Thinking about Science

Thinking about Science

Good Science, Bad Science, and How to Make It Better

FERRIC C. FANG

University of Washington School of Medicine
Seattle, Washington

ARTURO CASADEVALL

The Johns Hopkins University Bloomberg School
of Public Health and School of Medicine
Baltimore, Maryland

ASM PRESS
Washington, DC

WILEY

Editorial Correspondence:
ASM Press, 1752 N Street, NW, Washington, DC 20036-2904, USA

Registered Offices:
John Wiley & Sons, Inc., 111 River Street, Hoboken, NJ 07030, USA

For details of our global editorial offices, customer services, and more information about Wiley products, visit us at www.wiley.com.

Wiley also publishes its books in a variety of electronic formats and by print-on-demand. Some content that appears in standard print versions of this book may not be available in other formats.

Library of Congress Cataloging-in- Publication Data Applied for
Paperback ISBN: 9781683674344

Cover: © Naeblys/Shutterstock; KTSDesign/SCIENCEPHOTOLIBRARY/Getty Images
Cover design: Wiley

Set in 10.5/13pt ArnoPro by Straive, Pondicherry, India

SKY10071956_040524

Contents

enterprise and are symptomatic of a dysfunctional research culture in which incentives are misaligned with goals.

Boxes, Figures, and Tables

Boxes

Figures

Tables

Preface

We live in an age of science and technology. Science has allowed us to understand our place in the universe and our relationship to all life on Earth. Technology has provided a sophisticated computer on which to compose this book and allowed us to reshape our world to make life safer, more productive, and more comfortable. This has never been clearer than during this time of plague, as the COVID-19 pandemic enters its fourth year and finally shows signs of receding. The advanced technology of COVID-19 vaccines has greatly reduced the mortality of severe acute respiratory syndrome coronavirus 2 (SARS-CoV-2) infection and provided a vivid demonstration of the tangible benefits of science for society. As harrowing as the last few years have been, just think of what they would have been like *if we didn't have science.*

And yet, there are signs that not all is well with the scientific enterprise. The pace of transformative biomedical innovation appears to be slowing (1, 2). Fraud, sloppiness, and error have required the retraction of publications from the literature. Record numbers of research trainees are opting out of academic career pathways. Surveys report declining public confidence in scientists. With looming challenges from future pandemics, climate change, and shortages of food, water, and energy, it is vital for the world's scientific enterprise to be firing on all cylinders, to use an automotive metaphor (which will happily become anachronistic as electric vehicles displace those with internal combustion engines). This volume is a collection of essays exploring the nature of science and the way that it is performed today. In thinking about science, both good and bad, we will cast a light on contemporary scientific culture and practice, provide guideposts for young scientists, and propose a blueprint for reforming the way that science is done.

This book is written for scientists and science students, but also for technologists, engineers, mathematicians, teachers, journalists, administrators, policymakers, and anyone with an interest in science and how scientists think. The project began 15 years ago when we were editors at the journal *Infection and Immunity*. Our initial collaborative essay, called "Descriptive Science," was prompted by the tendency of many reviewers to dismiss work with the adjective "descriptive," despite the fact that *description* is the foundation of much of science (3). Encouraged by the positive responses from our colleagues, we subsequently collaborated on more than 40 articles, editorials, or commentaries. Many of the essays in this collection had their genesis in conversations or email exchanges, which eventually developed into editorials or commentaries. Each has been recently updated and supplemented with additional material for publication in this book. Nine of the chapters are completely new and have not been published elsewhere.

Our goal has been to create a volume that can be read either sequentially or as individual chapters, each constituting a freestanding essay that can be read and understood independently, although we have connected the themes through cross-referencing. Anyone reading the book from cover to cover will note some repetition, as certain issues arise again and again in various contexts. This is intentional and was necessary for the chapters to be able to stand on their own. We hope that this will help to reinforce these points.

Over the years, we have often commented to each other how writing these essays has improved our understanding of science and made us better scientists. We hope the same will be true for our readers. You will find that much of the material is slanted toward issues in the biomedical sciences, with a particular preference for the subdisciplines of microbiology and immunology. This reflects the fact that we are both active scientists with research programs focused on microbial pathogenesis. We make no apologies for writing about what we know best and note that other science essayists, such as Thomas Kuhn (4) and Eugene Wigner (5) writing about scientific revolutions and the unreasonable effectiveness of mathematics, have focused largely on examples from the physical sciences. In fact, we think that our biomedical emphasis makes sense since the 21st century is heralded to be the biological century. We subscribe to the view that science is a continuous discipline and observations made in one domain can apply to other domains as well. Nevertheless, we have attempted whenever possible to bring the physical sciences into the context of our essays, and readers will find numerous references to Newtonian physics, plate tectonics, and particle physics. We purposefully refer to some of the same scientific discoveries in multiple chapters in order to illustrate the continuity of themes across different aspects of science using familiar examples. Hence, some scientists, such as Alfred Wegener, Oswald Avery, and Rosalind Franklin, appear in more than one chapter, and we hope you will enjoy becoming more acquainted with them.

In many ways, *Thinking about Science* is a commentary on the current state of science in the early 21st century, with a particular emphasis on biomedical research. Although both of us are unabashed admirers of science and the scientific process, the reader may note a critical tone in many of these chapters. This, too, is intentional and reflects the fact that many chapters are written to highlight a problem in science in the hope of correcting it. The "Historical Science" chapter laments how often science ignores and neglects its history. In fact, we hope that the book provides an accurate snapshot, from the perspective of scientists working in the present day, for future historians of science. Similarly, we hope that chapters such as "Descriptive Science," "Mechanistic Science," "Reductionistic and Holistic Science," and "Important Science" have captured the tension of our time regarding preferred scientific approaches. "Impacted Science" describes a contemporary sociological malady that we hope will become obsolete in future years as science reforms its value system. "Dismal Science" delves into the economics of science, and we hope that more economists will take an interest in this important topic that remains largely unexplored. "Plague Science" feels unfinished, as every week brings a new development in the COVID-19 pandemic, and yet we hope that the words therein capture a sense of this moment in early 2023 by documenting successes and failures in confronting a novel viral scourge. In updating the early chapters of our collaboration, we have been both pleasantly surprised at the progress in certain areas, such as prepublication review, efforts to improve reproducibility, and efforts to improve equality and diversity in science, and dismayed by how little has been done in others, such as persistent problems with peer review and funding.

For us, this book has provided an opportunity to reflect and to gather and update our thoughts after 15 years of friendship and collaboration. This is, of course, a work in progress, and we will continue our work as practitioners, observers, and commentators of contemporary science who want to improve the scientific enterprise. We encourage readers to write to us with their comments, criticisms, and suggestions so that we can continue to think about science together.

January 2023

1. **Park M, Leahey E, Funk RJ**. 2023. Papers and patents are becoming less disruptive over time. *Nature* **613**:138–144. http://dx.doi.org/10.1038/s41586-022-05543-x.
2. **Casadevall A**. 2018. Is the pace of biomedical innovation slowing? *Perspect Biol Med* **61**:584–593. http://dx.doi.org/10.1353/pbm.2018.0067.
3. **Casadevall A, Fang FC**. 2008. Descriptive science. *Infect Immun* **76**:3835–3836. http://dx.doi.org/10.1128/IAI.00743-08.
4. **Kuhn TS**. 1970. *The Structure of Scientific Revolutions*, 2nd ed. The University of Chicago Press, Chicago, IL.
5. **Wigner EP**. 1960. The unreasonable effectiveness of mathematics in the natural sciences. *Commun Pure Appl Math* **13**:1–14. http://dx.doi.org/10.1002/cpa.3160130102.

Acknowledgments

The opinions expressed in this book are our own, and we take full responsibility for any errors (see chapter 25). We gratefully acknowledge the contributions of coauthors who have collaborated with us on previous publications, including Joan Bennett, Elisabeth Bik, Anthony Bowen, Erika Davies, Roger Davis, Daniele Fanelli, Michael Imperiale, Amy Kullas, Margaret McFall-Ngai, R. Grant Steen, and Andrew Stern. We also thank our many colleagues who have provided important insights and feedback on the topics covered in this book, including Gundula Bosch, Nichole Broderick, Lee Ellis, Sunil Kumar, Adam Marcus, Ivan Oransky, Liise-anne Pirofski, and Jessica Scoffield. We are grateful to the American Society for Microbiology for their longstanding support and for giving us a venue to publish our papers and ideas, in particular Stefano Bertuzzi, Christine Charlip, Shannon Vassell, Shaundra Branova, and our editor, Megan Angelini. We thank our mentors and students, who have taught us so much and renewed our passion for science. Finally, but certainly not least, we thank our families for their love, encouragement, and indulgence during the many late nights and weekends that we have spent thinking about science instead of other things. This book is dedicated to them.

About the Authors

Ferric C. Fang and Arturo Casadevall are physician-scientists and journal editors who have studied infectious diseases for more than three decades and have a longstanding interest in the culture and sociology of science. Dr. Fang is presently a Professor in the Departments of Laboratory Medicine and Pathology, Microbiology, Medicine, and Global Health at the University of Washington School of Medicine, and Dr. Casadevall is presently a Bloomberg Distinguished Professor in the Johns Hopkins Schools of Public Health and Medicine.

Ferric C. Fang   Arturo Casadevall

DEFINITIONS OF SCIENCE

1 What Is Science?

Science is not inevitable; this question is very fruitful indeed.

Edgar Zilsel (1)

Science is humanity's greatest invention. When difficult decisions are to be made, everyone says that they want to "follow the science." But what is science? The Merriam-Webster online dictionary defines science as "knowledge about the natural world based on facts learned through experiments and observation" (2). The word itself is derived from the Latin word *scientia*, which means "knowledge." However, as Carl Sagan observed, "Science is more than a body of knowledge. It is a way of thinking" (3). Thus, Great Britain's Science Council has defined science as "the pursuit and application of knowledge and understanding of the natural and social world following a systematic methodology based on evidence" (4). This is an improvement, but perhaps goes too far in emphasizing the process over scientific knowledge itself.

Thomas Huxley suggested that science is merely "common sense clarified" (5), although common sense tells us many things that science has shown to be untrue, such as that the Sun travels around the Earth (6). Science rises beyond mere observation, intuition, and association. Science is a way of acquiring knowledge that is progressive, cumulative, testable, and predictive. Fields that call themselves sciences share certain elements in common, including facts, theories, methods, practices, and predictions. The most persuasive characteristic of science is that it

Thinking about Science: Good Science, Bad Science, and How to Make It Better, First Edition.
Ferric C. Fang and Arturo Casadevall.
© 2024 American Society for Microbiology.

works. Science underlies all technology, from the light-emitting diode illuminating this room to the laptop on which this chapter is being composed, or the cellphone giving a reminder about an imminent meeting. Yet science is much more than technology, and its relationship to technology is complex (Box 1.1). Science allows the recognition of principles that make the natural world comprehensible. That doesn't mean that science is always right, not by a long shot. Scientific knowledge is always tentative and subject to change. But evidence of the power of the scientific method is all around us, and even when science leads to errors, the method itself embodies the means to correct its mistakes.

The scientific method was not invented all at once, but rather evolved over time with refinement from a range of sources. The scientific method has not arisen in every civilization. In fact, most scientific knowledge has been acquired only during the past 400 years, less than one-quarter of 1% of the time that our species has

---

Box 1.1 Science and technology

Science and technology are often mistakenly viewed as synonymous. Whereas a definition of science is elusive, the definition of technology is easier. Technology is "the application of scientific knowledge for practical purposes" (24). Hence, while science and technology are intimately associated, the two can exist independently. For example, the ancient world had the technology to construct majestic buildings and structures such as pyramids and the Great Wall of China without a formal understanding of the laws of physics. The Industrial Revolution was catalyzed by the invention of the steam engine, which was created by tinkering without any knowledge of thermodynamics. In fact, the field of thermodynamics emerged afterwards to explain phenomena observed in steam engines and in efforts to optimize their efficiency. On the other hand, major advancements in science often find no immediate technological applications. Einstein's theory of general relativity, formulated in 1916, did not find a clear technological application until the development of a geopositioning system in the 1970s required synchronization of clocks on Earth and in orbit, which run differently depending on the gravitational field that they experience. In 2016, gravitational waves were first detected using remarkable technology in the form of paired interferometers, constructed by highly exacting tolerances prescribed by physical laws, but these have yet to find a technological application. Today, much scientific research is dependent on technology made possible by our scientific knowledge.

inhabited the earth. In their books *The Unnatural Nature of Science* and *Uncommon Sense*, the embryologist Lewis Wolpert and the physicist Alan Cromer, despite their different perspectives, both trace the origins of science to ancient Greece (6, 7). Plato regarded reason as the most powerful capacity of human beings, Thales of Miletus attempted to describe the nature of the world, and Aristotle defined humans as rational animals. Aristotle distinguished *induction*, the inference of universal principles from particular observations, from *deduction*, in which general principles are used to make predictions in specific situations. Most of what Aristotle had deduced turned out to be incorrect, but his mode of thinking laid a foundation for others to follow. Modern scientists use induction to develop theories and hypotheses, which can then be tested experimentally to arrive at deductions (Fig. 1.1). Greek mathematicians developed the concept of mathematical proof, which allowed the systematic application of logic to deduce a level of knowledge that is regarded as the truth (Box 1.2). Another tradition that arose in ancient Greece was rhetoric, in which oratory was used for the purpose of persuasion. When modern scientists perform experiments and interpret results, they are carrying on the great ancient Greek traditions of reason (*logos*) and persuasion (*rhetor*).

During the so-called Dark Ages in Europe, Islamic scholars helped to preserve and further develop these concepts. Science and mathematics flourished in the Arab world in the Middle Ages (8), building upon earlier intellectual traditions to

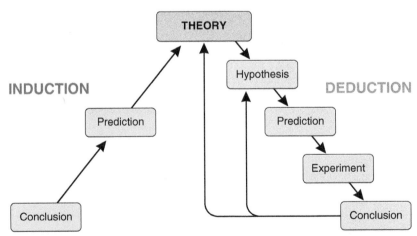

FIGURE 1.1 *Inductive versus deductive reasoning. In inductive reasoning, particular observations are used to infer universal principles or theories. In deductive reasoning, hypotheses lead to predictions that are tested experimentally. The results of experiments in turn may be used to revise hypotheses and theories.*

Box 1.2 Mathematics and science

In 1960, the physicist Eugene Wigner penned an influential essay titled "The Unreasonable Effectiveness of Mathematics in the Natural Sciences" (25), in which he noted how mathematical relationships pervade the natural sciences and, once identified, are predictive of new relationships and findings in nature. The relationship between science and mathematics may be viewed as essential, dependent, intricate, synergistic, and even symbiotic. Science depends on mathematics, and advances in science and technology further the development of mathematics, as evidenced by the ever-increasing reliance of mathematics on computers to probe its secrets, such as finding ever-larger prime numbers. At the heart of the matter is the fundamental question of whether the essence of the natural world is mathematical. The ancient cult of Pythagoras viewed the world as mathematical and promoted its understanding through mathematics, a world view with echoes in Plato's allegory of the cave, in which a perfect world lies just beyond the senses. The ability to express a scientific finding in the precise notation of mathematics is considered an apotheosis in modern science. The increasing recognition that we live in a probabilistic universe has reinforced the notion that both discovered and as yet undiscovered mathematical relationships underlie everything in the natural world, something that Pythagoreans would have embraced and appreciated. Although a detailed treatment of the relationship between mathematics and science is beyond the scope of this book, we encourage budding scientists to learn as much mathematics as they can.

---

create a body of knowledge that was communicated to Europeans through trade and contacts in the Iberian Peninsula. This eventually blossomed into what is recognizable as modern science during the Scientific Revolution in Western Europe. Two influential publications were Francis Bacon's *Novum Organum*, published in 1620 (9), and Rene Descartes' *Discourse on the Method*, published in 1637 (10). *Novum Organum* proposed an inductive method for understanding natural phenomena in which relevant facts were systematically assembled and categorized according to their association with a phenomenon of interest to generate axioms based on empirical data. *Discourse on the Method* proposed that problems be divided into smaller parts so that the simpler parts might be solved first and urged scientists to begin any inquiry from a skeptical perspective. Scholars continue to debate why the Scientific Revolution occurred in Western Europe rather than elsewhere. Contributing factors include the continuum with classical Greek philosophy, the increasing prominence of academic institutions, the development of printing and the increasing availability of books, a growing crisis between religious and humanistic world views, and the rise of capitalism, which lessened deference to authority and brought scholars and craftsmen together. The result was the emergence of a critical mass of practitioners of the scientific method who gave the revolution an unstoppable momentum (Box 1.3).

Box 1.3 Was science inevitable?

This chapter has emphasized the Western roots of modern science. That the Scientific Revolution occurred in 17th-century Europe in unquestioned, but contributions from many civilizations and cultures made this revolution possible (26). We have already mentioned the critically important contributions of Islamic scholarship. In addition, Chinese civilization developed science-enabling technologies such as the magnetic compass, the printing press, and papermaking, which allowed global exploration and efficient communication. As Bacon recognized, "Printing, gunpowder, and the compass ... changed the appearance and state of the world" (9). Chinese astronomy was also highly developed and precisely recorded a supernova in the year 1054, which created the Crab Nebula. It is noteworthy that there is no record of this event in Western records despite what must have been the spectacular event in the night sky, with the appearance of a new, very bright star that was visible during daytime. This curious and mystifying omission from European records may reflect that it conflicted with philosophical-religious consensus at the time, which held that the heavens were eternal and constant. Indian contributions to mathematics, such as the concept of zero, the decimal system, and advanced notation systems, were essential for later advances in theoretical physics (27). In the Americas, the Mayan civilization developed highly advanced astronomy and mathematics, along with the sophisticated engineering expertise to build magnificent cities. Ancient Africans developed advanced astronomy and metallurgy (28). In Oceania, ancient Polynesians mastered navigational skills that allowed them to travel to remote, isolated islands. Hence, the impulse to develop mathematics and science may be seen everywhere that humans settled and built civilizations and reflects the indomitable human curiosity. The will to do science, like the will to make music, can be viewed as a universal human trait. However, in contrast to the development of scientific concepts and mathematics in other

(Continued)

Box 1.3 (Continued)

societies, the Scientific Revolution gave rise to unique new insights, formalisms, and ways to investigate the world—the creation of the modern scientific method and scientific disciplines and institutions. The Scientific Revolution allowed humanity to overcome the limits of intuitive thinking, which serves us well in many situations but can lead us astray when trying to understand natural phenomena.

To return to Edgar Zilsel's question at the beginning of this chapter, we must consider the uniqueness of the Scientific Revolution. While there is abundant evidence of curiosity, ingenuity, creativity, and mathematics in many human civilizations, what we call modern science has arisen only once. This alone suggests that science was not an inevitable consequence of the evolution of human thought. Accordingly, science should not be taken for granted. Why it arose in 17th-century Europe, rather than in other scientifically and mathematically sophisticated societies, remains a fascinating and open question. It is probably not a coincidence that Europe during this period also witnessed new technologies like the printing press and telescope, brutal wars of religion, and the upheaval of medieval theology by the Protestant Reformation. Modern science may owe its existence to an unusual confluence of technological, historical, and sociological events.

In the 1920s, a school of philosophy known as *logical positivism*, with centers in the European capitals of Vienna and Berlin, asserted that truth must be demonstrated by direct observation or logical proof. Thus, scientific knowledge was favored over other forms of knowledge. In the classic formulation of the scientific method, science consists of careful observation and description, formulation of a hypothesis, and experimental testing of predictions. An implicit assumption is that experimental results can be replicated by others (chapter 7). Although this so-called "hypothetico-deductive" approach is not the only way of doing science, it is what many people think of when referring to the scientific method.

Logical positivism ultimately fell out of favor among philosophers of science, although its influence on 20th-century philosophy of science is undeniable. One reason for the decline of logical positivism is an inability to provide a clear demarcation between science and nonscience. Any definition of science must be able to distinguish it from pseudoscience, such as astrology, alchemy, creationism, and homeopathy. In fact, separating pseudoscience from science can be difficult since those disciplines have many of the trappings of science, including theory, method, and practice (see chapter 16). For the Austrian philosopher Karl Popper, the issue of demarcating science from nonscience was a central issue. He was concerned that a statement can never be definitively verified, only falsified (11). In his classic example, no amount of evidence can ever prove the assertion that all swans are white, but the assertion may be disproved by the sighting of a single black swan. Reliance on verification cannot distinguish science from pseudoscience, as pseudoscience may use inductive reasoning to generate false theories and selectively collect observations that appear to verify them. Instead, Popper proposed that scientific assertions should be falsifiable if shown to conflict with observable evidence (or even theoretically observable evidence). This concept has proven popular with scientists. However, falsifiability has its limitations and fails to describe much of what scientists actually do. Furthermore, the quest for falsification itself can be an impossible journey. For example, the statement that all iron has ferromagnetism cannot be falsified unless all iron in the universe is examined, which is an impossible task.

The physicist and philosopher Thomas Kuhn argued that much of "normal science" consists of puzzle-solving rather than testing for falsifiability, and that it is the feature of puzzle-solving that best provides a basis for the demarcation of science (12). Kuhn also proposed that scientists are guided by *paradigms*, broadly explanatory models or theories that can periodically undergo a radical change when they no longer conform to objective reality (see further discussion in chapter 11). Kuhn's views have also been challenged. The concept of paradigms implies an element of subjectivity in science and has been used by postmodernists to question whether science is superior to other forms of knowledge. In fact, Paul Feyerabend declared that a definition of science is impossible, and that the only principle is "anything goes" (13). Rather than specifying criteria to define science, the physicist John Ziman opted for a social definition consisting of a consensus among competent researchers (14).

While each of these efforts ultimately falls short of providing a universally satisfying definition of science, one can take a pragmatic approach to identifying core principles that distinguish science from nonscience (15). Conserved characteristics of science, such as an absence of supernatural explanations, measures to reduce the risk of bias, receptivity to new evidence, and support from independent forms of evidence, can help to distinguish science from pseudoscience. Although the world may indeed be "awash in bullshit," as some authors claim (16), science provides the tools to distinguish truth from bullshit. Nevertheless, as we will discuss in detail in later chapters, it can be challenging to distinguish good science from bad or fraudulent science, and the problem is even more daunting for nonscientists. Most nonscientists lack a sufficient understanding of the scientific method and the mathematics of uncertainty to critically evaluate scientific information, and people who trust science are not necessarily immune to misinformation (17).

Science as it is presently understood possesses several essential characteristics: (i) an assumption that nature can be understood without having to invoke supernatural forces; (ii) a requirement for supporting evidence; (iii) arguments that conform to logical chains of reasoning; (iv) accumulation of knowledge that can change and undergo refinement over time as a result of new ideas or observations and improvements in methodology; and (v) predictive power. Expertise is valued in science, but authority is not (chapter 28). A characteristic of science that sets it apart from other human endeavors is that it embraces the notion that all knowledge is provisional and that answering one scientific question inevitably leads to another (Box 1.4). Bias refers to any process that can improperly influence scientific inquiry and lead to erroneous conclusions. Accordingly, bias is to be avoided as much as possible, and potential sources of bias must be identified. For example, simple measures such as the inclusion of control groups, randomization, and blinding can help to control for selection bias, but there are many other potential forms of bias (18). As science has advanced, the development of probability, Bayesian analysis, and other statistical tools has allowed increasing rigor and the estimation of scientific certainty (Box 1.5).

Box 1.4 Science and uncertainty

The acknowledgment of uncertainty and that one might be wrong is as important for science as the quest for knowledge. Many ancient cultures were certain that the Earth was flat, with the heavens above and the ground below—until the 6th century BCE, when the pre-Socratic Greek philosopher Anaximander suggested that the Earth is suspended in space, with celestial bodies situated all around. Although he did not realize that the Earth is spherical or that it orbits around the Sun, his insight was nevertheless a revolutionary idea that enabled subsequent advances in Greek astronomy. His contemporary Pythagoras subsequently taught that the Earth is spherical, and this idea spread so that Pliny the Elder could state in the 1st century CE that this fact was universally accepted. In contrast, although Aristarchus of Samos proposed a heliocentric model of the universe in the 3rd century BCE, his notion was rejected in favor of the geocentric models of Aristotle and Ptolemy, and Aristarchus would not be vindicated until Copernicus published *De Revolutionibus Orbium Coelestium* 1,800 years later. Why did geocentrism persist for so long when it conflicted with the measured movement of the planets, in particular the periodic retrograde motion of Mars? Ptolemy was so certain of geocentrism that he created an extraordinarily complex explanation in which planets traveled in small epicycles that traveled around the Earth. When this also proved inadequate to account for planetary motion, additional epicycles were added, so that epicycles were traveling in epicycles around other epicycles. In retrospect, this model violates the principle of parsimony and is ludicrously wrong. However, it also illustrates the danger of certainty. As the theoretical physicist Carlo Rovelli has observed: "Science is not about certainty. Science is about finding the most reliable way of thinking at the present level of knowledge. ... In fact, it's the lack of certainty that grounds it" (29).

While science allows us to conclude some things with reasonable certainty, such as that the Earth orbits the Sun, even that knowledge is temporal and provisional, as the solar system is evolving. In a few billion years, the Sun will exhaust its fuel and is expected to expand into a red giant and swallow the Earth. An essential feature of science is the humility to remain open to new evidence and ideas.

Ancient Roman mosaic from 3rd century CE depicting Anaximander holding a sundial.
Courtesy of GDKE-Rheinisches Landesmuseum Trier, Thomas Zühmer.

Box 1.5 Where do formulas in science come from?

Laws in science are often stated as formulas that relate a fact to a set of variables. Examples are Newton's and Einstein's famous formulas for force ($F = ma$, or force = mass × acceleration) and energy ($E = mc^2$, or energy = mass × square of speed of light), respectively. So, a reasonable question to ask when thinking about science is: Where do these relationships come from? Certainly, they are not written down in nature until they come from the minds of scientists. Richard Feynman addressed this question in one of his famous Messenger Lectures, "The Character of Physical Law," given at Cornell University on November 9, 1964 (30). According to Feynman, the first step is to "guess," and then to compare the results of the equation to experimental data and see what happens. If the formula's predictions match the experimental results, the formula may be right. If the results are discrepant, then the formula is wrong and must be discarded. This reveals the insight that mathematical relationships can explain laws of nature. Scientists tinker with variables to see what works and validate formulas by careful observation and measurement. However, Feynman may have overstated things a bit when he said, "First, we guess it." An equation represents a highly educated guess, not just intuition. For instance, Newton's second law of motion can be inferred by the observation that a change in the momentum of an object varies directly in proportion to the force exerted on the object but inversely in proportion to its mass. All that's left is to collect observations and determine whether a coefficient is required. Ohm's law, in which current or flow varies in proportion to voltage but inversely in proportion to resistance, is analogous. Feynman was simply restating inductive inference from observation, which leads to hypothetical formulas or laws that can be tested experimentally (Fig. 1.1).

Hence, there may be different routes to finding an equation that describes nature. We could imagine that the sequence to finding a law of nature and representing it mathematically follows the sequence of observation, insight, tinkering with equations, and validating with experimental data. In this sequence, the first step, observation, can involve the collection of data that provide the basis for writing down formulas until one fits the data and making predictions that can be validated by subsequent experiments. The universe functions by unwritten laws that the human mind can discern, write down, and use to explain natural phenomena.

It is also important to remember that science aims to accurately describe physical reality but does not provide answers to questions of morality. In the 18th century, the Scottish philosopher David Hume discussed this as the "is-ought problem" (19); in other words, one cannot conclude what *ought* to be on the basis of what *is*. For questions of right and wrong, science and ethics may be viewed as separate *magisteria* (20), although science can certainly be informed by an application of ethical principles, and neuroscience may one day provide a biological explanation for human ethical behavior. In addition to adhering to principles of research integrity (see chapters 18 and 29), scientists have a responsibility to understand the ethical and social dimensions of their work and to ensure that scientific knowledge is used wisely and in the public interest (21).

The coexistence of science with religion has often been fraught, and that tension has led to conflict, as evidenced in the struggles between Galileo and the Inquisition and more recently between evolution and creationism. Stephen Jay Gould introduced the concept of "NOMA," or *nonoverlapping magisteria* (20), to argue that science and religion occupy different spheres of human action and thought. However, just as defining science is difficult, drawing a bright line of separation between science and religion is not always straightforward, and each may intrude on the other. Religion relies on received wisdom that is permanent and dependent on the supernatural, whereas science produces knowledge based on observable evidence, which is always provisional and subject to change as new information becomes available. The NOMA concept requires adherents of science and religion to agree to accept their limitations, e.g., Galileo's assertion that "the intention of the Holy Spirit is to teach us how one goes to heaven, not how the heavens go" (22). However, NOMA often fails in practice because science and religion encounter irreconcilable conflicts with regard to fundamental questions such as the origin of life, and science and faith are fundamentally incompatible at an epistemological level. Nevertheless, science and religion can find many areas of common ground (23), and frequently manage to coexist peacefully, even within the same individual. This is fortunate, as both science and religion have much to offer humanity.

In summary, science is a body of knowledge, a way of thinking, and a method for studying and understanding the natural world. Just as scientific knowledge has evolved over time, undergoing constant testing, reinforcement, and refinement from the contributions of innumerable scientists, so have the scientific method and the definition of science evolved over time. Although scientific revolutions occur only rarely (more on this topic in chapter 11), scientific knowledge and methodology continue to grow and advance with every passing year. With all that in mind, we shall now explore in detail how science can be done well or poorly and used for good or ill, and what that may mean for both science itself and society as a whole.

2 Descriptive Science

Certainly no developed science is merely descriptive in the narrower sense of the word—it seeks to explain.

Ernest Albee (1)

The word "descriptive" is defined as "referring to, constituting or grounded in matters of observation or experience" (2). Since practically all laboratory-based science is based on recording evidence from experimentation, it might be argued that all science is in some sense "descriptive." However, scientists distinguish between descriptive research, in which information is collected without a particular question in mind, and hypothesis-driven research, which is designed to test a specific explanation for a phenomenon. In this dichotomy, "descriptive" has numerous synonyms, including "observational," "inductive," or "fishing expedition," while "hypothesis-driven" may also be referred to as "hypothetico-deductive" or "mechanistic" (see chapter 3). When scientists favor hypothesis-driven science over descriptive science, they are really saying that they prefer work that is explanatory or provides insights into causation.

Journals commonly state that they will not consider papers that are "purely descriptive" (3), although what this means is usually not defined. When applied to science, the word "descriptive" has acquired dismissive or pejorative connotations and is frequently provided as justification for rejection of a manuscript or grant application. Given the widespread use of this adjective and its profound implications, it is worthwhile to reflect on what is right or wrong with descriptive science.

Thinking about Science: Good Science, Bad Science, and How to Make It Better, First Edition.
Ferric C. Fang and Arturo Casadevall.
© 2024 American Society for Microbiology.

In considering this issue, it is noteworthy that many esteemed scientific disciplines, such as astronomy, archaeology, and paleontology, are almost entirely descriptive sciences (4). Newton's laws of motion can be considered descriptive, and there is nothing mechanistic about the gravitational constant. Nevertheless, we hold these laws in great esteem because they are able to predict the behavior of the natural world. One cannot perform an experiment in which a stellar variable (Box 2.1) or a geological epoch is altered. Moreover, the descriptive sciences of taxonomy, anatomy, botany, and paleontology have been central to the development of evolutionary theory, which remains the linchpin of all biological sciences. Hence, there is nothing fundamentally wrong with descriptive research, with the caveat that a scientific field may demand more from an investigator once it becomes an experimental science.

Box 2.1 Careful description leads to cosmic understanding

As telescopes improved, astronomers discovered an astounding number of stars and cloudy patches known as nebulae. In the early 20th century, the nature of nebulae was unknown and there were no methods available to measure cosmic distances. The marrying of telescopes with photographic cameras allowed astronomers to take long-exposure photographs and see objects that were too dim for the naked eye. Soon, however, it was evident that analyzing celestial photographs to map stars was a very laborious process. Edward Pickering at Harvard College Observatory hired women known as "computers" to assist in the identification and cataloguing of stars according to their brightness and spectral characteristics. The Harvard Computers analyzed thousands of photographic plates. Among these women was Henrietta Swan Leavitt, whose careful observations would change our view of the universe and provide the first method for measuring intergalactic distances (14). Leavitt focused on a set of stars in the constellation Cepheus known as cepheid variable stars that appeared to change their brightness at regular intervals. In 1912, she reported a relationship between cepheid variable brightness and the period between intervals (15), which subsequently allowed Edwin Hubble to determine in 1924 that the Andromeda "nebula" was actually a spiral galaxy outside of our own Milky Way galaxy (16). Sadly, Leavitt died in 1921 of cancer and never knew how her insights changed our perception of the universe. Also sadly, but not uncommonly for women in science, Leavitt never received the recognition that was due from her discovery.

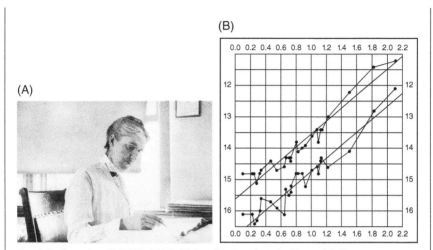

Henrietta Swan Leavitt (1868–1921). (A) Leavitt at work at the Harvard College Observatory. Photo by Margaret Harwood, courtesy of AIP Emilio Segrè Visual Archives, Physics Today Collection, Shapley Collection. (B) Plot showing a log-linear relationship between star magnitude and period length. From a 1912 paper prepared by Leavitt (15). Paper from John G. Wolbach Library, Harvard-Smithsonian Center for Astrophysics, NASA Astrophysics Datasystems.

In microbiology and related medical sciences, the transition from descriptive research to hypothesis-driven research has generally reflected the maturation of these fields. In the early stages of a field, descriptive studies may "represent the first scientific toe in the water" (5). Initial observation and induction give rise to novel hypotheses, which subsequently can be experimentally tested to provide a progressively detailed mechanistic understanding. Specific hypotheses allow a more discerning interrogation of complex data sets, something recognized by Charles Darwin when he noted, "Without speculation there is no good and original observation" (6). On the other hand, a descriptive approach may be less prone to bias (7). "It is a capital mistake to theorize before you have all the evidence," Sherlock Holmes once remarked, "it biases the judgment" (8). Microbiology and immunology have been transformed by a number of powerful technological advances; methods such as large-scale sequencing, microarrays, bioinformatics, and proteomics are generating enormous databases that provide invaluable resources for the research community. While these methods can certainly provide potent means to answer mechanistic hypotheses, in many cases they are initially used solely in a "descriptive" sense. In other words, some aspects of biological science have returned to an observational phase, in which research is primarily discovery-driven rather than hypothesis-driven (9). Such research is clearly important when it leads to the recognition of novel phenomena or the generation of novel hypotheses. However, microbiology and immunology are now experimental sciences, and consequently investigators can go beyond simply describing observations to formulate hypotheses and then perform experiments to validate or refute them.

Why, then, the proscription against descriptive science? Editors and reviewers distinguish between descriptive science that significantly advances the field and *mere* descriptive science that does not further understanding. The former might be appropriate for publication in a selective journal, but the latter will almost always be returned to the authors as too preliminary. An example of a rejected descriptive manuscript would be a survey of changes in gene expression or cytokine production under a given condition. These manuscripts usually fare poorly in the review process and are assigned low priority on the grounds that they are merely descriptive; some journals categorically reject such manuscripts (10). Although survey studies may have some value, their value is greatly enhanced when the data lead to a hypothesis-driven experiment. For example, consider a cytokine expression study in which an increase in a specific inflammatory mediator is inferred to be important because its expression changes during infection. Such an inference cannot be made on correlation alone, since correlation does not necessarily imply a causal relationship. The study might be labeled "descriptive" and assigned low priority. On the other hand, imagine the same study in which the investigators use the initial data to perform a specific experiment to establish that blocking the cytokine has a certain effect while increasing expression of the cytokine has the opposite effect. By manipulating the system, the investigators transform their study from merely descriptive to hypothesis-driven. Hence, the problem is not that the study is descriptive *per se* but rather that there is a preference for studies that provide novel mechanistic insights.

When a manuscript is rejected by a journal for being "merely descriptive," the reviewer is essentially saying that the work has not revealed novel phenomena, has failed to generate interesting novel hypotheses, or has failed to adequately follow up such hypotheses with further experimentation. The most common reason for a paper to be assessed as "merely descriptive" is that more in-depth investigation is required. A reviewer who recommends that a paper be rejected because it is "merely descriptive" can provide a great service to the authors by clearly and unambiguously explaining the additional studies required for the paper to become more significant and therefore more interesting (Box 2.2).

Box 2.2 "Descriptive" is a fraught word in the scientific lexicon

A major problem in scientific discourse is that the adjective "descriptive" has a different meaning for different scientists, which becomes even more confounded when the word is used in contrast to mechanistic science (chapter 3). For some scientists, descriptive means that the work is correlative and does not seek to establish cause-and-effect relationships. For others, it is an epithet for lack of depth or a survey-like treatment of a subject. This lack of precision is a major problem when labeling a scientific study as descriptive to justify rejection of manuscripts or grant applications. The origins of the use of "descriptive" as a pejorative term in the biological sciences is uncertain but may reflect the relatively recent transition of biology from a strictly observational science to one endowed with very powerful molecular and genetic techniques. Given this state of semantic confusion, it may be best to avoid using this word when trying to give criticism. Instead, scientists who review papers and grants should expand on their comments to be sure that the those who receive their criticism understand what is meant.

Descriptive observations play a vital role in scientific progress, particularly during the initial explorations made possible by technological breakthroughs. Certainly, Alfred Wegener could not have thought about the possibility of continental drift (11) without accurate mapping of continent coastlines, a descriptive endeavor, which gave Wegener the idea that these had once been joined (Box 2.3). Following up on this example, the subsequent theory of plate tectonics would find its strongest support in the mapping of magnetic orientation of the ocean seabed (12). Clearly, the phenomenally successful theory of plate tectonics, which revolutionized our view of the geology of the planet, could never have been developed without the support of a tremendous amount of descriptive data. In the area of infectious diseases, John Snow's painstaking spatial description of the cholera cases around a water well in London provided key evidence for the waterborne transmission of cholera and against the miasma hypothesis (Fig. 2.1).

At its best, descriptive research can illuminate novel phenomena or give rise to novel hypotheses that can in turn be examined by hypothesis-driven research. However, descriptive research by itself is seldom conclusive. Thus,

Box 2.3 Careful description can lead to revolutionary new insights

Wegener was a German meteorologist who suggested in 1912 that the earth's continents were once joined together and subsequently drifted apart. Wegener based his hypothesis on descriptive data including rock types, geological structures, fossil plants and animals, and coastline contours. The continental drift hypothesis was initially controversial and only became widely accepted after Wegener's death. Right panel from reference 17.

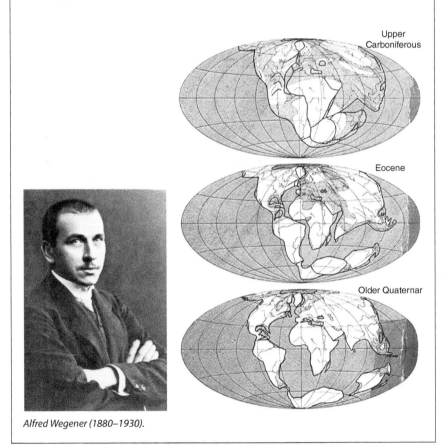

Upper Carboniferous

Eocene

Older Quaternar

Alfred Wegener (1880–1930).

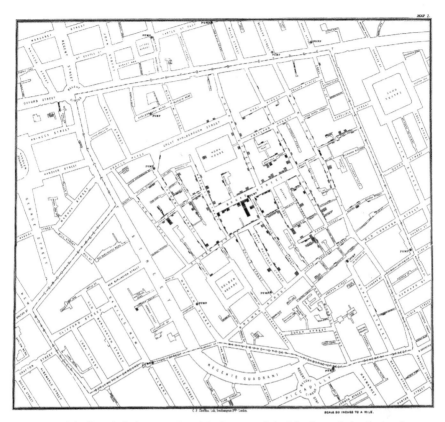

FIGURE 2.1 *John Snow's cholera map. An original map made by John Snow during the London epidemic of 1854 showing cholera cases (small rectangles) centered around the contaminated pump at the intersection of Broad and Cambridge Streets. This is considered a landmark event in the birth of modern epidemiology. Image from* On the Mode of Communication to Cholera *by C.F. Cheffins, Lith, Southhampton Buildings, London, England, 1854.*

descriptive and hypothesis-driven research should be seen as complementary and iterative (13). Descriptive science is neither good nor bad, and as we shall further explore in chapter 3, neither is mechanistic science. Nor is descriptive science the antithesis of mechanistic science. Observation, description, the formulation and testing of novel hypotheses, and the elucidation of mechanisms are all essential to scientific progress. The value of combining these elements is almost indescribable.

3 Mechanistic Science

Science is the knowledge of consequences, and dependence of one fact upon another.

Thomas Hobbes (1)

In contrast to the pejorative use of the adjective "descriptive" discussed in the preceding chapter, the term "mechanistic" is usually used to praise the quality of science. This is apparent in discussions when the phrases "mechanistic understanding" and "mechanistic approach" are used to convey a sense of high-quality research. Science educators attempt to instill "molecular mechanistic reasoning" into the curriculum through the use of graphics and models (2). In reviews of scientific manuscripts and grants, the words "mechanistic" and "descriptive" are often misused as synonyms for "good" and "bad," respectively (3, 4).

However, the extraordinary power of the words "descriptive" and "mechanistic" requires us to wield them carefully when critiquing science. In chapter 2, we considered the epithet "descriptive" and argued for an important role of descriptive studies in many scientific fields, while also acknowledging a general preference for studies that go further by including experimental work (5). Here we consider the more favored adjective "mechanistic" and explore its usage, meanings, implications, and limitations. Recognizing the centrality of mechanistic research to the history of science (6), we will explore what scientists mean when they use this term.

Thinking about Science: Good Science, Bad Science, and How to Make It Better, First Edition.
Ferric C. Fang and Arturo Casadevall.
© 2024 American Society for Microbiology.

DEFINITIONS

At first glance, one is struck by the fact that the terms "descriptive" and "mechanistic" are often used antagonistically as descriptors of scientific quality, yet they are not antonyms. "Descriptive" is defined as "referring to, constituting, or grounded in matters of observation or experience," while "mechanism" is defined as "the fundamental processes involved in or responsible for an action, reaction or other natural phenomenon" (7). From these definitions, "descriptive" can be seen as analogous to the interrogatives *who, what, where,* and *when,* whereas "mechanistic" in turn asks *how* and *why.* Hence, these terms collectively encompass the spectrum of inquiry. But if "descriptive" and "mechanistic" are not antonyms, what accounts for the general preference for mechanistic over descriptive work?

As "descriptive" and "mechanistic" are generally considered to denote different qualities in science, we must probe further to ascertain what these terms mean in the scientific vernacular. In chapter 2, we suggested that "descriptive" means different things to different people. Since practically all laboratory-based science is based on recording evidence from experimentation, it could be argued that all science is in some sense descriptive. However, this is unsatisfactory because every scientist intuitively knows that there are qualitative differences among scientific studies. Hence, the first problem we encounter is in the precision of language, as we try to understand and convey meaning in words. Similarly, the word "mechanistic" is used to refer to both complex natural phenomena and man-made mechanical devices (Box 3.1). The machine as an analogy for the natural world owes much to the writings of Hobbes and Descartes (even though the latter could not bring himself to ascribe the human soul to a mechanical process). Like Hobbes, the modern scientist makes the implicit assumptions that phenomena have rational explanations and that events may be connected as cause and effect. Scientists seeking mechanisms to explain the workings of the natural world are only the latest practitioners in a philosophical continuum extending back to the 17th century (6, 7).

The explanation for many biological phenomena requires a basic understanding of causal mechanisms. However, "mechanism" can mean different things in different fields. For example, in the late 1950s the problem of protein synthesis was central to biology, and "mechanism" to biochemists meant the formation of covalent bonds in polypeptides, whereas to molecular biologists, "mechanism" was the means by which the genetic code is translated into proteins (8). Although these approaches were eventually reconciled during the great synthesis of the mid-1960s (8), it is noteworthy for our discussion that the word "mechanism" can hold different meanings even in closely related fields like biochemistry and molecular biology. Furthermore, the meaning of the term "mechanism" with respect to science has

Box 3.1 The illusion of scientific processes as clockwork

The development of mechanical clocks such as pendulum clocks in the 17th century coincided with the Scientific Revolution and provided much more accurate measures of time than prior technologies, such as water and sand clocks. Mechanical clocks rely on gear mechanisms that operate at discrete time intervals to measure the passage of time. Such clocks were essential for measurements that unearthed basic physical laws and enabled much progress during the Scientific Revolution. Clocks are deterministic in the sense that the gears interact to produce a known outcome, which is reliable and produces an accurate measurement of time. Clockwork mechanisms have a powerful allure for the human mind, for the intricate gearwork and their interactions provide a perfect explanation for their function and output. The metaphor of clockwork as a representation of the universe has fueled philosophical debates about free will versus determinism, which led Pierre-Simon Laplace to hypothesize in 1814 that knowledge of all forces and matter at a certain moment would allow the prediction of all future events, a concept known as "Laplace's demon" (13). Hence, it is no surprise that scientists look for the mechanism of a process hoping that elucidation of its components will provide enlightenment as to both function and purpose. However desirable, clockwork mechanisms in nature are rare and illusory. In contrast to the deterministic mechanism of a clock, natural processes are often unpredictable. Quantum theory posits that the microscopic world is probabilistic and indeterminate, while at the other end of the size scale, gravitational interactions between bodies in the solar system are chaotic and unpredictable (14). Emergent properties reduce the predictability of complex biological systems (15). Thus, for atoms and planets and many things in between, the human desire for predictability and understanding collides with fundamental dynamic processes that limit what we can know for certain about the present and future. The scientific penchant for "mechanistic science" may reflect an anthropomorphic bias for clockwork processes that obscures a deeper understanding of the natural world. In other words, the quest for mechanisms can be thwarted by epistemological limitations, and there are limits to how precisely scientists can predict the future. We suggest that focusing on what can be known while remaining critical about what we know and how we know it is a more productive approach to science than attempting to judge science on the basis of such imprecise epithets as "descriptive" and "mechanistic."

(Continued)

Box 3.1 (Continued)

A pendulum clock designed by Galileo, as drawn by Vincenzo Viviani in 1659. Courtesy of University of Toronto Photographic services.

changed over time, from a version of philosophical materialism in opposition to vitalism to a stepwise explanation of how system components interact to produce an outcome (6).

WHERE IS THE LINE OF DEMARCATION BETWEEN "DESCRIPTIVE" AND "MECHANISTIC?"

Starting with the assumption that there is a difference between descriptive and mechanistic science and seeking a clear line of demarcation that can be expressed in words, one immediately runs into the problem that the description of a process can be considered the mechanism for another process. To further illustrate this point, let us consider a hypothetical situation. A scientist walks into a dark room and encounters impenetrable darkness. A candle is lit, and the scientist now perceives the outline of the room. The scientist decides to investigate the phenomenon of light. The mechanism responsible for the light is the candle. However, the scientist notes that only part of the candle emits light and determines that the mechanism for light is the flame. In describing the flame, the investigator establishes that the mechanism for the flame is combustion. Describing combustion, the scientist determines that the mechanism is a series of oxidation-reduction reactions that in turn are explained by electron transfer and, ultimately, quantum mechanics. At each step, the description of a process provides only a partial explanation in search of a deeper mechanism that must in turn be described (Table 3.1). What is striking in this hypothetical situation is that the difference between description and mechanism is one of proximate causation. Hence, the epithets "descriptive" and "mechanistic" are epistemologically related and differ quantitatively rather

TABLE 3.1 A scientist considers the illumination of a dark room

Description	Mechanism
Light	Candle
Candle	Flame
Flame	Combustion
Combustion	Chemical reaction
Chemical reaction	Oxidation-reduction
Oxidation-reduction	Electron loss and gain
Electron loss and gain	Quantum mechanics
Quantum mechanics	Metaphysics

than qualitatively. In other words, observations become regarded as progressively less descriptive and more mechanistic as one probes more deeply into a phenomenon. In fact, one might argue that there is no real line of demarcation between descriptive and mechanistic science but that the difference is rather a matter of depth and one's preferences.

The scientist in the dark room also gives us a model with which to explore the difference between descriptive and experimental science, a point that we emphasized in the preceding chapter (5). Note that the scientist can assign causality to the association between a lit candle and an illuminated room by extinguishing the candle, observing the return of darkness, and subsequently validating that the candle is responsible for light by reigniting the flame. Furthermore, our thought experiment illustrates the issue of "significance" of a scientific observation. The assessment of significance is a major criterion in grant or manuscript review, yet we have few tools for judging the significance of a finding in real time other than judgment and experience. One might argue that since the scientist needs light in order to see, the most significant finding is the association of the candle with light. The mechanistic details following subsequent questions may be important for understanding the related phenomena but are not essential in the context of a dark room unless the scientist decides to use the information to design a better candle or kindle a brighter fire. Hence, the significance of a finding is often related to the subsequent development and application of the revealed information and may become apparent only over time. Our hypothetical scenario also provides insight into the line of demarcation between science and nonscience. As when one peels away the successive layers of an onion or opens a series of nested Russian dolls, each revealed mechanism becomes a new description leading to a new mechanistic question, until the investigator arrives at a point where scientific inquiry cannot proceed without entering the realm of metaphysics.

Moving from the dark room into the world of science, it is possible to envisage similar scenarios in which it is difficult to identify a clear demarcation between descriptive and mechanistic research. For example, consider a disease characterized by a red, hot, painful, and swollen skin lesion. The investigator would note that this collection of signs and symptoms corresponds precisely to *rubor* (redness), *calor* (warmth), *dolor* (pain), and *tumor* (swelling), the Latin terms used to describe inflammation. These terms in aggregate represent descriptors that denote "inflammation," a process that provides a mechanism for the disease (Table 3.2). To investigate the mechanism of inflammation, the investigator employs a microscope and determines that the lesion is a result of an influx of white blood cells, specifically the subset known as neutrophils. While investigating the presence of neutrophils, the investigator discovers chemotaxis (migration of cells toward a signal), alterations in expression of chemokines (molecules that attract white blood cells), activation of signaling pathways, and perhaps the presence of microbeassociated molecular patterns responsible for chemokine elicitation. In essence, the boundary between descriptive and mechanistic science is moving and subjective and depends on both the depth of the experimental question and the technological sophistication of the investigator. In other words, one scientist's mechanism may become another's descriptive starting point.

When reviewers of scientific work express a desire for "more mechanistic" studies, some are probably asking for experimental work that establishes causality between the observations being reported. For example, a paper that reports a

TABLE 3.2 A scientist considers the cause of a skin lesion

Description	Mechanism
Pain, redness, swelling, and heat	Inflammation
Inflammation	Cellular infiltration and vascular leakage
Cellular infiltration and vascular leakage	Chemokines, cytokines, and arachidonic acid derivatives
Chemokines, cytokines, and arachidonic acid derivatives	Signaling cascade activation
Signaling cascade activation	Agonist-receptor interactions
Agonist-receptor interactions	Microbe-associated molecular patterns
Microbe-associated molecular patterns	*Staphylococcus aureus*

simple correlation between two phenomena might be criticized as insufficiently "mechanistic" during the peer review process because correlation does not necessarily imply causation (e.g., what happens if you blow out the candle?). In such instances, the reviewer can be most helpful by suggesting specific experiments to allow the inference of causation. In other cases, peer reviewers requesting mechanistic studies may desire more depth in ascertaining the explanation for a reported observation. One of the most frequent reasons for a paper or grant application to be labeled "descriptive" is that it fails to interpret its observations and tell a coherent story. As Peter Medawar observed, scientists tell "stories which are scrupulously tested to see if they are stories about real life" (9). A description followed by a hypothesis can still tell a story, albeit a tentative one, but a presentation of disconnected phenomena without a narrative to bind them together is likely to be poorly received and labeled as "merely descriptive."

"DESCRIPTIVE" AND "MECHANISTIC" IN THE SCIENTIFIC VERNACULAR

Both the scientist in a dark room and the example of inflammation suggest that there is no bright line of demarcation separating the terms "descriptive" and "mechanistic" as applied to science. Given the inexactitude of "descriptive" and "mechanistic" and the vagaries associated with their meaning, labeling research as descriptive or mechanistic is often not a productive exercise. Although we agree with the statement that many of the most important discoveries in the sciences relate to novel mechanisms (10), "descriptive" should not be used as a derogatory term, since description is a critical element of the scientific process, and elucidation of a "mechanism" always requires some form of description. Since "mechanistic" is not an antonym for "descriptive," and description can provide a mechanism in certain contexts, we are still left with the question of exactly what scientists mean when they use such terms in critiques. Probably the most honest answer to this question is that we do not always know, since the definitional boundaries are sufficiently fuzzy that these terms probably mean different things to different people.

The problem in demarcating "descriptive" and "mechanistic" is nicely illustrated by the field of crystallography. Solving the structure of a protein or nucleic acid may be considered a strictly descriptive exercise, since the output is often a series of atomic coordinates. However, describing a structure frequently provides key insights into function and mechanisms. In this regard, we are reminded that the description of DNA structure provided the critical insight for the mechanism of genetic replication, the conservation of information, and the deciphering of the genetic code. Hence, crystallographic studies that yield functional insights may be considered mechanistic despite the essentially descriptive nature of diffraction data.

"Descriptive" is closely related to "empirical," or that which is observed without regard to theory. However, the root of "empirical" is "experimental," which expands upon mere description by introducing perturbations into a system. This in turn may lead to novel hypotheses and predictions that can be tested, thereby completing the transition to "mechanistic" theory-driven models. The preference for "mechanistic" as a descriptor may be a result of the historical importance of elucidating mechanisms in science. Bechtel and Abrahamsen have noted that explanations based on mechanism are inherently attractive because they are able to avoid the limitations of linguistics by using diagrams and introduce directionality to the process of discovery and hypothesis testing (10). Another attribute of mechanisms is that they can point the way to targeted translational interventions, such as the targeted discovery of new drugs through rational drug design (Box 3.2). However, there are problems with attempting to reduce all biological sciences to a search for mechanisms. A description of a novel discovery or hypothesis can be of greater interest than the elucidation of a highly predictable or conventional mechanism (11).

Box 3.2 Mechanism leads to rational drug design

Until relatively recently, most drugs were discovered by trial and error, screening of libraries of natural products and synthetic compounds, or plain serendipity. For example, aspirin (acetylsalicylic acid) was synthesized to mimic salicylates found in willow bark and other plants, the antimalarial quinine was extracted from cinchona tree bark, and penicillin was obtained after the serendipitous finding of a mold that inhibited bacterial growth (chapter 14). Although the approach of screening and optimizing lead compounds has yielded many useful drugs, high costs and diminishing returns were raising concerns by the latter part of the 20th century (16). At the same time, a great deal of information was accumulating on molecular and cellular physiology, particularly regarding enzymatic reactions and signaling cascades, and there was growing interest in developing specific inhibitors, including molecules that might not exist in the natural world. This led to the field of rational drug design, where knowledge of enzyme structure and mechanism is used for the design of chemical compounds that act on specific biochemical targets to achieve a therapeutic effect. A giant in this field was Gertrude Elion (17), who was involved in the development of such major drugs as 6-mercaptopurine for leukemia, acyclovir for herpes virus infections, pyrimethamine for malaria, allopurinol for gout, azathioprine for prevention of transplant rejection, trimethoprim for bacterial infections, and azidothymidine for HIV (human immunodeficiency virus) infection (18). In her autobiography, Elion recounts how the loss of her beloved grandfather to cancer catalyzed her desire to apply chemistry for the design of new drugs (19). The death of her fiancé from bacterial endocarditis in 1941 intensified her desire to discover new therapeutic agents. If his infection had occurred just a couple of years later, he might have been treated with antibiotics (20). Although Elion never obtained a doctoral degree, she learned chemistry at Hunter College of the City University of New York and at New York University and spent her research career in the pharmaceutical industry. In 1988, Elion was recognized with the Nobel Prize in Medicine, which she shared with Sir James W. Black and her longtime collaborator Dr. George Hitchings. Today, rational drug design is an established approach to drug discovery that relies on mechanistic science.

Gertrude Elion (1918–1999). Photo credit: GSK Heritage Archives.

The words "descriptive" and "mechanistic" are so ingrained in the scientific psyche that we have no choice but to live with them. However, educating ourselves and the public as to their meaning and limitations might enhance scientific discourse, and by extension the scientific process. It would be best if the terms "descriptive" and "mechanistic" were not employed in scientific critiques unless accompanied by more-specific language to explain precisely what the critic means. In scientific conversations, it is preferable for scientists to provide context for their impressions and opinions rather than vague terms. The critique should specifically state what is required to make the content suitable. We suspect that for many scientists, the meanings of these terms are like Supreme Court Justice Potter Stewart's famous comment, "I know it when I see it," in reference to pornography (12). However, we argue that science, despite its potential to thrill, is not pornography, and that thoughtful and carefully chosen words can facilitate scientific progress. Science must describe not only what, when, and where events are occurring, but also the mechanisms of how and why they are linked together, in order to illuminate the darkness.

4 | Reductionistic and Holistic Science

Reductionism is one of those things, like sin, that is only mentioned by people who are against it.

Richard Dawkins (1)

Few scientists will voluntarily characterize their work as reductionistic. Yet reductionism is at the philosophical heart of the molecular biology revolution. Holistic science, the opposite of reductionistic science, has also acquired a bad name, perhaps due to an unfortunate association of the word "holistic" with new age pseudoscience. Fortunately, however, there is an increasingly popular euphemism that lacks the pejorative connotations of holism for scientists: "systems biology." Since its debut around two decades ago (2, 3), "systems biology" has appeared as a medical subject heading (MeSH) in the PubMed database more than 10,000 times, and a search of the database for the keyword "systems biology" yields nearly 200,000 publications. A fundamental tenet of systems biology is that cellular and organismal constituents are interconnected, so that their structure and dynamics must be examined in intact cells and organisms rather than as isolated parts. We recall that the late author Douglas Adams created a fictional detective named Dirk Gently who described his methods as "holistic" because he relied on the "fundamental interconnectedness of all things" to solve crimes (4). Gently used this to justify a large expense account, arguing that each of his personal expenses, like a beach holiday in the Bahamas, must be related to an ongoing investigation at

Thinking about Science: Good Science, Bad Science, and How to Make It Better, First Edition.
Ferric C. Fang and Arturo Casadevall.

some level. Although funding agencies are not likely to accept holistic accounting practices, holistic approaches have become increasingly popular in microbiology, sometimes advocated as superior to reductionistic ones (5). Researchers often adopt holistic or reductionistic approaches to study a problem without justifying their choice or explaining the advantages and limitations of such an approach. In this chapter, we consider the dichotomy between holistic and reductionistic approaches to science and their particular implications for microbiology. First, however, a few definitions are in order.

TYPES OF REDUCTIONISM

"Reductionism" can have epistemological, ontological, and methodological meanings (6). Epistemological reductionism addresses the relationship between one scientific discipline and another and is defined as "the idea that the knowledge about one scientific domain can be reduced to another body of scientific knowledge" (7). For instance, can one, as Francis Crick proposed, "explain all biology in terms of physics and chemistry" (8)? Perhaps, but today it is difficult to imagine how consciousness can be reduced to physics and chemistry based on our current understanding of those sciences. Certainly, different scientific disciplines are interrelated and share fundamental principles, but discrete disciplines continue to exist because phenomena are best understood at one level or another. In fact, it can be argued that in practice, disciplines such as physics and biology are epistemologically discontinuous, for science currently lacks a grand theory that allows us to connect such disparate phenomena as quantum mechanical states and the songs of birds. Epidemiology may be related to molecular biology, which in turn is related to chemistry and ultimately to physics, but the investigation of an ongoing cholera epidemic cannot be effectively carried out at the level of a molecule of cholera toxin or the quantum state of an electron around a single carbon atom within the toxin B subunit. In fact, the revolution in modern physics that replaced such bedrock assumptions of classical physics as continuity, separability, and determinism with discontinuity, entanglement, and the uncertainty principle has raised serious doubts about whether epistemological reduction can ever be realized. Exploring the epistemic relationships between different disciplines might be grist for the mill for a philosopher of science but does not seem like a particularly fruitful endeavor for a working scientist.

Ontological reductionism presents an even thornier issue. Ontological reductionism is defined as "the idea that each particular biological system is constituted by nothing but molecules and their interactions" (7), in other words, the centuries-old debate about whether physical matter is the only reality in nature. Instances in which esoteric mathematical knowledge has later been found to be perfectly suited for describing newly discovered physical phenomena have prompted contemplation of the "unreasonable effectiveness of mathematics" in describing the physical world and raised deep philosophical questions about the possibility of a Platonic reality beyond our measurements and senses (9). At this point, we find ourselves squarely within the realm of philosophy and feeling increasingly uncomfortable as we tiptoe gingerly through metaphysics.

The third category, methodological reductionism, describes the idea that complex systems or phenomena can be understood by the analysis of their simpler components. Methodological reduction is often traced back to Francis Bacon, who in the early 17th century proposed that principles derived from specific cases might be applied to make general predictions (10, 11). René Descartes soon afterward suggested that one should "divide each difficulty into as many parts as is feasible and necessary to resolve it" (12). As a contemporary example, a reductionistic approach would be to use a reporter fusion to the *ctxA* cholera toxin gene in order to identify environmental conditions responsible for regulating toxin production during infection (13). The experimenter would argue that regulation is most likely to occur at the level of transcription and that a simplified *in vitro* reporter system reduces the number of complicating experimental variables and facilitates analysis. An advocate of a more holistic approach could posit that cholera toxin gene expression is better studied during infection of a host and in the context of a genetic network of co-regulated loci monitored over time (14, 15). In this example, reductionistic and holistic methodologies can be viewed as alternative approaches to understanding a complex system, with each providing useful, but limited, information. This chapter will focus on the issue of methodological reductionism and leaves epistemological and ontological reductionism to the philosophers.

MOLECULAR BIOLOGY: A TRIUMPH OF REDUCTIONISM

If reductionistic methodology sounds familiar, that is because reductionism is implicit in much of molecular and cellular biology. Reductionism allows a microbiologist to explain that a bacterium fails to respond to therapy because it has acquired a gene encoding a beta-lactamase, which inactivates antibiotics, or that a patient exhibits enhanced susceptibility to infection because he has a mutant receptor for gamma interferon, which normally activates protective immune responses. Reductionism permits a microbiologist to screen *Salmonella* mutants for the ability to survive in cultured macrophages, knowing that this phenotype is predictive of the ability to cause mammalian infection (16). The successes of the reductionistic approach in biology during the latter half of the 20th century are undeniable, and yet limitations to methodological reductionism have been recognized. There are numerous examples of *in vitro* experimental observations made with isolated components of cells that are not directly applicable to the physiology of whole organisms. For example, mice deficient in Toll-like receptor 4 signaling are highly resistant to the effects of purified lipopolysaccharide but extremely susceptible to challenge with live bacteria (17, 18). The Instructions to Authors for the journal *Infection and Immunity* state that "papers that utilize conserved microbial constituents (e.g., lipopolysaccharide, peptidoglycan) to stimulate immune responses, unless accompanied by experiments demonstrating relevance to the interaction between intact microbes and hosts or host cells," are not within the scope of the journal. This is a tacit recognition of differences between pathogenic microbes and their parts and of the journal's preference for understanding the biology of whole organisms.

Some limitations of reductionism may reflect current technological capabilities rather than inherent shortcomings of the approach. An early triumph for reductionism was the discovery that one could separate tobacco mosaic virus (TMV) into its RNA and coat protein components, which could then self-assemble when combined (19). However, in contrast to TMV, the self-assembly of more-complex structures is often impossible. This underscores the relationship between the inherent complexity of a system under study and the limits of methodological reductionism. However, a 2010 report that a complete functional genome can be inserted into bacterial protoplasm through advances in synthetic biology (20) demonstrates that technological advancements can greatly empower and validate reductionistic approaches. The limitations of reductionism are a moving boundary.

EMERGENCE OF SYSTEMS BIOLOGY

The last decade has witnessed a backlash against the reductionism of molecular biology. The philosophical antecedents of holism can be traced back to Aristotle, who is said to have pithily observed that "the whole is more than the sum of its parts." Jan Smuts later coined the term "holism" as a tendency in nature to form wholes that are greater than the sum of their parts "through creative evolution" (21). Systems biology has increasingly been touted as a revolutionary alternative to molecular biology and a means to transcend its inherent reductionism (3, 22, 23). Theoretical biologists, such as Stuart Kauffman, have emphasized the ability of complex systems to give rise to emergent novel properties that are not predictable from the examination of individual components (24, 25). A humbling example is provided by the inability of detailed knowledge about the molecular structure of a water molecule to predict surface tension, a macroscopic phenomenon reflecting emergent behavior among multiple water molecules. The issue of emergence imposes a theoretical limit on the knowledge available from reductionistic methodology.

Systems biology has had a transformative effect on microbiology. An emphasis on pathways, networks, and systems has given rise to powerful new bioinformatic and experimental methods. Genomic, microarray, and proteomic analyses are now commonplace (26–28). Systems approaches can be "top-down," starting from "-omics" data and seeking to derive underlying explanatory principles, or "bottom-up," starting with molecular properties and deriving models that can be subsequently tested and validated (29). The first approach begins with data collection and a description of phenomena, while the latter is more mechanism based, but both produce models of system behavior in response to perturbation that can be tested experimentally, thus bringing together the descriptive and mechanistic approaches discussed in chapters 2 and 3. The construction of synthetic regulatory circuits, the modeling of complex genetic and metabolic networks, and the measurement of transcriptional dynamics in single cells are just some of the new ways of analyzing complex phenomena that have invigorated microbiology (30–34). Systems biology approaches are particularly attractive for analyzing the exceedingly complex events that occur as a host encounters a pathogenic microbe or a vaccine (14, 35, 36).

A FALSE DICHOTOMY

Methodological reductionism and holism are not truly opposed to each other (37). Each approach has its limitations. Reductionism may prevent scientists from recognizing important relationships between components or organisms in their natural settings, appreciating the evolutionary origins of processes and organisms, grasping probabilistic relationships underlying complicated and seemingly chaotic events, or perceiving heterogeneity and emergent multilevel properties of complex systems. Holism, on the other hand, is inherently more challenging due to the complexity of living organisms in their environment. Fundamental principles may be difficult to discern within complex systems due to confounding factors like redundancy and pleiotropy. Signal may be swamped by noise. The technology is seductive, but more data do not necessarily translate into more understanding. It is not yet certain whether current approaches to holism, such as systems biology, are adequate to cope with the challenges posed by emergent properties of complex biological systems. When fecklessly performed, systems biology may merely describe phenomena without providing explanation or mechanistic insight (38) or create virtual models that lack biological relevance.

It is difficult to imagine how a number of important scientific discoveries could have been made by any method other than a reductionistic approach. Without isolating DNA from other cellular constituents, Oswald Avery, Colin MacLeod, and Maclyn McCarty could not have conclusively demonstrated that DNA alone was responsible for the transformation of the pneumococcus (39). Similarly, the power of reductionism was shown when a single *Yersinia* gene could confer upon *Escherichia coli* K-12 the ability to invade eukaryotic cells in tissue culture (40) or when the replacement of the mouse E-cadherin protein with its human counterpart rendered transgenic mice susceptible to oral challenge with *Listeria* (41). Likewise, there have been important observations for which a holistic approach has been essential. The discoveries that high levels of expression are the predominant barrier to horizontal gene transfer between bacteria (42) and that *Helicobacter pylori* contains an unexpectedly large number of small untranslated RNAs and transcriptional start sites within operons (43) are but two examples. Confidence in these findings was critically dependent upon the researchers' ability to use holistic high-throughput methods to generate and analyze enormous data sets, in these instances, the attempted subcloning of nearly 250,000 genes and the sequencing of hundreds of thousands of cDNAs (complementary DNA molecules).

It should be emphasized that a combination of reductionistic and holistic approaches can be synergistic. In one example from the pathogenesis field, a holistic cRNA (complementary RNA) microarray analysis revealed that the RegIIIγ gene, encoding a C-type lectin, was strongly induced within intestinal Paneth cells following microbial colonization of germfree mice (44). The same lab subsequently went on to hypothesize that the RegIII lectin kills Gram-positive bacteria and demonstrated that it is able to bind the bacterial peptidoglycan carbohydrate backbone via a conserved molecular motif, confirmed by site-specific mutagenesis of a single amino acid in the tripeptide motif (45). In another example, a holistic genome-wide RNA interference (RNAi) screen was first used to identify host factors important for influenza virus replication (46). When the screen suggested that viral replication was dependent on the cell cycle regulatory protein p27, the investigators shifted to a reductionistic approach and were able to demonstrate reduced influenza virus replication in a p27-deficient mouse *in vivo*.

Reductionism and holism are in fact interdependent and complementary. Reductionism is most useful if observations made in a simplified system allow accurate predictions, or at least the generation of hypotheses, to be made when returning to the complex natural world. However, interpreting observations from holistic studies may require mechanistic insights gained from earlier reductionistic work or may generate hypotheses that are amenable to testing through reductionistic experimental approaches. Ironically, Hiroaki Kitano noted that systems biology became possible only once advances in molecular biology allowed the emergence of genomic analysis and high-throughput measurements (3). We conclude that one approach is not necessarily better than another. Observations made in test tubes that have no correlates in the real world may not be very useful biology, but the mere creation of large data sets without interpretation, or holistic cartoon models that fail to achieve concordance with empirical reality, is also of little value.

THE WAY FORWARD

How can these alternative ways of doing science be reconciled? Investigators employing a reductionistic approach should attempt to test the predictive power of their observations in a more complex setting. For example, a biochemical study of protein-protein interactions should obtain evidence that such interactions and their consequences occur in an intact cell. An *in vitro* study of microbial resistance to a stress condition could be enhanced by experiments to determine whether the mechanism applies to interactions with host cells in which the particular stress condition occurs. A study showing the behavior of a microbe infecting host cells in tissue culture might be fruitfully expanded to include a bona fide infection of an animal host. Similarly, investigators should attempt to determine the degree to which reductionist findings are generalizable to other systems. An immunological study that shows the importance of a certain response in mice should be tested in other animal models or, where possible, in humans, to ascertain whether general conclusions can be drawn. System-wide models, whether describing interactions of genes, proteins, small molecules, or organisms, should be rigorously tested and refined against real-life observations. Attempts should be made to identify the general organizing principles that underlie complex phenomena (47), and areas of discordance between predicted and observed results must be forthrightly addressed.

The recent focus on systems biology in microbiology is not a revolution or even a true paradigm shift (see chapter 11), in the sense that reductionistic and holistic methodological approaches have been coexisting and thriving for centuries. One can argue that Darwin's theory of evolution represents an early example in which many reductionist observations on finches and domesticated pigeons were synthesized into a system that unified all of biology. The real seismic event in the recent rise of systems biology arguably has more to do with the introduction of computer technology that allowed inexpensive calculation and the storage of prodigious amounts of information than with new conceptual approaches. Nevertheless, there is no denying the revolutionary impact of holistic thinking on the field, both in calling attention to situations in which reductionistic approaches have been deficient and in the generation of new experimental approaches for the analysis of

DNA → mRNA → protein (1958)

DNA → mRNA → protein → protein interactions → pathways
→ networks → cells → tissues → organisms → populations →
ecologies (2001)

FIGURE 4.1 *Modern extension of the central dogma of molecular biology (49–51).*

complex systems. Computer technology has permitted the development of sophisticated mathematical, engineering, and computational tools that have allowed new questions to be asked. The central dogma of molecular biology may not have been overturned, but it certainly has been extended (Fig. 4.1) (2).

Whether one's methodology is primarily reductionistic or holistic, it is wise to begin by considering the limitations of the approach. This will help to limit imprudent extrapolation and point the way for further experimentation. In the end, the test of both reductionistic and holistic paradigms is their ability to explain and make useful predictions about the real world. No one said it would be easy. As Douglas Adams said, "If you try and take a cat apart to see how it works, the first thing you have on your hands is a non-working cat. Life is a level of complexity that almost lies outside our vision" (48).

GOOD SCIENCE

5 Elegant Science

If one ignores contradictory observations, one can claim to have an "elegant" or "robust" theory. But it isn't science.

Halton Arp (1)

Among the qualities most highly valued by scientists is "elegance." Here we consider the meaning of elegance in the context of science. We humbly acknowledge that many who have come before us have found scientific elegance difficult to define. Interest in scientific elegance is not limited to scientists; the subject has been covered in the *New Yorker* (2) and *Atlantic Monthly* (3) magazines. In science, "elegance" and its cousin "beauty" have often been used in the context of physics. According to a 2002 *New York Times* article, the 10 most beautiful scientific experiments of all time were all in the field of physics (4). However, we suggest that elegance can also apply to biology, geology (Box 5.1), medicine (Box 5.2), and all fields of science.

The word "elegant" entered the English language in the 1400s, a French word (*élégant*) that was derived from the Latin *elegantem* (5). The Merriam-Webster online dictionary defines "elegant" as "tasteful richness of design or ornamentation, dignified gracefulness or restrained beauty of style," and in the specific context of science, as exhibiting "precision, neatness, and simplicity" (6).

Ian Glynn has argued that the essence of scientific elegance is simplicity and explanatory power, while at the same time noting that its appreciation requires

Thinking about Science: Good Science, Bad Science, and How to Make It Better, First Edition.
Ferric C. Fang and Arturo Casadevall.
© 2024 American Society for Microbiology.

Box 5.1 Elegance in the crust

To illustrate the Pentateuch of Elegance in scientific thought, we return to the theory of plate tectonics (see also chapter 2). Through the ages humans have pondered the following questions: How are mountains made? Why does the earth shake during earthquakes? Why do volcanoes occur in some places and not others? As mapmaking improved, cartographers noted the curious fact that some continents such as South America and Africa fit together almost as if they were parts of a jigsaw puzzle. As early as 1596, a Dutch cartographer proposed that these two continents were torn apart by earthquakes and floods. However, that idea went nowhere since people could not imagine how continents could move. Alfred Wegener revived the idea of continental drift in the early 20th century and began a systematic search for evidence focusing on the similarities in the geology and fossils along continental fissure lines (Box 2.3). He was widely ridiculed and destined to die on an ill-fated expedition to Greenland looking for more fossil evidence. However, evidence to vindicate Wegener's idea accumulated over the 20th century from such diverse fields as paleontology, geology, and sub-Atlantic and magnetic cartography, which ultimately led to the theory's acceptance.

Considering the five pillars of elegance, we note that plate tectonics is *parsimonious* and can be summarized as the earth's crust consisting of different plates that float on magma and move over time. Plate tectonics is *clever*, for it required a synthesis of very different types of evidence. It exhibits *clarity*, for it can be simply explained and understood by nonspecialists, is relatively simple, and is unencumbered by caveats or exceptions. For *correctness*, continental drift, which moves at the speed of nail growth, has now been measured by the global positioning system. Most spectacular is its *explanatory* power. Plate tectonics tells us that volcanoes and earthquakes occur at the edges of moving plates, that mountains rise when plates collide, that the earth's surface is in constant motion, that ancient epochs saw supercontinents like Pangaea and Rodinia, and that the Atlantic and Pacific Oceans are getting larger and smaller, respectively. Plate tectonics does not just explain geology but extends into space to explain why the Moon and Mars are cratered (they lack moving plates), the distribution of fossils, and the periodic glaciation of our planet as moving plates modified ancient climates. In microbiology, one of us (A.C.) has used this theory to explain the geographical distribution of fungal species belonging to the genus *Cryptococcus* (29). The elegance of plate tectonics illustrates the majesty of human thought and insight, which can provide a powerful antidote to the depression that can sometimes ensue from reading about the foibles of human history.

Box 5.2 Elegance in medicine

Clinical medicine is a subfield of biomedical science focused on treating disease and promoting health. Medicine differs from other sciences by its major practice component, which seeks to translate scientific advances into improved health outcomes. An example of elegance in medicine is the prevention of cervical cancer, a disease that is nearly always fatal if untreated. In the early 20th century, epidemiological studies revealed an association between cervical cancer and the number of sexual partners, suggesting the possible involvement of a sexually transmitted pathogen. Although technology at the time could not initially identify the microbial cause, a method developed in the 1920s called a "Pap smear" allowed the detection of abnormal cells in the cervix, which could often be treated before progression to cancer. In the 1970s and 1980s, molecular methods linked specific strains of human papillomavirus (HPV) to the development of cervical dysplasia and cancer (30). This in turn led to led to diagnostic tests for HPV and a vaccine that can prevent women from developing cervical cancer in the first place.

The physician Lewis Thomas described three stages in the evolution of medical technology (31). The first is nontechnological, in which clinicians care for patients but have nothing to alter the natural history of the disease. The second is "halfway technology," which can help to mitigate the course of a disease but not reverse it altogether. The third is transformative technology, which is so effective that the disease is quickly reversed or even eliminated altogether. Vaccines and antibiotics fall into the last category. Pap smears provided a halfway technology, which allowed the early recognition of cancer or precancerous lesions but were cumbersome and incompletely effective. The HPV vaccine now provides an elegant solution to the problem of a terrible disease, and it is an effective solution too, judging from recent data from Sweden (32).

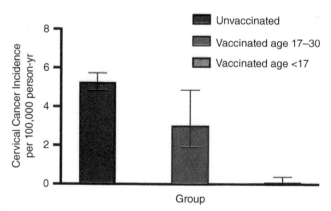

Rates of cervical cancer in Swedish women stratified by vaccination status. Vaccination against HPV early in life prevents the development of cervical cancer.

a historical context (7). An article celebrating the discovery of the C_{60} molecule buckminsterfullerene noted that an elegant theory or model must explain a phenomenon "clearly, directly and economically" (8). The *New Yorker* article adds that elegance in science requires "simplicity plus capaciousness" in explanatory power (2). Marco Nathan and Diego Brancaccio have gone further, arguing that elegance is "an intrinsic feature of successful scientific practice and observation, a benchmark that demarcates between good experiments and bad ones" (9). This formulation seems to imply that elegance could apply to all good science and that high quality is an intrinsic characteristic of scientific elegance.

The question of whether elegance is merely desirable or essential for good science is particularly relevant to current concerns about rigor and reproducibility in biomedical research (chapters 6 and 7) (10). If elegance in science is just an attractive attribute, then elegance is not a necessary goal but simply something to be admired when it happens. However, if elegance is a requisite feature of good science, then the characteristics defining elegance deserve the same attention given to scientific rigor (chapter 6). We will unpack these possibilities with examples provided from the biological sciences.

A search of the PubMed database for the word "elegant" yields the titles and abstracts of nearly 7,000 publications. Browsing through those publications reveals that the word "elegant" is used to describe models, experiments, methods, or theories. Furthermore, the appearance of the word "elegant" in PubMed publications has risen rapidly in the 21st century. In the mid-1990s, fewer than 100 articles used the word "elegant" in titles and abstracts, whereas in recent years the number of articles using this word now routinely approaches 500 per year (Fig. 5.1). Hence, biomedical scientists are increasingly using the word "elegant" in the context of their work.

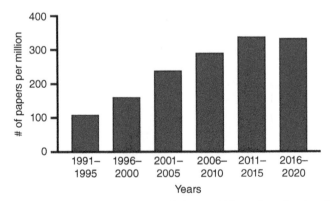

FIGURE 5.1 *Journal articles in PubMed containing the keyword "elegant" as a function of publication year. The number of papers has been corrected for the number of total journal articles in each year.*

EXAMPLES OF ELEGANT SCIENCE IN THE BIOLOGICAL SCIENCES

The structure of DNA is often referred to as an elegant model. In 1953, James Watson and Francis Crick proposed that the structure of DNA is a double helix (11). Linus Pauling had preceded Watson and Crick in proposing a DNA structure based on a triple helix (12). However, the Watson-Crick model was simpler, accounted for Erwin Chargaff's rules (stating that DNA from any organism must have a 1:1 ratio of purine and pyrimidine bases), and immediately suggested a mechanism for replication. In contrast, none of these criteria was met by Pauling's three-stranded structure. Hence, the double helix was more elegant than the triple-helical model because of its simplicity and greater explanatory power. Subsequent experimental work established the correctness of Watson and Crick's model. In this example, we note the qualities of simplicity, correctness, and explanatory power that help to define elegance.

The Watson-Crick model made the testable prediction that DNA replication would be semiconservative, providing the basis for the Meselson-Stahl experiment. In 1958, Matthew Meselson and Franklin Stahl isotopically labeled DNA bases and separated the products by ultracentrifugation to show that each new strand of DNA is built upon a previously existing strand (13). This was called the "most beautiful experiment in biology" by John Cairns (14). The experiment reinforced the Watson-Crick model and ushered in an era of experimentation that continues uninterrupted to this day (15). Again, the elegance of the Meselson-Stahl experiment is derived from its conceptual simplicity and broad explanatory power. Even so, it is noteworthy that the conclusions of the Meselson-Stahl experiment were not immediately accepted due to alternative explanations that were raised at the time (15).

Elegance also applies to the method of PCR developed by Kary Mullis to amplify DNA from minute quantities of single-stranded template (16, 17). PCR-related technologies have had a tremendous impact on numerous fields, including biomedical research, diagnostics, anthropology, and criminology, to name just a few. As we will discuss further in chapter 11, we believe that the development of PCR can be considered an example of revolutionary science (18) as well as elegant science. Although the fundamental idea of amplifying DNA by denaturing a DNA target and amplifying a segment with synthetic primers and polymerase was published a decade earlier (19), Mullis had the clever and transformative insight of using a heat-stable polymerase from a thermophilic microbe, which made the method simpler and faster. The elegance of PCR allowed the development of convenient and affordable automated platforms that could perform DNA amplification in a wide range of settings. The most elegant theory in the biological sciences is unquestionably Charles Darwin's theory of evolution by natural selection, which was all the more remarkable for being proposed before genes or DNA were known.

As with PCR, some of the basic principles underlying natural selection were proposed well before Darwin published *On the Origin of Species* (20); new species had been suggested to arise from existing species, and Jean-Baptiste Lamarck had hypothesized that speciation arose in response to environmental demands. Darwin's crucial insights were that differential fitness of individuals within a population could lead to differences in their survival and reproduction and that the ancestral relationships between species were reflected in the relatedness of their essential characteristics. This theory had the essential virtues of elegance: simplicity, clarity, and explanatory power. In addition, Darwin's theory had the virtue of being clever. An obvious idea could be simple, clear, and correct with great explanatory power, without necessarily being considered elegant. To be truly elegant, an idea, once proposed, should cause others to exclaim, as T. H. Huxley did, "How extremely stupid not to have thought of that!"

SIMPLICITY AND ELEGANCE

The theme of simplicity runs deep in elegant science, and definitions of elegance invariably contain such words as "simplicity," "neatness," and "economy." However, it is not immediately obvious that simplicity should be considered necessary or sufficient for scientific elegance. Elegance does not connote simplicity in fashion, as it is hard to imagine a bikini or a string tie being viewed as universally elegant. The association of simplicity with elegance in science may relate to the principle of parsimony also known as Ockham's razor, which states that the simplest explanation is the most likely to be correct. This in turn implies that for science to be elegant, it must also be correct. What William of Ockham, a 14th-century theologian, actually wrote was *"Pluralitas non est ponenda sine necessitate,"* which can be translated as "Plurality should not be posited without necessity." This principle is not without risk. As Crick observed, "Ockham's razor ... can be a very dangerous implement in biology. It is ... very rash to use simplicity and elegance as a guide in biological research" (21). He also said that "God is a hacker, not an engineer" (22), implying that evolution works without foresight and in doing so may solve problems in rather inelegant ways. An elegant theory can actually impede research progress if it delays the appreciation of conflicting observations due to confirmation bias (23).

Nevertheless, we suggest that parsimony is preferable to simplicity, as simplicity in science is relative. For example, when some physicists expressed frustration over the complexity of Albert Einstein's theories of relativity in comparison to Isaac Newton's formulas, he responded that "It can scarcely be denied that the supreme goal of all theory is to make the irreducible basic elements as simple and as few as possible," but he cautioned that this must be achieved "without having to surrender the adequate representation of a single datum of experience" (24). This in turn led to the aphorism that "Everything should be made as simple as possible, but no simpler." Hence, there are limitations on the relationship between simplicity and elegance in science.

A DEFINITION OF SCIENTIFIC ELEGANCE

From the foregoing discussion, we propose that to be considered elegant, science should meet the five criteria of being clear, clever, correct, explanatory, and parsimonious (Fig. 5.2). This concept of scientific elegance parallels our Pentateuch for improving rigor in science (chapter 6) (10) in its 5 essential components. The symmetry serves as a reminder that beauty is an important quality in the philosophy of science. Perhaps our definition of elegance also requires a measure of aesthetics—is the science beautiful? For example, imagine the complete dissection of a signaling pathway involving dozens of interacting components. Would this be considered elegant science? According to our proposed definition, such a pathway could be considered an example of elegant science, but the beauty of its construction may only be apparent to the cognoscenti. They might argue that such a system can be regarded as beautiful if one considers the challenge of communicating information to the cell rapidly, faithfully, and efficiently, in order to allow an appropriate homeostatic response. Our proposed definition can only provide boundary conditions that are necessary but not sufficient for scientific elegance. To be elegant, a scientific discovery or theory must also meet an aesthetic criterion in the mind of the observer. In essence, the discernment of elegant science is analogous to the appreciation of art, which means that the quality of elegance in science is ultimately a human judgment.

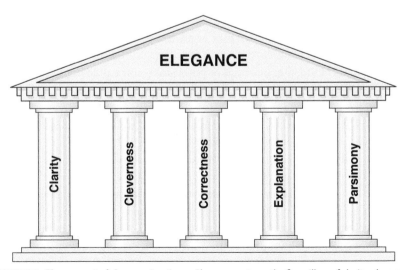

FIGURE 5.2 *The concept of elegance in science. Elegance rests on the five pillars of clarity, cleverness, correctness, explanation, and parsimony.*

STRIVING FOR ELEGANCE IN SCIENTIFIC RESEARCH

Although scientific rigor (10) is necessary but not sufficient for elegant science, it should be possible to make scientific work more elegant by considering the elements of elegance in research design. One may certainly strive for clarity, correctness, parsimony, cleverness, and explanation when designing experiments, perhaps leaving the more subjective quality of beauty to the appreciation of posterity. Emphasis on clarity and correctness can help to improve the reproducibility of science (see chapter 7). Achieving parsimony requires the generation and consideration of alternative theories, models, or methods that can enhance scientific work by opening a researcher's mind to other possibilities. Finally, asking whether a proposed experiment will definitively answer a question and explain the problem at hand can improve experimental design. Hence, the components of our definition of elegance (Fig. 5.2) are mutually self-supporting and represent important elements of good scientific work. The quest for elegance can make science better even when elegance is not achieved. Hence, this quest should be a goal of all scientists.

ELEGANT SOLUTIONS CAN TAKE TIME

In May 1905, Albert Einstein was still working as a young clerk in the Swiss patent office. As he headed home one night, he glanced at the clock tower in the center of Bern and suddenly had the thought, what if the streetcar in which he was riding was moving away from the tower at the speed of light? As Einstein recalled, "A storm broke loose in my mind" (25). He published a paper describing the special theory of relativity only 4 months later. This story fits neatly into the mythology of scientific "eureka" moments in which momentary flashes of insight quickly lead to elegant breakthroughs. But that may be an exception. In contrast, the theory of plate tectonics took centuries to develop from the curious observation that continental contours fit together like a jigsaw puzzle to the measurement of continental drift by global positioning systems (Box 5.1). In another example, an Italian physician named Antonio Rigoni-Stern first noted the rarity of cervical cancer in celibate nuns in 1842 (26), but a link between sexually transmitted human papillomavirus and cervical cancer was only established in the 1980s (27), and an effective vaccine to prevent cervical cancer took two more decades to develop (Box 5.2).

Despite the obvious appeal of elegance, one should not forget Crick's admonition mentioned above that nature sometimes works in inelegant ways; the seductiveness of an elegant theory may prevent a researcher from recognizing a less attractive truth. The elegance-seeking scientist would also do well to recall the words of the neuroscientist V. S. Ramachandran: "The quest for biological laws shouldn't be driven by a quest for simplicity or elegance. ... No woman who has been through labor would say that it's an elegant solution to giving birth to a baby" (28).

6 Rigorous Science

> *There is no rigorous definition of rigor.*
>
> Morris Kline (1)

Rigor is a prized quality in scientific work. Although the term is widely used in both scientific and lay parlance, it has not been precisely defined (1). Rigor has gained new prominence amid concerns about a lack of reproducibility in important studies (chapter 7) (2, 3), an epidemic of retractions due to misconduct (chapter 24) (4), and the discovery that the published literature is riddled with problematic images (chapter 17) (5). Insufficient rigor may be slowing the translation of basic discoveries into tangible benefits (chapter 12) (6, 7). New initiatives aim to understand deficiencies in scientific rigor and to make research more rigorous (8–10). Here, we consider the meaning of rigorous science and how it can be achieved.

The word "rigor" is derived from an old French word, *rigueur*, meaning "strength" and "hardness" (11). In scientific vernacular, the underlying concept of strength resonates in the expressions "hard data" and "solid work" used to convey a sense of reliable and trustworthy information. In common usage, the word "rigor" has evolved to mean the quality of being exact, careful, or strict (12). Although the words "exact" and "careful" also apply to science, additional definition is needed since practicing rigorous science means more than mere exactness and care in experimental design. An experiment in which all components were exact in their proportions and the procedures carefully executed would still not be considered

Thinking about Science: Good Science, Bad Science, and How to Make It Better, First Edition.
Ferric C. Fang and Arturo Casadevall.
© 2024 American Society for Microbiology.

rigorous in the absence of appropriate controls. Hence, the definition of scientific rigor requires a deeper exploration than can be provided by simple perusal of the dictionary.

The scientific literature adds surprisingly little to our understanding of rigor, with the term almost always used without definition, as if its meaning were self-evident. In 2016, the National Institutes of Health (NIH) defined scientific rigor as "the strict application of the scientific method to ensure robust and unbiased experimental design, methodology, analysis, interpretation and reporting of results" including "full transparency in reporting experimental details so that others may reproduce and extend the findings" (13). While we credit the NIH for providing a starting point for discussion, we find the NIH definition of rigor to be both excessively wordy and disconcertingly vague, as well as complicated by an insistence on transparency and reproducibility, which may be desirable but are arguably separate from rigor.

A WORKING DEFINITION OF SCIENTIFIC RIGOR

We suggest that rigorous science may be defined as theoretical or experimental approaches undertaken in a way that enhances confidence in the veracity of their findings, with veracity defined as truth or accuracy. Rigorous science could be entirely theoretical, as exemplified by a thought experiment used to illustrate a principle, such as (Erwin) Schrödinger's cat or (James Clerk) Maxwell's demon in physics, or entirely experimental, as illustrated by Henry Cavendish's measurement of the gravitational constant at the end of the 18th century. However, in the biomedical sciences, most research has both theoretical and experimental aspects.

A PENTATEUCH FOR SCIENTIFIC RIGOR

Different fields vary in the level of uncertainty that they are willing to accept with regard to conclusions. Certainty in science is often couched in terms of the probability that the null hypothesis may be rejected, which in turn depends on the methodologies employed. For example, the Higgs boson was announced when physicists were certain to "five sigma" or a P value of 3×10^{-7} (14). In contrast, many biological and medical studies accept a P value of 0.05, although more-stringent criteria have been advocated (15). Does this make physics more rigorous than biology? Not necessarily—differences in the complexity of physical and biological phenomena as well as limitations in methodology determine the level of certainty that is practically achievable in these disciplines. Hence, a definition of rigorous science cannot rely on strict and arbitrary levels of certainty.

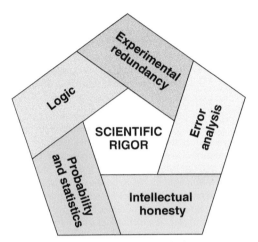

FIGURE 6.1 *A Pentateuch for improving rigor in the biomedical sciences.*

Traditional Chinese philosophy, Hinduism, Islam, and Judaism are each founded on five elements, pillars, or sacred texts. In Judaism, the first five books of the Hebrew bible are collectively referred to as the Pentateuch. Here, we humbly propose a Pentateuch for scientific rigor (Fig. 6.1).

1. **Redundancy in experimental design.** Good laboratory practices include proper controls, dose-response studies, determination of time courses, performance of sufficient replicates, and corroboration of major findings using independent experimental approaches and methods. It is important to establish whether a finding is generalizable, using a variety of cell types or species. New findings should lead to new predictions, which can in turn be experimentally tested. Experimental confirmation of predictions provides added assurance that the original findings are valid. Like rigor, redundancy is a multidimensional quality composed of many elements (Table 6.1). Redundancy in experimental design can enhance confidence in experimental results.

2. **Sound statistical analysis.** As we discussed in chapter 2, the progression of biology from a qualitative science focused on classification to a quantitative science based on numerical data represents the maturation of the discipline. However, this means that biologists must become conversant in the principles of probability and statistics. Attention to power calculations and other statistical considerations is essential for a more rigorous approach to science. A determination of certainty is not enough—the size of an observed effect is crucial. A large effect is more likely to be important and perhaps more reproducible as well.

TABLE 6.1 Some elements of scientific redundancy

Element[a]	Description and implementation
Replication	Carry out independent replicates, which provide information regarding the replicability and variability of an observation. This in turn influences the interpretation of the magnitude of the observed effects and the sample size required for statistical significance.
Validation	Validate the observation by an independent methodology. For example, the assignment of a protein on an immunoblot can be validated by immunoprecipitation using different antibodies, and the differential abundance of an mRNA by microarray or transcriptome sequencing (RNA-seq) can be validated by quantitative reverse transcription-PCR. This element also applies to purifications, which should use at least two independent methodologies. For example, chromatography can be complemented with differential sedimentation, electrofocusing, precipitation, etc. The element of validation is particularly important for experimental components such as antibodies and cell lines.
Generalization	Explore the generalizability of the findings. For example, in microbiological studies the use of different strains, cell lines, reagents, media, experimental conditions, etc., can be used to ascertain the generalizability of the finding. Findings that are generalizable are more likely to be robust.
Perturbation	Define the conditions where the observation occurs by perturbing the system. For example, before reporting a biochemical observation, perturb the system by changing the pH or the ionic strength of the experimental conditions. Knowledge of the perturbation boundaries introduces redundancy since it inevitably includes replication and generalization and reveals the degree of resiliency, which in turn enhances the likelihood of replication.
Consistency	Determine whether the various observations that define a scientific study are internally consistent. Although internal consistency does not necessarily imply validity, its absence may suggest the presence of uncontrolled variables.

[a] This list of elements is not exhaustive. The examples are provided to illustrate the principle of redundancy in experimental design. We note that some of these elements are interrelated and thus not independent. For example, any effort to validate or generalize a finding also involves replication. However, the elements listed are sufficiently distinct as to be considered independently when analyzing the redundancy of a scientific study.

3. **Recognition of error.** Error is pervasive in research (16), and any experimental technique ranging from simple pipetting to crystallographic analysis is subject to error. Errors can be random or systematic and tend to propagate with multiple experimental steps. Random errors may occur during any experimental procedure, are unpredictable, and can be estimated by performing replicates. Systematic errors tend to be associated with scientific instruments but can also be introduced by the use of impure reagents, contaminated cell

lines, etc. A rigorous approach to science must include a full appreciation of potential sources of error and of how error can affect experimental results and the incorporation of processes to authenticate key reagents and resources. In engineering, a sensitivity analysis examines how uncertainty or error in various parameters can influence the output of a system. Applying the principles of sensitivity analysis to scientific experiments can reveal the likelihood that a given result is true as well as identify potential sources of error. Results that remain robust despite variance in experimental conditions are more likely to be valid.

4. **Avoidance of logical traps.** Logical traps and fallacies lurk everywhere in experimental science, especially in the interpretation of results. The list of logical fallacies that can befall an investigator is lengthy and includes confirmation bias, congruence bias, affirming the antecedent, denying the antecedent, base-rate fallacy, etc. A bedrock principle of science is the possibility of falsification—confirmatory evidence cannot prove an assertion, but contradictory evidence may disprove it (17). Given that confirmatory evidence is not conclusive, scientists can enhance the rigor of their work by systematically challenging and attempting to falsify their hypotheses. Avoiding logical fallacies is therefore essential for a rigorous approach to science, but doing so requires training in critical thinking and avoiding illogical thought patterns that often come naturally to humans.

5. **Intellectual honesty.** Intellectual honesty is a mindset that encompasses diverse concepts ranging from ethics to proper scientific practice (18). Intellectual honesty is synonymous with objectivity, an essential requirement for scientific rigor (Box 6.1). Acknowledgment of nagging details that do not fit with one's hypothesis is often the first step to a new understanding and a better hypothesis. Implicit in the acknowledgment of earlier work is the need to reconcile one's observations with those made by others. Corroboration by independent researchers can enhance the confidence of the scientific community that a scientific finding is valid.

Box 6.1 Rigor, deception, and intellectual honesty

Our Pentateuch for rigor requires intellectual honesty, which is a broad term that includes honesty in considering multiple explanations to honest experimental work. At a time when there is concern about the prevalence of misconduct in science (see also chapter 18), the relationship between personal integrity and rigorous science raises an interesting question: Are rigor

(*Continued*)

Box 6.1 (Continued)

and deception compatible? Given that many fraudulent papers are published in leading journals despite careful peer review, one must conclude that deception can be done with *apparent* rigor. Lorne Hofseth has labeled the deliberate falsification of data as "insidious rigor" (26), but our definition of rigor is fundamentally incompatible with any sort of falsification. The aforementioned NIH definition of scientific rigor as "the strict application of the scientific method to ensure unbiased and well-controlled experimental design, methodology, analysis, interpretation and reporting of results" (27) leaves no room for deception. While it may be possible for a fraudulent investigator to manufacture results that include some of the ingredients of the Pentateuch of rigor, and simulate the appearance of rigor, fraudulent research cannot be rigorous, for such research lacks intellectual honesty. Careful application of the Pentateuch of rigor to a fraudulent scientific finding may reveal inconsistencies that help uncover the fraud. Many fraudulent studies are discovered when the findings cannot be replicated by other investigators, leading to questions that uncover fraud and demonstrating yet another connection between rigor and reproducibility (Box 6.1). Hence, rigor in science requires ethics and personal integrity on the part of scientists, and an insistence on scientific rigor can help to ensure the integrity of science.

As illustrated by these five principles, scientific rigor is multifaceted. No single criterion can define it. Even the most careful experimental approach is not rigorous if the interpretation relies on a logical fallacy or is intellectually dishonest. On the other hand, the principles of rigor can be synergistic, as when a logical approach and the awareness of error lead to greater purposeful redundancy in experimental design.

RIGOR AND REPRODUCIBILITY

Rigorous scientific practices enhance the likelihood that the results generated will be reproducible, and we will discuss reproducibility in greater detail in chapter 7. However, reproducibility is not an absolute criterion of rigorous science. One might rigorously characterize the mass, composition, and trajectory of a comet that collides with the sun, but those measurements could never be reproduced since each comet is unique. Nevertheless, improvements in scientific rigor are likely to improve reproducibility (Box 6.2).

Box 6.2 Rigor and reproducibility

The words "rigor" and "reproducibility" are often used together in reference to desirable qualities of science. The topic of reproducibility is covered separately in chapter 7. Although related, these qualities are independent. It is possible to do highly rigorous science that is not reproducible and reproducible science that is not rigorous. For example, in nature there are many unique events, as evident by the unique qualities of celestial objects such as asteroids, planets, stars, and galaxies. Are rigorous studies of these celestial objects reproducible? On reflection, the answer is both yes and no, but perhaps the question is not formulated properly. Mars is a unique object, and planetary insights learned from Mars might not be reproducible when applied to other planets, irrespective of rigor and accuracy. On the other hand, rigorous studies of Mars should lead to reproducible results, provided that all parameters remain constant. That last caveat is important. Consider the question of whether there is methane on Mars, which is relevant to whether microbial life exists there. Highly sophisticated instruments operated by highly competent scientists have produced evidence both for and against the presence of methane on Mars (28, 29). Explanations for the lack of reproducibility may be instrument error or variable methane emissions, or anything in between. Hence, the finding of methane on Mars may be real and variable or erroneous, and the only way to discriminate between these possibilities is to do more measurements, potentially with different instruments and at different times of year (30, 31). At the other extreme, highly reproducible results may be wrong (32). For example, measurements might be analyzed by a computer program that contains an error. For both the quest for methane on Mars and the possibility of software errors, incorporating redundancy into the experimental approach can improve the rigor of the experimental work and avoid error. Rigor increases the likelihood of reproducibility, while reproducibility enhances confidence in the rigor of the research.

ENHANCING RIGOR IN BIOMEDICAL RESEARCH TRAINING

The five principles outlined above provide a road map for increasing rigor in the biomedical sciences (Fig. 6.1). An obvious step is to strengthen didactic training in experimental design, statistics, error analysis, logic, and ethics during scientific training. We have previously called for improvements in graduate education, including areas of philosophy such as logic, as well as in probability and statistics (19). A 2016 survey of more than 1,500 scientists found that there is considerable support for such reforms (20). However, with the exception of statistics, none of these disciplines are formally taught in the training of scientists, and even statistics is not always a curricular requirement (21). Reforming scientific education will be a major undertaking since most academic faculty lack the necessary background to teach this material. A multifaceted approach would include greater attention in the laboratory environment and improvements in peer review and constructive criticism, as well as formal didactics. This would need to be accompanied by a cultural change in the scientific enterprise to reward rigor and encourage scientists to recognize, practice, and promote it. To initiate the process, we make the following five actionable recommendations.

1. **Develop didactic programs to teach the elements of good experimental practices.** Most biomedical scientists currently learn the basics of scientific methods and practice in their graduate and postdoctoral training through a guild-like apprenticeship system in which they are mentored by senior scientists. Although there is no substitute for good mentorship, there is no guarantee that mentors themselves are well trained. Furthermore, there is no guarantee that such individualized training will provide the basic elements of good experimental science. Today, training programs seldom include didactic programs to teach good research practices. Such courses could ensure a baseline background for all trainees.

2. **Require formal training in statistics and probability for all biomedical scientists.** As the biological sciences have become increasingly quantitative, they have become increasingly dependent on mathematical tools for all aspects of experimental design, implementation, and interpretation. Although the use of statistical tools has become widespread in biomedical research, these tools are often misused. Some authorities have gone so far as to issue a warning on the use of P values (22, 23), and the American Statistical Association (ASA) has declared six principles relating to P values (24) (Table 6.2). Clearly there is a need for more formal mathematical training in the biomedical sciences, a recommendation made previously by us and others (19, 21).

3. **Refocus journal club discussions from findings to methods and approaches.** Journal clubs are invaluable training formats that allow discussions of science in the context of a specific publication. However, articles selected for journal club discussions are often the flashiest papers from high-impact journals, which limit article length and truncate descriptions of methodology. Journal club sessions that focus on limitations of methodology, appropriateness of controls, logic of conclusions, etc., could provide a recurring venue for discussions of what constitutes scientific rigor.

TABLE 6.2 American Statistical Association principles for the use of *P* values[a]

P values can indicate if data are incompatible with a specified statistical model
P values do *not* measure the probability that a hypothesis is true or that the data result from random chance alone
Scientific conclusions should not be based solely on whether a *P* value passes a specific threshold
Proper inference requires full reporting and transparency
A *P* value does not measure the size of an effect or the importance of a result
By itself a *P* value does not provide a good measure of evidence regarding a model or hypothesis

[a] Modified from reference 24.

4. **Develop continuing education materials for biomedical sciences.** Fields in which basic information is changing rapidly, such as medicine, require continuing education to maintain competence. Science is arguably one of the most rapidly changing areas of human endeavor, and the biological sciences have experienced a revolution since the mid-20th century. Fields could develop continuing education materials for scientists that would allow them to gain proficiency in the latest techniques. Guidelines for the practice of science, analogous to those used for clinical practice and publication ethics, could help to establish standards that promote rigor.

5. **Develop teaching aids to enhance the quality of peer review.** The scientific process is critically dependent on peer review in publication and grant funding (see chapter 22). Remarkably, scientists are recruited into peer review activities without any training in this process, and the results can be uneven. As peer review is a process, it is amenable to study in order to identify best practices that can be promulgated to educate and improve reviewer performance. Strengthening peer review will help to ensure the quality of scientific information.

Enhancing scientific rigor can increase the veracity and reproducibility of research findings. This will require a tightening of standards throughout the scientific progress from training to bench science to peer review. Today's reward system in biomedical research is primarily based on impact (25) (see chapter 26), and impactful work is highly rewarded regardless of its rigor. Consequently, a scientist may obtain greater rewards from publishing non-rigorous work in a high-impact journal than from publishing rigorous work in a specialty journal. Perhaps it is time to rethink the value system of science. Prioritizing scientific rigor over impact would help to maintain the momentum of the biological revolution and ensure a steady supply of innovative, reliable, and reproducible discoveries to translate into societal benefits. Perhaps it will even become *de rigueur.*

7 Reproducible Science

Non-reproducible single occurrences are of no significance to science.

Karl Popper (1)

Reproducibility is a bedrock principle in the conduct and validation of experimental science. Consequently, scientific journal articles routinely include detailed information on the methodology used and on the reproducibility of experiments. Articles may describe findings with a statement that an experiment was repeated a specific number of times, with similar results. Alternatively, depending upon the nature of the experiment, the results from multiple experimental replicates might be presented individually or in combined fashion, along with an indication of experiment-to-experiment variability. For most types of experiment, there is an unstated requirement that the work be reproducible, at least once, in an independent experiment, with a strong preference for reproducibility in at least three experiments. The assumption that experimental findings are reproducible is a key criterion for acceptance of a manuscript, and instructions to authors typically insist that "the description of materials and methods should include sufficient technical information to allow the experiments to be repeated."

In this chapter, we explore the problem of reproducibility in science. In exploring the topic of reproducibility, it is useful to first consider terminology. "Reproducibility" is defined by the *Oxford English Dictionary* as "the extent to which consistent results are obtained when produced repeatedly." Although it is

Thinking about Science: Good Science, Bad Science, and How to Make It Better, First Edition.
Ferric C. Fang and Arturo Casadevall.
© 2024 American Society for Microbiology.

taken for granted that scientific experiments should be reproducible, it is worth remembering that irreproducible one-time events can still be a tremendously important source of scientific information. This is particularly true for observational sciences in which inferences are made from events and processes not under an observer's control. For example, the collision of comet Shoemaker-Levy with Jupiter in July 1994 provided a bonanza of information on Jovian atmospheric dynamics and *prima facie* evidence for the threat of meteorite and comet impacts. In paleontology, the discovery of a transitional fossil between fish and terrestrial tetrapod known as *Tiktaalik* in 2004 was of enormous significance in understanding how vertebrates moved onto land (2). The discovery of more *Tiktaalik* specimens in subsequent years yielded additional understanding of its anatomic characteristics (3) but was not required for validation of the original finding. Consequently, the criterion of reproducibility is not an essential requirement for the validation of scientific information, at least in some fields. Scientists studying the evolution of life on earth must contend with their inability to repeat that magnificent experiment. Stephen Jay Gould famously observed that if one were to "rewind the tape of life," the results would undoubtedly be different, with the likely outcome that nothing resembling ourselves would exist (4). (Note for younger readers: it used to be fashionable to record sounds and images on metal oxide-coated tape and play them back on devices called "tape players.") This is supported by the importance of stochastic and contingent events in experimental evolutionary systems (5).

Given the requirement for reproducibility in experimental science, we face two apparent contradictions. First, published science is expected to be reproducible, yet most scientists are not interested in replicating published experiments or reading about them. Many reputable journals are unlikely to accept manuscripts that precisely replicate published findings, despite the explicit requirement that experimental protocols must be reported in sufficient detail to allow repetition. This leads to a second paradox, that published science is assumed to be reproducible, yet only rarely is the reproducibility of such work tested or known. In fact, the emphasis on reproducing experimental results becomes important only when work becomes controversial, is called into doubt, or is of such enormous importance that replication is necessary to be certain of its implications. For example, in 1989 the announcement of cold fusion by Stanley Pons and Martin Fleischman was such a stunning development because of the possibility of limitless cheap energy. Unfortunately, the scientific community raced to repeat the findings but was unable to do so (6). Replication can even be hazardous. The German scientist Georg Wilhelm Reichmann was fatally electrocuted during an attempt to reproduce Benjamin Franklin's famous experiment with lightning (7). The assumption that science must be reproducible is implicit yet seldom tested, and in many systems the true reproducibility of experimental data is unknown or has not been rigorously investigated in a systematic fashion. Hence, the solidity of this bedrock assumption of experimental science lies largely in the realm of belief and trust in the integrity of the authors.

REPRODUCIBILITY VERSUS REPLICABILITY

Although many biological scientists intuitively believe that the reproducibility of an experiment means that it can be replicated, Chris Drummond, a researcher at the National Research Council of Canada, has made a distinction between these two terms (8). Drummond argues that reproducibility requires changes whereas replicability avoids them (8). In other words, reproducibility refers to a phenomenon that can be predicted to recur even when experimental conditions may vary to some degree. On the other hand, replicability describes the ability to obtain an identical result when an experiment is performed under precisely identical conditions. For biological scientists, this would appear to be an important distinction with everyday implications. For example, consider a lab attempting to reproduce another lab's finding that a certain bacterial gene confers a certain phenotype. Such an experiment might involve making gene-deficient variants, observing the effects of gene deletion on the phenotype, and, if phenotypic changes are apparent, then going further to show that gene complementation restores the original phenotype. Given a high likelihood of microevolution in microbial strains and the possibility that independently synthesized gene disruption and replacement cassettes may have subtly different effects, then the attempt to reproduce findings does not necessarily involve a precise replication of the original experiment. Nevertheless, if the results from both laboratories are concordant, then the experiment is considered to be successfully reproduced, despite the fact that, according to Drummond's distinction, it was never replicated. On the other hand, if the results differ, a myriad of possible explanations must be considered, some of which relate to differences in experimental protocols. Hence, it would seem that scientists are generally interested in the reproducibility of results rather than the precise replication of experimental results. Some variation of conditions is considered desirable because obtaining the same result without an absolutely faithful replication of the original experimental conditions implies a certain robustness of the original finding. In this example, the replicability of the original experiment following the exact protocols initially reported would be important only if all subsequent attempts to reproduce the result were unsuccessful. When findings are so dependent on precise experimental conditions that replicability is needed for reproducibility, the result may be idiosyncratic and less important than a phenomenon that can be reproduced by a variety of independent, nonidentical approaches.

REPLICABILITY REQUIREMENT FOR INDIVIDUAL STUDIES

Given the difference between reproducibility and replicability, which depends on whether experimental conditions are subject to variation, it is apparent that when most papers state that data are reproducible, they actually mean that the experiment has been replicated. On the other hand, when different laboratories report the confirmation of a phenomenon, it is likely that this reflects reproducibility, since experimental variability between labs is likely to result in some variable(s) being changed. In fact, depending on the number of variables involved, replicability may be achievable only in the original laboratory and possibly by the same experimenter. This accounts for the greater confidence one has in a scientific observation that has been corroborated by independent observers.

The desirability of replicability in experimental science leads to the practical question of how many times an experiment should be replicated before publication. Most reviewers would demand at least one replication, while preferring more. In this situation, the replicability of an experiment provides assurance that the effect is not due to chance alone or an experimental artifact resulting in a one-time event. Ideally, an experiment should be repeated multiple times before it is reported, with the caveat that for some experiments the expense of this approach may be prohibitive. Guidelines for experimentation with vertebrate animals also discourage the use of unnecessary duplication (9, 10). In fact, some institutions may explicitly prohibit the practice of repeating animal experiments that reproduce published results. We agree with the need to repeat experiments but suggest that authors strive for reproducibility instead of simple replicability. For example, consider an experiment in which a particular variable, the level of a specific antibody, is believed to account for a specific experimental outcome, resistance to a microbial pathogen. Passive administration of the immunoglobulin can be used to provide protection and support the hypothesis. Rather than simply replicating this experiment, the investigator might more fruitfully conduct a dose-response experiment to determine the effect of various antibody doses or microbial inocula and test multiple strains.

THE "REPRODUCIBILITY CRISIS" IN BIOLOGY

Until the early 2000s, there was little questioning of the reproducibility of biological research. Although many scientists in every field knew of some papers that reported work that was not reproducible, there was no general sense of an endemic replication problem in biology. That changed in 2011, when scientists from Bayer Pharmaceuticals reported that only about a quarter of published findings were reproducible in their hands (11). An even greater furor arose the following year when C. Glenn Begley of Amgen and Lee Ellis of the M.D. Anderson Cancer Center reported that when industry scientists tried to reproduce research on potential new drug targets in cancer research, most of the replication efforts failed (12). These papers and others that followed triggered an avalanche of concern about the reliability of basic research that forms the bedrock for health sciences and drug discovery. In 2016, a survey by the journal *Nature* reported that over half of researchers in the biomedical sciences thought that there was a reproducibility crisis, and more than 70% had been unable to reproduce another scientist's experiments (13).

These concerns led the Arnold Foundation to fund the Reproducibility Project: Cancer Biology, which sought to determine whether key findings in cancer biology could be reproduced (14). By 2021, results of this effort reported an overall 46% success rate in reproducing published research (15). Although this success rate is suboptimal, lack of reproducibility does not necessarily imply that the original report was in error. Over the past decade, numerous causes of irreproducibility and error have been identified in the biomedical sciences (Table 7.1), which have triggered efforts to improve methodology, reagents, and how findings are reported (Box 7.1). At the same time, it is important to keep problems with reproducibility in perspective. Not all are convinced that science is facing a "reproducibility crisis" (16). Irreproducible results can be misleading and waste time and resources on failed replication attempts. Nevertheless, they are a scientific dead end, and science progresses despite the existence of irreproducible results. Truly important

TABLE 7.1 Some causes of irreproducibility and error in biomedical sciences

Cause	Reference
Inadequate description of methods	35
Deficiencies in experimental design	36
Laboratory-generated reagents	35
Erroneous and contaminated cell lines	37
Poorly characterized antibody reagents	38
Differences in mouse breeding and rearing	39
P hacking[a] and statistical errors	40
Biological heterogeneity	41

[a] Inappropriate analysis of data in order to obtain a specific P value.

Box 7.1 Efforts to improve the reproducibility of biomedical sciences

In the past decade, various efforts have been undertaken to improve the reproducibility of experimental work in the biomedical sciences. Perhaps the most important development has been increasing awareness among biomedical investigators of the reproducibility problem, which has fostered national and local efforts to improve science. The realization that problems with cell lines and antibody reagents can lead to spurious results has translated into greater efforts in practice to maintain their integrity. In this regard, many journals now require precise listing of all reagents used in the experimental work. The Open Science movement strives to improve reproducibility through transparency, by making primary data, samples, software, etc.,

Factors that might improve reproducibility. From a survey of 1,576 researchers (13). Baker, M. 1,500 scientists lift the lid on reproducibility. Nature 533, 452–454 (2016). Reproduced with permission from SPRINGER.

freely available to allow subsequent investigators to better plan their experiments and understand how differences in results could have occurred. The U.S. government launched several initiatives in the past decade to enhance reproducibility in biological research through rigor and transparency (42). In addition, the subject of reproducibility has been incorporated into science education to make students aware of the potential for irreproducibility. In a 2016 *Nature* survey, respondents identified specific interventions that might improve reproducibility (figure).

findings tend to be quickly and independently replicated because scientists wish to build upon the findings of others.

IRREPRODUCIBILITY IN OTHER SCIENTIFIC FIELDS

The problem of poor reproducibility of published studies is not limited to the medical sciences. At the Center for Open Science in Charlottesville, Virginia, Brian Nosek and collaborators attempted to replicate 100 experiments in 98 psychology papers and found that only 39 were reproducible (17). In the new field of artificial intelligence, a major reproducibility problem was traced to the use of unpublished computer algorithms and differences in training conditions (18). With such diverse fields as biomedical sciences, psychology, and computer science grappling with problems of reproducibility, it is likely that similar issues exist in other scientific disciplines, and even fields such as systems biology modeling and chemical engineering do not appear to be immune (19, 20).

IRREPRODUCIBILITY SCIENCE

The realization that there may be a problem with reproducibility in science has led to efforts to understand its causes. Notable progress has been made in the last decade. The authors of this book carried out the first study of error in the biomedical literature (21) and found that the most common cause for retracting papers due to experimental error was the result of contamination and problems with molecular biology procedures (see also chapter 25). It has since become apparent that many experiments in the biological research literature suffer from low statistical power, and a significant proportion of irreproducible findings are likely to result from inadequate sample sizes and random variability of many assays (22). The recognized variation inherent in experimental systems can make it difficult to define the expected reproducibility of an experiment (22) or to estimate the power needed to accept or reject the null hypothesis (16). Nevertheless, efforts to replicate cancer research experiments as part of the reproducibility project have produced important insights into how irreproducibility comes about. For example, inability to replicate prior results concerning the effects of caveolin-1 on metastasis revealed experimental differences necessitated by changing mores in animal research, such as the need to terminate mouse experiments at earlier time points (23). Problems posed by inadequate replication have stimulated new questions that will require additional research, ushering in the emergence of a field focused on the irreproducibility of science. Learning about systemic causes of irreproducibility and limits of reproducibility should lead to more-robust science in the future.

LIMITS OF REPLICABILITY AND REPRODUCIBILITY

Although the ability of an investigator to confirm an experimental result is essential to good science, with an inherent assumption of reproducibility, we note that there are practical and philosophical limits to the replicability and reproducibility of findings. Although to our knowledge this question has not been formally studied, replicability is likely to be inversely proportional to the number of variables in an experiment. This is all too apparent in clinical studies, leading the epidemiologist John Ioannidis to conclude that most published research findings are false (24). Statistical analysis and meta-analysis would not be required if biological experiments were precisely replicable. Initial results from genetic association studies are frequently unconfirmed by follow-up analyses (25), clinical trials based on promising preclinical studies frequently fail (26), and a 2009 paper reported that only a minority of published microarray results could be repeated (27). Such observations have even led some to question the validity of the requirement for replication in science (28). Every variable contains a certain degree of error. Since error propagates linearly or nonlinearly depending on the system, one may conclude that the more variables involved, the more errors can be expected, thus reducing the replicability of an experiment. Scientists may attempt to control variables in order to achieve greater reproducibility but must remember that as they do so, they may progressively depart from the heterogeneity of real life. In our hypothetical experiment relating specific antibody to host resistance, errors in antibody concentration, inoculum, and consistency of delivery can conspire to produce different outcomes with each replication attempt. Although these errors may be minimized by good experimental technique, they cannot be eliminated entirely.

There are other sources of variation in the experiment that are more difficult to control. For example, mouse groups may differ, despite being matched by genetics, supplier, gender, and age, in such intangible areas as nutrition, stress, circadian rhythm, microbiota, etc. Similarly, it is very difficult to prepare infectious inocula on different days that closely mirror one another given all the variables that contribute to microbial growth and virulence. To further complicate matters, the outcomes of complex processes such as infection and the host response do not often manifest simple dose-response relationships. Inherent stochasticity in biological processes (29) and anatomic or functional bottlenecks (30) provide additional sources of experiment-to-experiment variability. For many biological experiments, the outcome of the experiment is highly dependent on initial experimental conditions, and small variations in the initial variables can lead to chaotic results. In such systems where exact replicability is difficult or impossible to achieve, the goal should be general reproducibility of the overall results. Ironically, results that are replicated too precisely are "too good to be true" and raise suspicions of data falsification (31), illustrating the tacit recognition that biological results inherently exhibit a degree of variation.

To continue the example given above, the conclusion that antibody was protective may be reproduced in subsequent experiments despite the fact that the precise initial result on average survival was never replicated, in the sense that subsequent experiments varied in magnitude of difference observed and time to death for the various groups. Investigators may be able to increase the likelihood that individual experiments are reproducible by enhancing their robustness. A well-known strategy to enhance the likelihood of reproducibility is to increase the power of the experiment by increasing the number of individual measurements, in order to minimize the contribution of errors or random effects. For example, using 10 mice per group in the aforementioned experiment is more likely to lead to reproducible results than using 3 mice, other things being equal. Along the same lines, two experiments using 10 mice each will provide more confidence in the robustness of the results than will a single experiment involving 20 animals, because obtaining similar results on different days lessens the likelihood that a given result was strongly influenced by an unrecognized variable on the particular day of the experiment. When reviewers criticize low power in experimental design, they are essentially worried that the effect of variable uncertainty on low numbers of measurements will adversely influence the reproducibility of the findings. However, subjective judgments based on conflicting values can influence the determination of sample size. For instance, investigators and reviewers are more likely to accept smaller sample sizes in experiments using primates. Consequently, a sample size of 3 might be acceptable in an experiment using chimpanzees while the same sample size might be regarded as unacceptable in a mouse experiment, even if the results in both cases achieve statistical significance. Similarly, cost can be a limiting factor in determining the minimum number of replicates. For nucleic acid chip hybridization experiments, measurements in triplicate are recommended despite the complexity of such experiments and the range of variation inherent in such measurement, a recommendation that tacitly accepts the prohibitive cost of larger numbers of replicates for most investigators (32). Cost is also a major consideration in replicating transgenic or knockout mouse experiments, in which mouse construction may take years. Hence, the power of an experiment can be estimated accurately using statistics, but real-life considerations ranging from the ethics of animal experimentation to monetary expense can influence investigator and reviewer judgment.

We cannot leave the subject of scientific reproducibility without acknowledging that questions about replicability and reproducibility have long been at the heart of philosophical debates about the nature of science and the line of demarcation between science and nonscience. While scientists and reviewers demand evidence for the reproducibility of scientific findings, philosophers of science have largely discarded the view that scientific knowledge should meet the criterion that it is verifiable. Through inductive reasoning, Francis Bacon used data to infer that under similar circumstances, a result will be repeated and can be used to generalize about other related situations (33). However, the logical consistency of such views was challenged by David Hume, who posited that inferences from experiences (or, in our case, experiments) cannot be assumed to hold in the future because the future may not necessarily be like the past. In other words, even the daily rising of the sun for millennia does not provide absolute assurance that it will rise the next day. The philosophies of logical positivism and verificationism viewed truth as reflecting the reproducibility of empirical experience, dependent on propositions that could be proven to be true or false. This was challenged by Karl Popper, who suggested that a hypothesis could not be proven, only falsified or not, leaving open the possibility of a rare unpredictable exception, vividly depicted as the metaphor of a "black swan" (34). One million sightings of white swans cannot prove the hypothesis that all swans are white, but the hypothesis can be falsified by the sight of a single black swan.

A PRAGMATIC APPROACH TO REPRODUCIBILITY

Given the challenges of achieving and defining replicability and reproducibility in experimental science, what practical guidance can we provide? Despite valid concerns ranging from the true reproducibility of experimental science to the logical inconsistencies identified by philosophers of science, experimental reproducibility remains a standard and accepted criterion for publication. Hence, investigators must strive to obtain information with regard to the reproducibility of their results. That, in turn, raises the question of the number of replications needed for acceptance by the scientific community. The number of times that an experiment is performed should be clearly stated in a manuscript. A new finding should be reproduced at least once and preferably more times. However, even here there is some room for judgment under exceptional circumstances. Consider a trial of a new therapeutic molecule that is expected to produce a certain result in a primate experiment based on known cellular processes. If one were to obtain precisely the predicted result, one might present a compelling argument for accepting the results of the single experiment on moral grounds regarding animal experimentation, especially in situations in which the experiment results in injury or death to the animal. At the other extreme, when an experiment is easily and inexpensively carried out without ethical considerations, then it behooves the investigator to ascertain the replicability and reproducibility of a result as fully as possible. However, there are no hard and fast rules for the number of times that an experiment should be replicated before a manuscript is considered acceptable for publication. In general, the importance of reproducibility increases in proportion to the importance of a result, and experiments that challenge existing beliefs and assumptions will be subjected to greater scrutiny than those fitting within established paradigms.

Given that most experimental results reported in the literature will not be subjected to the test of precise replication unless the results are challenged, it is essential for investigators to make their utmost efforts to place only the most robust data into the scientific record, and this almost always involves a careful assessment of the variability inherent in a particular experimental protocol and the provision of information regarding the replicability of the results. In this instance, more is better than less. To ensure that research findings are robust, it is particularly desirable to demonstrate their reproducibility in the face of variations in experimental conditions. Reproducibility remains central to science, even as we recognize the limits of our ability to achieve absolute predictability in the natural world. Then again, ask us next week and you might get a different answer.

8 Important Science

> It is clear that without the distinction between the important and the unimportant at our disposal, mankind could neither adequately understand, successfully teach, or effectively practice science.
>
> Nicholas Rescher (1)

Of all the epithets attached to the word "science," perhaps the most desirable is "important." In this chapter, we will consider the problem of "importance," recognizing that this most critical of descriptors is also among the most elusive, for it relies on judgment. The *American Heritage Dictionary* defines "important" as "strongly affecting the course of events or the nature of things" (2). This implies that the outcome of a scientific discovery or publication on subsequent events determines its importance. Hence, there is a separation between scientific quality and outcome, a distinction that raises interesting issues. For example, authors frequently complain that journals are biased against studies that yield negative results, a practice known as "publication bias" (3); this is unquestionably true and can be attributed to the greater perceived importance of positive studies. Moreover, the focus on outcome means that any judgment on assessing the importance of a scientific work is a function of time. This is apparent to any student of the history of science, which contains many instances in which the importance of a scientific finding was initially underappreciated or overvalued. An oft-cited example of this phenomenon is the work of Gregor Mendel, the importance of which was not

Thinking about Science: Good Science, Bad Science, and How to Make It Better, First Edition.
Ferric C. Fang and Arturo Casadevall.
© 2024 American Society for Microbiology.

recognized in the decades following its publication in an obscure journal (4) but was subsequently rediscovered to become a foundation of the nascent science of genetics.

The judgment of importance in science is a critical element of the peer review process. Reviewers are asked to assess the importance of a scientific manuscript or research proposal, which often becomes the critical determinant of the reviewer's decision. In the current scientific parlance, the word "priority" may be used as a synonym for importance. Although the two words do not mean precisely the same thing, both are comparative terms that may be used to rank one work relative to another. In the scientific publication process, reviewers are asked to rate manuscripts in terms of both general interest and significance to the field, suggesting a nuanced meaning to the criterion of importance. In every field, there are examples of papers that made little splash when first published but in retrospect were found to contain important information with a substantial impact on later work. One example is discussed in more detail below, regarding how DNA was found to transmit inheritance. Hence, reviewers are asked to judge the potential of a manuscript without the advantage of hindsight that only time can provide. On other occasions, the importance of a scientific finding is apparent immediately (Box 8.1).

It would seem at first glance that any assessment of importance must be a highly subjective and error-prone process. Nevertheless, even without a crystal ball one may systematically attempt to estimate the importance of scientific work. We propose four criteria by which one may assess scientific importance, forming the acronym "SPIN." Important science should be sizeable (S), practical (P), integrated (I), and new (N), and these criteria can be used individually or collectively to estimate the importance of a manuscript (Fig. 8.1).

Box 8.1 Importance in real time

Some scientific findings are recognized to be of enormous importance the moment that they are first announced. We can think of two recent such events, both in the field of physics: the demonstration of the Higgs boson at the Large Hadron Collider in 2012 (30, 31) and the detection of gravitational waves at the Laser Interferometer Gravitational-Wave Observatory (LIGO) in 2016 (32). Both discoveries were followed by Nobel Prizes, given in 2013 and 2017, respectively, a remarkably short interval between announcement and prize. The rapidity with which the Nobel Committee acted to recognize these discoveries underscores the fact that they were immediately appreciated as important. How do we reconcile these instances with the notion that importance in science often takes time to be recognized? In contrast to the report by Avery, MacLeod, and McCarty that DNA is responsible for pneumococcal transformation and therefore carries heritable information (13), both the Higgs boson and gravitational waves were anticipated by broadly accepted theoretical work that predicted their existence. Both discoveries required the assembly of expensive and sophisticated instruments with teams of thousands of physicists working collaboratively. Hence, when a finding is both new and predicted by generally accepted theory, it can be immediately considered important because it validates the correctness of the theory. The measurement of gravitational waves by LIGO offered a new practical way of visualizing the universe through gravity waves, which is likely to be of profound importance given that observations to date have relied on electromagnetic spectrum measurements, which include visual, radio wave, and infrared astronomy maps. Seeing the universe using gravitational waves promises to revolutionize astronomy, and revolutionary science (chapter 11) is by definition important.

S = size
P = practical
I = integration
N = novelty

FIGURE 8.1 *Scientific importance is all about the SPIN. S = sizeable, P = practical, I = integrated, N = new.*

SIZEABLE

In scientific investigations, the parameter of size can refer to the magnitude of a problem or the size of a field. Consider two pathogenic microbes that cause comparable levels of morbidity and mortality in susceptible hosts: *Mycobacterium microti* and *Mycobacterium tuberculosis*. Both belong to the *M. tuberculosis* complex. However, *M. microti* is found in rodents and llamas and only very rarely in humans, whereas approximately one-fourth of the world's population is infected with *M. tuberculosis* (5). This is paralleled by the roughly 281,000 publications on tuberculosis listed in PubMed, compared with fewer than 300 on *M. microti*, as of September 2022. Hence, following our criteria, a scientific observation would be considered more important if pertaining to *M. tuberculosis* rather than to *M. microti*, because the problem of tuberculosis and the number of tuberculosis researchers are so much more sizeable. Acknowledgment of the contribution of size to the overall importance of scientific work is evident in the introductions of many papers relating to microbes, in which one often finds descriptions of the prevalence of infectious diseases. In gauging importance, size matters, but those of us who work on rare microbes and/or inhabit small scientific fields need not despair, for other components of the SPIN factor can compensate for size.

PRACTICAL

Scientific findings with practical utility have great importance because they provide benefits to society. For the public and the political establishment, which currently supports most scientific research, the tangible rewards of science are the dominant measure of its importance. However, utility can come in many forms, and here the value system of the reviewer is paramount in assigning importance. For instance, a discovery can have utility in a theoretical context if it facilitates understanding, even if that information cannot be immediately translated into practical applications. Furthermore, history has repeatedly provided examples in which practical applications arise serendipitously from basic research initiated for other purposes (6).

Discoveries may have enormous practical utility even though they are not conceptually new. For example, earlier work showed that conjugation of the *Haemophilus influenzae* capsular polysaccharide to a protein carrier results in an effective vaccine (7, 8). This breakthrough had a dramatic impact on the incidence of invasive *H. influenzae* infections in the United States (9). Subsequent research led to the development of *Streptococcus pneumoniae* (pneumococcal) capsular polysaccharides conjugated to carrier proteins, allowing the prevention of invasive pneumococcal infections in infants and toddlers (10). From a definitional standpoint and public perspective, the pneumococcal research is clearly of high importance because it impacts many susceptible individuals. However, despite its importance, the finding lacks conceptual originality, and individuals who value newness over practical utility might dismissively label the latter work as derivative. However, the development of a pneumococcal conjugate vaccine was not without its challenges. Unlike *H. influenzae*, the pneumococcus comes in dozens of sero-types, and any single polysaccharide conjugate would protect against only one serotype. To get around the problem of pneumococcal antigenic diversity, vaccine developers combined conjugates of the 13 most prevalent serotypes into a vaccine, thereby meeting the "size" criterion for importance as well (see above). The pneu-mococcal conjugate vaccine has been a tremendous success for public health, and its formulation as a polyvalent antigenic mixture showed ingenuity. This example illustrates that although assigning importance to science inevitably involves some subjective judgment, one may nevertheless quantify importance by considering the different components of the SPIN factor independently.

INTEGRATED

Knowledge builds upon knowledge, and all scientific knowledge is interconnected. New information must become integrated with prior knowledge. The importance of a new discovery is therefore dependent on context and the readiness of existing knowledge to allow integration of the new information, which must sometimes await new tools or further understanding. Science that is initially underappreci-ated because it is ahead of its time has been referred to as "premature" (11). By "premature," we do not mean that a work was published before it was ready for publication but rather that the results were presented before the field could prop-erly grasp their significance. Svante Arrhenius' warnings of global warming in 1896 and Alfred Wegener's proposal of continental drift in 1915 are two classic exam-ples of premature science (12). When Oswald Avery, Colin MacLeod, and Maclyn McCarty identified DNA as the material responsible for heredity in 1944 (13), the initial reaction to this transcendent discovery was surprisingly muted (14). This is in part because prevailing concepts of DNA were not consistent with the coding of specific information, as DNA was then believed to consist of a monotonously repeating polymer of identical tetranucleotides (14). Only after Erwin Chargaff showed that DNA bases were not necessarily present in equal proportions (15) and Alfred Hershey and Martha Chase demonstrated internalization of phage DNA by bacteria (16) were the observations of Avery et al. fully appreciated and integrated with existing knowledge. Science may be prioritized on the basis of its level of integration. If one envisages scientific knowledge as a branching tree, then basic processes tend to occupy deeper positions within scientific branches and have broader implications. A discovery may be viewed as more important when it is situated more deeply within the tree of knowledge or possesses more interconnections with other findings. To illustrate this with an example, consider the discovery of nitric oxide biosynthesis from the amino acid L-arginine. This insight facilitated the elucidation of nitric oxide's role in vascular regulation, neurotransmission, signal transduction, and host defense (17). Knowledge of nitric oxide chemistry is located at a deeper level than knowledge of the actions of nitric oxide in a specific type of infection and, consequently, depending on the perspective of the reviewer, might be considered more important.

NEW

The parameter of newness reflects the time since information has come into existence. Newness plays a key role in the economics of science, where priority of discovery can confer great rewards on the scientist (see also chapter 19) (18). In most fields, there is no prize for coming in second or third, except for the knowledge that confirmation and validation reinforce the foundation of science. Although newness and novelty could be considered as synonymous, we will avoid the word "novel" and its derivatives since it is used so frequently in the current scientific literature that we regard it as tired and in need of rest (a search using the word "novel" in PubMed produced more than 1.5 million citations, whereas "newness" yielded only 250). In incorporating newness into our definition of importance, it is apparent that this parameter is different than size, practical utility, and integration because, unlike the other qualities, newness alone does not necessarily imply that a discovery affects "the course of events or the nature of things" (2). In fact, the newness of a discovery appears to be more important to the personal satisfaction of scientists than to the assessment of importance. This leads to the uncomfortable realization that new things are not necessarily important, and older findings may subsequently be judged of greater importance, as in the case of Mendel's work.

The emphasis on the term "novel" in manuscripts and grants as a measure of importance may therefore be somewhat misplaced. Nevertheless, the newness of information is relevant in judging the importance of a finding because immediacy can temper the other SPIN parameters. Alexander Fleming's initial discovery of penicillin production by the obscure saprophytic fungus *Penicillium rubens* was undeniably important (19), but largely in hindsight, as treatment applications were not realized until more than a decade later. Although other fungi and actinomycetes have subsequently been shown to produce antibiotics, Fleming's observations have retained special importance because they were the first. Ironically, newness can become more significant over time, as a greater value is placed on priority once the importance of a finding is recognized.

SURROGATE MEASURES OF IMPORTANCE

Most scientists would agree that the Nobel Prize represents the highest accolade for important scientific work (see also chapter 21). Hence, the analysis of science recognized by Nobel Prizes can provide insight into the validity of the SPIN parameters, their relative value, and their changes over time. Historically, the emphasis on size (S) is apparent in many Nobel Prizes in Physiology or Medicine that recognize progress against diseases affecting large numbers of people. For example, in 2008, Nobel Prizes recognized the association of human immunodeficiency virus and human papillomavirus with AIDS and cervical cancer, two diseases that afflict millions each year, while the association of human T-cell lymphotropic virus type 1 with the much rarer disease tropical spastic paraparesis did not receive comparable recognition, even though comparable associations with a specific virus were shown for all three conditions. Size is also undoubtedly a factor in the Nobel Prizes awarded for recognition of mechanisms of malaria transmission (1902), tuberculosis etiology and therapy (1905 and 1952), typhus control (1928), penicillin (1945), poliovirus (1954), and the association of *Helicobacter pylori* with peptic ulcer disease (2005), as each of these awards recognized progress against diseases affecting many individuals. No Nobel Prizes have been given for breakthroughs related to infections involving relatively small numbers of individuals, such as Kaposi's sarcoma, Whipple's disease, and Lyme disease, or the discovery of the first effective antifungal agent, amphotericin B.

Of the remaining parameters, integration (I) plays a dominant role, with the overwhelming majority of Nobel Prizes in Physiology or Medicine being awarded for basic discoveries that have had great consequences for diverse lines of life science research. The contribution of practical application (P) is apparent in Nobel Prizes awarded for techniques with wide applicability, such as the development of the radioimmunoassay (1977) and hybridoma technology (1984). In contrast to the Physiology or Medicine prize, technological advances have been recognized frequently by the Chemistry prize, including protein sequencing (1958), polarography (1959), DNA sequencing (1980), nuclear magnetic resonance (1991), PCR (1993), site-directed mutagenesis (1993), and green fluorescence protein applications in biology (2008). Newness (N) pervades all discoveries but is clearly apparent when given for the discovery of a specific phenomenon leading to a wholesale paradigm shift, such as "slow viruses" or prions (1976 and 1997), ribozymes (1989), and nitric oxide biosynthesis by vascular tissues (1998). We note that newness can sometimes make a finding more difficult to integrate with existing science. For example, the discovery of prions as infectious agents challenged the central dogma of DNA $\rightarrow$ RNA $\rightarrow$ protein and could not be integrated into mainstream biology until subsequent studies provided a mechanism by which a proteinaceous infectious agent could propagate itself by transmitting information, leading to an aberrant folding state.

The evolution of the Nobel Prize also provides some insights into the perception of importance over time. As examples of science no longer considered as important today, the 1926 prize to Johannes Fibiger for his erroneous "discovery" that the nematode *Spiroptera carcinoma* (*Gongylonema neoplasticum*) causes gastric cancer and the 1949 prize to António Egas Moniz for his work on prefrontal lobotomies come to mind (pun intended). Early in the 20th century, Nobel Prizes in Physiology or Medicine were often given for contributions that had a great impact on health, such as serum therapy (1901), malaria transmission (1902), phototherapy (1903), and surgical advances (1912), suggesting a great emphasis by the selection committee on practical application. In contrast, later in the century the majority of prizes were for very basic discoveries that often could not immediately be realized into useful applications. Regarding the significance of time in the ultimate judgment of importance, we note that Alfred Nobel stipulated in his will that the prize should be awarded annually "to those who, during the preceding year, shall have conferred the greatest benefit on mankind" (20). Although this sentence is sufficiently ambiguous to allow some room for interpretation, the emphasis on the preceding year does suggest an initial intention of rewarding immediacy in discovery. However, Nobel Prizes are nowadays almost always awarded many years after the original discovery because importance takes time to become evident. For less spectacular discoveries, one must rely on surrogate measures of importance, such as the number of times that a paper is cited by others. Nicholas Rescher has noted that the distribution of numbers of citations and scientific quality are described by similar exponential functions, suggesting that the former is a reasonable measure of the latter (1). However, he hastens to add, "Of course, no more than an estimate is at issue here. For it has to be acknowledged that in view of the ever-moving boundary lines of the frontiers of knowledge and the shifting ebb and flow of fashions in matters of theorizing have the unavoidable consequence that importance as best we can judge it is not a fixity but an ongoingly varying parameter."

OTHER MEASURES OF IMPORTANCE

With the advent of large computerized databases of bibliometric information, new quantitative measures of importance have emerged. Perhaps the most commonly used measure is the number of citations that a publication receives, with those accruing more citations considered to be more important or impactful. The obvious appeal of bibliometric indicators of importance is that they can be readily measured and followed over time, Nevertheless, one must be cautious with such assessments, as citations are a very imperfect measure of scientific importance. Papers are cited for various reasons, and citation does not imply an assessment of quality (21).

Mendel's paper on the laws of inheritance published in 1866 was cited only three times in the subsequent 35 years (22). Raw numbers of citations do not take an individual author's relative contribution to a work into account. Reviews and methods papers tend to be more widely cited than papers reporting original research findings (23). Larger fields have more authors who are publishing and citing manuscripts, which inflates citation numbers. To account for field size differences, the relative citation ratio has been developed (24) (see chapter 26, Box 26.3), which provides some improvement over simply counting citations. The number of citations for a publication is also dependent on time, with older papers accruing more citations than younger publications.

In addition to bibliometric information, there is also the time-honored tradition of asking scientists in a particular field for their opinion on important work. Although individual responses are likely to be highly subjective, a consensus in a field for the importance of particular work is probably a good indicator of its value. Polls can provide insights into how scientists feel about the importance of scientific work and sometimes provide surprising findings. For example, a poll of physicists attending a conference on quantum mechanics revealed that 27% thought quantum theory is wrong (25), providing a remarkable demonstration of divergent opinion regarding a foundational theory of modern physics. Yet another measure of importance is public interest in scientific findings, as evidenced by reporting in newspapers or other general media. However, caution is always a wise approach when judging the importance of a scientific report from publicity alone, which tends to reflect an instant assessment that may be undone with time. For instance, there was great public interest during the early days of the COVID-19 pandemic in the potential benefits of hydroxychloroquine, but subsequent work revealed that this drug was ineffective (chapter 30). Altmetrics are alternative measures of the impact of science based on online activity such as tweets or downloads. An empirical study has found that citations are a better correlate of scientific quality than altmetric score (26).

ASSESSMENT OF IMPORTANCE IS SUBJECTIVE AND IMPERFECT

Ultimately, the assessment of importance has a large subjective component that reflects the values, experience, interest, knowledge, and biases of the reviewer. Different individuals are inclined to give different weights to the SPIN parameters. For example, scientists concerned with global health and vaccines may have their SPIN algorithm altered to S^2P^2IN to reflect these priorities, while for scientists interested in basic research the algorithm may be SPI^3N. Although some important discoveries may meet all four SPIN criteria (Box 8.2), others will meet only two or three. The essential point is that a paper's importance may be assessed from different perspectives. Emphasis on S, P, I, or N is in the eye of the beholder, and it behooves one to be respectful of divergent views. The difficulties inherent in evaluating the importance of a manuscript should not deter reviewers from attempting their own assessments, with the important caveat that one should remain humble, for there is great uncertainty in the process. The journal *PLoS One* decided to eliminate the assessment of importance from its review process altogether: "*PLoS ONE* will ... publish all papers that are judged to be technically sound. Judgments about the importance of any particular paper are then made after publication by the readership (who are the most qualified to determine what is of interest to them)" (27). Since *PLoS One* is an open-access journal that is available to anyone with an Internet connection, this policy is revolutionary in the sense that it returns the assessment of importance from the editorial elite to the people. Søren Kierkegaard has reminded us, "Life must be understood backward, but ... it must be lived forward" (28), and as we have tried to emphasize, true importance can be fully recognized only in hindsight. It is also worthwhile to note that although authors, reviewers, and editors can disagree on the relative importance of a manuscript, the work was considered of sufficient importance by the investigators to warrant their time and resources. Hence, all manuscript submissions should be treated with respect, for they are important to someone.

Box 8.2 Important science in the 1890s: Anna Williams and diphtheria antitoxin

In the late 19th century, diphtheria, a terrible disease caused by the toxin-producing bacterium *Corynebacterium diphtheriae*, was a major killer of young children. Children with diphtheria would sometimes develop a tough pseudomembrane in their throat that interfered with breathing, resulting in death by asphyxiation. In 1891, Emil von Behring in Germany showed that antibodies elicited in animals immunized with diphtheria toxin neutralized the toxin and might be used therapeutically to treat the disease, a discovery for which he received the first Nobel Prize in Medicine in 1901 (chapter 21). However, production of antitoxin in the United States was difficult because of the lack of a bacterial strain that produced sufficient toxin to immunize animals. Working in the New York Department of Public Health, Anna Wessels Williams isolated such a strain from a patient with tonsillar diphtheria in 1893. This strain was named *Corynebacterium diphtheriae* Park-Williams number 8 (PW8), recognizing Williams and her collaborator William Park (33). This strain was remarkable for producing very large amounts of toxin and continues to be used as the universal strain for producing toxin in vaccine preparation to this day. Diphtheria toxoid has essentially eliminated diphtheria in most of the world, and we and our readers who are toxoid-immunized can trace this protection to the work of Williams. Her discovery of the PW8 strain meets all SPIN criteria for important science. The problem of diphtheria was sizable (S) since the disease was a major killer of children in all areas of the world. Identifying SW8 was eminently practical (P) since it allowed the immediate production of antitoxin within a year of its isolation. The identification of SW8 is integrated (I) since it gave a tool that allowed the production of antitoxin in the 1890s, was subsequently used to produce toxoids, and has been extensively studied in the ensuing 120 years to gain new insights into bacteriophages and the pathogenesis of diphtheria. PW8 was clearly new (N) in 1893 and remains a remarkable isolate.

Williams went on to a distinguished career that included numerous other discoveries in bacterial pathogenesis and public health and at her retirement in 1934 was lauded by the mayor of New York City as a "scientist of international repute" (34). Williams's contributions are all the more remarkable given the misogynist attitudes of late-19th- and early-20th-century medicine, which discouraged and minimized the contributions of women. Apart from being a trailblazer in medical research, she enjoyed riding with stunt fliers in vintage airplanes and motoring at high velocity, which resulted in many speeding tickets (35).

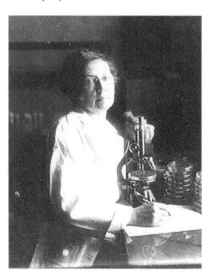

Anna Wessels Williams (1863–1954). Photo credit: Schlesinger Library, Harvard Radcliffe Institute.

The paucity of writings on the topic suggests that importance is considered to be self-evident by most scientists. However, the deconvolution of importance into its SPIN parameters suggests that a quantitative approach to the problem is possible and even worthy of future investigation. A departure from purely subjective assessments toward the establishment of quantitative criteria for importance could eventually provide the basis for a system of manuscript prioritization. Authors might also be tempted to use this approach to put the most favorable "SPIN" on their work. However, they should first recall the words quoted by Frank Harold (29): "Every novel idea in science passes through three stages. First people say it isn't true. Then they say it's true but not important. And finally they say it's true and important—but not new."

9 Historical Science

The history of science bores most scientists stiff.

Sir Peter Medawar (1)

One of the unique experiences of being human is to have a history. The ability to recount the past and pass it on to future generations is made possible by the symbolic language unique to our species. Most human history has been conveyed by oral narratives and legends. However, the invention of writing allowed history to acquire a new permanence. Herodotus, who lived in Greece during the 5th century BCE, is generally regarded as the first historian who attempted to systematically organize and analyze information. (There are others who regard his fellow Greek Thucydides as the first true historian and Herodotus as the "first liar" for getting so many of his facts wrong [2].) Personal histories define individuals, while communal histories define groups and nations. In some areas of human endeavor, such as law and politics, history is essential for interpreting and understanding the present, and competing versions of history are often critical points of contention. However, science is a human endeavor in which the study of its own history plays a less prominent role. This is evidenced by the scant attention paid to history during the scientific training process, the ahistorical style of most scientific literature, and the separation of science and the history of science as academic disciplines. In this chapter, we will examine the importance of history in the scientific process and the consequences of its neglect.

Thinking about Science: Good Science, Bad Science, and How to Make It Better, First Edition.
Ferric C. Fang and Arturo Casadevall.
© 2024 American Society for Microbiology.

Dictionaries describe history as a chronological record of significant events, often including an explanation of their causes (3). From such a definition, the history of science would include the Copernican revolution, Isaac Newton's *Principia*, the Darwin-Wallace theory of evolution, and the theory of relativity. Major events in the history of science are widely known and well documented, although the intellectual and experimental struggle required for discovery may not be as well appreciated. For example, while most scientists are aware of the Copernican revolution and Galileo's struggle with the Catholic Church, the scientific arguments made in favor of a geocentric universe, such as the inability to detect stellar parallax (4), are less common knowledge. Although major scientific discoveries eventually become accepted as fact, the hard-fought struggles to obtain this understanding tend to fade with the passage of time.

WHY DO MOST SCIENTISTS IGNORE THE HISTORY OF SCIENCE?

Assuming that Sir Peter is correct in saying that "the history of science bores most scientists stiff," it is perhaps not difficult to explain the limited interest that most scientists take in history. Science by its very nature seeks to push back the boundaries of the unknown—the border between the known and unknown is far more interesting to scientists than what happened in the past. Although most students in the biological sciences learn about the discoveries of Charles Darwin, Gregor Mendel, Marie Curie, and James Watson and Francis Crick, it is fair to say that historical training is not a major part of the undergraduate or graduate science curriculum. Very few scientific fields have an accessible historical literature to supplement scientific training. While some students may have learned additional science history from courses that consider classic papers, most learn the history of their chosen field of study from their laboratory mentor or from review articles that emphasize historical aspects of discovery. Human aspects of scientific discovery, such as scientific rivalries and their effect on science, are generally not discussed in formal articles. Rather, such information is maintained within fields by an oral tradition consisting largely of gossip, anecdote, and rumor, and is thus forgotten or lost when the participants depart from the scene through death or retirement. For example, we know many intimate details about the human drama surrounding the solution of DNA's structure because Watson decided to memorialize his experiences in a bestselling book, *The Double Helix*, in 1968. This book exposed the public to the competition that goes on behind the scenes in the process of scientific discovery and sparked controversy for diminishing the critical contributions of Rosalind Franklin. Edwin Chargaff, who discovered the rules of base composition that were essential for the DNA model, wrote a scathing review of the book (5), but whatever its virtues and vices, it provides an invaluable glimpse into the backstory of this discovery.

One can master a scientific topic without having the least idea of how the knowledge was obtained. For example, it is possible to describe the central dogma of molecular biology from transcription to translation in excruciating detail without having to mention a single scientist's name. In this regard, science differs from politics, law, economics, or most social sciences, in which the history of events is essential for understanding the field. It is impossible to understand the state of race relations in the United States without considering the history of slavery, civil war, reconstruction, segregation, and civil rights. In contrast to other intellectual pursuits, science can be viewed as being either privileged or disadvantaged because it has the luxury of neglecting its history.

THE SCIENTIFIC LITERATURE IS DELIBERATELY AHISTORICAL

In a lecture titled "Is the Scientific Paper a Fraud?" Medawar also noted that the format of a conventional scientific paper consisting of an introduction, description of methods, results, and discussion implies a logical inductive process that is completely alien to how most science is actually done (6). John Carmody expanded upon this point by observing that research papers not only idealize the scientific process but also drain it of the passion of discovery (7). Perhaps this has always been the case. When Elie Metchnikoff (Fig. 9.1) described his discovery of phagocytosis in starfish larvae in a research journal (8), he drily reported the following: "The reactive phenomena ensuing on artificial injuries may be readily observed in the much larger larvae, the *Binpinnaria asterigera*. ... If a delicate

FIGURE 9.1 *Elie Metchnikoff (1845–1916). He was the recipient of the 1908 Nobel Prize in Physiology or Medicine for his discovery of phagocytes. Photogravure after Henri Manuel. Image courtesy of the Wellcome Collection, ref # 13229i.*

glass tube, a rose-thorn, or a spine of a sea urchin be introduced into one of these larvae, the amoeboid cells of the mesoderm collect around the foreign body in large masses easily visible with the naked eye."

Yet the historical recollection of the events in his biography (9) paints a much different picture: "One day when the whole family had gone to a circus to see some extraordinary performing apes, I remained alone with my microscope, observing the life in the mobile cells of a transparent starfish larva, when a new thought suddenly flashed across my brain. ... I felt so excited that I began striding up and down the room and even went to the seashore in order to collect my thoughts. ... I was too excited to sleep that night in the expectation of the result of my experiment."

Medawar's criticism of the scientific literature resonates in the present day with additional profound and disturbing implications. As any working scientist knows, the process of scientific discovery is messy and often involves dead ends, chance, and being in the right place at the right time. At a minimum, the conventional format of a scientific paper distorts history by creating a narrative for scientific discovery that is different from what actually occurred. Susan Howitt and Anna Wilson revisited the question of whether writing a scientific paper in the current accepted style was itself a fraudulent act. These authors concluded that "doing science and communicating science are quite different things" and noted that little had changed since Medawar's provocative essay (10). Perhaps it is of even greater concern that the "winner take all" reward system of science and the pressure to demonstrate novelty may create perverse incentives for authors to overemphasize the novelty of their own work and fail to appropriately cite the contributions of others or selectively cite publications that support their conclusions (11, 12). Such historical neglect, whether inadvertent or purposeful, can misrepresent and even distort the scientific record.

In some respects, it is an advantage that science can convey its subject matter without having to consider history. This means that science, unlike other disciplines (or the legal system [13]), is not shackled to the misinterpretations of the past. While history demands that facts be interpreted in context, scientists are wary of interpretations that are difficult to validate or falsify. Instead, untethered scientific knowledge is independent of history and can serve as a platform for further research. Scientists do not need to consider the contentious emergence of the heliocentric theory to accurately deliver probes to Mars, Ceres, and Pluto. However, there are significant costs when science neglects its history. During the COVID-19 pandemic, the medical community first embraced and then abandoned the use of convalescent plasma after several clinical trials suggested that it was ineffective. Unfortunately, these trials were designed without taking into account the knowledge generated in the early 20th century that for antibody-based therapies to be effective, they needed to be used early in the course of disease (Box 9.1). This fiasco may have resulted in as many as 29,000 excess deaths in the United States

---

Box 9.1 Forgetting history cost lives in the COVID-19 pandemic

When the COVID-19 pandemic began spreading around the world in 2020, there were no specific antiviral therapies. As the crisis grew, physicians began using convalescent plasma as a treatment. Convalescent plasma had been used in many prior epidemics dating back to the 1918 influenza pandemic (45, 46). In the United States, the Food and Drug Administration allowed the use of convalescent plasma, first in a registry program and then under emergency use authorization. Over half a million patients received this therapy in the first year of the pandemic (47). The registry program produced the first evidence of efficacy when an analysis of the first few thousand patients showed that administration of units with high specific antibody content early in hospitalization reduced mortality by about 35% (48). Epidemiologic evidence showed a strong inverse correlation between convalescent plasma use and mortality in the United States, estimating that its deployment saved about 100,000 lives (14). In the meantime, many randomized controlled trials were begun, but most tested convalescent plasma late in the course of disease and hospitalization, when the pathogenic process had progressed into the inflammatory phase and antiviral therapies would not be expected to be effective. Most of these trials reported that convalescent plasma was ineffective, a result that could have been predicted if physicians were familiar with the history of medicine in the early 20th century when an enormous body of research established that the effectiveness of antibody-based therapies depends on the time of administration. In 1937, Russell Cecil, one of the giants of American medicine, wrote, "It is a fundamental principle of all serum therapy that to obtain the best results the serum must be given early in disease" (49). With physicians favoring evidence from randomized controlled trials over other types of medical evidence, the news of negative trials led to an abrupt abandonment of convalescent plasma in early 2021 that was associated with an increase in mortality, resulting in as many as 29,000 excess deaths (14). Eventually, other trials were completed, and analysis of the aggregate data showed what history would have predicted: that COVID-19 convalescent plasma is effective when used early in the course of disease (50).

---

alone (14). Although eventually other clinical trials showed that convalescent plasma is effective for COVID-19 when used early, that knowledge came late in the pandemic when other therapeutic options were available (15, 16). Hopefully, the knowledge of how to use antibody therapy, which was forgotten and then regained, will be remembered for the next infectious disease crisis when humanity may once again find itself without specific therapies.

The history of science is replete with instances in which facts and research were forgotten and later rediscovered. For example, the changes in cross-striated muscle during contraction were known in the 19th century but forgotten, only to be rediscovered in the mid-20th century (17). The vertical occipital fasciculus was described by the neuroanatomist Carl Wernicke in 1881 but later disputed and forgotten until the 2014 work of Brian Wandell and colleagues (18). Moreover, scientists who are concerned with only the facts and not the process miss out on the rich human drama of perseverance, serendipity, inventiveness, and conflict that characterizes the history of science. It is often such details that are most interesting to a nonspecialist, which in turn facilitates teaching and the engagement of the general public with science. The omission of the history of discovery from scientific papers may thus serve to perpetuate the barrier between scientists and the public whom they serve and depend upon for support.

To neglect history and accept the scientific literature as record is in fact to embrace a false narrative. The absence of a historical perspective of science can create a disconnect between perception and reality. In a seminal essay (19), Stephen Brush jokingly suggested that the history of science should be "X-rated": "Young and impressionable students at the start of a scientific career should be shielded from the writings of contemporary science historians ... (because of) violence to the professional ideal and public image of scientists as rational, open-minded investigators, proceeding methodically, grounded incontrovertibly in the outcome of controlled experiments, and seeking objectively for the truth."

However, the serious subtext of this statement is that "the history of science may be used to challenge the supposedly truth-seeking character of science" (20). This is a devastating criticism because it implies that scientists who ignore the discrepancies between the real and idealized views of science may also undermine their legitimacy as objective and trustworthy authorities on the realities of the natural world.

WHY SCIENTISTS SHOULD CARE ABOUT THE HISTORY OF SCIENCE

The history of science is important because it highlights the ingenuity of earlier scientists and provides a map to connect current pathways of discovery with the past. To this, we add five reasons why scientists should pay greater attention to history.

1. **Science is influenced by historical and social factors.** The great pathologist Rudolf Virchow initially rejected the germ theory of disease because his

passionate concern for social justice led him to attribute infectious diseases to poverty rather than to microbes. He actually had a point, but this example shows how science is not a purely objective endeavor that stands apart from society but rather that science and culture profoundly influence each other. This is most readily appreciated from a historical perspective. The historian may also be able to appreciate broad historical trends that are inapparent to a scientist. For example, the British philosopher Stephen Toulmin wrote of the "Alexandrian Trap," in which scientists in the 1st and 2nd centuries CE became increasingly specialized and focused on technology, losing sight of bigger questions (21). Historians can help scientists to avoid this conceptual trap in the modern era by illuminating the grand arc of scientific discovery and the importance of basic research.

2. **History allows scientists to learn from previous errors.** Errors are an inescapable part of science (22) (see also chapter 25). The history of science can help to show how investigators may be led astray and how the process of discovery can be improved. The historian James Atkinson has observed that scientists pay little attention to "the experiments that failed, the approaches that did not work out, the speculations without sound empirical support, and the metaphysical underpinnings of the work that did not appear in print" (23). However, such failures are the purview of historians, and scientists can learn a great deal from their insights.

3. **A historical perspective provides a greater appreciation of how discoveries occur.** Kuhn's seminal work on scientific revolutions used history to understand how discoveries occur and come to be accepted (24). In fact, history is essential for understanding how science advances, but the scientific literature does a poor job of documenting critical events in the process of discovery. For example, scientific papers seldom mention the critical role of chance in discovery. As a case in point, we consider the association of *Helicobacter pylori* with peptic ulcer disease, a discovery that changed the treatment of this common disease and was recognized by the Nobel Prize in Physiology or Medicine in 2005. In their landmark paper, Barry Marshall and J. Robin Warren paid tribute to the role of serendipity in a single sentence: "At first plates were discarded after 2 days, but when the first positive plate was noted after it had been left in the incubator for 6 days during the Easter holiday, cultures were done for 4 days" (25). Other than this casual reference to the religious calendar, the role of chance is not mentioned elsewhere in the paper. Marshall later acknowledged that prolonged incubation due to the holiday was a critical event leading to their landmark discovery. Decades of observations had suggested the presence of bacteria in stomach lesions, but these observations could not be validated experimentally because the slow-growing organism had not been successfully

cultivated. The ability to grow *H. pylori* from stomach tissue allowed Marshall to establish causality in his now-famous self-experimentation that fulfilled Koch's postulates. A greater appreciation of the role of chance and serendipity in discovery (26) (see also chapter 14) could eventually result in reforms to promote transformative, curiosity-driven research as opposed to an exclusive emphasis on hypothesis-driven and translational forms of research (27, 28).

4. **History can give credit where it is due.** Many alternative histories of science may emerge when scientists compete for rewards such as positions, prizes, and funding. Consider the discovery of the antibiotic streptomycin. Scientific papers tell us the origin of the compound, the properties of the molecule, and the spectrum of antimicrobial activity. However, underlying these cold facts is the struggle of a junior partner, Albert Schatz, for recognition and the efforts by a senior partner, Selman Waksman, to deny him that credit (29–31). Although the discovery of streptomycin was honored with a Nobel Prize, the committee never considered the contribution of Schatz, the graduate student who actually made the discovery while working in a basement laboratory. We have argued that the Nobel Prize often assigns disproportionate credit to certain individuals while neglecting the contributions of others (32) (see also chapter 21), and the Schatz-Waksman controversy is but one example. As professional recognition is the currency of science, history can play an invaluable role in setting the record straight.

5. **History reveals evolving ethical standards in science.** The history of science is essential for teaching about ethical behavior in science. The sanitized literature of scientific discovery often fails to detail ethical considerations, and it is striking to consider how scientific ethical standards have evolved over time. History has allowed us to see how Louis Pasteur's human trials, the Tuskegee and Guatemalan syphilis experiments (see chapter 29), and the unauthorized appropriation of Henrietta Lacks's cells are now considered ethical transgressions (33–36) (see also chapter 29), which underscores that the obligations of science to society must undergo continuing reevaluation to ensure that science remains a force for good in the world.

HOW TO BRING MORE HISTORY TO SCIENCE

We conclude by making a few recommendations to enhance the awareness of history among scientists.

1. **Recognizing science historians.** The scientific culture currently rewards priority and importance in discovery (37, 38) (see also chapters 8 and 19), but there is little recognition for those who chronicle and interpret the human stories behind those discoveries. Although historians of science are recognized

within their own field, they are too often regarded as curiosities by scientists. Scientific recognition that science historians and journalists have a critical role in the scientific enterprise will help to elevate the value of history in science and encourage students to take an interest in these fields.

2. **Promoting history in scientific societies.** Many scientific organizations, such as the American Society for Microbiology (ASM), contain groups that are focused on history, such as the Center for the History of Microbiology/ASM Archives (CHOMA). Such groups play a critical role in preserving the past and are largely maintained by a dedicated set of history-minded individuals. The efforts of such groups should be encouraged, supported, and made more visible. Meetings, conferences, and publications provide ample opportunities to provide historical perspectives on key scientific topics and ensure continuity between the scientific past and present. Science historians and scientists alike could benefit from greater interaction and cross-fertilization.

3. **Promoting history in scientific courses and literature.** The history of science can be a powerful tool to teach and promote science. In the early 20th century, Paul De Kruif's *Microbe Hunters* helped to inspire a generation of scientists to pursue problems in microbiology (39). One mechanism to enhance the appreciation of the history of science is to combine historical aspects of discovery with the didactic presentation of scientific information. For example, a course on nucleic acids could be supplemented by historical readings on the subject and include such material as Watson's *The Double Helix: a Personal Account of the Discovery of the Structure of DNA* (40), Horace Freeland Judson's *The Eighth Day of Creation: Makers of the Revolution in Biology* (41), and Edwin Chargaff's reminiscences on the critical discoveries that first elucidated DNA structure (42). The injection of history, with its inevitable human foibles and drama, can add interest to any course and help to stimulate discussions about how discoveries come about and what constitutes ethical behavior. Similarly, journals could encourage more historical articles, perhaps pairing historians with scientists to document the process of discovery and encourage interactions between these disciplines. Placing new findings in the context of historical questions and discoveries can help make science more interesting to the general public. Nonscientists are often more engaged by the human history of discovery than by stark scientific facts. A greater emphasis on the historical process of discovery could also enliven courses, journal clubs, seminars, and scientific papers.

4. **Assuring historical accuracy in scientific publications.** The scientific literature has been highly formulaic for many decades. In contrast to the papers of the early 20th century, which often provided considerable background on the problems being addressed, publications today are terse and often limited in word number and the space that they can occupy in journals. As research

publications are increasingly accessible in electronic format, space limitations have become less of a concern. This should allow journals to relax restrictions on word counts that prevent historical discussions and lead to inadequate citation of the relevant literature. Given that citations are increasingly used as a measure of scientific impact, removing artificial restrictions on reference list length will help to ensure that authors are appropriately credited for their work. Perhaps some journals could introduce a small "serendipity box" where authors could tell the reader how a particular discovery came about. For example, although the role of serendipity in the discovery of phenotypic switching in *Cryptococcus neoformans* (43) was briefly alluded to in the paper, more could have been said. For that paper, the serendipity box might have stated:

This project began when strange colony morphologies were observed on agar plated with a liquid culture that had been inadvertently forgotten in a walk-in refrigerator. Although contamination was initially suspected, the colonies were shown to be *C. neoformans*, which prompted a search for the conditions that promoted such phenomena. The precedent of phenotypic switching in *Candida albicans* led the authors to specifically test whether the unusual morphologies represented a similar mechanism in *C. neoformans*.

Those few words pay tribute to the importance of serendipity and chance and provide a truthful account of how the finding came to be recognized that also acknowledges critical prior observations made with *Candida albicans*. This anecdote illustrates Pasteur's quote that "chance favors the prepared mind," since the knowledge of the phenomenon in another system encouraged pursuit of the observation. There is a strong lore in microbiology about forgotten culture plates leading to discovery. We note that culture plates kept past their time led to Nobel Prizes for the discoveries of penicillin and *Helicobacter pylori*. Perhaps the role of serendipity is minimized in today's literature because it is contrary to the prevailing hypothesis-driven models of discovery, and crediting chance may be seen to take credit away from the investigators. In fact, investigators often acknowledge the role of serendipity in discovery once a finding is accepted as important and credit is assured. It is time for the scientific literature to more truthfully represent the process of discovery and to reinforce the notion that honesty is essential to the quest for truth in science.

Science is more than a disembodied collection of facts. It is a uniquely human construct, a detailed and interconnected understanding of the natural world based on innumerable observations and contributions from individuals spanning thousands of years. History can help to keep science honest, with a keen sense of where it has been and where it is going. As Darwin observed, "Great is the power of steady misrepresentation—but the history of science shows how, fortunately, this power does not endure long" (44).

10 Specialized Science

Every man gets a narrower and narrower field of knowledge in which he must be an expert in order to compete with other people. The specialist knows more and more about less and less and finally knows everything about nothing.

Attributed to Konrad Lorenz

Science is a highly specialized enterprise, but it was not always this way. In the early days of the Scientific Revolution, natural philosophers strived to learn about many fields. For example, Alexander von Humboldt (1769–1859) made tremendous contributions to cartography, geology, zoology, and botany. In fact, early scientists did not necessarily confine themselves to science. Johann Wolfgang von Goethe (1749–1832) is primarily known for his literary contributions, but he also made a significant contribution to physics by proposing a theory of colors. However, as scientific knowledge grew, the practice of science began to require a specialized knowledge base and a specialized approach to problems. In fact, today, scientists who stray from their narrow fields are often viewed with suspicion by their peers (Box 10.1). Consequently, science today comprises specialties and subspecialties that have evolved to define discrete fields of study. Scientists are now so highly specialized that their expertise is usually confined to a narrow segment of scientific knowledge.

Thinking about Science: Good Science, Bad Science, and How to Make It Better, First Edition.
Ferric C. Fang and Arturo Casadevall.
© 2024 American Society for Microbiology.

Box 10.1 Do scientific generalists pay a penalty today?

Alexander von Humboldt was a scientific superstar in the 19th century whose radiance dimmed in subsequent centuries. Johann Wolfgang von Goethe continues to be known as a literary giant, but few today would consider him a physicist. Similarly, Benjamin Franklin made major contributions to our understanding of electricity but is better known for his revolutionary politics. Scientists today are more comfortable with early scientists who stuck to science such as Nicolaus Copernicus, Johannes Kepler, and Galileo Galilei. Isaac Newton may have spent as much time on theology that he did on physics but is forgiven because his contributions to classical physics are considered so significant that his theological explorations of the Holy Trinity and biblical mysteries are given a pass. One explanation may be that today's scientists are so specialized and their success so dependent on specialization that they are uncomfortable with generalists. Today, we recognize the "Sagan effect," named after Carl Sagan to describe the decline in reputation that befalls scientists who achieve popularity among the general public (52).

Carl Sagan was probably the most famous living scientist in the latter decades of the 20th century. He was an accomplished physicist and astronomer who made major contributions toward explaining the greenhouse effect on Venus. He was also a generalist who popularized science

Carl Sagan (1934–1996). Courtesy of The Planetary Society.

and wrote many science books that were widely read by the public. Sagan eschewed specialization, featuring many scientific disciplines in his television show *Cosmos*. When criticized for straying from his field, he said, "The boundaries are arbitrary … . In the real world, these subjects flow into each other" (53). Unfortunately, his peers were not impressed. He was rejected for tenure at Harvard and never elected to the National Academy of Sciences as an astronomer. However, in 1996 he was given the Academy Public Welfare medal shortly before his death, in belated recognition of his work to popularize science. Today, the Sagan effect continues to weigh on science as young scientists are often told to focus on one area and avoid publicity, which contributes to a high degree of specialization and the difficulty that many scientists have communicating with the public.

For a field, specialization can be viewed as a sign of success. As disciplines mature and expand their knowledge base, specialization becomes inevitable as the amount of information becomes too large for any individual scientist to master. The major specialties of science are physics, chemistry, and biology, each of which comprises dozens of subspecialties ranging from astronomy to zoology. In the allied field of medicine, physicians long ago separated into surgeons and internists, each of which now includes over a dozen subspecialties. Surgeons specialize their skills primarily according to anatomical regions as they are required to master increasingly challenging technical procedures. More recently, medicine has developed specialists in pediatrics, women's health, radiographic techniques, and mental disorders, to name a few. Specialization is rife throughout society. For example, lawyers specialize depending on the type of law they practice, police officers specialize depending on the duties they perform, and the armed forces now include many branches that specialize according to the type of warfare in which they engage. Specialization is generally viewed in a positive light because it permits expertise in a subset of knowledge in a discipline and is encountered in all areas of human endeavor in which complexity emerges. Specialization can produce organizations that define themselves through technological prowess or the excellence of their trade, and this can be a source of pride that provides self-definition to specialists. Specialization emerges and is maintained because it confers obvious benefits to those who specialize.

The advantages and disadvantages of specialization have been studied primarily in the context of economic theory, finding forceful exposition in Adam Smith's 1776 treatise *An Inquiry into the Nature and Causes of the Wealth of Nations* (1).

Smith noted the advantages of a division of labor among workers to increase their efficiency and productivity. The guild system in Europe arose in the Middle Ages as artisans and merchants sought to maintain and protect specialized skills and trades. Smith promoted the view that specializing in certain types of labor, i.e., the division of labor, promotes efficiency and productivity by breaking down large jobs into smaller components that can be readily mastered by individuals, allowing the more rapid delivery of superior products. Smith famously used the example of a pin factory, in which the manufacturing process could be broken down into 18 discrete steps, each performed by a specialist. Through the division of labor, 10 workers could produce nearly 50,000 pins a day, whereas the same number of workers performing each step themselves could produce only 10 to 20 pins each day. Specialization can extend to entire countries, which develop specialized economies centered on those areas in which they have advantages, providing the basis for globalization and world trade.

Although scientific knowledge is quite different from a packet of pins, both have in common the delivery of goods, which for science consists of information, education, analysis, an improved understanding of the natural world, and the applications of that knowledge. Hence, the concepts developed from economics may have some relevance to analyzing the consequences of specialization in science. Like specialization in other fields of human endeavor, specialization in science has advantages and disadvantages.

Despite its benefits to those who practice it and to those who are served by it, specialization has its costs. Although guilds often produced highly trained and specialized individuals who perfected their trade through prolonged apprenticeships, they also encouraged conservatism and stifled innovation. Specialization in warfare has led to different services that compete for resources and prestige. Specialized services such as the Navy further subspecialize to create carrier, surface, submarine, and marine forces that may compete among themselves and fail to adapt to the changing nature of warfare. Interservice rivalry is a well-recognized problem in the military that can be detrimental to national interests. The United States Armed Forces require that officers rotate in other services prior to senior promotions in an effort to curb this problem (2). Hence, the benefits of specialization are tempered by the possibility that specialized groups become isolated, resist innovation, and engage in destructive competitiveness. Economists now recognize that one of the principal costs of the division of labor is the cost of coordinating the efforts of highly specialized workers, something that becomes increasingly important as the number of specialties and specialists increases (3).

Science is a highly specialized human endeavor, but the consequences of the divisions of labor found among scientists have not been examined systematically. From its beginnings in astrology, astronomy, alchemy, and classical medicine, science has generated a voluminous amount of information that has spawned the creation of dozens of disciplines that include the authors' fields of microbiology and immunology. Both microbiology and immunology are themselves sectarian, and each comprises many subdisciplines. For microbiology, these are generally microbe based, with a subdiscipline centered on researchers interested in specific microbes such that even within the larger groupings of bacteriology, mycology, and parasitology, there are mycobacterial, staphylococcal, chlamydial, candidal, and malarial communities, among many others. These groups tend to attend meetings focused on their favorite organisms and seldom interact collaboratively across microbial species. The immunological subdisciplines tend to focus on various components of the immune system, with adaptive (T and B cell), innate, and mucosal immunity constituting major affinity groups, and specialize in processes and functions of the immune system (4). Like the microbiologists, these constituencies are largely self-contained, although their boundaries are constantly challenged by the fact that the immune system is highly interconnected, rendering human-defined boundaries physiologically irrelevant when considering the system as a whole.

SCIENTIFIC FIELDS

Science is divided into areas that constitute fields. These areas range from the larger subdivisions, such as biology, physics, and astronomy, to smaller groups within those disciplines. For example, *Wikipedia* divides biology into dozens of subfields, including microbiology and physiology, with immunology considered a branch of physiology (5). Most scientists define themselves as belonging to a particular scientific field. As just one example, let us consider the field of microbiology. Here we explore the reasons that scientists self-organize into fields and how the shortcomings of fields might be ameliorated.

Dictionaries define "fields" as "particular branches of study or spheres of activity or interest" (6). This definition certainly applies to scientific fields, but there are other definitions more narrowly crafted for science. Thomas Kuhn argued that a field of science should ideally have a paradigm: "Acquisition of a paradigm and of the more esoteric type of research it permits is a sign of maturity in the development of any given scientific field" (7). Lindley Darden had a more extensive list of requirements: "A central problem, a domain of items taken to be facts related to that problem, general explanatory factors and goals providing expectations as to how the problem is to be solved, techniques and methods, and concepts, laws and theories related to the problem which attempt to realize the explanatory goals" (8). A special vocabulary is often characteristic of a field (9). Although these definitions apply to some scientific fields, they do not work very well for microbiology. Perhaps the greatest limitation is that microbiological fields tend to be defined by common interests rather than by theoretical concepts. In microbiology, field definitions are often microbe-centric rather than focused on specific problems. For example, the authors of this book are members of the *Salmonella* and *Cryptococcus* fields, respectively, which include individuals working on very different sorts of problems. These communities are centered around a microbe of choice. Membership in one of these fields requires contributing in some manner, but the contribution may range widely from structural biology to physiology to clinical medicine. Paradigmatic classification is perhaps more applicable to immunology, which contains some fields organized according to processes, e.g., antigen recognition, tolerance, autoimmunity, etc., but even there, one finds groupings that fail to conform to specific paradigms, such as the B cell, T cell, antibody, innate immunity, and mucosal immunity fields. Hence, there are many alternative ways in which scientists can be grouped. Apart from classification by phylogeny and process, fields may be defined by the meetings, journals, and organizations created to serve scientists with common interests. Determining field membership can sometimes be difficult. For instance, the authors of this book are both in the field of microbiology, but one is in the field of bacteriology and the other in the field of mycology. Since both authors study how microbes cause disease, they also belong to the fields of infectious diseases and microbial pathogenesis. However, one predominantly studies questions in molecular biology, genetics, and biochemistry, while the other is focused on immunology. To further complicate matters, both are interested in the workings of science, an interest that relates to the fields of philosophy, sociology, and even economics, as reflected in this book.

The point of this exercise involving your authors is to illustrate the difficulty of assigning individuals to particular fields with any degree of certitude, as individual scientists typically belong to multiple fields. In fact, field membership appears to be determined largely by self-definition by individuals according to their interests and is also dependent upon their acceptance by the other scientists working in a particular area. This implies a sociological dimension to field membership, which in turn suggests that field definition depends not only on scientific content but also on the other scientists with whom a scientist associates. From this vantage point, one can argue that associations and meetings define field membership. For example, those who attend the annual meetings of the American Society for Microbiology (ASM) and the American Association of Immunology may be labeled "microbiologists" and "immunologists," respectively. Similarly, individuals may be classified by where they publish their scientific work, with those publishing in microbiology or immunology journals being assigned to those respective fields. In U.S. academic institutions, microbiologists and immunologists are often housed in the same department based on the rationale that many microbiologists work on pathogenic organisms and immunologists are interested in mechanisms of immunity. However, this union is not always satisfactory because microbiologists tend to focus on individual microbes, while immunologists tend to be more interested in processes of the immune system (4).

Clearly, scientific field delineation, membership, and boundaries do not lend themselves to easy definition. Nevertheless, nearly everyone would agree that fields are important. We propose that a scientific field is a collection of individuals with a common interest in some aspect of science who interact on a regular basis. The interaction may be social, professional, and/or through the act of publication. This definition of a field differs from earlier ones by focusing on the human element as the key to field composition and yet incorporates interests, common goals, etc.

SCIENTIFIC FIELDS AS SOCIOLOGICAL UNITS OF SCIENCE

A scientific field viewed as a group of interacting individuals sharing a common interest is distinct from earlier definitions because it is based on human choices rather than on mere subject matter. The human interactions required by fields provide them with a social dimension, and fields may therefore be regarded as the sociological units of science. Scientific fields provide opportunities for professional friendships, collaborations, and interactions that promote science, but they are not immune to the problems found in society at large, such as gender and racial discrimination. Alice C. Evans, the first woman to become president of the ASM, had to overcome gender discrimination when she proposed that brucellosis could be transmitted by unpasteurized milk. She persevered and was eventually vindicated despite fierce opposition from men who "did not want any woman scientists" (10). Although women are now well represented in graduate training programs in the life sciences, the underrepresentation of women and members of historically underrepresented groups at senior academic levels remains an ongoing concern, and women continue to face significant obstacles in science (see chapter 15).

THE NORMATIVE ROLE OF SCIENTIFIC FIELDS

A scientific field, once formed, plays an important role in achieving consensus on matters large and small, including the following. What are the next important questions to pursue? What are the appropriate methods to study a problem? What are the standards for data acquisition and analysis? In this manner, fields establish the rules by which science is done. However, the answers to these questions have a sociological dimension. The decision of what problems to pursue may be driven as much by the charisma and personalities of individuals in a field as by more objective factors. Fields may develop cultures based on accepted customs and ways of thinking. Some fields may encourage the open sharing of information and reagents, while others may foster secrecy and internecine competition. For fields that foster the sharing of newly developed resources such as strains and reagents, the refusal of an investigator to share could lead to ostracism with negative consequences for the investigator. Such individuals might receive fewer invitations to give presentations and have more difficulty obtaining collaborations or funding as a result of a poor reputation. Hence, fields can enforce their norms by punishing those who transgress them, a topic that will be further explored below when we consider the social dimension of scientific fields.

To understand the customs of a field, one must be in the field or have close contact with those who are. Fields are also the keepers of specialized knowledge, which is not always accessible to outsiders. For example, it may be common knowledge within a microbiological field that there are peculiarities associated with a particular microbial strain, but that information may be shared informally and may not be available to a scientist who relies on the published literature. In this regard, fields share some characteristics with guilds, which regulate competition by strictly controlling admission and the sharing of information. Those who seek to work in a particular field must obtain insider knowledge in order to publish and become accepted by a field. As repositories for specialized knowledge, fields carry the risk that the knowledge and beliefs within fields can develop into dogma, which then becomes normative and can impede progress. The history of science is replete with examples in which prevailing dogma has impeded scientific progress. For example, Alphonse Laveran had to overcome the dogma that malaria was caused by noxious air to show that it was actually the result of a parasitic bloodstream infection (11). A nearly universally held belief in the tuberculosis field for the last half-century was that antibody-mediated immunity had no role (despite many observations that suggested otherwise), which delayed appreciation of the importance of humoral immunity in tuberculosis (12, 13). More recently, the central dogma of molecular biology, in which information flows from DNA to RNA to proteins, has been challenged by the discoveries of reverse transcription and prions (14, 15). Although the central dogma remains essentially intact, scientists now appreciate that there are exceptions to the rule. The failure of reverse transcription and prions to conform to existing dogma initially delayed their acceptance (16, 17). It can be very difficult for a scientist to overcome established dogma, as publication and funding are controlled by established members of a field.

The requirement for peer review in publication can be a two-edged sword that on the one hand helps to enforce rigorous standards of quality but on the other hand may stifle innovation and lead to stagnation. Peer review makes it difficult for outsiders to publish in certain areas, as the reviewers may see them as naïve interlopers. However, with persistence, outsiders can join fields and bring in new ideas that lead to changes in paradigms. The infusion of new members can also invigorate a field, hence Max Planck's observation that "a new scientific truth does not triumph by convincing its opponents and making them see the light, but rather because its opponents eventually die and a new generation grows up" (18), a concept now referred to as "Planck's principle" (see also chapters 19 and 28).

FIELDS AS COMMUNITIES

Given the interdisciplinary nature of science, the dangers of dogmatism, and the fuzzy nature of fields, one might then ask, as Victor DiRita has done (19), why have fields at all? One important reason is that fields provide a sense of community. Humans are social animals, and fields provide venues for socialization in professional areas. The anthropologist Robin Dunbar has observed that the size of social groups in primates is related to the size of the brain's neocortex ($r^2 = 0.61$) (20). He hypothesizes that this is the result of cognitive limits on the number of stable social relationships that can be maintained. A regression analysis of data obtained from various primate genera yields an estimated limit on human group size of 148 (95% confidence interval, 100 to 230) (21). Dunbar's number has been applied in settings as diverse as companies, government agencies, and social networks. Groups that exceed Dunbar's number may experience a loss of cohesion, which could be a factor in the emergence of new fields as old ones expand and diversify. The social organization of human enterprises that includes scientific fields may therefore be a direct reflection of the tendencies and limitations of the human brain. At one time, the ASM considered consolidating its 40,000 members into 4 large groups in place of the existing 25 divisions. However, the divisional structure fulfills the need of scientists, like all primates, for intimacy and cohesive communities. Perhaps as a compromise, 8 broad groups were created within the society, but the divisional structure was also maintained.

THE EMERGENCE OF FIELDS

New fields in science emerge when new problems are recognized as worthy of study or new technologies require a high degree of specialization for their use. An example of a new field is exobiology, which marries elements of astronomy with biology. Although life has not been discovered outside of our planet, there is intense interest in whether life exists in other worlds. This curiosity, along with a space program that allows the placement of robots on other planets in our solar system and the discovery of exoplanets, has stimulated research on how extraterrestrial life can be detected and what its biosignatures may be like. An example of a new technology that requires considerable technical expertise is nuclear magnetic resonance (NMR). Scientists and physicians specializing in radiology who utilize NMR have organized themselves around a body of knowledge that is tied to the instruments and its specialized uses (in medicine, this technology is called magnetic resonance imaging). Alternatively, a specialized technology may evolve to become available to scientists belonging to many fields. An example of the latter is molecular biology, which was limited to a few laboratories in its early days but is now used in fields as diverse as forensic science, genealogy, animal husbandry, and cancer research. Adaptation of a technology to a new use can itself foster further specialization, such as forensic scientists who specialize in DNA fingerprinting. Hence, technology can foster specialization and the emergence of new fields when first developed and disseminated to other fields as it matures, where the cycle of new specialization is repeated.

THE DECLINE OF FIELDS

Fields may decline or disappear depending on the interest level of scientists and the opportunities for investigation. A field can also decline once the scientific method exposes problems with its epistemic basis. Once-thriving fields such as alchemy, astrology, phrenology, and numerology are now recognized as pseudosciences (chapter 16). Other fields have seen their fortunes rise or fall as knowledge evolves. Anatomy was once a thriving field as scientists analyzed the structure of the human body to gain insight into how it functioned. Interest in anatomy waned as the structure of organs, vessels, nerves, etc., became well-known but has been reinvigorated by the development of new imaging technologies and the discovery of new structures (22, 23). The field of phage biology emerged following the discovery of phages and expanded markedly during the mid-20th century when it attracted many of the leading luminaries of molecular biology, Interest in phages declined during the latter part of the 20th century as individuals became more interested in animal viruses, only to be reinvigorated in recent years by the genomics revolution, an appreciation of the important role of phages in gene transfer, the development of phage-based therapies for antibiotic-resistant pathogens, and the transformative CRISPR (clustered regularly interspaced short palindromic repeat)/Cas9-based technology (24). The health of a field is proportional to the number of people interested in it, and fields can wither and die if abandoned by individuals who find other areas more worthy of study. Alternatively, the success of a field can also bring its own demise as it becomes incorporated by other fields. Consider, for example, the field of molecular biology, which emerged from the realization that proteins and DNA were molecules that could be studied, characterized, and manipulated. Progress in molecular biology spawned a wide variety of powerful techniques that went on to revolutionize other areas of biology. Whereas molecular biologists were once at the cutting edge of a revolution in biology, practically everyone in the biological sciences today can be considered a molecular biologist of one sort or another. Molecular biology has become so pervasive that the field of molecular biology has lost its unique identity. Perhaps fields that are incorporated into multiple other fields have achieved the ultimate level of success.

ADVANTAGES OF SPECIALIZATION IN SCIENCE

The advantages of specialization in science mirror those delineated by Smith for the division of labor, including efficiency, reduced time to production, improved quality, and the partitioning of vast quantities of knowledge into more-manageable units. In fact, there is no alternative to specialization in science, for the subject matter is so vast that progress requires a concentrated focus on a narrow problem for a protracted period of time. Consequently, scientific training has become highly specialized, with graduate programs channeling students into ever narrower fields.

Gaining recognition as a specialty or field can be important to establish legitimacy and to compete for resources. The medical subspecialty of infectious diseases originally arose from an increasing demand for expertise in the administration of antibiotics. The inaugural meeting of the Infectious Diseases Society of America (IDSA) took place in 1963 (25), and subspecialty board certification was first offered in 1972. However, demand and reimbursement for the expertise of infectious disease specialists was tenuous at first, leading the IDSA president to observe in 1978 that "I cannot conceive the need for 309 more infectious disease experts unless they spend their time culturing each other" (26). This observation has not aged well. The subsequent emergence of the AIDS epidemic changed the equation, and today there are estimated to be 8,500 board-certified infectious disease specialists in the United States alone. The complexities associated with treating a chronic multiorgan disease have led to the further subspecialization of some infectious disease specialists into those who focus primarily on HIV, and this has led to the formation of the HIV Medicine Association (HIVMA), closely allied with IDSA. Hence, success, complexity, and need are powerful forces in promoting specialization.

Given the success of science in the past two centuries and the fact that this success has occurred in the setting of increasing specialization, it is likely that the process is beneficial to the enterprise. The advantages of reducing the amount of information that must be mastered by any individual are largely self-evident. Given that specialization will remain the status quo in the foreseeable future, we will devote more attention to the disadvantages, especially as they apply to specific fields.

DISADVANTAGES OF SPECIALIZATION IN SCIENCE

Determining the boundaries of a field can sometimes be difficult, for no two individuals in a field have identical interests, and many spheres of interest overlap. The interdisciplinary interests of scientists provide human links that interconnect fields. As Richard Feynman observed, "We make no apologies for ... excursions into other fields, because the separation of fields, as we have emphasized, is merely a human convenience, and an unnatural thing. Nature is not interested in our separations, and many of the interesting phenomena bridge the gaps between fields" (27). Fields contribute to the balkanization of science and create artificial barriers that can impede interdisciplinary interactions and transdisciplinary thinking.

Some of the disadvantages of specialization in science also mirror the problems resulting from the division of labor in the economic sphere, including monotony, lack of mobility, monopoly, isolation, and the costs of coordination. Monotony was a major problem in optimizing the efficiency of industrial production once individuals became dedicated to specific tasks. The extent to which monotony is a problem among scientists is unknown, but given human nature, it is likely that some scientists become disenchanted with their chosen areas of expertise and may wish to move to other pastures. The industrial solution to monotony involved rotating jobs, but that is not readily applicable to science, for the development of scientific expertise and the maintenance of specialized laboratories require enormous expenditures of personal and financial resources. Consequently, many scientists live and die in their chosen fields of expertise, for it is simply too difficult to change fields. Adding to the cost of changing fields is the fact that most scientists are identified with their fields and develop social connections accordingly. For example, an individual who has specialized in *Salmonella* pathogenesis or T cell function would have to make a major effort to change to work on cryptococcal pathogenesis or B cell function and vice versa, despite the fact that all of these specialties are subfields within the parent fields of microbiology and immunology, respectively. As discussed earlier, fields become social units that define norms and are essential for advancement. For example, funding proposals are reviewed by established members in a given field, and in a similar fashion, awards and honors are generally bestowed by those who constitute the "establishment" in a field. In this regard, acceptance into a field carries some of the benefits of the medieval guild system, whereby accepted scientists are considered experts and given considerably more latitude in their work than newcomers, especially if their contributions contribute to the status quo or reinforce prevailing paradigms in the field. Conversely, it is very difficult for newcomers to break into fields and achieve the acceptance accorded to longstanding members, especially if they bring new ideas that are contrary to the accepted views in that field. Hence, specialization in science has the immediate disadvantage for an individual that the chasm can be too deep for movement to another field and that the benefits of field membership are too great. Once an individual becomes established in a certain field, changing fields carries a disproportionate cost that results in a *de facto* lack of mobility for most scientists.

Is lack of scientific mobility good or bad for science? The fact that most scientists become wedded to their fields of study has the advantage of providing continuity and stability to their respective fields, including the maintenance of specialized knowledge and normative standards for research. However, these advantages carry potential disadvantages, since continuity and stability can also exclude new ideas and promote the phenomenon of groupthink, whereupon fields may stagnate. The ability of Louis Pasteur to radically transform the fields of microbiology and immunology has been attributed to his "outsider" status as a chemist and nonphysician taking a fresh look at infectious diseases and strategies for their prevention (28).

One paradox is that all fields want to be recognized outside their fields and most desire growth, yet those desires are often thwarted by the same forces that bring cohesion to a field. For example, there is ample historical precedent that great progress can be made at the interface between fields, where each field can cross-fertilize the other, resulting in synergistic interactions. Unfortunately, scientists who strive to bridge two fields do so at their peril for they run the risk of being considered outsiders and thus fail to accrue the benefits that come with field membership.

Monopoly is another potential disadvantage of specialization. In science, a monopoly can emerge with regard to information, access to reagents, access to facilities, or collaborative interactions. Specialization in an area can lead to the generation of unique reagents, such as certain microbial strains, transgenic mice, etc. Most journals have strict policies requiring the sharing of reagents that are described in the instructions to authors. However, not all individuals with unique reagents are free and generous with their distribution, which creates a situation akin to a monopoly. Monopolies can also arise in the context of working with dangerous microbes, such as those requiring stringent biosafety level (BSL) 3 or 4 containment. In those situations, the monopoly arises from the regulatory requirements that the experimental work be performed in containment facilities that are available only in certain institutions, thus constituting a scarce resource. Fields focused on research of microbes that need high-level containment define norms for publication that often require work with the wild-type virulent strain and thus effectively exclude investigators who lack such facilities from entering the field. This exclusion can find many expressions. For example, in fields of research in which attenuated organisms that allow work with BSL2 containment exist, research papers involving such strains may find little acceptance by the established group, who demand validation of the data using fully virulent strains before accepting the findings. This, in turn, requires that any investigator who wishes to contribute to such a field must find the means to carry out experiments under conditions of high-level containment, often with the collaboration and to the benefit of established investigators who have a monopoly on production by virtue of access to the required facilities. Although we are clearly not advocating the relaxation of rules put in place to ensure the safety of investigators and the public, we merely use this example to point out that such rules may serve to create monopolies.

The mania around the journal impact factor (see chapter 26) that has proven so problematic in the biological sciences (29, 30) may have some of its roots in the increased specialization and intellectual isolation of working scientists. As scientists specialize, they tend to lose their capacity to critically evaluate the importance and quality of work in other areas of science and may increasingly look for surrogate markers. In this context, the journal impact factor has emerged as a means to judge the quality of individual research articles, in stark contrast to the impact factor's origin as a bibliographic tool to help librarians gauge the relative importance of journals (31). Consequently, many scientists have begun to judge the value of a scientific paper based on the venue in which it is published rather than on the importance, quality, and novelty of its content (32). This has introduced a major distortion in the practices of scientists as they seek to publish their work in higher-impact-factor journals that increasingly restrict publication (in order to maintain their high impact factors), thereby creating an environment conducive to questionable research practices (33, 34).

Given the enormity of scientific knowledge and the dispersed nature of the modern research enterprise, it is not surprising that the costs of coordinating specialized researchers can be substantial. A study of nearly 500 multi-institutional research projects supported by the National Science Foundation revealed an inverse relationship between the number of institutions involved and the achievement of project outcomes, suggesting that group heterogeneity reduced the efficiency of research when members belonged to different fields and/or institutions (35). Yet, as noted in numerous instances (examples provided below), the benefits of transdisciplinary research can be considerable once scientists leave their intellectual silos. Understanding a complex phenomenon typically requires a combination of approaches. Just as economists have documented the critical role of generalists on innovation teams (36), scientific leadership may benefit from individuals with broad vision and an ability to synthesize observations from diverse fields.

TRANSDISCIPLINARY RESEARCH AND TEAM SCIENCE

Two landmark scientific discoveries that transformed microbiology in the past century were the development of antibiotics and the discovery that heredity is conferred by DNA. Both were made possible by transdisciplinary research. Although the bacteriologist Alexander Fleming made his famous seminal observation in 1928, more than a decade elapsed before the chemists Ernst Chain and Edward Abraham, working with the immunologist Howard Florey, were able to purify sufficient quantities of penicillin to demonstrate its antimicrobial activity in mice. Further refinements by the biochemist Norman Heatley played a crucial role in making the industrial production of penicillin a reality, just in time for victims of the 1942 Cocoanut Grove nightclub fire to receive this lifesaving treatment (37). In other words, the bench-to-bedside translation of Fleming's observation required contributions from multiple scientific disciplines. Elucidating the structure of DNA and recognizing its potential to encode genetic information similarly emerged from multiple lines of inquiry, including crucial contributions by physician-scientists (Oswald Avery, Maclyn McCarty, and Colin MacLeod), physicists (Maurice Wilkins, Francis Crick, and Rosalind Franklin), a biochemist (Erwin Chargaff), and a molecular biologist (James Watson). Another example of fertilization across fields was provided by the enormously influential "phage group" organized by Max Delbrück, a theoretical physicist who teamed up with the molecular biologist Salvador Luria and the bacterial geneticist Alfred Hershey to promote the use of bacteriophages in exploring fundamental biological questions. A revolution linking the microbiome to many aspects of human health is in full swing, and multiple fields, including microbiology, immunology, metagenomics, clinical medicine, chemistry, physiology, and bioinformatics, are playing a major role. Microbiome-related research itself has already become highly specialized, with subgroups focusing on health, disease, specific anatomical regions, host species, computational tools, bacteria, fungi, phages, etc.

It is therefore not surprising to see an emergent consensus that transdisciplinary research and team science integrating the biological and physical sciences with engineering are critically important for the future of science (38, 39). The American Academy of Arts and Sciences has proposed numerous recommendations for achieving synergy across disciplines (40). However, it is also evident that the implementation of this vision will need to overcome significant barriers, including the physical segregation of scientists working in different disciplines, the current reward system of science, and the increasingly anachronistic organizational structure of academic institutions (41, 42), as well as deeply rooted epistemic differences between fields (43).

STRATEGIES TO AMELIORATE THE CONSEQUENCES OF SCIENTIFIC SPECIALIZATION

Specialization in science is a necessity due to the enormity of scientific information, and specialization clearly confers significant advantages to the scientific community. However, although scientific specialization will clearly remain a fact of life, we can envisage some measures to retain the benefits of scientific specialization while mitigating its disadvantages.

1. **Give latitude to those who wander.** Fields can be hard on those who choose to relocate to other pastures. Such individuals may lose the benefits of field membership and may be perceived as not being seriously interested in the core subject matter of a field. However, allowing scientists to wander can benefit a field by sharing the specialized knowledge of the field with others (28) and fostering cross-fertilization. Being gentle on those who wander also increases the probability that those individuals will one day return with new knowledge and experiences gained from other fields, which in turn could enrich their original field.

2. **Promote interactions among fields.** Recognizing the importance of interdisciplinary, multidisciplinary, and transdisciplinary interactions, institutions, funding agencies, associations, journals, and meeting organizers should strive to create mechanisms to mingle and reassort individuals and concepts in ways that transcend conventional field boundaries and get scientists out of their comfort zones. This might involve research centers, conferences, sessions, or collaborative projects, just to name a few possibilities. Greater efforts can be made to bring together researchers with complementary expertise through transdisciplinary work-in-progress meetings and centers, such as the ASM General Meeting (now called the "Microbe Meeting") and FASEB Science Research Conferences, which actively encourage exchanges between fields. We acknowledge that the tribal organization of microbiology and immunology is unlikely to change in the foreseeable future, but there are encouraging efforts to forge transdisciplinary links. Specialized meetings are likely to remain very popular. Nevertheless, it is possible for fields to benefit from advances in

other fields and to reduce the problems associated with groupthink. Perhaps the online meeting options that have arisen in response to the COVID-19 pandemic can allow researchers to more easily attend meetings outside of their fields. Other mechanisms to reduce isolation can include inviting speakers from other fields to specialized meetings, encouraging cross-field visitations, and actively supporting interface research. However, the success of initiatives is critically dependent on efforts by the participants to reach out to other groups. For example, inviting speakers from other groups to specialized meetings will succeed only if each speaker makes an effort to integrate his perspective with that of the audience, which usually requires the creation of a new type of presentation.

Seminars, journal clubs, and scientific meetings are often structured around individual departments or fields. Physical isolation of scientists is an important contributor to the development of intellectual silos within institutions. One mechanism for promoting transdisciplinary research is the creation of institutes within institutions that include individuals from diverse fields and provide opportunities for interactions outside of specialized fields. The development of institutional criteria to recognize the contribution of individuals to team science projects when there are appointment and promotion assessments should also be encouraged.

3. **Articulate dogmas.** A dogma is defined as "a principle or set of principles laid down by an authority as incontrovertibly true" (44). Certainly, there is no problem when a dogma is true, such as the theory of heliocentrism. However, when dogmas are false or incomplete, they can impede scientific progress. As discussed earlier, all fields have dogmas, which influence the type of work that can be done and published through the mechanism of peer review. It is therefore useful for a field to articulate its existing dogmas. This exercise can prompt a healthy reevaluation of the evidence supporting current dogmas and focus work on outstanding questions in the field.

4. **Define important problems.** In parallel, the most important unsolved problems in a field should be defined as the first step toward finding solutions. At the beginning of the 20th century, the German mathematician David Hilbert laid out 23 important problems, which helped to catalyze many novel solutions (45). This is relatively easy to do, as demonstrated by one of the authors of this book at the conclusion of a meeting in 2012 (46). The mere exercise of listing problems can lead to discussions that promote new scientific directions.

5. **Welcome outsiders.** Fields wish to grow and attract members but ironically put up many barriers that make it difficult for outsiders to join. Field membership includes publications and invitations to meetings, which in turn are determined by peer review. It behooves fields to be generous in welcoming outsiders

by not creating undue obstacles for those wanting to enter a field. Growth is a sign of health, and all fields should strive to recruit new members. Established members can assist newcomers by educating them with regard to unstated dogmas, esoteric protocols, and the prevailing social dynamics of a field. Fields can promote diversity by being more inclusive of women, members of historically underrepresented groups, and young investigators when selecting speakers for meetings, recruiting journal editors, and electing society officers.

As discussed above, scientists who want to switch fields or diversify encounter many obstacles. However, scientists need not become terminally differentiated. One mechanism for breaking down barriers confining scientists to their specialized fields would be to design fellowships and awards to be used specifically for cross-field training. Although such fellowships and awards already exist, they are relatively rare, narrowly focused, and designed primarily to recruit investigators to certain fields rather than provide scientists with freedom of movement. For example, the Burroughs Wellcome Fund has an interface award to recruit young scientists trained in the physical sciences and mathematics to biology (47), and several institutes of the National Institutes of Health offer training awards to encourage work in specific fields (48). Many universities continue to permit a sabbatical leave as a mechanism for established scientists to visit other laboratories and become familiar with new fields of study. However, sabbaticals are increasingly difficult to obtain, as scientists are burdened with the immense efforts needed to keep their laboratories operational at times of scarce funding and to meet administrative responsibilities. An increase in dedicated career development awards with the goal of diversifying scientists' expertise may have a salutary effect on the increasing specialization of science.

6. **Avoid tribalism.** As fields have many similarities to tribes, it is unsurprising that fields can lead to tribalism. Tribalism is detrimental for science, for it encourages conformity and creates boundaries between groups. Resource scarcity may enhance tribalistic behavior that can interfere with the proper functioning of review groups, scientific organizations, meetings, etc. Tribalism can be avoided by focusing on what is good for science rather than what is good for a field and by promoting the fluid exchange of ideas and individuals between fields.

7. **Broaden postgraduate training.** Postgraduate training today is designed to deliver young scientists into narrow fields of study, such as microbiology, immunology, or cell biology. It is noteworthy that Ph.D.'s are doctorates in philosophy despite the fact that most graduates today have no training in philosophy. Current doctoral programs are designed to teach students more and more about less and less. As discussed in chapters 6, 9, 26, 31, and 32 of this book, current Ph.D. training programs are too narrowly defined; the first-year

curriculum could incorporate fields of philosophical and historical knowledge that bear directly on the scientific method (e.g., ethics, logic, epistemology, and metaphysics) together with increased training in quantitative skills, such as probability and statistics (34, 49). Greater facility with philosophical concepts may facilitate transdisciplinary thinking by broadening the young scientist's intellectual tool kit, and enhanced quantitative skills will facilitate synergy with the physical sciences and improve experimental design. More-broadly-trained scientists have a better chance of appreciating other fields and benefiting from their knowledge while retaining the possibility for further specialization later in their training and careers. In response to these needs, the Johns Hopkins University has developed the R3 program (Rigor, Reproducibility, and Responsibility), with the goal of providing trainees with generalist tools to supplement their specialized education (50).

8. **Provide plain-language summaries of journal articles.** One seemingly inevitable consequence of the specialization of science is that fields develop increasingly arcane nomenclature. This, in turn, reduces interdisciplinary communication, promotes further specialization, and increases the isolation of fields. One mechanism to encourage communication would be to require plain-language summaries of scientific papers, and several journals are already using this approach.

Adam Smith rightly foresaw the benefits of specialization in complex human endeavors. However, specialism carries a price, and a healthy enterprise, whether a factory, a laboratory, or a global community, requires both specialist expertise and generalist thinking. The chemist Leo Baekeland, whose invention of Bakelite ushered in the era of plastics, foresaw this over a century ago (51): "If specialization may be advantageous for increasing our productiveness in a given field of activity, over-specialization, on the other hand, may develop one-sidedness; it may stunt our growth as men and citizens; even for persons engaged in scientific pursuits it may render impossible the attainment of true and general philosophic conceptions."

11 Revolutionary Science

Revolution doesn't have to do with smashing something; it has to do with bringing something forth.

Joseph Campbell (1)

Any discussion of revolutionary science must begin with Thomas Kuhn (Fig. 11.1), who popularized the notion of paradigm change in his enormously influential treatise *The Structure of Scientific Revolutions*, first published in 1962 (2) and subsequently expanded. Kuhn argued that scientific revolutions occur when a crisis in normal science resulting from unresolved anomalies causes a paradigmatic shift in a world view. In developing his arguments, Kuhn drew heavily on examples from the physical sciences, such as the Copernican revolution with its shift from a geocentric to a heliocentric viewpoint. Over the past half-century, some of Kuhn's concepts have been criticized, including his distinction between normal and revolutionary science (3, 4). One of the greatest problems with the Kuhnian view of revolutionary science is that it does not account for other types of scientific revolution, particularly those in biology (5). Neither the theory of evolution, nor the germ theory of disease, nor the discovery that DNA carries genetic information seems to have been triggered by the type of crisis in normal science that he envisioned.

Thinking about Science: Good Science, Bad Science, and How to Make It Better, First Edition.
Ferric C. Fang and Arturo Casadevall.
© 2024 American Society for Microbiology.

FIGURE 11.1 *Illustration of Thomas Kuhn (1922–1996). Image courtesy of Davi.trip, under license CC BY-SA 4.0.*

WHAT IS REVOLUTIONARY SCIENCE?

In considering revolutionary science, we begin with the definition of the word. "Revolutionary" is derived from the Latin word *revolutionem*, which referred to a turning motion and was originally used in relation to celestial bodies (6). The *Oxford English Dictionary* gives several definitions of the word, of which the most useful for our purposes is "a dramatic or wide-reaching change" or the "overthrow of an established … order by those previously subject to it." Kuhn did not provide specific criteria to distinguish revolutionary from normal science and often refers to revolutionary and "extraordinary" science as if they are synonymous; others have similarly discussed revolutionary science without explicitly defining it

(3, 7). Bruce Charlton proposed that major prizes be used to define and measure revolutionary science (8–10), but we are concerned that not all great discoveries are recognized by awards (11), nor are all award-winning discoveries revolutionary (12), as will be further discussed in chapter 21.

It is noteworthy that revolutions in politics not only change the political system but also affect other areas of human endeavor. Both the American and French revolutions in the late 18th century and the Russian and Chinese revolutions in the 20th century replaced prior systems of government with new forms and affected other human endeavors, including the relationship between church and state and the social order. Furthermore, each revolution also had major and immediate repercussions for other nations. In science, the Copernican revolution similarly signaled the end of a geocentric view of the world and its replacement with a heliocentric model. The success of the heliocentric model, together with other observations, broke the notion that received wisdom from antiquity was certain and reliable, thus opening the way for additional questioning in other areas of natural philosophy that ultimately ushered in the Scientific Revolution of the 17th century. Heliocentric theory also affected other important disciplines, such as theology, astronomy, and astrology, and directly impacted the calculation of the calendar.

We propose a definition of revolutionary science as a conceptual or technological breakthrough that allows a dramatic advance in understanding that launches a new field and greatly influences other fields of science. The Darwin-Wallace theory of evolution therefore qualifies as a revolution because it spawned the new field of evolutionary biology and profoundly influenced diverse fields, including anthropology, theology, sociology, and political science, soon after its publication in 1859. The discovery that DNA is the transforming principle of heredity and the subsequent elucidation of its structure also meet our criteria for revolutionary science because they launched the field of molecular biology while transforming the fields of genetics, medicine, and biochemistry. Eventually, this revolutionary discovery would even resonate in the fields of criminology, anthropology, and history, where it was used to solve crimes, clarify relationships between human populations, and determine the paternity of Thomas Jefferson's descendants. This discovery also had an immediate impact on numerous fields of scientific inquiry and within a generation had given rise to technologies to produce new pharmacological products such as human insulin. However, seven decades after publication of the structure of DNA, we may still be only at the beginning of what promises to be a continuing revolution in the uses and applications of polynucleotides. For example, DNA has tremendous potential for data storage, and DNA-based computer technologies may one day revolutionize the field of computer science (13).

Revolutionary science is not synonymous with extraordinary or important science (see chapter 8) (14). For example, the discovery of reverse transcriptase upended the central dogma of molecular biology, explained how RNA viruses replicate, and provided a target for some of our most important antiretroviral drugs. The discovery was recognized with a Nobel Prize. By any measure, this is extraordinary and important science. And yet, in our view, the discovery of reverse transcriptase does not qualify as a scientific revolution, as the discovery occurred within the established field of molecular biology and did not launch a new discipline or significantly affect disciplines outside virology and closely related fields. Similarly, targeted genome editing with CRISPR (clustered regularly interspaced short palindromic repeat)/Cas9 is a technology of tremendous importance that is rapidly having an impact in many fields associated with molecular biology. The development of CRISPR/Cas9 genome editing has recently also been recognized with a Nobel Prize given to Emmanuelle Charpentier and Jennifer Doudna. However, at this time, this discovery does not yet meet the criteria for revolutionary science, although it remains possible that subsequent events will establish it as a revolutionary finding. The shift in world view from a geocentric to a heliocentric model took more than a century, and both reverse transcriptase and CRISPR/Cas9-based technologies may yet evolve to meet our revolutionary criteria. In fact, it would be nice to be wrong, because such evolution will mean new fields and future uses that can benefit humanity. In contrast, the invention of PCR meets our criteria for revolutionary science because it spawned new fields such as forensic DNA analysis and provided a transformative tool for such unrelated fields as anthropology, archeology, criminology, and historical analysis. We note that other accepted revolutions in science also appear to meet our criteria (Table 11.1).

Microbiology is currently undergoing a microbiome revolution (15). This is another example of how normal science can progress to revolutionary science without a Kuhnian crisis. We have known for more than a century that microbial communities occupy all ecosystems from intestinal tracts to soils, but their diversity and function was not fully appreciated until development of the revolutionary technologies of DNA sequencing and PCR (Table 11.1), combined with high-throughput metagenomic next-generation sequencing, which has allowed the exploration of complex microbial communities in tremendous detail. Genomic sampling reveals that much of the microbial world is not culturable with current methods. Today, microbiome research includes the description of microbial diversity and the role of microbial communities in human, animal, and plant health, as well as in natural environments, including but not limited to extreme settings such as thermal pools, geologically active zones, and ocean depths. The implications of the microbiome include climate change, as there is now belated recognition that climate prediction models must include microbial elemental flows (16).

TABLE 11.1 Characteristics and impact of scientific revolutions[a]

Revolution	Year	Nobel Prize[b]	Type	New field(s)	Affected field(s)	Time (yr) to impact[c]
Heliocentric solar system	1543	NA	Conceptual	Astronomy	Theology	~100
Light microscopy	1600s	NA	Experimental	Microbiology, cytology	Biology, anatomy, physiology	70
Newtonian mechanics	1687	NA	Conceptual	Classical mechanics, calculus	Physics, astronomy, mathematics	10–20
Vaccination	1796	NA	Experimental	Vaccinology	Medicine, public health	Variable
Computers	1822	NA	Experimental	Computer science	All fields	>100
Thermodynamics	1824	NA	Experimental	Classical thermodynamics	Chemistry, physics, engineering, geology	30
Electromagnetism	1820	NA	Experimental	Electrodynamics	Physics, engineering	10
Urea synthesis	1828	NA	Experimental	Organic chemistry	Biology, biochemistry	5
Natural selection	1859	NA	Conceptual	Evolution	Biology, political science, theology	10
Germ theory	1850s–1870s	NA	Experimental	Infectious diseases, epidemiology	Public health, immunology	20–30
Mendelian inheritance	1866	NA	Conceptual-experimental	Genetics	Biology, botany, medicine	35
Phagocytosis, antibodies	1882–1890	Y	Experimental	Immunology	Medicine	5–10
Filterable viruses	1890s	Y	Experimental	Virology	Microbiology, medicine, public health	5

(Continued)

119

TABLE 11.1 (Continued)

Revolution	Year	Nobel Prize[b]	Type	New field(s)	Affected field(s)	Time (yr) to impact[c]
X-rays	1895	Y	Experimental	Radiology, X-ray spectroscopy, X-ray crystallography	Astronomy, medicine, dentistry	15–20
Radioactivity	1896	Y	Experimental	Radiation biology, radiometric dating, nuclear medicine, nuclear engineering	Anthropology, archaeology, history, military science, medicine	10
Quantum theory	~1900	Y	Conceptual-experimental	Quantum mechanics, quantum chemistry, quantum information	Classical physics, chemistry, electronics, biology	10
Relativity	1905–1920	Y	Conceptual	Relativity	Atomic physics, nuclear physics, quantum mechanics, astronomy, cosmology	10–20
Continental drift	1912–1970	N	Conceptual	Plate tectonics	Geology, evolutionary biology	10
Laser physics	1917–1960	Y	Conceptual-experimental	Nonlinear optics	Astronomy, biology, chemistry, medicine, physics	5
Transistor	1947	Y	Experimental	Solid-state electronics	Computer science	5–10

Heredity from DNA	1944–1953	Y	Experimental	Molecular biology	Genetics, medicine, biochemistry	10
Prions	1960s–1980s	Y	Experimental	Prion biology	Biochemistry, microbiology, neurology, veterinary medicine	20–30
DNA sequencing	1970	Y	Experimental	Genomics	Biology, medicine, forensics	5–10
Molecular cloning	1972	Y	Experimental	Recombinant DNA	Biology, medicine	5
Three domains of life	1977	N	Experimental	Archaeal biology, molecular taxonomy	Microbial ecology, evolutionary biology	10
PCR	1987	Y	Experimental	Molecular forensics, molecular diagnostics, synthetic biology	Molecular biology, medicine, anthropology, archaeology, forensics, history	5
Microbiome	1990s–present	N	Experimental	Metagenomics	Medicine, ecology	5

[a] Not a complete list.
[b] NA, not applicable; Y, yes; N, no.
[c] Estimates based on history of the field and the historical record.

Two other very recent examples of extraordinary and important scientific findings, the Higgs boson in 2012 and the detection of gravitational waves in 2016, also do not yet meet our definition of revolutionary science (see also Box 19.1). Finding the Higgs boson is part of a scientific revolution that began with the discovery of radioactivity (Table 11.1). The detection of gravitational waves is a prediction from the revolution initiated by Albert Einstein's general theory of relativity (Table 11.1). Whereas neither the Higgs boson nor gravitational wave detection have yet impacted other fields, we would not be at all surprised if in the future they make the transition from extraordinary and important science (chapter 8) to revolutionary science, for these are fundamental discoveries at the heart of physics, and all other fields depend on physics.

We suspect that some readers may quibble with inclusions and omissions in Table 11.1. Some of the scientific revolutions in Table 11.1, like the Copernican, Newtonian, Einsteinian, and molecular biology revolutions, are already well established in the pantheon of revolutionary science. Others, such as plate tectonics and the development of PCR, may not quite be chiseled into the pantheon walls but clearly meet our criteria. Some extraordinary and important discoveries, such as Fritz Haber's process for ammonia synthesis, which greatly facilitated the production of munitions and fertilizer, have had a major societal impact but fail to meet our definition of revolutionary science, because they relied on conventional concepts and did not spawn new fields. We acknowledge that our list is subjective, and some inclusions or omissions can be debated. One area of potential controversy is whether a revolutionary technology qualifies as revolutionary science. The discerning reader may note that Table 11.1 includes the microscope but not the telescope, both optical devices that function on the same principles of light refraction. This is because the telescope was not an absolute requirement for astronomical measurements, something that dates to antiquity, whereas the microscope is essential for becoming aware of the microscopic biological world. For others, the invention of the transistor might be considered a technological innovation rather than a scientific revolution. This discovery was catalyzed by an industrial scientific effort to improve upon vacuum tube technology at Bell Laboratories. However, the transistor led to the new scientific discipline of solid-state physics, and its importance for modern electronics had an enormous impact on many aspects of society, which argues for its inclusion as revolutionary science. The microscope, the laser, and the PCR are other examples of technologies that have allowed scientists to ask new questions that have transformed our understanding of the world. We consider other technologies such as DNA sequencing and recombinant DNA as having emerged in the context of laboratory science and include them in Table 11.1. The interested reader is welcome to add to or subtract from our list. In fact, doing so can be fun and necessitates learning about the history of science (chapter 9).

Our definition provides a straightforward means to demarcate revolutionary and nonrevolutionary science. However, it is interesting to note that the pace of scientific revolutions can vary widely. The Copernican revolution took almost a century to unfold, whereas the molecular biology revolution occurred within a decade and the PCR revolution began to influence criminology within a few of years of publication. Gregor Mendel's findings, despite their fundamental importance to genetics, were extremely delayed in their influence, as illustrated by the merely three times that his work was cited in the 19th century. The dissemination of scientific information has accelerated markedly since Mendel's time, but revolutions may still be delayed if there are no adequate experimental tools to test the predictions of a novel theory. Alfred Wegener's theory of continental drift failed to spark a revolution when proposed in 1912 because he lacked an explanatory mechanism; the plate tectonics revolution only became possible half a century later, when technological advances allowed the demonstration of sea floor spreading. Nevertheless, the theory of plate tectonics is truly revolutionary science, as it created a new field with tremendous explanatory power that has also profoundly influenced paleontology, evolutionary biology, oceanography, and even astronomy, with regard to our understanding of crustal dynamics on other moons and planets. Plate tectonics is essential for understanding great climatic events in deep time such as Snowball Earth, when the planet froze over (17). One of us (A.C.) has even used the concept of continental drift to explain the speciation of *Cryptococcus neoformans*, which is a subject of his research (see also chapter 10) (18).

Confusion may occur when words have different meanings in common parlance and in science, such as the words "chaos," "error," and "significant," and here we note the limitations of "revolution" as a metaphor when used to describe a transformative scientific discovery. As Stephen Jay Gould memorably observed, "Great revolutions smash pedestals" (19), and political revolutions destroy one social order to allow its replacement with another. Scientific revolutions, in contrast, do not necessarily destroy or invalidate earlier work but rather place it in a new light. Old observations can be newly understood in the context of a novel paradigm. Einstein's theory of relativity did not destroy Newtonian mechanics but rather demonstrated their limitations. Newtonian mechanics were still used to get a man to the Moon. Moreover, although some have interpreted Kuhn's analogy of revolution to indicate that science is merely a social construct that does not make cumulative progress (20), science exhibits a strong tendency to build upon, not to discard, what has come before, particularly as fields mature and coalesce around a consensus paradigm.

ON THE RELATIONSHIP BETWEEN REVOLUTIONARY SCIENCE AND NORMAL SCIENCE

Kuhn posited that extraordinary science, which could lead to revolutionary science, differs fundamentally from normal science, and this distinction has taken hold in the zeitgeist, as evidenced by its frequent mention by essayists. "Normal science" was characterized as routine day-to-day research focused on what Kuhn called "puzzle-solving." Although he later denied any feelings of condescension (21), Kuhn also compared scientists engaged in normal work to "the typical character of Orwell's *1984*" (2). Karl Popper, in many respects Kuhn's adversary, was even more dismissive of normal science, describing it as "the activity of … the not-too-critical professional: of the science student who accepts the ruling dogma of the day; who does not wish to challenge it; and who accepts a new revolutionary theory only if almost everybody else is ready to accept it—if it becomes fashionable by a kind of a bandwagon effect" (22). Kuhn's depiction of normal science was controversial even in its time (3, 4). His and Popper's descriptions appear to be caricatures of science when they describe science at all. While scientists certainly do attempt to solve problems, they are constantly testing existing dogmas and, indeed, hoping to find evidence that current thinking may need to be revised, even if the revisions are more modest than a full-fledged scientific revolution.

The way in which we have defined revolutionary science provides a new perspective from which to view Kuhn's claim. According to our definition, revolutionary science cannot be identified at the moment of discovery since the implications and consequences of a finding are only evident after the passage of time. Accordingly, it follows that if revolutionary science cannot be distinguished from nonrevolutionary science at the moment of discovery, there must be no fundamental qualitative or quantitative difference between the two. By way of illustration, let us examine the discovery of the structure of DNA and the scientists working on the problem using similar techniques, including James Watson, Francis Crick, Rosalind Franklin, Maurice Wilkins, and Linus Pauling. These researchers were attempting to incorporate Erwin Chargaff's observations regarding the relative amounts of purine and pyrimidine bases into a chemical structure for DNA, which is, in essence, a puzzle that fits within Kuhn's view of normal science. The technological breakthrough of X-ray fiber diffraction allowed investigators to produce structural models that integrated the X-ray data with biochemical constraints such as Chargaff's ratios and the acidic pH of DNA. Although Pauling published first, his model of a three-stranded structure was not consistent with chemical observations and was rapidly discarded. Rosalind Franklin obtained the best diffraction data, which were used by Watson and Crick to propose their double-helix model. Watson and Crick's efforts involved false starts and a remarkable piece of luck—Watson shared an office with

Jerry Donahue, a chemist who noticed that Watson was using the wrong tautomeric structures for the bases and provided a key insight by providing base structures that allowed complementary pairing. Watson and Crick also benefited when Wilkins shared Franklin's unpublished diffraction data with them. It is difficult to view the sinuous trail of discovery punctuated by false starts and serendipity and regard this as epistemically distinctive from what was being done in other laboratories. Had not Watson and Crick proposed their double-helix model, it is virtually certain that another group would have eventually stumbled onto the correct model of DNA. Watson and Crick were honored for their discovery with a Nobel Prize in 1962, the same year Kuhn's work was published. This biological revolution resulted from puzzle-solving, or "normal science," without any paradigmatic crisis. Hence, the Kuhnian notion of a separation between normal and revolutionary science does not apply to what is perhaps the most important biological finding of the 20th century.

THE TRUE NATURE OF SCIENTIFIC REVOLUTIONS

A survey of scientific revolutions (Table 11.1) suggests that scientific revolutions have developed in a variety of ways. Some, indeed, seem to correspond to Kuhn's description, such as the discoveries of filterable viruses and prions, in which the progressive accumulation of anomalous observations led to a crisis that culminated in the generation of a new paradigm—the existence of submicroscopic life. However, many others do not correspond to such a scenario. The theory of evolution was an intellectual synthesis suggesting an explanation for biological variation, which made its debut without a mechanism. In contrast to Wegener's theory of continental drift, which initially failed to gain traction because no one could imagine how continents could move, the Darwin-Wallace theory of evolution captured the public imagination and gradually gained acceptance among scientists despite the initial absence of a mechanism. The germ theory of disease began with speculation and eventually emerged in mature form from the contributions of researchers in multiple countries who were investigating such disparate phenomena as silkworm disease, cholera, childbed fever, and ringworm. Such observations were eventually able to conclusively link specific diseases to certain microbes. Acceptance of the germ theory required decades of work involving both observation and experimentation, as exemplified by John Snow's investigation of a London cholera outbreak and the transmission of anthrax by Robert Koch. In contrast to continental drift, the germ theory of disease was accepted despite the lack of a mechanism to explain why some microbes could be pathogenic to some individuals yet harmless to others, a problem that continues to vex the field of microbial pathogenesis (23). It is interesting how mechanism plays a role in the acceptance of some theories and not of others.

The molecular biology revolution required both intellectual and experimental advances that culminated in the identification of DNA as the agent of heredity and the determination of its structure, which in turn provided a mechanism to explain the transfer of information. The invention of the transistor at Bell Laboratories arose from experimental observations and transformed the field of electronics. The PCR revolution was a technological innovation that allowed the amplification of small segments of DNA, which was enabled by the availability of a thermostable polymerase. Although Kary Mullis was recognized for the discovery of PCR (24), it is noteworthy that the concept of denaturing and replicating DNA with synthetic primers had been published over a decade earlier (25), and the concept could not have been successfully realized without the preceding isolation of a thermophilic microorganism with a thermostable DNA polymerase (26), which might not have occurred had Tom Brock not decided to stop at Yellowstone to break up a long car trip in 1964 (27). Hence, PCR emerged from normal science using established facts that were assembled into an extraordinary idea, which ushered in a revolution. Kary Mullis has stated that his inspiration came during a night drive on California State Highway 128, when the air was redolent with flowering buckeye (28). Although other aspects of PCR required careful attention to detail to become facile and useful, the PCR revolution appears to have begun by inspiration. It is also noteworthy that the PCR method relies on natural components put together to create a reaction that does not occur in nature. PCR reflects human ingenuity and was made possible only by the previous revolutions in DNA, heredity, recombinant DNA, and DNA sequencing. PCR required remarkable sophisticated engineering in the form of thermal cycler instruments that can translate the PCR concept into an affordable reality. Thermal cyclers in turn relied on earlier revolutions in thermodynamics and semiconductors. Revolutionary science is not an island—most if not all revolutionary science stands on the shoulders of prior revolutions.

The most striking aspect of the revolutions in Table 11.1 is the absence of any common structure to explain their occurrence or to define their nature. Revolutionary science can emerge from careful observation and description, experimentation, thought, or inspiration and often requires a combination of these elements, seasoned with a touch of serendipity. The only common denominator of all scientific revolutions is that they resulted from human curiosity, ingenuity, and an unceasing drive to understand the natural world.

SOCIETY AND REVOLUTIONARY SCIENCE

Scientific revolutions have had tremendous practical benefits for society (Table 11.2). The human population has increased exponentially since the mid-18th century, around the time when scientific inquiry became firmly

TABLE 11.2 Some practical societal benefits from scientific revolutions

Revolution	Societal benefit(s)[a]
Heliocentric solar system	More-accurate calendars
Newtonian mechanics, thermodynamics	Industrial Revolution, mechanical transportation
Vaccines, phagocytosis, antibodies	Vaccines, passive antibody therapies
Computers	Enhanced computational power
Natural selection	Comparative anatomy
Light microscopy, germ theory, viruses	Antibiotics, epidemic prevention
X-rays	Diagnostic tests
Radioactivity	Cancer treatment, energy, weaponry, radioactive dating
Quantum theory	Improved electronics
Relativity	Global positioning system
Continental drift	Seismological forecasting
Laser physics	Medical applications, printing, information management, communications
Transistors	Solid-state electronics
Heredity from DNA, molecular cloning, DNA sequencing	New therapeutic agents, genetically modified plants and animals, molecular diagnosis of birth defects
PCR	Biotechnology, forensic analysis, diagnostic tests

[a] Not a complete list.

established as a foundation for many human activities. Humanity has repeatedly avoided a Malthusian crisis by increasing the efficiency of food production, a direct result of the Industrial Revolution coupled with advancements in farming, crop varieties, and food preservation, which themselves have benefited from other revolutions. For example, the ability to preserve food by canning was made possible by the industrial and germ theory revolutions, which in turn allowed food to be consumed long after it was produced and stored in sealed containers that were free of botulism. Since the late 19th century, the pursuit of science began to be supported by public funds, first in Germany and then in numerous other countries, including the United States, particularly during and after World War II. We note that more than half of the scientific revolutions listed in Table 11.1 occurred with public support. Furthermore, the linkage between revolutionary scientific findings and the emergence of measurable public goods (Table 11.2) provides a direct refutation of the viewpoint that public spending on basic science is not associated with technological advances that benefit humankind (29). We recommend to our

Box 11.1 Revolutionary science as an antidote for depression

It is difficult not to be depressed after reading about current events or human history. Violence, war, inequity, injustice, oppression, starvation, ecosystem destruction, etc., are humankind's constant companions. These foibles are products of the human mind, which bring much human suffering. However, the human mind is also responsible for revolutionary science. Learning how humans figured out that the Sun was the center of the solar system (heliocentric theory), and that the Earth moves around the Sun, despite our senses telling us that it the other way around, is an uplifting journey. The more one learns about how scientific discoveries have been made, the more impressed and amazed one becomes with the extraordinary potential of the human mind. Together with music, literature, and the arts, and human acts of love, kindness, and generosity, which also flow from the mind, knowledge of how humans solve scientific and mathematical problems can provide great optimism for the human condition. Until intelligent extraterrestrial life is discovered, the thought that the universe is aware of itself through the scientific discoveries of the human mind brings a mystical connection to scientific discovery. For many of revolutions listed in Table 11.1, there are books written for the general public and scientists alike. We suggest that reading such books after digesting the daily news can be an antidote for depression.

readers that learning about revolutionary science provides an antidote for the depression that can result from thinking about the human condition (Box 11.1).

FOSTERING REVOLUTIONARY SCIENCE

As society is both the beneficiary and major sponsor of revolutionary science (Table 11.2), it is important to consider how revolutionary science can best be encouraged. Are there lessons from earlier scientific revolutions that can hasten the pace of scientific discovery? Although 25 revolutions are too small a sample from which to draw firm conclusions, some themes are discernible. First, although scientific revolutions are often associated with individual scientists, a closer inspection reveals that each of these individuals required a community of scientists making observations and raising questions that contributed to a revolutionary discovery. Second, a striking interdependence of scientific disciplines is evident in the genesis of certain scientific revolutions. For example, the molecular biology revolution depended upon advances in physics applied to molecular structures (X-ray diffraction), microbiology (pneumococcal transformation), chemistry (bases, amino acids, pH), biochemistry (Chargaff rules), mathematics (fiber diffraction analysis), and a well-supported academic system to provide scientists with sufficient time and resources to pursue their curiosity. Third, scientific revolutions are reliant on both routine scientific pursuits and moments of brilliant insight. The theory of natural selection was critically dependent on the assembly of many descriptive observations on species variation obtained through the mundane actions of specimen collection, characterization, classification, and archiving, activities that no biological scientist would consider extraordinary. However, when the information gathered from these mundane activities is illuminated and unified by an extraordinary thought, a larger synthesis emerges that constitutes revolutionary science. Similarly, neither the isolation of thermophiles from hot springs not the report that their enzymes are thermostable might seem, on the surface, to be extraordinary, yet without these findings, the PCR revolution could not have taken place. Fourth, most of the scientific revolutions listed in Table 11.1 emerged from inquiries into problems of basic science. This suggests that society must support research in broad fields of inquiry, with a major emphasis on basic science, to create the fertile substrate from which tomorrow's scientific revolutions will arise.

In closing, we emphasize that the thoughts developed in this chapter are not intended as a rejection of the seminal contributions of Thomas Kuhn. Although our definition of revolutionary science implies that there is no essential distinction between revolutionary and normal science and our argument that scientific revolutions lack a common structure challenges the conclusions of *The Structure of Scientific Revolutions* (2), we have had the benefit of an additional half-century of scientific history for analysis, including four biological revolutions, as well as access to many thoughtful discussions and criticisms of Kuhn's views. Our consideration of revolutionary science was made possible by the intellectual spaces that Kuhn created. In a sense, we are following Kuhn's directive by challenging his paradigm, as we have found it to be insufficient to explain the multifarious nature of scientific revolutions. The philosophy of science, like science itself, is a work in progress, and the analysis of the nature of science can be anticipated to evolve with additional human experience. We encourage a continuing dialogue among historians, philosophers, sociologists, economists, and working scientists in an ongoing effort to understand this essential human institution that constitutes science and to foster an environment conducive to revolutionary science.

12 Translational Science

Poetry is what is lost in translation.

You have probably already gathered that this chapter will not be about mRNA translation for protein synthesis or the science of translating languages. Rather, we here consider translational research, defined as "the process of applying ideas, insights, and discoveries generated through basic scientific inquiry to the treatment or prevention of human disease" (2), sometimes abbreviated as "from bench to bedside" (3). Although there is no universal agreement on what the word "translational" really means (4), there is no denying that translational research has become an important buzzword. In 2006, the National Institutes of Health (NIH) launched the Clinical and Translational Science Awards consortium of nearly 50 centers throughout the country (5). In 2009, the American Association for the Advancement of Science announced the publication of a new journal, *Science Translational Medicine,* and other journal publishers soon followed suit. The National Center for Advancing Translational Sciences (NCATS) was established by the NIH in 2012 and presently has a budget exceeding $500 million annually. Graduate programs have begun to offer programs focusing on translational research, and dedicated translational research institutes and centers have been popping up everywhere.

Thinking about Science: Good Science, Bad Science, and How to Make It Better, First Edition.
Ferric C. Fang and Arturo Casadevall.
© 2024 American Society for Microbiology.

THE ALLURE OF TRANSLATIONAL RESEARCH

There are obvious reasons for the emphasis on translational research. One is political. With ever-present political pressure on NIH administrators to demonstrate the tangible public benefit from the billions of dollars invested in scientific research, translational research is an easy sell—the testing of new treatments, vaccines, and diagnostic tests. Some of the impetus toward translational research comes from an impatient public speaking through its political leaders, who are the ultimate source of support for most scientific investigation through federally supported research. Another reason is to fill a genuine need. A number of factors have combined to impede the flow of information between basic science and clinical medicine, perhaps most notably a lack of sufficient resources to support early-stage investigation and the challenges involved in organizing clinical trials. A focus on translational research aims to remove these obstacles and facilitate and expedite the practical application of scientific discoveries. A third reason is an increasing impatience with the pace with which basic scientific discovery has resulted in new products and cures. Although translation of the molecular biology revolution into genetically modified crops, recombinant drugs, molecular forensics, and nascent gene therapy within a mere generation has been rapid by historical standards, the age of instant communication and fast-forward remote control buttons has created even greater expectations. For feared diseases such as cancer, AIDS, and Alzheimer's disease, progress toward prevention or cure has not been as rapid as many would like, although the remarkable progress made in treating human immunodeficiency virus (HIV) infection with antiviral drugs nevertheless provides a demonstration of how basic research with no obvious initial utility can lead to important clinical advances (Box 12.1).

Box 12.1 How basic research allowed scientists to meet the HIV challenge

The story of human immunodeficiency virus (HIV) and the development of HIV therapeutics illustrates the nonlinear and cross-fertilizing relationship between basic and translational science. Retroviruses were discovered in 1908, when Danish physicians showed that a poultry disease called chicken leukosis was caused by a virus now known as avian leukosis virus (ALV). In 1911, Peyton Rous showed that a closely related retrovirus could cause tumors in chickens (he received the Nobel Prize for this discovery—but not until 1966) (30). Over subsequent years, dozens of animal retroviruses were discovered, each associated with veterinary diseases that usually involved cancer. Despite a lack of evidence at the time that retroviruses caused disease in humans, funding agencies wisely supported retroviral research because of the high likelihood that such work would provide major biological insights. That premise paid off, as retroviral research led to the discovery of a reverse transcriptase enzyme that generates complementary DNA from an RNA template, a major modification of the central dogma of molecular biology that was recognized by another Nobel Prize (31). Rous's discovery also stimulated research on how retroviruses cause cancer, leading to the discovery of oncogenes, yet more Nobel Prizes, and revolutionary advances in cancer therapy (32). Work on mechanisms of retroviral replication would demonstrate the formation of a large polypeptide that is cleaved by a viral protease, a finding that would be exploited decades later in the development of highly effective therapies for AIDS and hepatitis C. The first human retrovirus associated with disease was human T-cell lymphotropic virus 1 (HTLV-1), reported in 1980, just 1 year before the first report of U.S. patients with acquired immunodeficiency syndrome (33, 34). In 1984, AIDS was shown to be caused by the retrovirus HIV (35). Hence, when AIDS cases exploded across the world, a large body of basic science knowledge was already in place that could be immediately utilized in response to the emergency. Independent of the basic retrovirology research was a major effort in the 1960s to identify new anticancer agents among nucleotide derivatives. This produced many new compounds that were not suitable as cancer therapeutics, but among these compounds was azidothymidine (AZT), which proved to be the first effective therapy for AIDS. In subsequent years, protease inhibitors were developed that together with reverse transcriptase inhibitors transformed AIDS from a uniformly lethal disease into one that could be chronically managed, even if not cured. Rapid advances in AIDS diagnosis and treatment were made possible by a wise decision to support basic science research on animal retroviruses many years and decades earlier, which produced important and extraordinary science that would save many lives.

OBSTACLES TO TRANSLATION OF BASIC SCIENCE

Not all promising biomedical discoveries are successfully converted into medical advances. In fact, there is a growing disparity between investment and research deliverables such as drugs (6). We seem to be investing more in research for fewer tangible benefits, and the pace of biomedical innovation and disruptive breakthroughs in science and technology appears to be slowing since the mid-20th century (7, 8). To be sure, one can still point to recent life-saving biomedical innovations such as mRNA vaccines, checkpoint inhibitor cancer therapies, and other novel medicines that were developed by targeting precise metabolic pathways identified by basic research. The question is whether the rate of novel applications has slowed relative to what is known and what is being discovered by current research. The translation of basic research into biomedical deliverables has encountered many challenges. For example, we are now aware of problems with the reproducibility of basic science findings that hinder their commercialization (chapter 7). For other discoveries there may be insufficient funding to follow up on promising observations, as large clinical trials required to license therapeutics are extremely expensive to perform. This problem is often referred to as the "valley of death" in drug development, as a metaphor for the fact that many basic discoveries fail to achieve fruition as a commercial product (9). Other factors contributing to the difficulty in translating basic research into deliverables include regulatory complexity, insufficient clinical researchers and clinical research infrastructure, pharmacology industry priorities, and the structure of academic research incentives (7). Finally, some scientific fields, such as immunology, are recognizing that animal models do not always precisely extrapolate to humans (10). For this reason, more translational research in humans is believed to be essential, despite the complexities and logistical hurdles posed by such research. Clearly, the attrition encountered when attempting to translate basic research into deliverables (11) has many complex causes, some of which are rooted in the systems used for discovery while others reflect hurdles at the level of academia, industry, and government.

SHOULD SOCIETY INVEST IN BASIC OR TRANSLATIONAL RESEARCH?

In a different era, Vannevar Bush argued strenuously to President Truman that the federal government should invest in basic science. He presciently observed the following:

> Discoveries pertinent to medical progress have often come from remote and unexpected sources and it is certain that this will be true in the future. It is wholly probable that progress in the treatment of … refractory diseases will be made as a result of fundamental discoveries in subjects unrelated to those diseases, and perhaps entirely unexpected by the investigator. …
>
> Basic research is the pacemaker of technological progress. …

New products and new processes do not appear full-grown. They are founded on new principles and new conceptions, which in turn are painstakingly developed by research in the purest realms of science. (12)

In Bush's view, basic research should be performed by academia, and applied research is performed largely by industry and government facilities (13). The conceptual dichotomy of basic and applied research has proven to be an enduring one. The late Daniel Koshland viewed basic and applied science as "revolutionary" and "evolutionary," respectively, summarizing the difference thus: "Basic research is the type that is not always practical but often leads to great discoveries. Applied research refines these discoveries into useful products" (14). Some discoveries in basic research, such as semiconductors and the structure of DNA, have revolutionized electronics and biology (chapter 11), making possible the laptop computer on which this chapter was written and the molecular research to which so many of us have devoted our careers.

The consensus forged after the Second World War that basic and applied research were the domains of academia and industry, respectively, began to fade to in the 1980s when the Bayh-Dole act allowed universities to patent knowledge obtained with federal funding. Universities ascertained that certain discoveries were enormously lucrative, and academic scientists began to emerge in a new role: that of the discoverer-entrepreneur. Within a decade, all major universities developed offices specializing in intellectual property to promote the protection and commercialization of scientific discoveries. Whatever the merits of this approach, one outcome was the blurring of the intellectual boundaries between academia and industry. Hence, scientists who formerly worked solely on basic biological mechanisms found greater freedom to develop their research along more practical lines, with the encouragement of their institutions. Furthermore, universities learned that it was much easier to connect with the public as well as with potential benefactors by highlighting their translational advances rather than their basic science discoveries. Translational research generated revenue, brought publicity, and enhanced public relations. In the evolving zeitgeist, academia is no longer viewed as an impartial champion for basic research.

In chapter 8, we considered the definition of "importance" in science and concluded that this quality is a function of four parameters: size, practicality, integration, and newness (15). From this perspective, basic and translational science differ primarily in integration and practicality, respectively. The importance of basic science derives from its contribution to knowledge deeper within the tree of information and, consequently, its greater potential for integration with other facts. In contrast, the importance of translational science lies in its practicality. Hence, we do not view basic and translational science as one being more important than the other but rather as complementary areas of human endeavor, with the important distinction that basic science findings often precede advances in translational science (see Box 12.2 for a discussion of why this is not always the case). We also note that observations in translational or applied science can generate new questions for fundamental research, as illustrated from the fact that vaccination preceded the field of immunology. Hence, the epistemological flow is bidirectional, and investments in both types of science are needed. As we scan a recent issue of the journal *Nature*, we see a stimulating mix of basic research, such as the mechanism by which giant meteorite impacts led to the formation of the earth's continents three and a half billion years ago (16), and applied or translational research, such as the mechanism by which a novel antibiotic called teixobactin targets a unique structure on bacterial membranes to compromise their integrity (17)—which seems just as it should be.

If the current emphasis on translational research leads to more scientific applications that benefit human society, that will be all for the better. However, it will be critical not to allow our impatience for translational applications

---

Box 12.2 Reverse translation: drug toxicity triggers bedside-to-bench research

Acetaminophen (paracetamol; Tylenol) is a small-molecule aniline derivative with potent antipyretic and analgesic properties and a widely used drug that does not require physician prescription in the United States. After acetaminophen was marketed, it became apparent that some users could develop serious liver injury. This prompted basic investigations into the mechanism of hepatotoxicity, which was shown to relate to the generation of N-acetyl-p-benzoquinone imine, a highly reactive metabolite. Unexpectedly, some of the damage was found to be mediated by the immune system, advancing our understanding of how drugs can damage organs (36). Another example of reverse translation is the thalidomide tragedy of the late 1950s, when this compound was used in Europe as an antiemetic in pregnancy, which produced major developmental birth defects affecting as many as 10,000 children (37). Although the compound was withdrawn from clinical use, it continued to be studied in the laboratory and was found to have potent immunomodulatory actions, which are now useful for the treatment of leprosy, tuberculosis, and cancer (38). The examples of acetaminophen and thalidomide show how unexpected clinical events can stimulate new basic research and unanticipated medical applications.

to skew resources and researchers away from the open-ended exploration of the natural world that has provided the foundation for so many translational successes and remains as essential as ever. In other eras, generations elapsed between the discoveries of electricity and the invention of the light bulb or the laws of thermodynamics and construction of the internal combustion engine. In our own impatient times, we note that retroviruses were discovered long before they were associated with human disease and that the dizzyingly rapid development of effective antiretroviral therapy against HIV was possible only because basic science had already provided the framework for rapid diagnosis and drug development (Box 12.1). One might argue that the absence of an effective vaccine against HIV despite considerable efforts at translating what we currently know reflects an inadequate basic knowledge of immunology as it relates to retroviruses.

Writing in *Newsweek*, Fareed Zakaria expressed concern that America may have lost its innovative edge (18). He noted that the rest of the world is rapidly catching up and that bright young foreign scientists are no longer flocking to the United States as they once did. Citing the robust federal funding of basic science that began in the 1950s, he observed the following: "Government funding of basic research has been astonishingly productive. Over the past five decades it has led to the development of the Internet, lasers, global positioning satellites, magnetic resonance imaging, DNA sequencing, and hundreds of other technologies." Whatever the need for further investment in translation, one must acknowledge that basic research has served us extremely well.

It is somewhat discomfiting that every grant application to the NIH must be evaluated on its practical merits, as if an obvious practical application is an essential requirement of all research (19). It is difficult to imagine how one might have justified the practical applications of basic research into the DNA polymerase of *Thermus aquaticus*, a thermophilic microbe with no medical or agricultural consequences, without the 20/20 hindsight provided by the development of PCR, an example of revolutionary science (chapter 11). Similarly, how would one justify the practical applications of fungal metabolism research that led to the discovery of statins, without the hindsight that these molecules can lower cholesterol? Of course, all basic scientists have become proficient at writing justifications that "X will lead to improved strategies to prevent or treat …" in the introductions to their grant applications, but the words can sometimes feel disingenuous. Furthermore, such efforts neglect the marvelous "childish curiosity" of unfettered exploration that led to such unanticipated discoveries as antibiotics and statins (20). While one can be confident that increased scientific knowledge can lead to improvements for society, it is impossible to accurately predict from whence key breakthroughs will come. As Harold Varmus insightfully noted:

> Just investing in clinical trials and things that are very disease-specific would be a huge mistake. ... Look at what pride people take now in advances made in diabetes and cancer research and infectious disease research. Almost all of it is based on recombinant DNA technology, genomics and protein chemistry. These are methods that grew out of basic science that was funded for years and years in a noncategorical way. Even if you move a little closer to the disease borders, you find that predictability is not the mantle under which we fund this stuff. (21)

PUBLIC COMMUNICATION

The scientific community must educate politicians and the public about how science really works, emphasize the complementary relationship between basic and applied research, and advocate more stable and sustained support of the nation's scientific enterprise. Our goal in this chapter is not to pit translational versus basic research but rather to draw renewed attention to the tenuous present condition of basic research, which will continue to be the engine driving humanity's hopes for curing disease, increasing productivity, eliminating poverty, developing renewable sources of energy, sustaining agriculture, and ameliorating climate change, to mention only a few immediate challenges. In the current enthusiasm for translational research, we must not forget that basic science is under threat. Medically related basic science research is particularly vulnerable because the NIH is the only source of support for much of this work, whereas applied research may be supported by a mixture of government, commercial, and private foundational sources. Moreover, in recent decades, industry has replaced the federal government as the leading source of support for research and development (Fig. 12.1). Funding trends have shown stagnant federal support for basic science (22, 23).

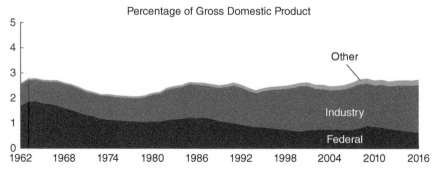

FIGURE 12.1 *U.S. research and development spending as a percentage of the gross domestic product, 1953 to 2016. Reprinted from reference 39.*

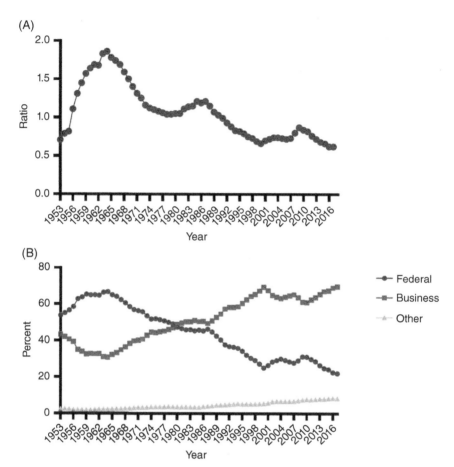

FIGURE 12.2 *Trends in U.S. investment in research and development. (A) U.S. research and development intensity. Changes in federal R&D in proportion to total gross domestic product over time is shown. (B) Percentage of R&D expenditure by funding source. Percentage of total U.S. R&D funding from federal government, business/industry, and other sources are shown (40). From Mandt R, Seetharam K, Cheng CHM. 2020. Federal R&D funding: the bedrock of national innovation. MIT Sci Policy Rev 1:44–54, CC BY 4.0.*

Success rates for individual investigator-initiated grant applications have declined (24). The declining relative contribution of the federal government to R&D (Fig. 12.2) is critical for basic science support because industry sources are less willing to invest in higher-risk research with uncertain applications and prolonged and unpredictable timelines for application.

Translators need something to translate. The time is ripe for a massive new national investment in science that includes basic research. Until the pendulum swings and basic science reemerges as a national priority, basic scientists will have to be imaginative in promoting the potential translational applications of their research, develop new methods to "humanize" their work (25), integrate their basic studies as components of larger translational programs, and hope that grant reviewers will continue to support good science even when it is not immediately apparent what the practical applications will be.

THE CIRCUITOUS PATH TO UTILITY

History has taught us that the path from basic discoveries to scientific and technological applications is seldom a straight line. Marie Curie described how her discovery of radium, which presaged the therapeutic use of radioisotopes, was purely serendipitous (chapter 14): "When radium was discovered no one know that it would prove useful in hospitals. The work was one of pure science. And this is a proof that scientific work must not be considered from the point of view of the direct usefulness of it" (26). In fact, the direction of knowledge and inquiry is not a one-way street. There are numerous instances in medicine when an unexpected clinical observation has required bedside-to-bench research in order to understand the phenomenon (Box 12.2).

More recently, we have seen studies of insect embryogenesis lead to a revolution in innate immunity (27), resulting in innumerable applications in drug and vaccine development. In her Nobel banquet speech, Christiane Nüsslein-Vollhard recalled her discovery of the Toll gene in *Drosophila*: "We started out in our research with a deep interest in understanding the origin and development of pattern during embryogenesis. None of us expected that our work would be so successful or that our findings would ever have relevance to medicine" (28). And when Carol Greider learned that she had been awarded the Nobel Prize for her groundbreaking work on telomeres, which promises to lead to advances in the treatment of cancer and the amelioration of aging, she emphasized the following: "We didn't know at the time that there were any particular disease implications. We were just interested in the fundamental questions. ... [This] is really a tribute to curiosity-driven basic science" (29).

Her words require no translation.

13 Moonshot Science

Shoot for the moon, because even if you miss, you'll land among the stars.

Leslie Brown (1)

On January 12, 2016, President Barack Obama called for a moonshot to cure cancer. The word "moonshot" served as a metaphor for a grand and epic endeavor (Fig. 13.1), echoing the successful space program that took humans to the moon in the 1960s (Box 13.1). Obama's call for a moonshot elicited mixed responses from medical researchers. Many praised it, because any increase in funding for biomedical science was welcome after more than a decade of diminishing research dollars (2), but others worried that the problem of cancer is too complex and too difficult to promise a cure (3). Critics were concerned about what could realistically be accomplished against cancer and feared that failing to deliver a cure in a timely fashion would risk eroding public confidence in science. Today, as we begin to recover from the devastating COVID-19 pandemic, there are calls for a pandemic prevention moonshot targeting the likely causes of future pandemics (4), as well as calls to reignite the Cancer Moonshot (5). Hence, the moonshot metaphor retains a powerful hold in the minds of the public and scientists alike as an example of something grand that can be accomplished with a concerted effort. However, others fear that the moonshot metaphor, like the words "roadmap" and "initiative," have been overused in the context of big and expensive science (6).

Thinking about Science: Good Science, Bad Science, and How to Make It Better, First Edition.
Ferric C. Fang and Arturo Casadevall.
© 2024 American Society for Microbiology.

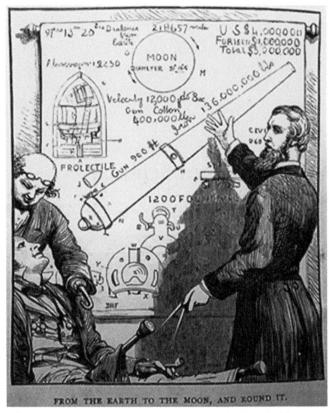

FIGURE 13.1 *Illustration from Jules Verne's* From the Earth to the Moon, and Round It *(40). Image courtesy of Andrew Cox, http://andrewcoxbooks.co.uk/contact-us.html.*

Aside from raising doubts about curing cancer, the 2016 Cancer Moonshot initiative raised the larger questions of how science works and how scientific progress can be fostered. History provides several examples of large public expenditures that produced spectacular results, including the Manhattan Project in the 1940s, the Moon landing in the 1960s, the development of AIDS therapies beginning in the 1980s, completion of the Human Genome Project in the 1990s, and more recently the accelerated development of vaccines during the COVID-19 pandemic. On the other hand, President Nixon's war on cancer of the 1970s failed to deliver a cure. Similarly, decades of effort to harness nuclear fusion and develop a vaccine for human immunodeficiency virus (HIV) have not yet delivered a commercial fusion reactor nor an effective vaccine.

Why are some large research endeavors successful, while others fail to deliver? As discussed in the preceding chapter, translational science builds upon previous advances in basic science. One major difference between successful and

Box 13.1 Science and the moonshot

In analyzing the moonshot metaphor for grand and epic science projects, it is worth reflecting upon the actual science that took humans to the Moon and returned them safely to Earth. The basic physics needed to know what it takes to lift a rocket and deliver a payload to the Moon was developed by Isaac Newton at the dawn of the Scientific Revolution. Advances in rocketry, propellants, and communications in the early 20th century allowed the design of the Saturn V rocket. Hence, the successful 1969 moonshot relied to a large extent on known scientific facts and principles. The success of the moonshot was a political bonanza for the United States, but some have argued that the scientific gains from visiting the Moon have been limited (34). One outcome of the moonshot was bringing lunar rocks to Earth, which provided convincing evidence that the Earth-Moon system resulted from the collision of a Mars-sized object with an earlier version of Earth. Another was the placement of laser reflectors at lunar sites, which allowed the precise measurement of the rate at which the Moon is receding from Earth. But perhaps most importantly, many technological advances developed by the Apollo program have impacted our everyday lives, ranging from digital fly-by-wire systems that guide modern airliners to advances in satellites, miniaturization, food safety, insulating materials, shock absorption, and energy storage (35, 36).

Regarding the direct impact of the moonshot on science, the physicist Freeman Dyson commented that "All was good science, but it was not great science. For science to be great it must involve surprises, it must bring discoveries of things nobody had expected or imagined" (34). Although we can understand Dyson's sentiments, we would quibble with his statement, as the knowledge that the Moon came from Earth was uncertain prior to having the actual rocks in hand, which allowed the precise measurement of oxygen isotopes (37). Prior to the moonshot, a variety of theories were advanced to explain the origin of the Moon, including gravitational capture of a wandering body, but this was disproven by the isotopic analysis. The lunar rocks contained other surprises for geologists as well, including feldspar-rich rocks indicative of magma and a hot origin of the Moon (38, 39). The interested reader may wish to compare the contributions of the moonshot effort with discoveries described in our chapter on revolutionary science (chapter 11). Although no Nobel Prizes have been directly associated with the moonshot effort, it has unquestionably had many benefits for humankind.

unsuccessful scientific moonshots is the extent to which the fundamental basic science underlying the goal is understood. The Manhattan Project, which delivered the atom bomb in 1945 to end World War II, was based on a solid theoretical understanding of nuclear fission, which had been understood since the late 1930s (7). Similarly, the Moon landings of the 1960s relied on 17th-century Newtonian physics and advances made in the German ballistic missile programs of the 1940s (8). Scientists combating the AIDS epidemic benefited from decades of research on retroviruses dating back to the early 20th century (9), so that the first cases of AIDS in 1981 were quickly followed by the discovery of HIV in 1983 (10) and the advent of antiretroviral therapy in 1987 (11). Even the first antiretroviral drug, zidovudine (AZT), was a repurposed anticancer compound that was first synthesized in 1964 (12) (see Box 12.1). The Human Genome Project of the 1990s was a triumph that relied on decades of fundamental science (13). Scientists had already established that DNA was responsible for heredity, had a double-helical structure, and could be sequenced by various techniques. Technological advances greatly accelerated their efforts to sequence the human genome.

Contrasting these successes with the efforts to harness nuclear fusion or an HIV vaccine, we can see how the slower pace of the latter projects reflects an insufficient understanding of the underlying basic science. To harness nuclear fusion, one must ignite and then control the same process that powers the Sun within a container that allows the heat to be harnessed for industrial purposes. Although this task poses formidable scientific and technological challenges, steady progress has fed optimism that fusion can become a source of unlimited clean energy in the 21st century (14–16). For HIV vaccine development, scientists don't yet understand how to direct the immune system to defend against a rapidly changing virus that hides inside immune cells. Nevertheless, here too incremental progress is being made. Of six HIV vaccine trials, one has shown signs of efficacy, hinting that a successful vaccine may require immunogens that can elicit certain specific antibodies that target the virion (17, 18). For both nuclear fusion and HIV vaccine development, unsuccessful efforts have not extinguished hopes for eventual success.

The same rules have applied to four recent public health emergencies: the severe acute respiratory syndrome coronavirus 1 (SARS-CoV-1) epidemic of 2003, the West African Ebola virus outbreak in 2012, the 2015 emergence of Zika virus in the Americas, and the COVID-19 pandemic of 2019. In 2003, a new coronavirus rapidly spread from Asia throughout the world (19). However, the SARS-CoV-1 epidemic was contained after only a few months as the result of a rapid response that relied on well-established principles for the epidemiological control of infectious diseases, including rapid identification of cases and the isolation and quarantine of infected individuals. The full SARS-CoV-1 genome sequence was known within weeks of the identification of the infectious agent (20). After just a few months, neutralizing human monoclonal antibodies (MAbs) were available to provide specific treatment and prophylaxis (21). In 2014, the world experienced the largest Ebola virus outbreak in history, killing thousands of individuals in multiple countries (22). The Ebola outbreak was contained by implementation of strict infection control protocols, including the isolation of infected patients and the use of full personal protective equipment. Once again, new therapeutics in the form of passive transfer of MAbs and immune sera were employed, and a vaccine was developed so rapidly that it could be tested in the final stages of the outbreak in 2016 (23). In both the SARS-CoV-1 and Ebola emergencies, the development of antibody therapies relied on decades of basic science studies on antibody-mediated immunity, immunoglobulin structure, and the development of MAb technology.

More recently, the world has faced the threat of new emerging viruses: Zika virus, which reached in the Americas in 2015 to cause a constellation of diseases ranging from microcephaly to Guillain-Barré syndrome (24), and of course, SARS-CoV-2, the cause of the COVID-19 pandemic (see chapter 30. Although SARS-CoV-2 has posed an unprecedented challenge, the international scientific community has responded to the pandemic with remarkable vigor, including rapid mobilization of clinical trial networks to rapidly test antiviral compounds and other therapeutic interventions; production of neutralizing antibodies for prophylaxis and therapy; and development of vaccines, utilizing a new mRNA platform made possible by a robust existing scientific infrastructure that built upon prior knowledge to tackle a new microbial threat (25). In fact, rapid vaccine development for COVID-19 was feasible only because basic coronavirus biology was already understood and the technical problems of delivering and expressing mRNA in human cells had been solved through decades of basic research (26).

Although the challenges associated with landing on the Moon and controlling a viral epidemic are very different, each of the above examples shares a common denominator: projects can ultimately succeed when earlier generations have invested in basic science research, often without necessarily knowing where it will lead. Is the cancer field ready for a moonshot? Perhaps. In the five decades since Nixon's war on cancer, there has been tremendous progress in our understanding of cancer, including the discovery of oncogenes, cellular growth factors, and mutations associated with carcinogenesis. Although cancer remains a major killer throughout the world, the death rates for a number of major cancers, including gastric, breast, uterine, lung, prostate, and colorectal cancer, have been significantly declining (27), while average survival times have improved. Improvements in prevention, screening, and treatment have resulted in a remarkable 33% decline in the U.S. cancer death rate since 1991, translating into nearly four million deaths averted (28). One of the most important conceptual advancements in recent decades has been the realization that each type of cancer is a different disease, such that any war on cancer must involve a multitude of separate wars to understand and target the specific mechanisms involved in each type. Even if a cure is not forthcoming, the risk of shooting for the moon is low. Earlier moonshots have produced benefits that could not possibly have been envisaged when the projects began. The Manhattan Project generated a vast amount of spin-off information that found its way into civilian nuclear power, radioisotopes for medical use, and plutonium-based batteries for exploratory spacecraft. The space program of the 1960s improved weather forecasting through satellite observation, enhanced telecommunication, and gave us the global positioning system that empowers our cell phones. Successes in HIV treatment showed that it was possible to effectively treat chronic viral infections, and today the same technology is being applied to many different viruses, some of which cause cancer. The human genome project of the 1990s led to the development of rapid sequencing technologies that have brought molecular biology into routine clinical use, including the use of sequence information to guide cancer therapy. That effort ultimately led to the sequencing of the genome of *Homo neanderthalensis* (29, 30), our extinct relative. This revealed differences that may have led to the success of our species, clarified our relationship to other members of the *Homo* genus, and won the 2022 Nobel Prize in Physiology or Medicine (31). Although Nixon's war on cancer failed to deliver a cure in the 1970s, that effort improved our understanding of the molecular causes of cancer, which is now bearing fruit in the form of new drugs. The efforts to contain SARS-CoV-1 provided new information about coronaviruses, and that experience was applied to the later coronavirus threats Middle East respiratory syndrome (MERS) (32) and SARS-CoV-2. Even moonshots that do not reach the moon can provide tremendous benefits for society.

The 2016 Obama cancer initiative reopened the debate on the optimal approaches for investment of public funds in biomedical research. While acknowledging the complex and formidable challenges posed by cancer, we are confident that if money for the Cancer Moonshot is spent on good projects, those projects are likely to yield many benefits to society, even if cures for some cancers continue to be elusive. In fact, in the first few years since the Cancer Moonshot was launched, remarkable progress has been made, for example, in the areas of immunotherapy and biomarker discovery (33). Just as money spent on cancer in earlier years allowed rapid progress against HIV, knowledge generated from this initiative is also benefiting many fields in addition to oncology. We urge the public, policymakers, and administrators to recall that the successful moonshots of the past each involved a solid foundation of basic science that was translated into practical applications that were useful to society. A broad research effort that balances advances in basic knowledge, the development of novel technologies, and robust clinical trials will give us the best chance to succeed. Moonshot science is grand, optimistic, and well worth the risk.

14 Serendipitous Science

> *One of the major ingredients for professional success in science is luck. Without this, forget it.*
>
> Leon Lederman (1)

An ancient Persian tale describes three princes from the country of Serendip (the ancient name of Sri Lanka) who are wandering through a foreign land in search of a lost camel (Fig. 14.1). As recounted in a letter by the 18th-century writer Horace Walpole, the princes "were always making discoveries, by accident and sagacity, of things which they were not in quest of" (2). This story has given rise to the word "serendipity," defined as "an aptitude for making desirable discoveries by accident" (3). Serendipity is now widely acknowledged to play an important role in scientific discovery (see also discussion of the role of serendipity in chapter 9).

Scientists love serendipitous anecdotes. Discoveries rooted in chance have a special allure, for they seem to belie the careful methodologies and intellectual approaches that form the basis of science. Serendipity celebrates the human element in discovery, in which seemingly random and unpredictable events intrude upon the prosaic world of everyday scientific inquiry to produce something sublime. As Isaac Asimov quipped, "The most exciting phrase to hear in science, the one that heralds new discoveries, is not 'Eureka' but 'That's funny.'"

In fact, the history of science is replete with stories of chance observations that led to great discoveries. One of Louis Pasteur's many discoveries was the chirality

Thinking about Science: Good Science, Bad Science, and How to Make It Better, First Edition.
Ferric C. Fang and Arturo Casadevall.
© 2024 American Society for Microbiology.

FIGURE 14.1 *Three Princes of Serendip. From a 19th-century engraving (45).*

of organic molecules, i.e., that they exist as enantiomers that are mirror images of one another. Less than a year after completing his doctoral studies, Pasteur was studying tartaric acid, which is an important determinant of the flavor, color, and stability of wine. A German chemist had concluded that two forms of tartaric acid crystals were identical despite their different interactions with polarized light in solution. However, Pasteur discovered that the crystals were composed of a 1:1 mixture of enantiomers whose crystals could be separated and shown to exhibit opposite effects on the rotation of linearly polarized light. It is now recognized that the precipitation of tartaric acid into enantiomeric crystals is temperature dependent, and Pasteur noted that this was best achieved in the morning before the warming of the day (4, 5). Hence, it appears that Pasteur had a serendipitous ally in the cool French climate that promoted tartaric acid crystallization into enantiomeric crystals.

French weather also played a role in the discovery of radioactivity. In 1896, Henri Becquerel was interested in phosphorescence and began conducting experiments to determine whether sunlight could trigger elements to emit X rays (6, 7). He wrapped photographic plates in black paper to protect them from sunlight, placed uranium crystals on top of the plates, and placed them outside in the sun. On the first attempt, he observed a crystal outline on the plates, appearing to support his hypothesis. However, an overcast sky prevented him from confirming his observations, so he placed the uranium and photographic plate in a drawer. He subsequently developed one of the plates that hadn't been exposed to sunlight to confirm that his chemicals were still working and was surprised to see that uranium radiation could be detected even in the absence of sunlight (8). His doctoral student Marie Curie went on to discover radium and polonium, and in 1903 Becquerel shared the Nobel Prize in Physics with Pierre and Marie Curie for the discovery of radioactivity.

In the field of bacteriology, no stain is as ubiquitous or useful in clinical medicine as the Gram stain, named after the Danish pathologist Hans Christian Gram, who discovered that bacteria could be divided into two groups, depending on whether they retained crystal violet after the addition of iodine and decolorization with alcohol. Such bacteria are known today as Gram positive and Gram negative. Gram is reported to have made the discovery after accidentally spilling Lugol's iodine on infected tissue samples while working late one night in 1884 (9). He modestly wrote to a friend that the stain is "very defective and imperfect, but it is to be hoped that it will turn out to be useful." His accidental stain is still proving useful more than 130 years later.

Perhaps the best-known example of serendipity in microbiology led to Sir Alexander Fleming's discovery of antibiotics. Fleming was studying colony variants of pathogenic staphylococci at St. Mary's Hospital in London. After returning from vacation in the summer of 1928, he noticed that some of his culture plates had become contaminated by mold. He set them aside to be discarded but noticed that some of the bacterial colonies adjacent to the mold were undergoing lysis. The mold turned out to be *Penicillium rubens* (10), originally designated variously as *P. notatum* or *P. chrysogenum*, which was producing the antibiotic penicillin, and the rest, as they say, is history. Through the brilliance, resourcefulness, and perseverance of Ernst Chain, Howard Florey, and Norman Heatley at Oxford University, Fleming's chance observation was translated into the antibiotic revolution. But the discovery was even more serendipitous than initially realized. For *Penicillium* and staphylococci to grow together on the same plate, laboratory temperatures had to be sufficiently low to permit the growth of the environmental mold. A review of temperature records in London during the summer of 1928 found that only nine days were sufficiently cool for this to occur, providing a narrow window of opportunity during which Fleming just happened to set up his cultures before departing on vacation (11). And what was the source of the mold? Fleming himself believed that it must have arrived through an open window. However, Ronald Hare, a bacteriologist who was a contemporary of Fleming at St. Mary's, later determined that Fleming's penicillin-producing strain of *P. rubens* was identical to a mold isolate that was being studied in the lab of C. J. La Touche, a mycologist who worked one floor downstairs from Fleming's laboratory (12). There can now be little doubt that the lifesaving fungus drifted upstairs into Fleming's lab as an airborne contaminant from La Touche's lab.

Innumerable other inventions owe a debt to serendipity, including corn flakes, Vaseline, Post-it notes, safety matches, saccharin, Teflon, the microwave oven, X rays, the anticancer drug cisplatin, Gore-Tex Vascular Grafts, and Viagra (13–26). Serendipity has also facilitated basic science advances such as the discovery of electromagnetism and the demonstration of background cosmic radiation left over from the big bang event that created our universe (27, 28). Serendipity is sufficiently common that most scientists can recall their own serendipitous experiences. For example, in the lab of one of the authors of this book, agar plates containing *Cryptococcus neoformans* colonies that were allowed to sit undisturbed for months revealed unusual morphologies that led to the discovery of phenotypic switching (29). In another example, a videomicroscopy experiment of fungal-macrophage interactions was inadvertently allowed to run longer than anticipated, revealing the unsuspected phenomenon of nonlytic exocytosis (30). The other author recalls once searching for paperwork in the notoriously disorganized office of his mentor and finding the latest issue of a scientific journal that had fallen to the floor—where by coincidence it was open to an article describing a recent discovery of growth-regulated gene expression that turned out to be pivotal for his research (31, 32). Serendipity touches scientists in ways both large and small.

What can we learn from these anecdotes? Recurring stories of unexpected discovery illustrate the contingent nature of serendipity, the importance of a receptive mind, and the potential value of undirected inquiry. The philosopher Samantha Copeland has noted that serendipity is profoundly contingent on the context in which it occurs (33). Many factors may prevent a serendipitous event from being appreciated (34). Fleming was primed to recognize the potential importance of penicillin because he had been studying antibacterial substances and had discovered the host antibacterial protein lysozyme 6 years earlier (35). Moreover, the translational application of Fleming's discovery could not have been realized but for the context of an international community of ingenious and determined researchers who could take Fleming's initial observation and perform the painstaking work required to develop it into a therapeutic agent capable of being produced on an industrial scale (36).

Pasteur famously observed that "Chance favors the prepared mind" ("*La chance ne sourit qu'aux esprits bien prepares*") (37). Becquerel, Gram, and Fleming certainly illustrate Pasteur's description of a prepared mind, and one can imagine many scenarios in which their serendipitous discoveries might have been overlooked by a less astute or attentive observer. Thus, in their seminal review of serendipity in science, James Shulman, Elinor Barber, and Thomas Merton concluded that "chance discoveries ... depend on an impressive list of estimable qualities in a scientist: enterprise, courage, curiosity, imagination, determination, assiduity and alertness" (2).

Like the princes of Serendip, scientists often seem to be struck by serendipity when they are in search of something else. Pasteur's contemporary, the great physiologist Claude Bernard, compared his scientific approach to a journey: "We take a walk, so to speak, in the realm of science, and we pursue what happens to present itself to our eyes." Just as microbes forage in random patterns when nutrients are few and far between (38), scientists can increase their encounters with serendipity by exploring new areas and asking original questions. As Gandalf reminded Frodo in *The Lord of the Rings*, not all those who wander are lost (39).

Importantly, the ability to take advantage of serendipity requires a scientific community. As serendipity relies on unpredictable occurrences, a large and diverse community of researchers is more likely to encounter, recognize, and follow up on serendipitous events. Chance events capable of revealing new scientific insights undoubtedly occur every day, but only a few are discerned. Joseph Henry, the first director of the Smithsonian Institution, remarked that "the seeds of great discoveries are constantly floating around us, but they only take root in minds well prepared to receive them" (40). Although scientific history is replete with examples of serendipitous discoveries, we cannot say precisely how many potentially serendipitous events have been missed. Scientists regularly confront unexpected observations in their daily work. Some may lead to novel insights or discoveries (see Box 14.1), while others are simply anomalous results due to error or experimental variation. How can one recognize the unexpected findings that are

---

Box 14.1 The cultivation of *Mycobacterium ulcerans*

Buruli ulcer is a chronic debilitating disease characterized by slowly expanding skin ulcers that may eventually involve the bone. The disease is caused by an environmental bacterium called *Mycobacterium ulcerans* and was first described by Sir Albert Cook in Uganda in 1897. However, the cause was not known until Peter MacCallum managed to culture the bacterium in Australia in the 1930s (43). The organism is only able to grow at a temperature of 33°C (91.4°F) or below, but conventional clinical laboratory incubators are set at 35 to 37°C. MacCallum was only able to grow *M. ulcerans* because his incubator was serendipitously malfunctioning and unable to maintain its usual growth temperature (44).

conveying important new information? Perhaps a systematic approach to such events could help to prioritize the ones worth pursuing. Confronted with the unexpected, a scientist must consider whether a finding is likely to be a false negative or a false positive and to consider the implications of the finding, if it is proved to be true, and whether it is consistent with what is known in the field. After trivial errors are excluded, scientists might apply the criteria discussed in chapter 8 to decide whether the finding is potentially important and worth pursuing (41). In science, knowing what to ignore is as valuable as knowing what to pursue.

We shall give the last word on serendipitous science to Sir Alexander Fleming, who remarked in a lecture at Harvard University: "If I may offer advice to the young laboratory worker, it would be this—never neglect an extraordinary appearance or happening. It may be—usually is, in fact—a false alarm that leads to nothing but may on the other hand be the clue provided by fate to lead you to some important advance" (42).

BAD SCIENCE

15 Unequal Science

This is the time when American higher education understands that our strength as a country will be inextricably tied to our success in bringing people from all backgrounds into the problem solving as we face the future.

Freeman Hrabowski III (1)

For whatever reason, I didn't succumb to the stereotype that science wasn't for girls. I got encouragement from my parents. I never ran into a teacher or a counselor who told me that science was for boys. A lot of my friends did.

Sally Ride, physicist and astronaut (2)

The Merriam-Webster online dictionary defines "inequality" as "disparity of distribution or opportunity" and "inequity" as "injustice and unfairness." Science is full of inequalities and inequities (3, 4), which pose a big problem for both science and humanity. Whereas subsequent chapters will cover unfairness in prizes (chapter 21) or credit (chapter 19), here we focus on the inequalities and inequities in science with regard to race, gender, and geographic origin. While there are certainly other types of disparity and injustice in science, the three discussed here are distinguished by a large body of evidence documenting major problems in these areas. We recognize that other forms of discrimination undoubtedly exist in

Thinking about Science: Good Science, Bad Science, and How to Make It Better, First Edition.
Ferric C. Fang and Arturo Casadevall.
© 2024 American Society for Microbiology.

science as well and note recent scholarship reporting that lesbian, gay, bisexual, transgender, and queer (LGBTQ) professionals face significant hurdles in STEM careers including career limitation, harassment, and devaluation of their work (5).

GENDER

A scientific career can be stressful for any scientist, as a result of demands for funding, promotion, publication, and success, but there is unequivocal evidence that life in science is more difficult overall for women than for men. One need only read James Watson's recollection of the discovery of the structure of DNA in *The Double Helix* to see how Rosalind Franklin was marginalized and undermined by the men in the story, despite her seminal contributions as an X-ray crystallographer. Unfortunately, Rosalind Franklin's experience was not atypical, as many if not most woman scientists encounter some form of gender discrimination during their careers (6). As Rita Colwell, former director of the National Science Foundation (NSF), recently recalled: "When I applied for a graduate fellowship to study bacteriology, a professor told me the department didn't waste such positions on women" (7).

Women continue to be underrepresented in many STEM fields in academia, industry, and government (8). Although this is changing as increasing numbers of women are choosing STEM careers, underrepresentation remains a particular problem in certain fields and among senior scientific ranks. The causes for gender disparities and underrepresentation in STEM fields are complex and incompletely understood. Evidence shows that women collectively have less funding (9–12), fewer publications (13), fewer collaborators (14), and lower citation rates than men (15). In a study in the field of social psychology, men were more likely to share research results with other men, suggesting the involvement of male-exclusive social networks (16). Women are less likely than men to be credited as collaborators in publications and patents when working on teams (17). Women are underrepresented in review panels that determine the funding of scientific grant applications (18), which could contribute to gender-related funding inequalities and also reduces their visibility. This could hinder career advancement, since such participation is viewed as a sign of expertise and authority. Such problems for women scientists are not necessarily independent, since, as one example, less funding can lead to fewer publications, which collectively can exert a negative synergistic impact on career progress. Interestingly, despite all these obstacles, an analysis of data from the National Institutes of Health (NIH) Office of Research Integrity found that women were underrepresented at all career stages among life science researchers who committed scientific misconduct (19).

Sometimes referred to as a "leaky pipeline," the gender gap between men and women becomes progressively apparent at more-advanced career stages. Women who remain in STEM fields are more likely to change career goals to pursue positions that require fewer years of postgraduate education (20). A 2019 report found that more than half of pre- and postdoctoral trainees receiving National Research Service Award support were women, but a widening gap began to appear in the transition from K to R01 awards, so that only one-third of R01 grant recipients were women (21). In fact, there is not a single "pipeline," as gender differences vary among scientific fields (22). Surveys indicate that the decision to have children impacts the career goals of women more than men, and marriage and childbirth may account for the largest career disruptions between the receipt of a Ph.D. and the acquisition of tenure for women scientists (23). The European Molecular Biology Organization analyzed its selection process for fellowships and awards and concluded that the decreased observed productivity and competitiveness of women applicants resulted from women taking on the majority of childcare responsibilities (24). A recent study by the National Academies found that the COVID-19 pandemic has amplified gendered expectations for women in STEM positions, adversely affecting their productivity and leading to a loss of personal well-being as well (25).

RACE

The concept of race has no justification in biological science (26, 27) (see also Chapter 16). In fact, race is a social construct and a social reality. In considering the effect of race on a scientific career, we should first mention that some non-White historically underrepresented and excluded racial and ethnic groups are designated as "historically underrepresented groups," or HUGs. It is important to note that the term HUGs can often include various diverse groups spanning from race, ethnicity, gender, physical disability, sexual orientation, neurodiversity, and more. However, for the sake of this chapter, we will associate the term HUGs with historically underrepresented racial and ethnic groups. Furthermore, the term "HUG" conveys the information that certain groups are underrepresented in science relative to their numbers in the population. For example, only 3 to 4% of medical school faculty members come from HUGs, although they constitute more than 30% of the general population (28). As the proportion of non-White people in the United States increases each year and they are expected to collectively constitute the majority in coming decades, there have been increasing efforts to increase the participation of HUGs in STEM fields. However, despite success in recruiting more students from HUGs into Ph.D. programs, resulting in a nearly

10-fold increase in STEM graduates from 1980 to 2013, this increase has not correspondingly increased their numbers in academic faculties (29).

Like women, individuals from HUGs seeking STEM careers face a wide variety of obstacles. Sixty-two percent of Blacks and more than 40% of Hispanics and Asians working in STEM jobs report experiencing discrimination at work, compared to just 13% of Whites (6). An analysis of NIH funding trends revealed that White applicants for NIH grants were 70% more likely to succeed than Black/African-American applicants after controlling for numerous variables that could influence success, including educational background, country of origin, training, previous research awards, publication record, and employer characteristics (30). Applications from Hispanic and Asian applicants were also less likely to be successful. A racial gap in funding success for Black and Hispanic applicants has persisted despite overall changes in funding rates (31, 32). Reduced funding parity dates to the end of the NIH budget doubling in 2003 and worsened following budget sequestration in 2013 (33), suggesting that inadequate public research support may have a disproportionate impact on women scientists and scientists from HUGs of both genders. Similar funding disparities have been seen at the NSF and have persisted across all directorates as recently as 2019 (34).

As with gender, career outcomes for Black and White researchers begin to diverge at the postdoctoral stage (35). Gender, race, and geographic origin can have an intersecting impact on inequality (22). In a survey of college seniors, White men were most likely and women from non-White racial and ethnic groups least likely to feel a sense of belonging in a STEM field (36). "Imposter syndrome," or the feeling that one is not deserving of one's accomplishments, is more prevalent in women and HUGs and may contribute to higher career attrition rates (37).

GEOGRAPHIC ORIGIN

There are strong impediments to performing scientific work in many countries. For example, the difficulties encountered by scientists in Latin America, including political instability, inadequate equipment, and inconsistent funding, are well-known (38). Funding cycles in many countries follow boom-and-bust dynamics, whereby periods of generous funding are followed by periods of funding scarcity, and the combination of insufficient funds and high publication fees can pose a major obstacle to scientists to publish their work (39). Journals from low- or middle-income countries are often omitted from international citation indices, which means that scientific work published in those journals will not be seen or acknowledged, creating a vicious cycle that can consign good science to oblivion (40). Discrimination and neglect can occur even within geographically proximal regions, as exemplified by the historical tendency for Northern European scientists to ignore the scientific contributions of Iberian scientists to the Scientific Revolution, which may have its roots in the schism between Catholicism and Protestantism (41).

INSIDIOUS INFLUENCES

Prejudices and biases are ingrained into the fabric of science. For example, an analysis of the etymology of new species names revealed a gender bias for species named after eminent scientists (42). In another instance, a randomized controlled study asked faculty to judge student applicants with identical credentials and found that men were favored, showing how conscious and unconscious biases work against women scientists (43). The analysis of paper co-authorship has been particularly informative. Researchers co-publish with other authors of the same gender more often than predicted by chance (44). Women are underrepresented as first or senior authors of manuscripts, and women-authored papers are less frequently cited (45, 46). Analysis of author order among scientists who shared the first author position showed that men were more likely to be placed first despite statements that both had made equal contributions (47, 48). Since credit for scientific work is often inferred from author order, a higher prevalence of women as second authors even when equal contributions are acknowledged can bias evaluators against women authors when applicants are evaluated for hiring and promotion. A higher likelihood of man-only associations is also observed in authorship sharing, providing further evidence that men preferentially cooperate with other men in sharing authorship credit (48). Surveys find that women are more likely to be involved in authorship disagreements and to feel that their contributions are devalued (49). In an effort to promote greater fairness in authorship assignment, journals published by the American Society for Microbiology (50) and the American Society for Clinical Investigation (51) now require explanations of how author order was decided upon. While these journals are under no illusion that requiring an explanation will redress all author order inequalities and inequities, the hope is that forcing a discussion among authors can perhaps lead to more equitable outcomes. Furthermore, having an explanation may be helpful to authors listed in the second position on the byline if it clarifies how the decision was made. For example, if the order was decided based on alphabetical order or a toss of a coin, then such an explanation would confirm that the first and second authors truly made equal contributions to the work.

ORIGIN OF INEQUALITIES AND INEQUITIES IN SCIENCE

Modern science (see chapter 1 and Box 1.3) is a social human enterprise that operates within the societies that support it. A sociological perspective can help to explain why the scientific enterprise reflects the inequalities and inequities of the society that created it, which is supported by the historical record. To some extent, the persistence of inequality and inequity in science may be a consequence of the longstanding predominance of White men in senior positions and the tendency of social networks to stabilize existing group structures (52). As Yuh Nung Jan observed, when attempting to explain the underrepresentation of Asians as awardees of biomedical research prizes, "People tend to choose people they are familiar with" (53).

However, there is also evidence that the scientific enterprise itself has been a willing collaborator in policies that have maintained gender, identity, and racial inequity (3). Science has entertained theories of biologically based differences between racial groups in the past. Louis Agassiz, a founding member of the U.S. National Academy of Sciences, was a vocal proponent of the existence of racial differences and the inferiority of Africans (54). Agassiz and many other scientists lent the weight of their reputations and work to justify the emergence of scientific racism (see also chapter 16), by which the prestige and accoutrements of science gave credence to despicable racial views. This has continued into present times as evidenced by the comments of Nobel laureate James Watson, who has espoused scientific racism by stating that there are genetic differences between Black and White populations that are responsible for differences in intelligence, despite abundant evidence to the contrary (for further discussion, see chapter 16). In contrast to Agassiz, Watson was criticized and shunned by the scientific establishment and forced out of his honorary academic positions (55). Similarly, the history of modern science is littered with supposedly scientific arguments to demonstrate the inferiority of women in intellectual endeavors (56). As I. I. Rabi, the 1944 Nobel laureate in Physics, explained:

> "(Women are) temperamentally unsuited to science. ... It's simply different. It makes it difficult for them to stay with the thing. I'm afraid there's no use quarreling with it, that's the way it is. Women may go into science, and they will do well enough, but they will never do great science" (57).

One suspects that Donna Strickland, a recipient of the 2018 Nobel Prize in Physics, might have disagreed: "I know there is certainly a lot of effort right now being placed on equity, diversity and inclusivity. ... But I don't see myself as a woman in science. I see myself as a scientist" (58).

To this day there are persistent stereotypes about gender and racial differences in STEM ability that lack any scientific validity but undermine efforts to correct the problems of inequality. Hence the origins of inequalities and inequities in science can be traced not only to the societies where science operates but also to the scientists who have promulgated ignorant stereotypes.

An example of how science and culture can mix to sustain a stereotype occurred in 2005, when the economist Larry Summers, as the president of Harvard University, stated in a speech discussing inequalities in STEM that "there is relatively clear evidence that whatever the difference in means—which can be debated—there is a difference in the standard deviation and variability of a (man and woman) population" (59). This comment ignited a firestorm of controversy and condemnation given his prominent position in academia because he seemed to attribute the problem of gender inequality and inequities in academic STEM fields to biological differences between men and women while ignoring all the other contributory factors cited in this chapter. Summers was alluding to the fact that while studies had found no differences in average ability between White men and women, greater variability is observed in men at the extremes of high and low ability, such that boys were overrepresented in the top and bottom percentiles. However, this finding did not apply to Asian students, which immediately raised questions about a biological explanation (60). The selective use of gender differences to explain gender inequalities and inequities in STEM was at best simplistic and incomplete, and at worst an example of how a soft finding in science can be used to promulgate a stereotype. Certainly, there is no evidence to support the notion that gender inequalities and inequities are the result of biological determinism.

Inequality and inequities date back to the beginning of science as a human field of endeavor. With few notable exceptions, such as the Cult of Pythagoras, which welcomed women, and the example of Hypatia of Alexandria, who was renowned for her knowledge of mathematics and astronomy, the paucity of women in the STEM of antiquity likely reflected the misogyny of those societies (61–63). Misogyny and gender discrimination remain a problem in the present even in secular westernized societies and undoubtedly contribute to existing inequalities and pose a barrier to full equality. Similarly, in the United States, the paucity of Black and African American scientists in the blossoming of American science that occurred around the turn of the 20th century reflects the racism of society and the legacy of slavery (64). Despite this tremendous handicap, we note the remarkable contributions of many Black and African American scientists (65–67), including George Washington Carver, who was born into slavery, Ernest Everett Just, Percy Julian, Alice Ball, and Katherine Johnson, among others, in the areas of agriculture, development, sterol chemistry, leprosy therapy, and orbital mechanics, respectively.

Despite the structural barriers alluded to above, both women and underrepresented racial and ethnic groups of both genders have made great inroads into successful participation in science (68). However, equality remains an elusive goal, and some fields continue to struggle to increase representation. For example, in physics, the proportion of women in 2018 was increasing by only 0.1% per year; at that rate, it was estimated that it would take 258 years to close the gender gap (44). Persistent gender and racial gaps in research funding also continue to be of great concern (33). Success in science is intimately linked to success in obtaining funding. One recent study concluded that topic choice is a factor in the lower rates of NIH funding success observed among Black scientists (69). In a possibly related vein, an analysis of more than five million published articles found that U.S. Asian, Black, and Latinx first authors tend to concentrate in specific fields, whereas White authors are ubiquitous (70). White scientists may have more latitude in the range of scientific fields available to them. In addition, scientists from HUGs may favor topics that impact their communities, while many reviewers prefer to support basic research (71), resulting in the continued devaluing of other areas of research.

THE BENEFITS OF DIVERSITY, EQUITY, AND INCLUSIVITY FOR SCIENCE

In addition to the injustice experienced by individuals who are affected by discrimination and prejudice in the scientific enterprise, the persistence of inequality and inequity has a tremendous cost for both science and humanity. The narrowing of gender and racial gaps in academic achievement over time suggest a strong role of environmental influences (72, 73). We believe that there are no innate gender, racial, or geographic differences in scientific potential. Given that scientific talent is a finite resource in human populations, any barriers to specific groups from participating in science only make the scientific community poorer and slow scientific progress (74). Historically, U.S. science has relied excessively on non-U.S.-born citizens to populate its scientific ranks. American society has been a magnet for immigrant scientists because it has provided opportunities not found elsewhere; approximately one-third of its Nobel Prize winners were foreign-born (75). However, importation of scientific talent from other countries creates a brain drain in those countries and exacerbates geographic inequalities. Instead, expanding participation by HUGs in the scientific enterprise of the United States will reflect the evolving demographics of the country and can expand the talent pool, enhance innovation, and improve the country's global economic leadership (76). There is a growing body of evidence to show that greater participation by HUGs translates into progress that benefits all of humanity. For example, women are more likely to patent innovations related to women's health, and given that women's health affects all in society from the women themselves to children, spouses, and family, having more women in STEM could mean more therapies that benefit everyone (77, 78). Team collaboration is improved by the presence of women in a group, due to a positive effect on group processes (79). Ethnic or gender diversity of authors, as inferred by names, correlates with higher citation impact and publication in higher-impact journals (80–83). Finally, a more diverse scientific community will be less permissive to harassment, which remains all too common in academic sciences, engineering, and medicine (84).

REDRESSING INEQUALITY AND INEQUITY IN SCIENCE

Obstacles to diversity of the scientific workforce exist at the levels of education, recruitment, retention, and promotion, including institutional culture, bias, admission criteria, salary disparities, a lack of role models, low rates of funding success, and concerns about work-life balance (85). These factors can work synergistically to worsen existing inequalities and inequities. To effectively address deep-seated patterns of inequality and inequity will require concerted efforts at systematic, institutional, and individual levels.

Intersectionality provides a framework to understand the interrelationship between such categories as race, gender, sexual orientation, and geographic origin and can help to understand the particular challenges faced by historically underrepresented women (86). Furthermore, gender differences observed within racial categories vary according to scientific discipline (86). Having less diversity in the workforce reduces the range of perspectives informed by social identities and thereby diminishes innovation.

At a systemic level, more research funding will be essential to achieve equity. It will not be possible to overcome historic disadvantages simply by eliminating discrimination and instituting a strict meritocratic approach to funding and promotion. Even in the absence of discriminatory practices, established senior investigators, who are predominantly White and men, have many advantages including reputation, experience, track record, collegial networks, robust infrastructural support, and an ability to recruit the best trainees. When funding availability is severely limited, meritocracy can serve to preserve the status quo rather than to ensure that all talent rises to the top. It is therefore unsurprising that inadequate funding has a disproportionate adverse impact on the very researchers who could enhance the diversity of the scientific workforce. In addition, women faculty and faculty from historically excluded racial and ethnic groups bear a disproportionate share of what has come to be known as "cultural taxation," the burden of being expected to serve on institutional committees to promote diversity and provide mentorship to women and trainees from HUGs (87–90). Although they feel an obligation and often personal gratification regarding their service in these roles, they are also acutely aware that such commitments divert time and effort from research activities that more directly impact promotion and tenure decisions. Despite having long hours and competing work-life issues in common, attrition is more common for women in research career pathways than for those in clinical medical training. The crucial difference may be the lack of job security and the precarity of funding for researchers (91). An increased public investment in research can help to lessen inequality and inequity in science. Diversifying the peer review process may also be beneficial.

Institutions also have an important role to play in reforming the scientific culture, which women and HUGs often find to be unwelcoming, if not frankly hostile. An NSF report found that a substantial proportion of individuals from HUGs who ultimately obtained doctoral degrees received their undergraduate education from institutions whose student bodies comprise a high percentage of a particular racial or ethnic group, such as historically Black colleges and universities (HBCUs), which attribute their success to providing a supportive environment, diverse faculty, and high academic standards despite chronically inadequate funding support (92–94). Achieving equity will require increasing the number of hires who are women and members of HUGs and providing them with support throughout the career timeline, as well as initiating a frank self-examination of institutional cultures and incentives from the perspectives of HUGs and women (3). Family-friendly policies such as paid parental leave, flexible hours, childcare, tenure clock adjustments, and workplace accommodations can provide crucial support and a welcoming environment for scientists who are pregnant or have young children (95). Studies in the private sector indicate that typical diversity training programs fail to yield lasting results (96). Employees resist what they perceive as coercion and indoctrination. More effective are measures targeting women and historically excluded racial and ethnic groups in recruitment, mentoring, increasing contact between groups, and efforts to promote social accountability, including diversity managers and task forces (97). Diversity statements can backfire and should be aspirational rather than coercive, emphasizing autonomy and the value of human differences (97). The counterargument made by some in academic circles that pro-diversity measures threaten excellence is not supported by data or rigorous analysis. Instead, we endorse the view that there can be no excellence in science without diversity (98).

Finally, individual scientists too have a responsibility for addressing inequality and inequity. Some specific steps that scientists can take include:

1. **Becoming informed about the problems of unequal science.** Scientific studies detailing the extent of inequalities and inequities in science are a relatively recent phenomenon, and many scientists are not fully aware of their findings. In this regard, we are reminded of the acclaimed Black American writer James Baldwin, who said, "Not everything that is faced can be changed, but nothing can be changed until it is faced."

2. **Redoubling efforts to be a good mentor.** Good mentorship is critical for the training of all scientists, but there is evidence that for women and historically excluded racial and ethnic groups a good mentor can make a huge difference. One study found that having a woman peer mentor increased the likelihood that women trainees would stay in engineering (99), but others have reported that active mentorship by men in academics can also play an important role in retaining women and other underrepresented groups in the pipeline (100). Another study of successful underrepresented racial and ethnic groups in science found that those in academia ranked good mentorship first in importance, in contrast to those in nonacademic positions, who ranked it fifth (101).

3. **Opposing the use of science or pseudoscience to promote inequality and inequity.** As noted earlier, some have used science and pseudoscience to argue for the inferiority of women and certain racial groups in STEM (see also chapter 16). Although the scientific racism of Louis Agassiz took place in the middle of the 19th century, such theories have continued to resurface in more recent times. In 1994, the book *The Bell Curve* analyzed IQ scores in populations and continued to promote the notion of racial differences in intelligence, even though racial categories have no biological basis. Scientists must be prepared to rebut any attempts to justify inequality and inequity on the basis of science by familiarizing themselves with contemporary data on human genetic diversity.

4. **Promoting diversity in teaching and lectures.** Scientists who teach can include and emphasize scientific findings from women, members of HUGs, and scientists working in disadvantaged geographical regions in their lectures (Box 15.1). For example, when teaching the germ theory of disease, lecturers can highlight contributions to understanding disease transmission from pioneering South American scientists such as Carlos Chagas and Alberto Barton. Other microbiology lectures could highlight the seminal contributions of women scientists such as Fanny Hesse, Anna Williams, and Alice Ball (Fig. 15.1).

5. **Advocating diversity in meetings, organizations, and selection and promotion committees.** Advocating diversity can make an enormous difference in ensuring that committees don't overlook women and candidates from HUGs for training, faculty appointments, speaking opportunities, and awards. Small changes can make a large difference. For example, increasing the participation of women in planning the American Society for Microbiology General Meeting significantly increased the proportion of lectures given by women and improved the gender balance of scientific sessions (102, 103). Broader criteria for promotion and tenure that give due credit for service and administrative contributions to promote institutional diversity, equity, and inclusion can reduce the career costs for faculty members who devote time to these activities.

Box 15.1 Using science teaching to highlight inequality and recognize diversity in science

Science courses are an important venue for the teaching of historical inequalities in science. Past injustices can be addressed by recognizing scientists whose contributions were ignored or marginalized in their time and by using historical vignettes to highlight the contributions of women and underrepresented groups to scientific discovery. Figure 15.1 provides an example of how important contributions of women could be incorporated into a course on medical microbiology or microbial pathogenesis. Figure 15.1A shows some of the scientific giants who are recognized for major contributions to the germ theory of disease and featured on national postal stamps, all White men of European ancestry. Figure 15.1B depicts five outstanding women scientists, including three Black microbiologists, and some of their accomplishments. Note that Dr. Anna Williams is featured in chapter 8 (Box 8.2) for her contributions as an example of important science. Another example could be Dr. June Almeida, a skillful electron microscopist who first noted the structural similarities of coronaviruses and gave them their name. Highlighting unsung heroes of science can enliven any course while also enhancing inclusiveness and recognizing diversity.

June Dalziel Almeida (1930–2007). Photo courtesy of Joyce Almeida.

(A)

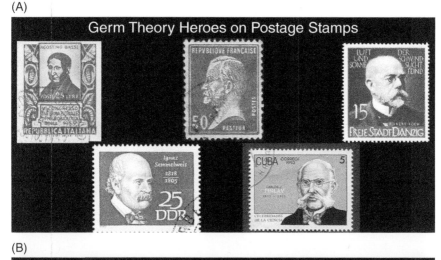

(B)

FIGURE 15.1 *Important contributors to the germ theory of disease. (A) Postage stamps commemorating scientists and medical researchers. Photo credits: Agostino Bassi (courtesy of Solodov Aleksei/Shutterstock), Ignaz Semmelweis (courtesy of wantanddo/Shutterstock), Carlos Finlay (courtesy of Shan_shan/Shutterstock). (B) Five pioneering women scientists who are also deserving of this honor. Photo credit: Jessie Price (courtesy of the Division of Rare and Manuscript Collections, Cornell University Library).*

Regrettably, science is rife with inequality and inequity, both as a reflection of the human society that created it and as a result of dubious studies that have suggested innate gender and racial differences in STEM ability. To its credit, the scientific community is presently trying to embrace diversity as well as to produce scholarship that seeks to better understand the scope and causes of the problem and to identify effective interventions. Although we are a long way from the ideal of equality in science, that day can be hastened if all scientists embrace the goal of creating a scientific community that is open and welcoming to all.

16 Pseudoscience

The demarcation between science and pseudoscience is not merely a problem of armchair philosophy: it is of vital social and political relevance.

Imre Lakatos

To begin this chapter, we would like to bring your attention to a remarkable medical therapy that has been shown to be effective in multiple clinical trials. Applications of the therapy range from chronic pain to rheumatoid arthritis, cardiovascular disease, and cancer. Several devices based on these principles are in use worldwide, and National Institutes of Health (NIH)-supported studies are ongoing. Dr. Mehmet Oz, the well-known surgeon, television show host, and, most recently, politician, referred to this therapy as being at the forefront of healing. The therapy is called "energy medicine" but has also been known by many other names including "biofield therapy" and "therapeutic touch." The central concept of energy medicine is that therapists believe that they can channel healing energy to a patient, and this can be achieved without even physically touching the patient or being in the same location. It is a form of *pseudoscience*.

Despite decades of study, randomized clinical trials of energy medicine have been inconclusive (1, 2). Various concerns have been raised about these studies, including methodological flaws, selection bias, and marked heterogeneity, not to mention the absence of biological plausibility. Like many complementary

Thinking about Science: Good Science, Bad Science, and How to Make It Better, First Edition.
Ferric C. Fang and Arturo Casadevall.
© 2024 American Society for Microbiology.

and alternative therapies, energy medicine is based on a nonsensical theoretical foundation, and any clinical benefits are likely to be the result of the placebo effect. Nevertheless, advocates of energy medicine couch their endorsements in the language of science and point to positive results that are selectively cited from the many studies that have been performed.

In 1796, the historian James Pettit Andrew referred to alchemy as a "fantastical pseudo-science." This is the first recorded use of the term "pseudoscience" (3). The Merriam-Webster online dictionary defines pseudoscience as "a system of theories, assumptions and methods erroneously regarded as scientific," while the *Oxford English Dictionary* provides the definition of "a pretended or spurious science; a collection of related beliefs about the world mistakenly regarded as being based on scientific method or as having the status that scientific truths now have."

Pseudoscience refers to an issue within the domain of science that is supported by assertions falsely represented to be reliable and based on scientific evidence (4). Defining the demarcation between science and pseudoscience has posed a challenge for philosophers of science. The Hungarian philosopher Imre Lakatos noted that Copernican heliocentrism and Mendelian genetics were held at one time to be pseudoscientific, which was used as a basis for persecution of the adherents of these theories (5).

Pseudoscience is not the same thing as *junk science*, which is sometimes used to refer to "methodologically sloppy research conducted to advance some extrascientific agenda or to prevail in litigation" (6). On the other hand, some forms of so-called *pathological science*, coined by the Nobel laureate Irving Langmuir to describe "the science of things that are not so," might be viewed as examples of pseudoscience (7). Langmuir was concerned about the potential for scientists to become deceived into believing in false phenomena, citing examples such as N-rays and extrasensory perception (ESP). N-rays were hypothesized by Prosper-René Blondlot to be a form of radiation, but their existence was ultimately disproven. This perhaps represents an example of observer bias and self-delusion rather than pseudoscience (8). ESP, popularized by J. B. Rhine at Duke University, more closely fits the definition of pseudoscience, referring to a range of paranormal abilities of the mind including telepathy, clairvoyance, and telekinesis. Other suggested examples of pathological science include polywater, a hypothetical polymerized form of water, and infinite dilution, the homeopathic principle in which a substance can be diluted a nearly unlimited number of times without losing its essential properties (9, 10).

Pseudoscience is used more generally to refer to entire disciplines, such as alchemy, astrology, or homeopathy. A common feature of pseudoscientific disciplines is that proponents claim to base their beliefs on evidence and often couch their arguments in scientific language, but the beliefs cannot be readily falsified.

TABLE 16.1 Warning signs of pseudoscience (11)

1. A tendency to invoke *ad hoc* hypotheses
2. An absence of self-correction
3. An emphasis on confirmation rather than refutation
4. A tendency to place the burden of proof on skeptics, not proponents of claims
5. Excessive reliance on anecdotal and testimonial evidence to substantiate claims
6. Avoidance of peer review
7. Failure to build upon existing scientific knowledge
8. Use of scientific-sounding jargon
9. An absence of boundary conditions, i.e., settings under which claims do not hold
10. A failure to progress over time despite additional research

Proponents of pseudoscience may appeal to authority, rely on evidence that cannot be replicated, cite handpicked examples, disregard conflicting information, or fail to subject their theories to rigorous tests that might disprove them. Pseudoscience is often represented as a complete and closed doctrine rather than as a methodology for open-ended inquiry (3).

Scott Lilienfield has assembled a useful list (Table 16.1) of warning signs that one may be encountering pseudoscience: the invocation of *ad hoc* hypotheses, a lack of self-correction, an emphasis on confirmation rather than refutation, placement of the burden of proof on skeptics rather than proponents, reliance on anecdotal evidence, evasion of peer review, lack of integration with other scientific knowledge (see chapter 8), use of scientific-sounding jargon, and the lack of boundary conditions defining the limits of the phenomena described (11). Another important characteristic of pseudoscience is that it fails to progress over time despite additional research.

Examples of pseudoscience are particularly abundant in clinical medicine, where they are collectively referred to as "quackery." Historical examples of quackery such as phrenology, humoral medicine, and miasma theory are often cited to demonstrate how far modern medicine has come. But contemporary forms of quackery abound and cloak themselves in the rubric of "complementary and alternative medicine" (12). With legitimacy conferred by this dignified title, pseudoscientific forms of therapy such as Ayurveda, chiropractic, and homeopathy have made their way into respected medical institutions and publications, gaining support from federally supported research (13). But as John Diamond observed many years ago, "There is really no such thing as alternative medicine—just medicine that works and medicine that doesn't" (14).

An instructive example of the destructive potential of pseudoscience may be seen in the history of scientific racism. The original concept of what we refer to as human racial groups is often attributed to the German philosopher Immanuel Kant, who published *On the Different Human Races* in 1775, dividing humanity into the "noble blond" of northern Europe, the "copper red" of America, the "black" of Africa, and the "olive-yellow" of Asia and India. The concept of races was quickly taken up by scientists—including such luminaries as Carl Linnaeus, Charles Darwin, Johann Blumenbach, and Louis Agassiz. Darwin's cousin Francis Galton advocated deliberate sexual selection on the basis of race, which he called "eugenics." Policies based on eugenics led to the involuntary sterilization of persons considered to be unfit and ultimately provided a foundation for the genocidal policies of Nazi Germany.

Galton was a statistician, and many statisticians whose names remain familiar today held racist views, including Karl Pearson, the inventor of the chi-squared test, and Ronald Fisher, considered to be one of the founding fathers of modern statistics. Purported genetic differences between racial groups have been inappropriately used to justify social and economic disparities. Scientists holding these views expected that genetics would ultimately confirm that racial groups fall into clusters based on distinctive genetic traits. However, modern genetic studies have shown no such thing.

The Harvard geneticist Richard Lewontin challenged the idea of a genetic basis for racial classification in the 1970s when he reported that more genetic variation occurs within groups rather than between groups (15), a conclusion that has since been corroborated by many others using more-extensive data sets (16, 17). Although considerable human genetic diversity is observed, this diversity does not correlate with traditional concepts of race but rather reflects local adaptation and the uniqueness of individuals (18). The division of humanity into racial groups has been exposed as a social construct without a biological basis, one that is contingent on specific cultural and historical context (19). Moreover, even if races were valid clusters with distinct genetic traits, the heterogeneity of these groups would preclude the application of generalizations based on group characteristics to individuals.

Although science may have helped to promote racist ideas for more than a century, it at least eventually provided the tools to expose these concepts as erroneous. Unfortunately, scientific racism now lives on as pseudoscience, with cherry-picked or manipulated data from population genetics studies disseminated via the Internet and marketed as "race realism" in the service of ethnic nationalistic agendas, just as retracted and discredited studies continue to be touted by antivaccine groups. This may resemble science to the casual observer, but it is merely bigotry and confirmation bias running wild.

Science denial, whether pertaining to evolution, HIV/AIDS, climate, vaccines, or COVID-19 origins, may also be viewed as a form of pseudoscience. Importantly, science denial shares essential epistemic features with pseudoscience: reliance on cherry-picked evidence, neglect of refuting information, fabrication of controversies, and demands of unreasonable criteria for assent (20). By using forms of argument that superficially resemble science, supported by fabricated or flawed evidence, science denialists negate the potential benefits of science for humanity and cause immeasurable harm by discouraging people from taking actions that are in their best interest.

Surveys show that magical thinking and pseudoscientific beliefs are widely prevalent among the public (21). Why are people so susceptible to pseudoscience? Pseudoscience exploits human intuition and innate distrust of recognized experts. Pseudoscientific arguments that appear to confirm intuitive beliefs are easier to accept, while rejection by the scientific establishment is seen as a courageous defiance of authority (22). Pseudoscience is dangerous because it hijacks the language and formalisms of science to mislead, just as fraudulent science uses false evidence to achieve the same ends.

It is all too easy to reinforce pseudoscientific beliefs because nothing comes more easily to human nature than the desire to confirm that one is right. The fundamental difference between science and pseudoscience boils down to a difference in philosophical stance: a search for truth, no matter where the evidence leads, or a search for evidence to confirm one's predetermined conclusions. The willingness to be proven wrong, the essence of what the philosopher Lee McIntyre calls the "scientific attitude" (23), gives us an important tool to distinguish science from pseudoscience and reminds us that practicing good science requires an open mind. As Lakatos observed, "A statement may be pseudoscientific even if it is eminently plausible and everybody believes in it. … The hallmark of scientific behavior is a certain skepticism, even towards one's most cherished theories" (5).

17 Duplicated Science

Journal editors were possibly embarrassed and perhaps even overwhelmed when I started to send them dozens of cases of duplicated images. Over half of the cases I sent to them in 2014 and 2016 have not been addressed at all, which has been frustrating. Some editors refused to respond to me, and others have told me that they did not see any problems with those papers.

Elisabeth Bik (1)

Our journey into the world of duplicated science began in 2013, when Dr. Elisabeth (Elies) Bik (Fig. 17.1) contacted us to relate her frustrations at having journals ignore her reports of problems with figures in published papers. At the time she was working in Dr. David Relman's laboratory at Stanford University, performing microbiome-related research and scrutinizing figures in published articles on the side. Elies decided to reach out to us since we had recently published a study of retractions (see chapters 18 and 24) (2), which concluded that most were the result of misconduct. After a few emails were exchanged, the three of us decided to team up to look at the problem systematically. Thus began a fruitful collaboration that brought in additional scientists and ultimately led to four publications (3–6). Before further exploring the problem of image duplication, we should place the topic into the broader context of the problem of scientific error, which is covered in more detail in chapter 25.

Thinking about Science: Good Science, Bad Science, and How to Make It Better, First Edition.
Ferric C. Fang and Arturo Casadevall.
© 2024 American Society for Microbiology.

FIGURE 17.1 *Dr. Elisabeth Bik. Photo courtesy of Gerard Harbers, under license CC BY-SA 4.0.*

ERRORS, MISCONDUCT, REPRODUCIBILITY, AND PROBLEMATIC IMAGES

Inaccuracies in scientific papers have many causes. Some result from honest mistakes, such as incorrect calculations, use of the wrong reagent, or improper methodology (7). Others are intentional and constitute research misconduct, including situations in which data are altered, omitted, manufactured, or misrepresented in a way that fits a desired outcome. The prevalence rates of honest error and misconduct in the scientific literature are unknown. One review estimated the overall frequency of serious research misconduct, including plagiarism, to be 1% (8). A meta-analysis by Daniele Fanelli, combining the results of 18 published surveys, found that 1.9% of researchers admit to modification, falsification, or fabrication of data (9).

There is little firm information on temporal trends regarding the prevalence of errors and misconduct. Research errors and misconduct have probably always existed. Even scientific luminaries such as Charles Darwin, Gregor Mendel, and Louis Pasteur have been accused of manipulating or misreporting their data (10, 11). However, the perception of error and misconduct in science has been recently magnified by high-profile cases and a sharp rise in the number of retracted manuscripts (12). In recent years, retractions have increased at a rate that is disproportionately greater than the growth of the scientific literature (13). Although this could be interpreted as an increase in problematic papers, the actual causes may be more complex and could include a greater inclination by journals and authors to retract flawed work (13). Retractions are a poor indicator of error because most retractions result from misconduct (2) and many erroneous studies are never retracted (7). In fact, only a very small fraction of the scientific literature has been retracted. As of this writing, the PubMed bibliographic database lists 11,789 retracted publications among more than 32 million articles (0.036%).

Concerns about misconduct have been accompanied by increasing concerns about the reproducibility of the scientific literature (see chapter 7). An analysis of 53 landmark papers in oncology reported that only 6 could be reproduced (14), and other pharmaceutical industry scientists have also reported low rates of reproducibility of published findings, which in some cases led to the termination of drug development projects (15). In the field of psychology, less than half of experimental and correlational studies are reportedly reproducible (16). Inaccurate data can result in societal injury. For example, a now-retracted study associating measles vaccination with autism continues to resonate and contribute to low vaccination rates (17). Corrosion of the literature, whether by error or misconduct, may also impede the progress of science and medicine. In this regard, false leads may be contributing to increasing disparities between scientific investment and measurable outcomes, such as the discovery of new pharmacological agents (18).

AN IMAGE PROBLEM IN THE BIOMEDICAL SCIENCES

Catalyzed by Elies Bik's inquiry, we attempted to estimate the prevalence of a specific type of inaccurate data that can be readily observed in the published literature, namely, inappropriate image duplication. This study was only made possible by Elies's uncanny ability to spot duplicated images in scientific papers. The study could not have been performed without her remarkable visual memory and passion for exposing dodgy data. By the time our study was completed in 2016, she had personally screened 20,000 research papers from 40 different journals (Box 17.1). Each of her findings was independently verified by each of us.

Image duplications were classified as simple, with repositioning, or with alteration (Fig. 17.2 to 17.4). Additional types of image modification were detected and not necessarily considered to be problematic, although they may signify questionable research practices by current standards and would not be accepted by some journals. Such modifications included "cuts," in which abrupt vertical changes in the background signal between adjacent lanes in a blot or gel suggested that lanes not adjacent to each other in the original gel had been spliced together, and "beautification," in which part of the background of a blot or gel where no band of interest was expected showed signs of patching, perhaps to remove a smudge or stain. When powerful image modification software programs were introduced in the late 1990s, the line between acceptable and unacceptable was initially uncertain. Many well-meaning investigators believed that it was reasonable to enhance the appearance of

Box 17.1 A study of problematic images in published papers

Dr. Elies Bik (Fig. 17.1) examined more than 20,000 research papers containing the search term "Western blot" from 40 different journals and 14 publishers for the presence of inappropriate duplications of photographic images, with or without repositioning or evidence of alteration (3). Of these, 40% were published by a single journal (*PLoS One*) in 2013 and 2014; the other articles were published in 39 journals spanning the years 1995 to 2014. Overall, 782 (3.8%) of these papers were found to include at least one figure containing inappropriate duplications. Problematic images were classified into three major categories: simple duplications, duplications with repositioning, and duplications with alteration. Simple duplications contained two or more identical panels, either within the same figure or in different figures within the same paper, purporting to represent different experimental conditions. The most common examples in this category were beta-actin protein loading controls that were used multiple times to represent different experiments or identical microscopy images that purported to represent different experiments. For papers with such figures, the methods and results were reviewed to establish that the duplicated figures were indeed reused for different experiments. The reuse of loading controls in different figures obtained from the same experiment was not considered to be a problem. Examples of simple duplication are shown in Fig. 17.2. Duplication with repositioning included microscopic or blot images with a clear region of overlap, where one image had been shifted, rotated, or reversed with respect to the other. Figure 17.3 shows examples of duplicated figures with repositioning. Duplication with alteration consisted of images that were altered with complete or partial duplication of lanes, bands, or groups of cells, sometimes with rotation or reversal with respect to each other, within the same image panel or between panels or figures. This category also included figures containing evidence of "stamping," in which a defined area was duplicated multiple times within the same image; "patching," in which part of an image was obscured by a rectangular area with a different background; or fluorescence-activated cell sorting (FACS) images that shared conserved regions and other regions in which some data points had been added or removed. Examples of duplicated images with alteration are shown in Fig. 17.4.

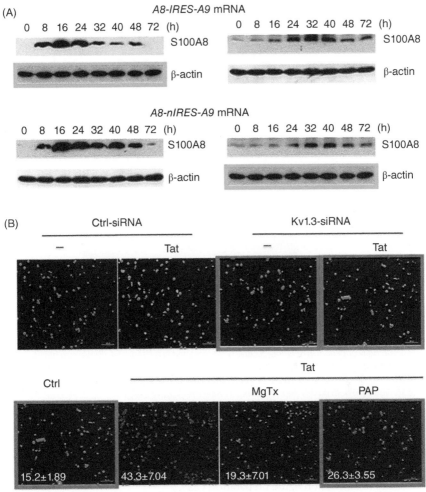

FIGURE 17.2 *Examples of simple duplication. (A) The beta-actin control panel in the top left is identical to the panel in the bottom right (green boxes), although each panel represents a different experimental condition. This figure appeared in reference 49 and was corrected in reference 50. Reproduced with permission from the publisher. (B) The panels shown here were derived from two different figures within the same paper (published in reference 51; corrected in reference 52). Two of the top panels appear identical to two of the bottom panels, but they represent different experimental conditions (red and blue boxes). Reproduced under the Creative Commons (CC BY) license. All duplications might have been caused by honest errors during assembly of the figures.*

their data by enhancing signals, reducing background noise, etc. The standards for acceptable and unacceptable image modification took several years to develop, but by the early 2000s were being taught in scientific ethics courses. The *Journal of Cell Biology* was among the first to call attention to the problem of figure alteration in

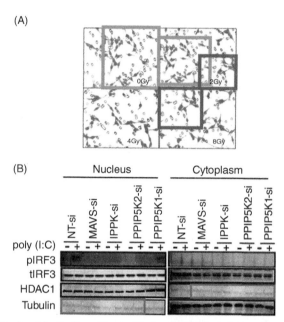

FIGURE 17.3 *Examples of duplication with repositioning. (A) Although the panels represent four different experimental conditions, three of the four panels appear to show a region of overlap (green and blue boxes), suggesting that these photographs were actually obtained from the same specimen. These panels originally appeared in reference 53 and were corrected in reference 54. (B) Western blot panels that purportedly depict different proteins and cellular fractions, but the blots appear very similar, albeit shifted by two lanes (red boxes). Panels originally appeared in reference 55 and were corrected in reference 56. Figures in both panels were reproduced under the Creative Commons (CC BY) license.*

manuscripts (19), and that journal instituted a policy to carefully inspect all manuscripts for image manipulation prior to publication (20). By 2007, a consensus had emerged, and journals such as the *Journal of Clinical Investigation* were providing guidelines to authors on figure preparation (21).

THE PROBLEM OF INTENT

When confronted with a problematic image, the next question is whether it represents an error or a deliberate action intended to deceive. Although researcher intent could not be definitively determined in our study with Elisabeth Bik, the three categories of duplicated images were felt to have different implications regarding the likelihood of scientific misconduct. Simple duplication is most likely to result from honest errors, in which an author intended to insert two similar images but mistakenly inserted the same image twice. Alternatively, simple duplications may result from misconduct, for example, if an author intentionally recycled a

(A)

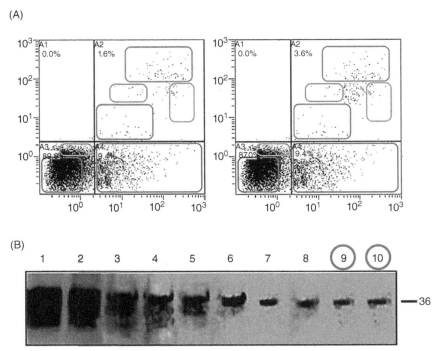

(B)

FIGURE 17.4 *Examples of duplication with alteration. (A) The left and right FACS panels represent different experimental conditions and show different percentages of cell subsets, but regions of identity (colored boxes) between the panels suggest that the images have been altered. (This illustration originally appeared in reference 57; the article was retracted in reference 58.) Reproduced with permission from the publisher. (B) The figure shown here displays Western blotting results for 10 different protein fractions isolated from a density gradient. The figure appears to show a single blot, but the last two lanes (red circles) appear to contain an identical band. Exposure was altered to bring out details in reference 59; the figure was corrected in reference 60. Figure reproduced under the Creative Commons (CC BY) license.*

control panel from a different experiment because the actual control was not performed. Duplication with repositioning or alteration may be somewhat more likely to represent misconduct, as conscious effort would be required for these actions. Among the 782 problematic papers found in our study, approximately 30% contained simple duplications and 45% contained duplicated images with repositioning, while the remainder contained duplicated figures with alteration. A definitive determination of author intent requires discussion with the author and inspection of the original data; intent cannot be reliably determined by looking at published figures alone. For this reason, questions about published images that require a determination of intent are usually referred to the institution where the work was done and where the original data are (or at least, should be) retained.

THE PREVALENCE OF PROBLEMATIC IMAGES

Plotting the percentage of papers containing inappropriate image duplication over time (Fig. 17.5) revealed that the percentage of papers with image duplications was relatively low from 1995 to 2002, with no problematic images found among the 194 papers screened from 1995. However, a sharp rise in the percentage of papers with duplicated images was observed in the year 2003 (3.6%), after which the percentages remained close to or above 4%. This pattern remained similar when only a subset of 16 journals for which papers were scanned from all 20 years was considered, other than a decline in the duplications found in 2014, the final year of our screening period.

Our data set of 782 problematic papers contained 28 papers (i.e., 14 pairs of papers) written by a common first author. To determine whether authors of papers containing inappropriate image duplication were more likely to have published additional papers containing image duplication, we screened other papers written

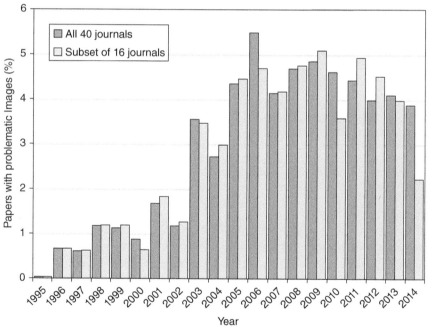

FIGURE 17.5 *Percentage of papers containing inappropriate image duplications by year of publication. No papers with duplications were found in 1995. The dark gray bars show the data for all 40 journals. The light gray bars show a subset of 16 journals for which papers spanning the complete time span of 20 years were scanned. Reproduced with permission from mBio (3) © 2016 Bik et al., under license CC BY 4.0.*

by the first and last authors of 559 papers (all from unique first authors) identified during our initial screening. This analysis encompassed 2,425 papers, or a mean of 4.3 additional papers for each primary paper. In 217 cases (38.8%), at least one additional paper containing duplicated images was identified. In total, 269 additional papers containing duplicated images (11.1%) were found among the 2,425 papers in the secondary data set. The percentage of papers with duplicated images in the secondary data set was significantly higher than that of the first data set (11.1% versus 3.8%), a difference that was statistically significant, indicating that other papers by first or last authors of papers with duplicated images are more likely to also contain duplicated images.

Our study represented a major effort to empirically determine the prevalence of just one type of problematic data, inappropriate image duplication, by visual inspection of electrophoretic, microscopic, and flow cytometric data from more than 20,000 recent papers in 40 primary research journals. We showed that figures containing inappropriately duplicated images can be readily identified in published papers through visual inspection without the need for special forensic software methods or tools, and that approximately 1 out of every 25 published papers contains inappropriately duplicated images. We further showed that the prevalence of papers with inappropriate image duplication rose sharply after 2002 and has since remained at increased levels. This coincides with the observed increase in retracted publications (2) and provides empirical evidence that the increased prevalence of problematic data is not simply a result of increased detection, as has been suggested by others (22).

These findings have important implications for the biomedical research enterprise and suggest a need to improve the literature though greater vigilance and education. The finding that figures with inappropriate duplications can be identified by simple inspection suggests that greater scrutiny of publications by authors, reviewers, and editors might be able to identify problematic figures prior to publication. The low prevalence of problematic images in the *Journal of Cell Biology* (0.3%, or about 1 out of every 333 papers) suggests that the measures employed by that journal to discourage and detect inappropriate image manipulation have been effective. The *EMBO Journal*, which was not part of our study, has also instituted a manual screening process for anomalous images (23). In addition, some journals have started to insist that original data such as pictures of Western blots and gels be deposited at the time of manuscript submission. For example, in 2012 the *Journal of Clinical Investigation* instituted a policy requiring authors of accepted manuscripts to submit uncropped, unedited blots for review and archiving (24). In 2019, that journal published an accounting of their observations on 200 manuscripts accepted in the prior 6 months. Remarkably, even after peer review, 21% had issues with blots and 27.5% had issues with images when the manuscript figures were compared to the unedited material (24). Although this study confirmed that an overwhelming majority of figure issues result from errors in figure construction, for 1% of manuscripts, the problems led the journal to rescind the acceptance decision. Having authors submit original and primary data for archiving along with manuscripts can be an important safeguard for maintaining the integrity of data and addressing questions that arise after publication. Although the situation may have improved in recent years, analysis of published blots shows that there is still much room for improving data presentation and methodology, which can help to improve scientific reproducibility (25). Our findings and those of the *Journal of Clinical Investigation* (24) suggest that greater scrutiny by journals can reduce the prevalence of problematic images. However, this is likely to require a concerted effort by all journals, so that authors of papers with problematic data do not simply shop around for journals that fail to employ rigorous screening procedures.

A large variation in the prevalence of papers containing inappropriately duplicated images was observed among journals, ranging from the *Journal of Cell Biology* (0.3%) to the *International Journal of Oncology* (12.4%), a more-than-40-fold difference. The differences among journals are important because, as noted above, they suggest that journal editorial policies can have a substantial impact on the problem. Alternatively, the variable prevalence of duplication could be partly accounted for by variations in the average number of figures and the number of panels per figure, which is likely to differ among journals but was not determined in our study. In nearly 40% of the instances in which a problematic paper was identified, screening of other papers from the same authors revealed additional problematic papers in the literature. This suggests that image duplication results from systematic problems in figure preparation by individual researchers, which tend to recur.

One out of every 25 published papers containing Western blotting results or other photographic images could contain data anomalies, which we found to be a surprisingly high number. Moreover, this is likely to be an underestimate of the extent of problematic data in the literature for several reasons. First, only image data were analyzed; thus, errors or manipulation involving numerical data in graphs or tables would not have been detected. Second, only duplicated images within the same paper were examined; thus, the reuse of images in other papers by the same author(s) would not have been detected. Third, since problematic images were detected by visual inspection, the false-negative rate could not be determined; we readily acknowledge that our screen may have missed problematic papers. It should be noted that our findings contrast with a recent small study by Oksvold examining 120 papers from three different cancer research journals, which reported duplicated images in 24.2% of the papers examined (26). Many of the reported image duplications in the Oksvold study involved representation of identical experiments, which we do not regard as necessarily inappropriate, as this form of duplication does not alter the research results. For comparison, we screened 427 papers from the same three journals examined by Oksvold and found the average percentage of problematic papers in these journals to be 6.8%, which is closer to our findings for other journals. Moreover, our study included more than 20,000 papers from 40 journals; in addition to more-rigorous inclusion criteria, we required consensus between three independent examiners for an image to be classified as containing inappropriate duplication, ensuring a low false-positive rate.

THE *MOLECULAR AND CELLULAR BIOLOGY* EXPERIENCE WITH PROBLEMATIC IMAGES

To follow up on our initial study and examine the consequences for journals, we collaborated with the American Society for Microbiology (ASM) in an in-depth analysis of one of the journals that was formerly owned by the society. In 2013, the journal *Molecular and Cellular Biology* (*MCB*) instituted a program to analyze the figures in all accepted manuscripts before publication (27), modeled after a similar program used by the *Journal of Cell Biology* (19, 28). For our follow-up study, we applied our previous approach (3) to papers in the journal *MCB* and followed up the findings with a process that included contacting the authors of the papers (Box 17.2). This allowed us to better understand how inappropriate image duplications occur and to measure the time and effort spent on following up papers with such duplications. The results provided us with new insights into the prevalence, scope, and seriousness of the problem of inappropriate image duplication in the biomedical literature.

By focusing on a single journal, we were able to determine the outcome of image duplication concerns. The reassuring conclusion of our study is that most image duplications result from errors during figure construction that can be easily corrected by the authors. However, discerning whether an apparently duplicated image warrants correction, retraction, or no action requires consultation with the authors and review of the primary data. The finding that 5.5% of *MCB* articles had inappropriate image duplications is a result consistent with our earlier findings. This confirmation is noteworthy because the approach used in the current study differed from prior work in that it focused on a single journal, with a 120-paper sample for each of six publication years. Of concern is that approximately 10% of the papers containing problematic images required retractions after the adjudication process, due to apparent misconduct, an inadequate author response, or errors too numerous for an author correction.

Box 17.2 The *Molecular and Cellular Biology* study

A set of 960 papers published between 2009 and 2016 by the journal *Molecular and Cellular Biology* (MCB), including 120 randomly selected papers per year, was screened for inappropriate image duplication (4). Of these, 6% were found to contain inappropriately duplicated images. The annual incidence showed a decline since 2013, when the screening of accepted manuscripts was introduced. From 2009 to 2012, the average percentage of inappropriate image duplication was 7%, which declined to 4% after the introduction of screening in 2013. The 59 MCB papers with inappropriate image duplications were investigated by contacting the corresponding authors and requesting an explanation. This in turn led to 41 corrections, 5 retractions, and 13 instances in which no action was taken. The reasons for not taking action included origination from laboratories that had closed, resolution of the issue in correspondence, or occurrence of the event more than 6 years earlier, consistent with ASM policy and federal regulations established in 42 CFR 93.105 (45) for pursuing allegations of research misconduct. Of the retracted papers, one contained multiple image issues for which correction was not an appropriate remedy, and for another retracted paper, the original and underlying data were not available, but the study was sufficiently sound to allow submission of a new paper for consideration, which was subsequently published. Authors who were contacted about image irregularities most frequently reported errors during assembly of the figures. The most frequent error was the accidental inclusion of the same blot or image twice. Other commonly reported mistakes were the selection of the wrong photograph, the assembly of figure panels with mock photographs that were not properly replaced, etc.

For the 59 papers published with potential inappropriate image duplication concerns, ASM publication staff recorded approximately 580 emails pertaining to these cases, or an average of ~10 emails per case (range, 4 to 103). In addition, at least two phone conversations with authors took place, each lasting approximately 1 hour. The production editor and assistant production editor handled approximately 800 emails in their folders regarding these corrections. In addition, for 20 papers the editor in chief was involved in communications with the authors, which involved a total of 244 emails (range per paper, 4 to 29), or an average of about 12 messages per paper. Including this time commitment added another 61 hours (~15 min × 244 emails). Hence, the problem of inappropriate image duplication after publication imposed a large time burden on the journal, with an average of 6 hours of combined staff time spent to investigate and follow up each paper.

THE ESTIMATED SIZE OF THE COMPROMISED LITERATURE

Other efforts to investigate causes of inappropriate image duplication for papers published in two other ASM journals produced retraction rates ranging from 3 to 21%. If the three ASM journals are representative, an estimated 35,000 papers are candidates for retraction due to inappropriate image duplication. These numbers might be an overestimate, since not all papers in the literature have images of the type studied here and *MCB* publishes many articles with figures involving photographic images. On the other hand, we screened only for visible duplications, and papers might contain additional problems in graphs, tables, or other data sets that are less easy to find, suggesting that this could also be an underestimate. Whatever the actual number, it is evident that the number of compromised papers in the literature is considerable. The continued presence of compromised papers in the literature could exert pernicious effects on the progress of science by misleading investigators in their fields. Nevertheless, even the most liberal estimates of the total number of papers that are candidates for retraction still represent a small percentage of the literature.

JOURNAL PRACTICES CAN IMPROVE THE LITERATURE

Our study also documented the potential value of increased journal vigilance for reducing inappropriate image duplications in published papers. A reduction in the number of inappropriate images identified in *MCB* papers was observed after initiation of dedicated image inspections by the journal in 2013 (27). Increased vigilance reduces problematic images by identifying and correcting errors before publication and by heightening awareness among authors to prevent such problems. However, such efforts come at considerable time and financial costs to the journal. The time invested in inspecting manuscripts prepublication was approximately 8.3 minutes per paper, and the identification of a problematic image resulted in additional time investment in communicating with authors and deciding whether a problem raised an ethical concern. Additional costs to science include the time taken by the authors to correct figures and the delays in publication. However, these costs are likely lower than the overall cost associated with discovery of inappropriate image duplication after publication, which triggers an investigation by the journal that consumes considerable time, as is evident from the average of 10 emails per case, and outcomes including publication of corrections and retractions. In our analysis, we found that following up on problematic images before publication costs about 30 minutes per problematic paper, whereas the time spent to follow up similar issues after publication, not including editor-in-chief time, was 6 hours per paper, which is 12 times greater. Hence, even though most inappropriate image duplications result from simple errors in assembling figures, their occurrence, once identified, imposes considerable costs to journals and authors and, by extension, to the scientific enterprise. Perhaps this is why many of the journals contacted by Elise Bik failed to respond to her inquiries regarding problematic figures. Identifying image problems before publication, even though this requires additional time for journal staff, might save journals time in the end by preventing problematic images from appearing in published papers. These time estimates do not include the time required when instances of inappropriate images are referred to an author's institution and trigger an ethics investigation. Identifying potential problems before publication protects authors' reputations and prevents collateral damage to the reputations of all authors of a retracted paper (29).

PROBLEMATIC IMAGES AND THE PEER REVIEW PROCESS

Peer review is a cornerstone of science (30, 31), but it is primarily designed to look for fundamental errors in experimental setup and data analysis. Most peer reviewers do not have the expertise to analyze papers for scientific misconduct. Consequently, the responsibility of screening for plagiarism, falsification, fabrication, and other forms of scientific misconduct often lies with editors (32). This underscores the critically important roles and responsibilities of journals in maintaining the integrity of the scientific record, which include both the detection and correction of problematic data (33). Although carelessness and misconduct in science have always existed, the problem may be becoming more acute because of advances associated with the availability of programs that allow authors to prepare figures easily. The ability to cut and paste text or images combined with availability of software to manipulate and generate photographic images gives authors powerful tools that can be misused. Our earlier study noted that the problem of inappropriate image duplications was largely a 21st-century phenomenon temporally associated with the proliferation of software for image construction (3). However, software advances have also provided tools to reduce error and abuse. Some publishers, including ASM, already perform routine screening of manuscripts using plagiarism detection software. Combined with manual curation and supervision, these tools work reasonably well (33, 34). Identifying image duplication of the types reported here and in our prior study (3) is more challenging and depends on individuals capable of spotting suspicious patterns. We noted that the prescreening process for *MCB* is quite good at picking up spliced images but poor at finding image duplications of the type reported in our study. Hence, without routine screening by individuals who are gifted at identifying image duplications and modifications, it is likely that the type of image problems identified in our study will continue (3). Although detecting image problems is difficult, the recent development of improved software tools offers a potential solution (more on this below) (35).

CAUSES AND SOLUTIONS FOR THE EPIDEMIC
OF PROBLEMATIC IMAGES

Although the causes of the increased frequency of image duplication since 2003 are not known, we have considered several possible explanations. First, older papers often contain figures with lower resolution, which may have obscured evidence of manipulation. Second, the widespread availability and usage of digital image modification software in recent years may have provided greater opportunity for both error and intentional manipulation. Third, the increasing tendency for images to be directly prepared by authors instead of by professional photographers working in consultation with authors has removed a potential mechanism of quality control. One possible mechanism to reduce errors at the laboratory level would be to involve multiple individuals in the preparation of figures for publication. A fourth consideration is that increasing competition and career-related pressures may be encouraging careless or dishonest research practices (36).

The finding that most inappropriate image duplications result from carelessness and error during figure construction is reassuring with regard to misconduct and fraud. Nevertheless, this problem still imposes large costs to authors and journals for their correction and indicates that greater efforts to prevent such errors should be instituted by researchers. Much of the problem appears driven by powerful new technologies used to generate figures and the changing nature of scientific publications. Prior to the availability of image editing software, figures for research papers were usually made by individuals who specialized in this activity and were not involved in data collection. In our initial study, we found no instances of inappropriate image duplication prior to 1997 (3). We suspect that prior to the availability of software that allowed authors to construct their own figures, the discussions between photographers or illustrators and authors combined with the separation of data generation from figure preparation reduced the likelihood of these types of problems. For example, when we completed our scientific training in the early 1980s, figures were usually assembled by professional photographers who worked closely with scientists who generated the primary data to assemble figures that were photographed before submission to a journal. Since figure construction and photography was costly and labor-intensive, this inevitably meant that figure mockups were circulated among authors prior to a final printing, and many errors were caught at an early stage. However, most labs today have software programs to allow rapid construction of sophisticated images. Figures are assembled by authors themselves, which reduces the chance that errors are caught early. In addition, the complexity of figures has increased tremendously in recent decades, and most figures today include multiple panels that show pictures, graphs, and diagrams. For example, the number of data items in the average paper in the biomedical sciences increased by 100% between 1993 and 2013 (37). Greater figure complexity combined with a doubling of data items inevitably increases the likelihood of error. The prevalence of image problems might be reduced by asking others in the laboratory who are not directly involved with a research project to participate in figure construction or review. In addition, providing clear guidelines for the preparation of photographic images as part of a journal's instructions for authors is helpful. For example, instructions might include rules about how to disclose cuts in Western blots, the requirement of each experiment to have its own control (e.g., β-actin or globin) protein blots (with no reuse of these blots allowed), etc. Examples of such guidelines currently exist (38). The ASM maintains an ethics portal in its website with information that can be helpful to authors (39).

The high prevalence of inaccurate data in the literature should be a finding of tremendous concern to the scientific community, as the literature is the record of scientific output upon which future research progress depends. Papers containing inaccurate data can reduce the efficiency of the scientific enterprise by directing investigators to pursue false leads or advance unsupportable hypotheses. Although our findings are disturbing, they also suggest specific actions that can be taken to improve the literature. Increased awareness of recurring problems with figure preparation, such as control band duplication, can lead to the reform of laboratory procedures to detect and correct such issues prior to manuscript submission. The variation among journals in the prevalence of problematic papers suggests that individual journal practices, such as image screening, can reduce the prevalence of problematic images (19, 23). The problems identified in our studies provide further evidence for the scientific establishment that current standards are insufficient to prevent flawed papers from being published. Our findings call for the need for greater efforts to ensure the reliability and integrity of the research literature.

Collectively, the results of our studies and those of others provide both reassurance and concern regarding the state of the biomedical literature. We are reassured that most duplication events result from errors that do not compromise the validity of the scientific publication and are amenable to correction, notwithstanding the considerable burden of correction on journal staff, editors, and authors. Also reassuring is the fact that only 0.5% of the papers screened in the *MCB* study had image problems of sufficient severity to require retraction. However, even this low percentage suggests that the current biomedical research literature contains many publications that warrant retraction. At the very least, our findings suggest the need for both authors and journals to redouble their efforts to prevent inappropriate image duplications. The emergence of pre- and post-publication review mechanisms offers additional checks to identify papers with problematic images (Box 17.3).

Box 17.3 Post-publication review

Whereas the information revolution has brought the problem of inappropriate duplication of figures, primarily caused by image preparation software programs, other technological advances have given us new ways to correct, avoid, and identify errors in science. The increasing use of preprints in the biological sciences allows the sharing of manuscripts even before submission to journals, a process that can lead readers to identify errors. Comments can be posted directly at the preprint site or communicated to the authors. Recent decades have also seen the emergence of post-publication review, whereby anonymous individuals can raise criticisms about published papers in public sites such as PubPeer, which was launched in 2012. Most comments left at PubPeer reflect the subject of this chapter—concerns about image manipulation (46). PubPeer has also seen the emergence of "super commenters," individuals who regularly police the literature by posting about issues with published papers. The PubPeer platform functions mainly as a public reporting site rather than a vehicle for scientific discussion. Some have criticized PubPeer as encouraging "vigilante science," where anonymity provides a shield for making unsupported accusations (47). On the other hand, some PubPeer comments have catalyzed investigations that have led to corrections or retractions and raised awareness among scientists that the review process does not necessarily end with publication. The PubPeer experiment in post-publication review is barely a decade old, and as with all innovations, it will take time to learn how to use it correctly. In the meantime, PubPeer provides a mechanism for unearthing problems in a rough-and-tumble way that uses shame in the public square as a mechanism to enforce quality control in science.

Whereas the problem of inappropriate figure duplication is mostly a consequence of the development of powerful software that allows investigators to prepare their own figures in increasingly complex ways, there is hope that image checking programs will one day help both authors and journals to reduce this problem. In 2022, the *Journal of Clinical Investigation* reported that it had implemented additional checks on its accepted manuscripts using the Proofig (40) software program (41). After 1 year of screening, the journal tripled its rejection rate of papers that had passed peer review and were on their way to publication from 1 to 3% (41). This experience suggests that computer-assisted review can increase the detection of figure duplications and provides hope that over time such software can reduce the frequency of problematic images in the published literature.

Finally, we note that Elisabeth Bik has continued her efforts to root out error and misconduct in science and remains very active in the field. In 2021, she was awarded the John Maddox Prize for her efforts on behalf of good science. During the COVID-19 pandemic, she took on a public role in criticizing and exposing problematic papers, leading to threats of a lawsuit by the French investigator Didier Raoult after she found irregularities in his paper reporting the use of hydroxychloroquine for the treatment of COVID-19 (42, 43). This in turn led more than 1,000 scientists to sign a petition in support of her efforts (44) (see also chapter 30). Bik's efforts provide a shining example of how a single individual can make a difference in the responsible conduct of science.

18 Fraudulent Science

The real question is not why a few scientists commit fraud, but why more don't do it.

T. M. Fenning (1)

Science is a search for truth. There is no place for fraud in science, except possibly to understand the underlying psychopathology, as research fraud is obviously counterproductive to the goals of science. Like many scientists, we used to believe that research fraud was exceedingly uncommon. Daniel Koshland, when he was the editor in chief of the journal *Science*, estimated that "99.9999% of (scientific) reports are accurate and truthful" (2). In 1981, Philip Handler, president of the National Academy of Sciences, testified to a Senate subcommittee considering research misconduct that "one can only judge the rare such acts that have come to light as psychopathic behavior originating in minds that . . . may be considered deranged" (3). We were also reassured by the widespread belief that science is self-correcting, so that any fraudulent findings would be exposed as soon as other scientists failed to confirm them and purged them from the scientific literature. Unfortunately, that belief has turned out to be a comforting fiction.

Admittedly, there have been some highly publicized instances of scientific fraud. In 1912, Charles Dawson claimed to have discovered fossil remains of a "missing link" between apes and humans in a quarry near Piltdown, England. His hoax was not conclusively exposed until more than 40 years later. In more-recent times,

Thinking about Science: Good Science, Bad Science, and How to Make It Better, First Edition.
Ferric C. Fang and Arturo Casadevall.
© 2024 American Society for Microbiology.

William Summerlin, a researcher at the Memorial Sloan-Kettering Institute for Cancer Research, was found to have faked the results of his mouse experiments in 1974, in a fraud so comically amateurish that it was featured in a book (4). In 1981, Marc Spector, a graduate student in the laboratory of the renowned biochemist Efrain Racker, iodinated proteins to simulate ATP in a much-publicized episode of fraudulent science (5). Subsequently, Harvard cardiologist John Darsee was found to have fabricated research findings in numerous publications and grant applications in 1983 (6), University of California, San Diego radiologist Robert Slutsky was removed from his faculty position after an institutional committee discovered repeated instances of data fabrication in 1985 (7), and Bell Labs physicist Jan Hendrik Schön was required to retract numerous publications in 2002 after his semiconductor research was found to be faked (8). In the late 1990s, the question of fraud in science became public drama when the U.S. Congress held hearings to investigate problems with a publication that included the Nobel laureate David Baltimore as an author (Box 18.1). That episode catalyzed the National Institutes of Health (NIH) to develop formal guidelines and regulations for its funded research. In 2005, a Korean researcher was required to retract two high-profile stem cell papers due to apparent data falsification (9). In 2006, University of Vermont metabolism researcher Eric Poehlman was sentenced to prison for falsifying data in grant applications (10). However, the newsworthiness of such cases seemed to only reinforce the perception that research fraud was rare, and that its consequences could be readily ameliorated.

Our experiences as editors of scientific journals have led us to question these assumptions. Let us recount just one example. In March 2010, a reviewer of a manuscript submitted to the journal *Blood* by a prominent Japanese virologist became concerned that some of the data looked strangely familiar (the data contained in a new manuscript would have been expected to be completely original). The reviewer recruited a student to assist her by pulling earlier publications by the same author for comparison and found a match with the new submission. The journal rejected the manuscript and notified the author's institution of suspected data falsification. The institution began a systematic review of all articles published by this faculty member during the previous 8 years. Four months later, they notified approximately a dozen different journals of additional instances of data fabrication or falsification, consisting primarily of the reuse of various images and their misrepresentation. In all, 50 articles were reviewed by the institutional committee, and two-thirds were found to contain "anomalous" data.

Figure 18.1 shows one example of this author's fraudulent data from two published articles (11, 12). The data represent the results of an electromobility shift assay, a routine molecular biology method to demonstrate specific interactions

Box 18.1 A problematic paper triggers a congressional hearing

In 1989, a dispute about the accuracy of data and the possibility of misconduct triggered a series of events that brought the problem of scientific fraud into unprecedented public scrutiny. Congressman John Dingell, chair of the House Committee on Oversight and Appropriations, held hearings on potential misconduct involving a 1996 paper in the journal *Cell* (34). Complaints from a laboratory whistleblower led to institutional investigations that found errors in the interpretation of the data but concluded that no misconduct had occurred. Thereza Imanishi-Kari, the senior scientist on the study, had been accused of fabricating data. She was initially found by the NIH Office of Scientific Integrity to have deliberately falsified and fabricated experimental data and results but was subsequently cleared by institutional and government review committees after review of 6,500 pages of testimony from 35 witnesses (35–37). One of the paper's authors was Nobel laureate David Baltimore, who was at that time the president of Rockefeller University. The involvement of Baltimore raised the profile of the hearings and led to a public confrontation between Dingell and Baltimore. The dispute morphed from a consideration of the reliability of certain data into a broader debate over Baltimore's vigorous defense of Imanishi-Kari (38–40) and became renamed in the media as the "Baltimore affair" (41) or the "Baltimore dispute" (42). The episode included drama not generally associated with scientific research, including congressional hearings, secret service investigations of laboratory notebooks, and whistleblowers. In the end, the paper was retracted and Baltimore resigned his position at Rockefeller (43). The most lasting consequence of the entire incident was the development of formal guidelines by the NIH to address problems of scientific and professional misconduct arising at institutions receiving federal grants.

between proteins and DNA. Even those lacking familiarity with the method can easily see that the panels on the left and right are identical. And yet the author had claimed these to represent the results from completely different experiments involving different types of human cells and bacteria. In electromobility shift assays, random variations in experimental conditions invariably produce subtle and minor differences in results obtained from experiment to experiment. Yet no such differences can be seen here—the images are identical to the last pixel.

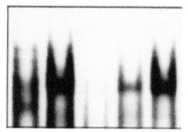

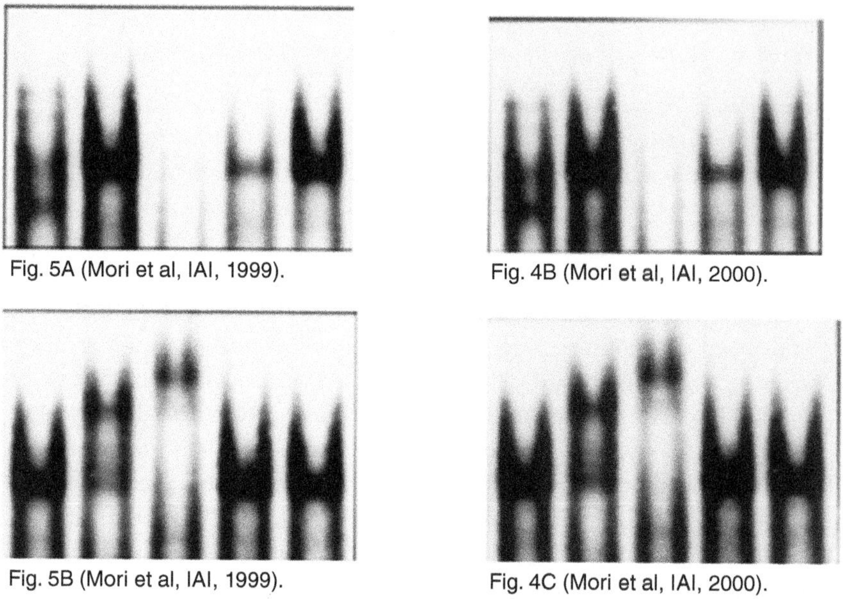

Fig. 5A (Mori et al, IAI, 1999).

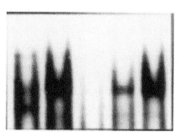

Fig. 4B (Mori et al, IAI, 2000).

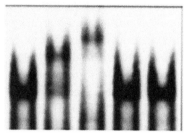

Fig. 5B (Mori et al, IAI, 1999).

Fig. 4C (Mori et al, IAI, 2000).

FIGURE 18.1 *Example of data fabrication. Electromobility shift assay (EMSA) data from different papers are shown (13, 14). The author claimed that the panels on the left were obtained from* Pseudomonas aeruginosa *infection of respiratory epithelial cells and that the panels on the right were obtained from* Helicobacter pylori *infection of gastric epithelial cells. This is not possible because the images are identical and do not exhibit the natural variation that one would expect from data obtained in different experiments.*

This degree of replicability is impossible to achieve in real life. The results must be fraudulent.

One of us (F.F.) was the editor in chief of the journal that had published these two articles. The articles were retracted immediately (13, 14). At one level, it seems as if the system worked. Fraud was uncovered by the peer review process, the journals and institution fulfilled their obligations, and the fraudulent publications were retracted. No harm, no foul. However, we found the experience to be deeply disquieting. The author was Naoki Mori, a professor at the University of the Ryukyus. At the time he had written more than 135 research articles and 17 book chapters and received multiple awards. If not for an unusually perceptive reviewer, his perfidy might not have been discovered. More than 30 of Mori's articles were retracted, but only a fraction of his career output had been investigated. Mori was fired by his institution but was successfully reinstated after a lawsuit. Why had his fraud not been discovered earlier? How many other fraudulent publications remained unrecognized in the scientific literature? What could prevent more fraud from occurring in the future?

In the cartoon view of science, Mori's fraudulent work would fail to be confirmed by other scientists, who would report their findings, thereby correcting the scientific record and directing others away from the flawed publications. However, this did not happen, for several reasons. First, Mori's research was not groundbreaking, so his claims were not particularly surprising, and some were likely to be true even though they were based on fabricated or falsified data. Thus, they did not attract attention. A second factor is that scientists seldom attempt to precisely repeat an experiment that has already been published, because they are trying to do something original. Third, failure to replicate a finding would not necessarily imply fraud (chapter 7). Thus, many of Mori's findings were not directly tested by others. A fourth possibility is that some scientists may have been unable to reproduce Mori's results but decided not to publish their results because they were uncertain of the reasons for the discrepancy.

Sobered by this episode, in 2012 we and our colleague Grant Steen decided to perform a systematic analysis of more than 2,000 retracted scientific articles that were indexed by PubMed, a database comprising more than 20 million publications at that time, primarily from the life sciences. Our study revealed a number of findings, some of which were unexpected (see chapter 24). After reviewing secondary sources, we found that retraction notices were sometimes opaque or misleading, giving an impression of honest error when in fact deliberate fraud had occurred. Taking the new information into account, we concluded that most retractions resulted from scientific misconduct. After correction for the number of publications, we found that the rate of retraction for fraud had increased nearly 10-fold between 1975 and the time of our study.

Although only 1 out of every 10,000 papers in the PubMed database was retracted at the time of our study, extensive surveys of working scientists indicate that misconduct and questionable research practices are considerably more common, with misconduct acknowledged by 1 to 8% of survey respondents (15, 16). Of even greater concern is the frequency of other questionable research practices, such as failing to disclose conflicting data or dropping data points based on a "gut feeling," which were observed by as many as a third to half of respondents. An analysis of individuals sanctioned by the NIH Office of Research Integrity (ORI) found that misconduct occurs throughout the academic spectrum, from students to postdoctoral fellows to faculty, and that men are overrepresented (17). Of concern for the notion that science is self-correcting are numerous studies showing that flawed research frequently persists in the literature, even after being shown to be fraudulent or completely erroneous (7, 18–23). The myth of the self-correction of science has been laid bare; science is only corrected when scientists, institutions, and journals make a concerted effort to correct it (24). Misconduct is very costly, leading to time and resources wasted on misinformation, careers destroyed, and the undermining of public trust in the research enterprise. With Andrew Stern and Grant Steen, we analyzed the costs of cases of misconduct that resulted in retracted publications. Papers retracted for misconduct following investigations by the ORI between 1992 and 2012 accounted for approximately $50 million in NIH research funding. The authors experienced a greater than 90% reduction in publication output and large declines in funding following ORI censure (25). Although only a small proportion of scientists commit fraud, they cause disproportionately large damage to the scientific enterprise, the costs of which affect everyone. When scientists are caught cheating, their careers, in the vast majority of cases, are over (25).

Why do some scientists cheat? First it must be noted that cheating is ubiquitous in biology, in organisms ranging from bacteria to humans, wherever there is competition for limited resources (26). Moreover, some scientists have suggested that the neocortical region of the primate brain may have evolved specifically in response to increasing complexity of social interactions, including deceptive behavior and its detection (27, 28). Thus, cheating can be viewed as an unfortunate natural impulse that drives individuals to act under certain circumstances unless constrained by other influences. Researchers such as Dan Ariely at Duke University have found that the potential for cheating is a basic feature of human nature, but whether an individual will decide to cheat in a given situation depends on many factors (29).

Our collaborations with Daniele Fanelli and Elisabeth Bik (chapter 17) analyzing duplicated images have produced additional insights into factors that may affect the likelihood of one form of research misconduct. Although figure duplications that were likely to result from simple error did not show an association with the country where the research originated, duplications suggestive of fraud were more likely to be observed in countries with cash-based publication incentives that lack an academic culture of mutual criticism and national policies to promote research integrity (30). Although we caution that association does not necessarily imply causation, these associations suggest that the country-level scientific culture can influence the likelihood of misconduct. A follow-up study of inappropriately duplicated images confirmed an association of misconduct with cash incentives, but also found that such individual-level "misaligned incentives" were more evident in some countries than in others (31).

The behavioral economists Scott Rick and George Loewenstein have observed that dishonesty is more often motivated by the fear of loss rather than by a desire for gain. We have observed this tendency in science as well. Scientists who have committed misconduct typically do so out of fear of losing research funding, employment, or prestige. Potential losses create what Rick and Loewenstein call a "hypermotivation" to cheat, which may overcome the desire of scientists to behave in an ethical manner (32). Intense competition for funding and jobs (chapter 20) can create a perfect storm of incentives for some individuals to engage in fraudulent science. Few scientists found to have committed fraud have spoken publicly about the reasons for their misconduct. An exception is Eric Poehlman, who, as mentioned earlier, was found to have fabricated data in grant applications. At his sentencing hearing, Poehlman attempted to explain himself:

> I had placed myself ... in an academic position (in) which the amount of grants that you held basically determined one's self-worth. Everything flowed from that. With that grant I could pay people's salaries, which I was always very, very concerned about. I take full responsibility for the type of position that I had that was so grant-dependent. But it created a maladaptive behavior pattern. I was on a treadmill, and I couldn't get off. ... Certainly there is this point of having a grant because it raises your esteem and raises your standing vis-á-vis your colleagues. ... The structure ... created pressures which I should have, but was not able to, stand up to. I saw my job and my laboratory as expendable if I were not able to produce. (10, 33)

Psychologists have taught us that cheating is a natural human behavior, and that the fear of loss is a hypermotivation for dishonesty. Viewed in this light, fraudulent science is a predictable outcome of too many scientists chasing too few positions and research dollars in academic cultures that reward productivity rather that scientific truth, using misaligned incentives ranging from promotion to cash bonuses. Arresting the rising tide of research fraud will require reforms to both the structure and the culture of the contemporary scientific enterprise (chapter 32). In the ensuing decades since the 1980s, when the studies of Spector and Darsee shocked the biomedical community, we have learned much more about fraudulent science. Research fraud is far more common than was once believed, and the guardrails of science based on peer review and replication are often insufficient to detect it. There is now a greater awareness of the problem, and efforts to minimize it include mandatory ethics training for all trainees, although the efficacy of such training is uncertain. What is certain is that all scientists must be vigilant about the possibility of fraudulent science. Perhaps with increased vigilance, the frequency of fraudulent science can be minimized, although given that science is a human endeavor and that cheating is part of human nature, it is unlikely that fraud can be completely banished from the scientific enterprise. Nevertheless, more still needs to be done to reduce the incentives for dishonesty in science. Perhaps the tools of science can be turned inwardly to the subject of research misconduct in the hope of better understanding its underlying causes and effective strategies for its prevention.

19 Dismal Science

Science is not only about fame; it is also about fortune.

Paula Stephan (1)

[Economics is] not a "gay science" ... no, a dreary, desolate, and indeed quite abject and distressing one; what we might call, by way of eminence, the dismal science.

Thomas Carlyle (2)

Economics is dismal, but is it science?

Joseph Epstein (3)

Although neither of us is an economist, our experiences as working scientists have given us a strong sense that the economic system of science is ripe for analysis and, perhaps, reform. We begin this chapter with a title and two quotes that include the word "dismal," but we note with some irony (and more dismay) that Carlyle described economics as dismal because it failed to justify the value of slavery (4). It seems to us that rather than being dismal, the arguments made by 19th-century economists on behalf of human liberty and equal opportunity attest to its greatness as a scientific discipline.

Economic systems, defined as "systems of production, resource allocation, exchange, and distribution of goods and services in a society or a given geographic

Thinking about Science: Good Science, Bad Science, and How to Make It Better, First Edition.
Ferric C. Fang and Arturo Casadevall.
© 2024 American Society for Microbiology.

area" (5), emerge whenever humans have to deal with the allocation of limited resources, and science is no exception. Scientific resources include scientists themselves, trainees, funding, laboratories, jobs, ideas, information, publications, and everything related to scientific products and infrastructure. The resources essential for the conduct of science exist in finite quantities and are thus the object of intense competition among scientists. Although economics plays a vital role in the functioning of the scientific enterprise, it is scarcely considered by most working scientists. Yet the economics of science is of fundamental importance to society, as scientific advances drive technological ones, which in turn drive productivity and societal growth (6). In recent years, the scientific enterprise has been beset by problems ranging from inadequate funding to problems of reproducibility and integrity. In this chapter, we consider how the problems of present-day science are a consequence of its economic system.

THE ECONOMIC SYSTEM OF SCIENCE

The economics of science cannot be simply summarized in classical macroeconomic terms such as capitalism, socialism, and communism. Nevertheless, some parallels with these systems are evident. Funding is provided by a mixture of government and private sources, giving science the characteristics of a mixed economy. Roughly 30% of total R&D funding in OECD (Organization for Economic Co-operation and Development) market economy countries is performed by government and academic institutions (7), and this is particularly important for the support of basic research, which is felt to be too risky for private companies since its conversion to profitability is always uncertain and has an unpredictable timeline. Nevertheless, industry occasionally makes basic science contributions of great importance during the course of applied and technical research. An example of this was the discovery of cosmic background radiation by Arno Penzias and Robert Wilson at Bell Laboratories in 1965 while researching methods to improve telecommunications (8). Although industry is a substantial and important source of research funding, this chapter will focus on the essential role of government-sponsored basic research, as industry tends to focus its investment on the development of commercial applications (9, 10) and operates according to a largely independent incentive system (11). Private foundations are an additional source of funding for targeted research, but their contribution is usually dwarfed by government spending.

Individual scientists search for knowledge, but on a more personal level they also seek income and prestige (12). The freedom of scientists to compete with one another for resources on the basis of merit is a characteristic of science economics shared with a capitalist economy. However, the ability of government to set aside funding or award contracts on the basis of societal priorities is more analogous to a managed socialist economy. As the amount of funding or the number of positions is fixed at any point in time, competition among scientists for these resources is a zero-sum game. This would tend to discourage cooperation and the sharing of ideas and information that could benefit one's competitor, except for the priority rule, which awards credit to the first scientist to make a claim (13–16). Thus, scientists are incentivized to share their findings with the scientific community in order to receive credit. Scientists are strongly motivated to receive credit, as the resulting prestige leads to tangible rewards such as academic positions, grants, and admission to honorific societies, which in turn confer recognition and prestige that can translate into better jobs, more funding, and more trainees (Fig. 19.1) (17). Hence success in science can beget further success, creating positive feedback cycles. Competition for credit is essentially a winner-takes-all system, as the benefits for those who fail to come in first are meager. Although the origins of the priority rule in science are uncertain, it has clearly been around for a long time, as Isaac Newton engaged in bitter priority disputes with Robert Hooke and Gottfried Wilhelm Leibniz and was able to parlay his scientific success into a prestigious academic appointment.

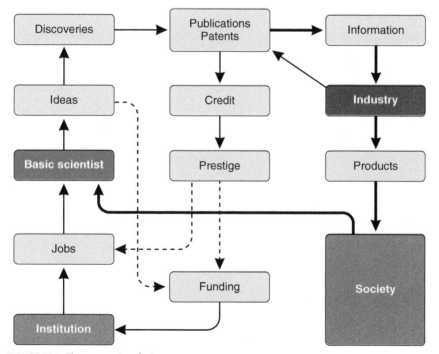

FIGURE 19.1 *The economics of science.*

Scientific credit is allocated according to complex informal rules that vary with the field. As scientific fields are the sociological units of science (18) (see chapter 10), the mechanisms for credit allocation reflect social conventions. Although credit for a discovery is given to a research team based on the priority rule, credit must also be allocated within a research group and among collaborating groups. In the biomedical sciences, credit allocation within a research group is often denoted by the order of authorship in a publication. By this convention, the first and last positions identify the individuals who carried out the bulk of the work and head the research group, respectively. This is in keeping with the winner-takes-all economics of science by allocating the lion's share of credit to the first and last authors, who are often junior and senior investigators, respectively. As the first and last authors occupy different circles in the hierarchy of science, both benefit from the credit generated by the study. However, allocating credit on the basis of authorship order may be unfair if a scientific discovery results from team science or if middle authors make critical contributions that are not that different from those of the first and last authors. Some publications attempt a more equitable distribution of credit by using asterisks to denote individuals who "contributed equally" to a project, although this practice has been questioned (19). Another way to share credit is to include a contributor in a patent, which also confers the benefit of potential financial remuneration. Informal credit may also be assigned by mention in letters of recommendation, public presentations, or other forms of acknowledgment. Ideally, credit allocation for scientific work is proportional to the importance of a discovery, but in practice the distribution of credit may be delayed or misallocated. The importance of a scientific discovery is not always appreciated at the moment of discovery and may be assessed on the basis of subjective and variable criteria (20).

The assignment of scientific credit by peers has several implications. First, the recognition of a scientific achievement is dependent on appreciation by other scientists and the ability of a scientist to persuade others of a discovery's importance. Second, the time elapsed between discovery and credit is highly variable, as many discoveries are either overlooked at the time or initially acclaimed only to be later discredited. Third, over time the responsibility for credit assignment passes from peers to historians of science, but this only occurs for major discoveries. In real life, scientists must be recognized by peers in a timely fashion if they are to enjoy the credit from their labors, which can then be transformed into prestige.

CREDIT AND PRESTIGE

Credit may be awarded for a variety of scientific contributions, including discoveries, ideas, theories, reagents, methods, or instruments. Publication is typically a necessary condition for the awarding of credit and prestige, but publication quantity and quality are also taken into account (21). Peers who are familiar with the relevant area of study are primarily responsible for making a value judgment regarding the importance of a discovery. Hence, fields may be viewed as reservoirs of credit. Once conferred, credit can be converted into prestige, an important currency of science. Prestige requires an accumulation of credit, but the correlation between credit and prestige is complex and does not follow a linear relationship. Some individuals receive a disproportionate amount of credit for a discovery, while others who have made major contributions are forgotten. In this regard, the Nobel Prize is notorious for recognizing some scientists and ignoring others (see chapter 21) (22). Further complicating the allocation of credit are the sometimes capricious and unpredictable ways in which credit is given, with the naming of the Higgs boson and its accompanying Nobel Prize as just one example (Box 19.1). For an individual scientist, prestige can translate into employment, academic advancement, acquisition of resources, and the ability to influence the course of science. Prestige differs from credit in that the

Box 19.1 The Higgs boson as a case study in the capriciousness of credit allocation in science

The Higgs boson is named after Peter Higgs, who proposed it in a research paper in 1964 and shared the 2013 Nobel Prize in Physics for its discovery. Certainly, naming a fundamental particle after a scientist is a remarkable event that immortalizes the individual, especially when this boson has also been called the God particle. According to lore, the name "God" was first invoked in association with the Higgs boson by the physicist Leon Lederman, who wrote a book that he called *The Goddamn Particle* to indicate how difficult the Higgs boson was to detect (68); the publisher changed the title to *The God Particle* to be less offensive, and the nickname stuck (69). In addition to Higgs, there were other scientists who proposed the existence of such a particle, including François Englert, Gerald Guralnik, Tom Kibble, Robert Brout, and Carl Richard Hagen. In fact, some suggested that if the particle was to be named after a scientist, it might more appropriately be called the Englert-Guralnik-Kibble-Brout-Hagen-Higgs boson, abbreviated as the BEHGHK boson and pronounced "berg" (70). Defenders of Higgs pointed out he was the only one who actually proposed such a particle, in a 1964 publication (71, 72). But even that claim of priority was somewhat ironic since Peter Higgs's original paper did not propose a new particle when it was first submitted and rejected by *Physical Review Letters* on the grounds

(Continued)

Box 19.1 (Continued)

that it did not warrant rapid publication (73). In his revised paper, Higgs decided to "spice it up" by adding more paragraphs, including a prediction of the particle that would come to be named after him (74). Had the original paper been accepted, it is doubtful that his name would have figured so prominently. Finding the Higgs boson in 2012 would eventually require decades of work by an untold number of physicists and the construction of the approximately U.S. $4.5 billion Large Hadron Collider at CERN (Conseil Européen pour la Recherche Nucléaire, or the European Organization for Nuclear Research). By all accounts, Peter Higgs is a private and modest man who is embarrassed by all the credit he has been given over colleagues in his field (74). We use this example solely to illustrate the capriciousness of credit in science and its dependence on chance and happenstance, but in no way intend any criticism of Peter Higgs or his monumental contributions to science.

Peter Higgs, 2013 Nobel laureate in Physics. Courtesy of Hans G. Andersson, under license CC BY-SA 2.0.

latter is theoretically timeless, whereas prestige is more ephemeral and limited to the lifetime of the scientist.

In addition to the accomplishments leading to the accrual of credit and prestige, there is an important social dimension. Highly influential scientists belong to networks that include their associates and trainees. These associations form scientific cartels that can promote each other's ideas and careers, and all experience a boost in prestige when any one of its members achieves recognition. This has a pervasive effect as senior members achieve positions as editors and reviewers who control access to resources such as funds and publications that can be used to obtain more prestige, creating a positive feedback loop. Similarly, all experience a decline in influence when their leader passes from the scene (23)—an illustration of Planck's principle in action, named after Max Planck, who stated, "A new scientific truth does not triumph by convincing its opponents and making them see the light, but rather because its opponents eventually die and a new generation grows up that is familiar with it" (24) (see also chapter 28).

BIBLIOMETRICS

Although publication record is an important determinant of prestige, there is no perfect means to measure publication output. The number of publications is a crude measure that reflects quantity but without regard to quality or importance of the work. Nevertheless, the number of publications should not be denigrated, as it is a measure of scientific output that reflects effort and participation in the scientific process. In contrast, the number of publications in prestigious, highly selective journals contributes disproportionately to prestige but suffers from the imprecise relationship between journal and article impact (see chapter 26). The number of times that a researcher's publications have been cited provides somewhat more information but does not reflect an author's contribution to a published work and may be skewed by a small number of papers that are extremely well cited. The h-index was proposed by Jorge E. Hirsch in 2005 as an estimate of the "importance, significance, and broad impact of a scientist's cumulative research contributions" (25); the value is derived as the number (h) of papers that have at least h citations each. However, like the raw number of citations, the h-index gives equal weighting to primary, senior, or coauthorship and fails to account for different citation practices in different fields. Recently, a citation metrics ranking algorithm has been proposed, which adjusts for coauthorship and includes annotation for scientific field and subfields (26). Yet each of the used and proposed bibliometric indices remains vulnerable to Goodhart's law: "When a measure becomes a target, it ceases to be a good measure." Number of papers, number of citations, h-index, and impact factor have become targets, and as targets, they are subject to manipulation and are no longer good measures (27).

DIFFERENCES BETWEEN PRESTIGE AND MONEY

In science, prestige may be viewed as analogous to money and used to attain certain ends (Table 19.1). However, prestige and money are not fungible commodities. Prestige alone cannot be used to obtain research funding or laboratory

TABLE 19.1 Competition for money versus credit

	Money	Credit
Actionable unit	Money	Prestige
Source	Government	Peers
Zone	National	International field and community
Value	Fixed	Variable
Amount	Variable	Variable
Stability	Variable	Variable
Inflation	Yes	Yes
Deflation	Yes	No
Regulation	Yes	No
Regulator	Central bank	None
Use	Goods and services	Positions, grants, honors
Fungibility	Yes	Limited
Encourages cooperation	Not necessarily	Yes
Investment	Yes	Yes
Inheritability	Yes	No

equipment, although it may increase the likelihood of obtaining research funding. Unlike money, which can be spent or saved, scientific prestige does not diminish when used for career advancement but may slowly erode over time or be lost precipitously by unsavory actions. Prestige may even beget more prestige through the positive feedback loop that occurs when highly honored scientists receive additional recognition. Like money, prestige must be converted in order to be used across international borders. The convertibility of prestige depends on whether the work is recognized by the international scientific community. Jewish scientists fleeing Nazi Germany found ready appreciation for their work while in exile, but Trofim Lysenko's pseudoscientific theories in support of Communist orthodoxy earned no prestige outside of the Soviet Union despite his exalted position within the Soviet Academy of Sciences. International scientific prestige requires acceptance according to the norms of the international scientific community and publication in respected international journals. Another difference between prestige and money is that prestige cannot be bequeathed to scientific progeny as a form of inheritance, and as has already been noted above, its value for associates rapidly declines with the death of the scientist.

PRESTIGE SUPPLY AND DEVALUATION

Like money, prestige is in limited supply. In fact, the value of prestige is defined by its scarcity. As Groucho Marx famously observed, "I don't want to belong to any club that will accept me as a member." The creation of new honors and titles can only increase total prestige up to a point, since such expansion may result in prestige devaluation. For example, efforts to expand the prestigious U.S. National Academy of Sciences have been resisted, partly due to the fear that less exclusivity might diminish the prestige of membership (28). Attempts to create science prizes to rival the prestige of the Nobel Prize have thus far had only limited success (29, 30), suggesting that the economy of prestige applies to prizes as well as individuals.

THE RELATIONSHIP BETWEEN FUNDING AND PRESTIGE

If a society invests generously in science, the increased availability of funding allows an expansion of the scientific workforce and a corresponding increase in the infrastructure needed to sustain the scientific enterprise. An increase in the job supply reduces the amount of prestige and extramural funding required to obtain a position. Conversely, when funding is tight, a greater amount of prestige is required to obtain these assets. Hence, the value of prestige is not only proportional to the credit awarded for scientific discovery but also depends on the supply of resources for which scientists compete. Research institutions must maintain their infrastructure even when funding availability diminishes, and therefore must place greater pressure on their scientists to compete for a shrinking pool of funds. Heavily leveraged institutions must focus on meeting their fiscal obligations and too often give priority to revenue generation over the needs of their scientists. Consequently, leveraged institutions can become a source of inertia in the system. The sustained plateau in government funding during recent decades, combined with an imbalance between the size of the research workforce and the number of available positions (31), has created a hypercompetitive environment in which scientists attempt to enhance their prestige to remain viable. Many young scientists today are prolonging their postdoctoral training to accumulate additional credit in the form of publications in prestigious journals, which they hope will allow them to secure a senior research position. This in turn forces their peers to follow suit as they compete for the few jobs and grants available. However, there are indications that the pressure on scientists is reaching a breaking point for the system, as we are witnessing an exodus of young scientists from seeking academic careers (32). The graying of the workforce and a rightward shift in the age when investigators obtain their first independent grant award (33) are predictable consequences of these trends. Or, as the Red Queen observed, "Here, you see, it takes all the running you can do, to keep in the same place" (34).

The quest for prestige also contributes to the phenomenon of impact factor mania (see chapter 26), in which the value of scientific work becomes secondary to the prestige of the journal in which it is published (35). Publication in the most highly selective journals becomes akin to admission to a "Golden Club" that rewards its members with employment and funding opportunities (36). The tremendous competitive pressure to publish in the small number of elite journals with limited publication space sets the stage for many of the problems that plague contemporary science. Just as income inequality inversely correlates with societal happiness (37), the maldistribution of scientific resources has correlated with a decline in job satisfaction among scientists, with adverse consequences for research integrity and productivity. The "Matthew effect" (38) allows a small minority of scientists to exert a disproportionate influence on the direction of research. As the elite continue to attract more funding and more of the brightest trainees, they are able to dominate their fields, the journals, and the study sections that prioritize grant applications for funding.

PROBLEMS IN SCIENCE REFLECT ITS ECONOMIC SYSTEM

In recent years, the life sciences have been beset by a host of problems including an epidemic of retractions and research misconduct (see chapter 24) (39), reproducibility concerns (see chapter 7) (40, 41), and poor quality control in the published literature (see chapters 17 and 18) (42). These problems, along with a misplaced emphasis on journal impact factor (see chapter 26) (35, 43), can be traced in part to a reward system that disproportionately rewards publication quantity and publication in prestigious venues. A pathological obsession with seeking credit can lead to suboptimal research practices and damage the public reputation of the scientific enterprise (17). The sustained shortage of research funding and academic jobs has created a hypercompetitive environment that places the next generation of scientists in jeopardy and stifles creativity (44–47). An emphasis on impact rather than rigor perversely incentivizes scientists to care more about the prestige of the journals where they publish than about the quality of their research. Promotion decisions increasingly rely on superficial metrics instead of scientific rigor and comprehensive contributions to a field (48–50). Of course, the limited availability of funding is also a constraint on productivity. A study of National Institutes of Health funding levels in 2006 found publication productivity to be proportional to funding levels up to $750,000 per laboratory, indicating that funding was rate-limiting for most laboratories (51).

ALTERNATIVE ECONOMIC SYSTEMS

As the economic system of science is a social construct devised by scientists themselves, it may be subject to change. Certainly, scientists operating in companies function under a different system that values discoveries leading to successful products over publications and prestige. It is interesting to imagine how other alternative economic systems could affect science. For example, imagine the replacement of the priority rule by a system where credit is given for a discovery that has been confirmed by others. Such a scheme would create incentives for scientists making new discoveries to spread word of their findings to ensure that someone reproduces them and for other scientists to reproduce the new discoveries and thereby share in the credit. Similarly, one can imagine a system that rewarded fields rather than individuals, a reform that we have suggested for the Nobel Prize (22). Such a system might incentivize scientists to cooperate more within their fields to achieve common scientific goals. Another consideration is how the economic system can influence scientists' behavior. There are already some encouraging signs that scientific prestige depends on more than just novel discoveries. For example, a number of well-known scientists have recently been expelled from the National Academy of Sciences or stripped of other honors based on poor behavior, such as racist statements or sexual harassment (52–54).

AN IMPROVED ECONOMIC SYSTEM FOR SCIENCE

As noted in our first paragraph, Thomas Carlyle called economics the "dismal science" because he disagreed with its conclusions (although Carlyle was wrong, the epithet stuck) (4). However, economics describes not how things must be, but rather why things are the way they are, given certain choices and incentives. Many of the problems of contemporary science result from its economic system, which contains elements both within and outside the scientific enterprise that have resulted from choices made by society, scientists, and their institutions in response to specific incentives. These economic conditions are not inevitable. A new economics of science calls for an increased public investment, increased incentives for scientists to work cooperatively, and alleviation of the barriers to information and technology transfer between academic institutions and industry: "The institutional machinery which has been performing these vital functions for our society is intricate, jerry-built in some parts, and possibly more fragile and sensitive to reductions in the level of funding for open science than often may be supposed" (13).

In an optimally functioning economy, success should beget greater investment, creating a non-zero-sum dynamic that encourages cooperation. Private rates of return on investment in research and development approach 30%, and social returns exceed that by 2- to 3-fold (55). According to one analysis, a greater investment in the U.S. research enterprise could sustain the recruitment of half a million more scientists and engineers and raise the gross domestic product (GDP) by half a percent (56). However, in contemporary science, success in the form of new knowledge and technology does not directly influence investment, which is susceptible to competition with other spending priorities and political considerations. It is therefore essential for information regarding the successes of the scientific enterprise to be effectively communicated to policymakers who are responsible for budget decisions. For basic research, it becomes particularly important to document the ways in which fundamental discoveries play an essential role in the development of tangible societal benefits, as the connections are frequently indirect and unobvious (57–59). New methods to measure the collective productivity of science are required (60, 61). Creating a more sustained and stable stream of research funding in proportion to productivity and the tangible benefits of science to society would stabilize the scientific workforce and alleviate current bottlenecks in jobs and funding that drive excessive competition and create inefficiencies.

Of course, some competition would continue even in an adequately funded system, and the allocation of credit will continue to be based on priority (14, 16). However, there is no reason to persist in the outdated assignment of credit to an individual scientist on a winner-takes-all basis when the pursuit of science has evolved into an undertaking by research teams and communities (45). In addition to recognizing scientists based on the priority rule, credit may also be awarded to scientists who verify scientific discoveries and establish their reproducibility (62, 63), which would increase the confidence of the community in these findings and enhance their value. Increasing the value of collaborative contributions to publications in promotion and tenure decisions would incentivize cooperative behavior and improve the dynamics of research communities. It is important to recognize that competition for funding and prestige generates different incentives. While competition for funding may discourage cooperation, collegiality, and the sharing of reagents and ideas, competition for prestige can incentivize more-positive social behavior, as prestige is dependent on the judgments of colleagues (Table 19.1). Thus, the alleviation of funding shortages could drastically alter the interactions between scientists and improve the efficiency of the system. Although awarding credit in a winner-takes-all fashion promotes early disclosure, conferring disproportionate benefits on winners can also foster haste, secrecy, and dishonesty (44), and may disincentivize revolutionary high-risk, high-reward research that tends to flourish in the absence of competition (64). Shifting recognition to fields rather than to individuals (22) would more fairly recognize the many contributions that go into a typical scientific breakthrough. The following specific steps could be undertaken to improve the economics of science:

1. **Replace the current system of annual research funding allocations with stable, long-term support indexed to productivity.** With regard to economic returns, scientific research and development is one of the most reliable investments a government can make (55). Science is a non-zero-sum game, as research breakthroughs translate into greater societal wealth. However, the current system of research funding is excessively prone to political influence, and the artificial scarcities created by inadequate funding hinder long-range planning and investment in transformative lines of research. The imbalance between funding, jobs, and the scientific workforce at the root of today's dysfunctional economics is a direct consequence of the failure of the political system to adequately fund the research enterprise in relation to its output.

2. **Replace the winner-takes-all economic model of science with the fairer allocation of credit.** Science is saddled with a priority rule established centuries ago, when scientists were isolated geniuses working alone and the only reward was the personal glory of discovery. This is no longer appropriate for the needs of the scientific enterprise and the society it serves. Science today is a communal undertaking that involves research teams and collaborations building upon the discoveries of many others.

3. **Reward scientific rigor rather than publication output.** Like all people, scientists respond to incentives. To obtain robust and reliable scientific findings, one must reward scientific rigor rather than impact (49). To obtain useful research, one must reward scientific quality rather than publication quantity and journal impact factor.

4. **Recognize team science and collaborative contributions.** The current system allocates the most credit to the first and last authors of a publication. In the era of team science, this is increasingly obsolete and disincentivizes collaboration and cooperation.

5. **Abolish prizes to individual scientists for specific discoveries.** As most discoveries have multiple contributions and credit allocation is problematic (65), it is inherently unfair to reward individual scientists for specific scientific discoveries. This is highlighted by the all-too-common situation in which a senior scientist is credited for work largely carried out by junior scientists (66, 67). Prizes should be given to groups and fields rather than to individuals (22).

An improved economic system for science could allow science to more effectively fulfill its promise to serve society by providing reliable and accurate information with which to meet future challenges and create new technologies that improve the quality of life.

20 Competitive Science

Science would be ruined if (like sports) it were to put competition above everything else.

Benoit Mandelbrot (1)

As discussed in the previous chapter, in the winner-take-all economics of science, scientists compete above all for priority, the recognition that they are the first to make a discovery (2). The "priority rule" gives scientists an incentive to share knowledge of their discoveries with the community (3) but also ensures that individuals and research teams must compete with one another. The primary currency of science is the prestige conferred by peers on the basis of one's discoveries (chapter 19). Prestige in turn can lead to employment, funding, prizes, and membership in honorific societies.

It is often taken for granted that competition is beneficial for any enterprise because it provides incentives for individuals to excel. In his classic 1957 essay, the sociologist Robert K. Merton viewed competition as a favorable influence on the scientific enterprise, which promotes the rapid dissemination of research discoveries and motivates scientists (2). There are many historical examples of intense scientific rivalries. Whether one is talking about Isaac Newton and Gottfried Wilhelm Leibniz disputing the invention of calculus, Charles Darwin's anxiety on learning that Alfred Russel Wallace was converging on ideas similar to his own, or

Thinking about Science: Good Science, Bad Science, and How to Make It Better, First Edition.
Ferric C. Fang and Arturo Casadevall.
© 2024 American Society for Microbiology.

James Watson and Francis Crick sparring with Linus Pauling over the structure of DNA, the popular notion is that competition spurs scientific progress.

It has also been suggested that scientific rivalry can provide a corrective for confirmation bias, the tendency to favor evidence that supports one's preexisting beliefs. Although scientists may be reluctant to disprove their own ideas, their rivals are unlikely to show the same restraint (4). Thus, in theory, competition may help to protect science from stagnation and dogma.

THE DARK SIDE OF COMPETITION

Not all historians of science have viewed competition in a uniformly favorable light. Warren Hagstrom suggested that competition could be inefficient and wasteful because it leads to duplication of effort, although he acknowledged that competition may be stimulatory and can help to diffuse new ideas (5). While Hagstrom suggested that competition might encourage scientists to publish their results, Daniel Sullivan suggested that competition could actually lead to greater secrecy, as scientists fear being scooped by their rivals (6). Katherine McCain also observed that competition reduces the tendency of researchers to share materials, information, and methods, thus impeding scientific progress (7).

More recently, focus group discussions have confirmed that competition discourages sharing and may even lead some scientists to sabotage competitors, perform biased peer review, and engage in questionable research practices (8). Scientific leaders have decried the detrimental effect of today's hypercompetitive environment on science (9, 10). Hypercompetition may be driving some young people away from careers in science, and this may be particularly true for young women scientists (11–13).

In 2014, the author Margaret Heffernan wrote a book called *A Bigger Prize: How We Can Do Better Than the Competition*, which explored the detrimental effects of competition in all aspects of society, particularly in the business world (14). She noted that science has a tournament structure with a steep drop-off in rewards for those who do not come in first: "Even economists now concede that tournaments have perverse outcomes. These are especially costly in an activity that depends on collaboration—like science."

A vivid real-world example of the costs of naked competition arose in 2006 when a prominent senior scientist discouraged a younger researcher from joining the faculty at his institution (15). The young researcher received an email stating that the problem was that their research interests were too similar:

> I have a strong reservation about having you as a faculty colleague in the same building here at this time because of a serious overlap in research and approach. ...

We briefly discussed the possibility of a collaboration. But this is complex. ... An additional drawback in logistics is about the shared resources and facilities. ... I, as Director of the Institute, took the major role in securing and designing rodent holding, behavior and transgenic facilities. ... I am afraid that accommodating your lab would be difficult. ... I am sorry, but I have to say to you that at present and under the present circumstances, I do not feel comfortable at all to have you here.

Soon afterward, the younger scientist took another position, and an opportunity for two researchers with common interests to work together was lost. Perhaps the senior scientist merely wanted to avoid a duplication of effort. Nevertheless, the episode had collateral costs. After an institutional committee found that the senior scientist had "behaved inappropriately," he stepped down as the institute director.

A relationship between competition for funding and scientific misconduct is increasingly recognized (16), with the important caveat that most scientists are able to maintain their integrity despite difficult circumstances. Eric Poehlman, a University of Vermont researcher who served 12 months in federal prison for falsifying data, attributed his actions in part to a system "in which the amount of grants basically determined one's self-worth. ... I was on a treadmill, and I could not get off" (17) (see also chapter 18). Many scientists have noted that the "publish or perish" culture of contemporary science can foster bias in the scientific literature (18). John Ioannidis explicitly stated that "competition and conflicts of interest distort too many medical findings" (19). Large surveys of scientific faculty in the United States have shown a significant association between pressure to obtain external funding and soft money salary support with questionable research practices and neglectful or careless behavior (20). In a 2014 Belgian survey, half of scientists agreed that the competitive scientific climate led them to publish more articles, and more than half felt that publication pressure is detrimental to the validity of the scientific literature and to their relationships with fellow researchers (21).

COMPETITION IS NOT ESSENTIAL FOR SCIENCE

History has repeatedly shown that competition is not required for seminal discoveries. Emil von Behring developed the concept of humoral immunity in his studies of the diphtheria toxin. When Shibasaburo Kitasato joined the same laboratory, the two men worked together to show that protection against tetanus could be achieved in a similar fashion, and they published their results jointly (22). Similarly, Frederick Griffith's discovery of the transforming principle of heredity and Kary Mullis's invention of PCR occurred in the virtual absence of competition (23, 24). We suggest that the development of such complex concepts benefited from the long intervals of time in which scientists could develop their ideas without the pressure of competition. In fact, it is not unusual for transformative scientific discoveries to be made in the absence of competition, as such events often involve serendipity and wholly unanticipated findings (25). Even the classic examples of scientific competition may be somewhat misleading. Newton and Leibniz developed calculus independently, using different approaches, and their notorious rivalry over credit occurred only afterwards (26). Similarly, Darwin and Wallace developed their theories of evolution by natural selection independently. When Wallace sent a manuscript describing his ideas to Darwin, the latter hastily prepared a paper to be presented simultaneously to the Linnean Society of London, an action that in itself acknowledged the contributions of Wallace. To his credit, Wallace never contested the priority and greater depth and influence of Darwin's work (27). As much as competition might have provided an added incentive for Watson in his search to solve the structure of DNA, concern over priority is likely to have led Pauling to prematurely publish a three-stranded helical structure of DNA that proved to be embarrassingly wrong (28). The researchers' competitive drive may also have led to some actions that were ethically questionable (29, 30). Thus, competition may at best be seen as a two-edged sword.

Competition also yielded mixed results in 1894 when Kitasato and Alexandre Yersin converged on Hong Kong seeking to discover the causative agent of plague. Although Kitasato had successfully collaborated with von Behring a few years earlier, he was no longer in a collaborative mood. He reportedly paid the local authorities to deny Yersin access to the bodies of plague victims (31). In his haste to beat Yersin, Kitasato rushed an announcement of the discovery of the plague bacillus into print (32). However, his description of the bacillus was contradictory in a number of respects to subsequent publications and gave rise to a persisting controversy about whether his cultures may have been contaminated (33, 34). Yersin's report came out six days later, but it is his name that is immortalized as *Yersinia pestis* (35).

And what of the argument that competition reduces the danger of confirmation bias? While it is true that competing scientists are less likely to be invested in corroborating a theory, fierce competition might actually reinforce confirmation bias by encouraging scientists to dig in their heels and defend their positions rather than lose face. Collaboration with others who do not precisely share your views might be a more effective safeguard against confirmation bias.

HOW COMPETITION EMERGES

Although transformative discoveries leading to entirely new fields can occur in the absence of competition, such discoveries typically spawn intense competition afterwards. The widespread appreciation of the importance of a new discovery can stimulate competing efforts to build upon the finding. Once goals are clearly defined, scientists recognize that achieving the goals can lead to rewards and that the first to solve the next problem will reap the greatest reward.

An illustrative example is the discovery that heredity is transmitted by DNA. Initial speculation as to the chemical nature of genes was restricted by technological limitations of the time. Key research on the transforming principle came from a laboratory interested in understanding the relationship between different bacterial strains in the hope of developing better therapies and vaccines. DNA was thus linked to heredity, but several more years were required for enough scientists to become persuaded. At this point the race was on, with participants including Crick, Maurice Wilkins, Rosalind Franklin, Erwin Chargaff, Pauling, and Watson. In the end, the race was won by Watson and Crick, who reported their findings in a landmark 1953 publication and shared the 1962 Nobel Prize with Wilkins. The elucidation of DNA's structure led to new fields focused on replication, transcription, and translation, which in turn spawned their own goals, competitions, and prizes.

COMPETITION VERSUS RIVALRY

When competition occurs between specific individuals or teams, it is referred to as rivalry. Scientific competition is different from scientific rivalry, although one may evolve into the other. Rivalries often involve competing interests that can range from commercial interests to feuds over scientific views and theories. Nikola Tesla and Thomas Edison were rivals in the late 19th century over the relative merits of direct versus alternating current, and that rivalry had a strong commercial dimension since the winner stood to make a lot of money. Edison claimed that alternating current was more dangerous and went as far as to use an alternating current generator to power the first execution by electrocution in the United States to make point that it could kill. Eventually alternating current prevailed because of its intrinsic ability to be delivered far from the generating source. (Interestingly, direct current is preferred over alternating current to defibrillate the heart during cardiac arrest [36].) The race to find a polio vaccine provides another example of how scientific competition can degenerate into rivalry (Box 20.1).

Scientific rivalries sometimes further degenerate into feuds. While scientific feuds nominally involve a scientific disagreement, their continuation is often a function of the personalities involved and the inherent uncertainties in research methods and data. Given human nature and the myriad of personality types represented among scientists, it is unlikely that feuds can be avoided entirely. However, debilitating disputes may be ameliorated, and science itself has provided tools for conflict resolution (Box 20.2).

Box 20.1 Polio vaccine wars: Albert Sabin versus Jonas Salk

In the years following World War II, infantile paralysis was a serious public health problem that instilled fear in parents and communities. Drs. Sabin and Salk both graduated from New York University School of Medicine before engaging in a race to develop a vaccine against infantile paralysis. Even though both ultimately succeeded, their quests degenerated into a bitter controversy about the relative virtues of their vaccine approaches and various personal disagreements. At the time, vaccines available against such viral diseases as smallpox, yellow fever, and rabies were all based on live attenuated virus strains that caused a mild infection and elicited long-lasting immunity. Sabin favored this approach, but Salk worked to develop an inactivated virus vaccine. Both approaches had potential advantages and disadvantages. A live attenuated vaccine that could be given orally would make administration easier and provide long-term protection but carried a small risk of causing the disease it was intended to prevent. A killed vaccine required immunization by injection and was safer but offered only transient protection. Sabin criticized Salk's approach on the grounds that formalin inactivation was unreliable and inactivated virus was a weak vaccine antigen. Despite these criticisms, a large clinical trial of the Salk vaccine was completed in 1955, which established its ability to prevent paralytic polio.

Albert Sabin (left) and Jonas Salk (right).

Although Salk won the race to develop an effective vaccine, Sabin's oral vaccine eventually found widespread use throughout the world, where it was favored because of its ease of administration. Sabin's concerns seemed to be validated in 1955 when a batch of Salk vaccine prepared at Cutter Laboratories was inadequately inactivated, leading some children to acquire poliomyelitis and prompting a cessation in vaccinations. However, Salk was also vindicated when Sabin's live attenuated vaccine was subsequently found to cause polio in some recipients as a result of regaining virulence. As Paul Offit has observed, the Cutter incident ironically led to the greater use of a vaccine that was even more dangerous (77). In time, the sequential use of both vaccines was found to best balance safety and efficacy (78), and both are used today. Both Sabin and Salk received great recognition for their efforts, but neither was awarded the Nobel Prize, which went instead to John Enders, Thomas Weller, and Frederick Robbins, scientists who figured out how to grow poliovirus in culture, a critical step in vaccine production.

Box 20.2 Conflict resolution in science

Scientific conflict is largely the result of competition for resources and ideas. Unlike politics and religion, science has effective mechanisms for conflict resolution. One is the time-honored method known as Planck's principle (79) (chapters 19 and 28), which is to wait for combatants to die and be replaced by a new generation of scientists who are less consumed by the old disagreements. This mechanism does not seem to apply to politics and religion, where long-standing conflicts are often perpetuated by succeeding generations. Another mechanism is inherent in the ethos of science itself, which holds that all knowledge is provisional and subject to change in the face of new information. New experimental work has proven to be a remarkably powerful method for conflict resolution. If the uncertainty that underlies a scientific rivalry can be precisely stated, then it should be possible to obtain additional evidence that supports or refutes one of the competing views. Scientific conferences where diverging views are aired can catalyze the amicable resolution of such conflicts. Even when no experimental path forward can be immediately discerned, the exercise of articulating differences as a point-counterpoint can highlight areas in need of further study and depersonalize the conflict. Although some participants may remain unconvinced, as in case of AIDS denialism (chapter 16), the greater scientific community and public at large can usually form a new consensus that emerges from conflict resolution.

LIMITS OF COMPETITION

It is unclear whether competition is an important incentive in science. The Nobel Prize is the most prestigious honor in science, but few laureates have had the prize in mind when they made their award-winning discoveries, and most recipients have already received ample recognition by the time Stockholm calls (37) (see chapter 21). Competition probably works best when the goals are clearly defined and a field is technologically ready. Although the space race rivalry between the United States and the Soviet Union probably hastened the Moon landing, the basic laws of physics and rocketry were already well-known. Another example in which competition probably hastened progress was in the completion of the Human Genome Project, in which a race between publicly and commercially supported teams resulted in success years ahead of schedule (38). However, competition by itself cannot necessarily lead to progress if the goals are too ambitious, as demonstrated by the many unclaimed prizes in science and technology (39).

With the shortcomings of competition becoming more evident, its counterpart collaboration appears to be in ascendancy. The successful demonstration of the Higgs boson by research teams at CERN (Conseil Européen pour la Recherche Nucléaire, or the European Organization for Nuclear Research) involving thousands of scientists working over a period of decades is an example of successful scientific teamwork (see also Box 19.1). The need for diverse research approaches and transdisciplinary teams to address complex problems is increasingly recognized (40). However, a greater emphasis on "team science" will require a radical reconsideration of how scientists are organized, supported, and rewarded (41–43).

DETRIMENTAL EFFECTS OF COMPETITION ON CREATIVITY

By channeling research efforts along defined paths, competition may constrain the creativity required for transformative breakthroughs. Moreover, there may be an even more insidious effect of competition on scientific creativity. At its best, science is a creative process on par with art, music, and literature (44). Like creative disciplines in the humanities, science involves imagination, intuition, synthesis, and aesthetics (45). Patterns of brain activation observed by functional magnetic resonance imaging during word association tasks show similar patterns of activation in association cortex and socioaffective processing areas among artists and scientists (46). The emergence of new ideas is associated with the default mode network, or random episodic silent thought (REST), in which an individual primed by a long incubation period is allowed to perform a relaxing activity such as watching television, reading a book, taking a shower, driving, or exercising (47).

Studies by psychologists have shown that intense competition and stress can actually stifle creativity. Teresa Amabile is a psychologist on the faculty of the Harvard Business School who has been studying the social psychology of creativity for 40 years. Among her findings are that creativity is more likely to respond to intrinsic rather than extrinsic motivation and requires sufficient periods of time for ideas to incubate (48). Creativity flourishes when an individual is allowed to pursue a subject about which he or she cares passionately in an environment that feels more like play than work (49). Experimental subjects motivated by external rewards are less likely to produce creative results (50). In an article entitled "How To Kill Creativity," Amabile describes a hypothetical work environment that is antithetical to creativity, one that relies on external financial rewards, creates relentless deadlines, and subjects any proposals to "time-consuming layers of evaluation . . . and excruciating critiques" (51)—pretty much a dead-on description of what it means to be a scientist today. Amabile notes that "when creativity is under the gun, it usually ends up getting killed" (52), and "job security appears to be extremely important in fostering creativity" (49). This underscores the fundamental difference between the positive motivation of competition and the negative "hypermotivation" that comes from the fear of loss of funding or employment faced by scientists today (53). The intensity of the stress is also an important factor; a meta-analysis of 76 experimental studies found that although mild levels of stress can stimulate creativity, high stress levels impede creative thinking (54).

Amabile has also cautioned that the detrimental effects of competition on creativity may affect men and women differently (49). This has been borne out in subsequent studies by others. Experimental studies of groups of undergraduates and scientists in the real world have revealed important differences between how men and women tend to respond to competition. In one such study, intense competition between teams enhanced creativity in teams composed of men but impeded creativity in teams composed of women, whereas women thrived in a collaborative environment in which teams worked side by side (55). Although we caution about generalizations and stereotypes, some have suggested that this might relate to deep-seated gender differences in cooperation and competition that have been designated the "male-warrior hypothesis" (56). The ability of women to broaden the perspective of research teams and promote collaboration has been one factor used to advocate for their greater inclusion at all levels of the scientific hierarchy (57). Although the extent to which these concerns apply to problems currently encountered by women in science is uncertain, it is worth considering whether a competitive culture could be contributing to gender inequity and inequality in science (see chapter 15).

Does cooperation work? In contrast to competition, there are many examples of the benefits of cooperation and collaboration. Empirical evidence suggests a synergy between networking of individuals in noncommercial settings. The author Steven Johnson analyzed 135 major innovations in science and technology that emerged during the 19th and 20th centuries and found that 40% of these discoveries arose from networks in nonmarket settings, in comparison to 26% from networks in market settings or from individuals in nonmarket settings and just 8% from individuals working in market settings (58). The products of nonmarket networks have included such innovations as aspirin, magnetic resonance imaging, plate tectonics, atomic reactors, penicillin, and quantum mechanics. Johnson concluded the following:

> Openness and connectivity may, in the end, be more valuable to innovation than purely competitive mechanisms. ... We are often better served by connecting ideas than we are by protecting them. ... When one looks at the innovation in nature and in culture, environments that build walls around good ideas tend to be less innovative in the long run than are open-ended environments. Good ideas may not want to be free, but they do want to connect, fuse, recombine. They want to reinvent themselves by crossing borders. They want to complete each other as much as they want to compete.

The history of science repeatedly shows that important scientific findings arise from unfettered exploration, the passion of individual scientists to understand a problem, and research environments that foster interaction. Although our current scientific enterprise could hardly be less conducive to creativity, these principles come as no surprise to scientists. When asked how to build a motivated research group, Uri Alon recommended providing young scientists with challenging problems that engage them, giving them the autonomy to seek their own solutions, and placing them in an environment in which they can readily network with others (59).

HOW TO CHANNEL COMPETITION AND FOSTER COOPERATION

There remains a role for competition in science. Competition appears to work best for algorithmic tasks rather than heuristic tasks that require great creativity. Thus, defining specific goals that are technologically feasible can help to advance a field, just as David Hilbert's definition of 23 unsolved problems in 1900 helped to galvanize the attention of mathematicians.

However, most science today would benefit from a radically different structure that promotes cooperation, collaboration, and creativity. Useful measures may include changing the criteria for professional advancement, with an emphasis on common rather than individual goals and a reduced emphasis on publication in prestigious venues (60). Unselfish scientific acts such as mentoring and making useful reagents and information available to the community should be recognized, along with more-effective policing of scientists who behave selfishly. Another strategy to reduce the detrimental effects of competition is for competing groups to cooperate by publishing their findings at the same time so as to not "scoop" one another. We and others have done this on several occasions (61–73). When groups simultaneously present their findings, there are no losers, and the scientific community benefits by having immediate corroboration of a new finding. Furthermore, knowledge that their work will be published simultaneously may allow rival groups to complete their studies with greater care. Simultaneous publication requires open communication, which in itself is beneficial to science.

A major change in the economic structure of science with a renewed national investment in research and development is required to alleviate hypercompetition for grants and jobs. (Imagine the efficiency of the armed forces if only one out of every five soldiers were issued weapons and the rest were asked to spend all of their time writing applications to explain what they would do if they had one.) While it is often stated that more funding alone will not be adequate to fix science, more funding is an essential part of any effective solution. In this regard, a system that funds people instead of projects may be more rational given studies showing that this approach fosters higher-impact science (74) and that track record rather than project reviews is predictive of future researcher productivity (75, 76). A greater emphasis should be placed on open-ended, investigator-initiated research and less on targeted programs. Institutions should reduce their dependence on soft money to provide researchers with more-stable salary support. Larger research teams to increase numbers of senior scientist positions can enhance intragroup networking and ameliorate competition among trainees.

Scientists today must work in an environment of relentless stress, time pressure, and insecurity, factors that are counterproductive to good science. Fortunately, research in neurobiology and social psychology has provided a clear prescription. Creativity thrives on freedom and interactivity. It is time to apply these principles to reform the scientific enterprise itself.

CONCLUSIONS

The progress of science does not require competition, and in most circumstances, the costs of competition are likely to exceed whatever short-term benefits it can provide. Improving the efficiency of science by reducing competition and promoting collaboration will require major changes to the incentive structure of science (chapter 19) and ensuring more-stable funding and research resources. As declines in funding have intensified competition among scientists, the detrimental effects of hypercompetition on creativity, efficiency, communication, collegiality, and integrity have become increasingly evident. With the emergence of team science and the need for multidisciplinary approaches to address challenging problems, a more cooperative and collaborative scientific culture is sorely needed. However, it is not going to be easy to make science more collaborative. We are going to have to work together.

21 Prized Science

The Nobel medallion is etched with human frailties.

Robert Marc Friedman (1)

Each fall, the attention of the world's scientific community focuses on Stockholm, where the Nobel Prizes are announced. The Nobel Prize is the most prestigious award recognizing scientific achievement and confers instant fame and monetary rewards on recipients. The Nobel Prizes are given to no more than three scientists. Even when shared, the Nobel Prize represents the culmination of a winner-take-all strategy that is the basic economy of science (2) (see chapter 19). In this chapter, we discuss the notion of prizes in science, with particular emphasis on the Nobel Prize because it is so prominent and influential, and we will specifically consider the benefits and liabilities of prizes for recipients, science, and society.

WHY HAVE PRIZES IN SCIENCE?

It is worthwhile to consider the purpose of prizes in science. A prize recognizes an achievement; awarding a prize for a scientific finding or advance is recognition of progress made and commemoration of the accomplishment. From this perspective, prizes provide visible landmarks of progress in science. For those scientists fortunate enough to be recognized, a prize is a source of prestige (see chapter 19) that can further their careers and perhaps even reserve a seat in the pantheon of

Thinking about Science: Good Science, Bad Science, and How to Make It Better, First Edition.
Ferric C. Fang and Arturo Casadevall.
© 2024 American Society for Microbiology.

science. Organizations that confer prizes also benefit from the publicity that often accompanies the prize and from association with an accomplished scientist. Thus, the publicity and attention that the Nobel Prizes have brought to the nation of Sweden have led to other international prizes such as the Asturias, Japan, and Kuwait prizes awarded by Spain, Japan, and Kuwait, respectively. Some prizes are given to stimulate progress and solve problems. When used in this manner, a prize focuses attention on a problem that needs to be solved and offers a tangible incentive. A famous example was the Longitude Prize sponsored by the British government in 1714 to award £20,000 for a method to determine the longitude of a ship at sea. The prize ultimately required the invention of a clock suitable for deployment on ships and was claimed by John Harrison, who invented the naval chronometer (3). Hence, prizes in science are given to recognize accomplishment, bringing attention to societies or countries, and to focus attention on specific problems.

The role of prizes in science is an important question for scientists to consider. Prizes are rapidly proliferating, increasing from around 20 a century ago to more than 350 today (4). The number of recipients is not escalating as rapidly as the number of prizes, with 64% of winners having won multiple prizes (4). Some have questioned the purpose, equity, fairness, inclusiveness, and effect of prizes in science (5). In the discussion that follows, we focus on the Nobel Prize because it best-known and most prestigious prize in science. Although we primarily focus on the Nobel Prize, many of our critiques apply to other science prizes as well.

DID A MISTAKEN OBITUARY LEAD TO A PRIZE?

Over the course of his life, Alfred Nobel amassed a fortune from selling explosives. According to one account, when his brother Ludvig died of a heart attack in France in 1888, a newspaper confused him with Alfred and published an obituary calling him the "merchant of death" who made money by developing new ways to "mutilate and kill" (6). This misdirected criticism reportedly had a deep impact on Alfred, who began to worry about his legacy and decided to bequeath his fortune to the creation of prizes for science, literature, and peace. Whether this tale is true or not, Nobel's will (7, 8) clearly states:

> The whole of my remaining realizable estate shall be dealt with in the following way: the capital, invested in safe securities by my executors, shall constitute a fund, the interest on which shall be annually distributed in the form of prizes to those who, during the preceding year, shall have conferred the greatest benefit on mankind. The said interest shall be divided into five equal parts, which shall be apportioned as follows: one part to the person who shall have made the most important discovery or invention within the field of physics; one part to the person who shall have made the

most important chemical discovery or improvement; one part to the person who shall have made the most important discovery within the domain of physiology or medicine; one part to the person who shall have produced in the field of literature the most outstanding work in an ideal direction; and one part to the person who shall have done the most or the best work for fraternity between nations, for the abolition or reduction of standing armies and for the holding and promotion of peace congresses. The prizes for physics and chemistry shall be awarded by the Swedish Academy of Sciences; that for physiological or medical work by the Caroline Institute in Stockholm; that for literature by the Academy in Stockholm, and that for champions of peace by a committee of five persons to be elected by the Norwegian Storting. It is my express wish that in awarding the prizes no consideration whatever shall be given to the nationality of the candidates, but that the most worthy shall receive the prize, whether he be a Scandinavian or not.

The Nobel Prizes are given out each year on December 10, on the anniversary of Alfred Nobel's death on December 10, 1896. His will explicitly stipulates that the prize should go to "the person who shall have made the most important discovery" in the fields of physics, chemistry, and physiology or medicine. The qualifier "important" was undefined and left to a committee from the Royal Swedish Academy of Sciences. As we detailed in chapter 8, importance in science is often a matter of judgment (9), and the relative importance of scientific discoveries is very much dependent on the perspective taken by the evaluators. In this regard, the Nobel Committee has shifted over the years from focusing on practical discoveries to more basic science. A problem inherent in Nobel's will was the stipulation to recognize a single person, given that scientific discovery is almost always a collaborative exercise. Since the prize was established, the committee has attempted to deal with this problem by allowing sharing of the prize by up to three scientists, but, as will be discussed below, this solution is increasingly inadequate and represents a major source of controversy. Another problem created by the will was the condition that the contribution had to have been made in the preceding year, a condition that was soon disregarded after the realization that importance in science requires time for recognition and the need to be validated by other investigators. The fact that the Nobel Committee has chosen to disregard some of the stipulations in the will is important because it establishes a precedent that they are not bound to the letter of the will and gives hope that the selection process may one day be reformed and made fairer.

The restriction of the Nobel Prize to living scientists means that some discoveries are not recognized with this award, emphasizes the personal nature of the award to individuals, creates a need to ensure that the individual is recognized in his or her lifetime, and disqualifies those who die early. The list of deserving scientists who have missed the prize because of untimely death is long. Henry Moseley missed his chance for a Nobel for the relationship between atomic number and X-ray spectra when he was killed in the Gallipoli campaign in 1915 (10). The development of the radioimmunoassay was recognized with a Nobel Prize in 1977 to Rosalyn Yalow, but not to Solomon Berson, who had died in 1972, despite the fact that the two had collaborated closely for many years (11). Similarly, the contributions of Rosalind Franklin were not recognized in the 1962 Nobel Prize for the discovery of DNA structure, because she had succumbed to cancer four years earlier. In 2011, the Nobel Committee for Physiology or Medicine announced that Ralph Steinman had been awarded the prize for his research on dendritic cells, unaware that Steinman had died three days earlier from pancreatic cancer. After an emergency meeting, the Nobel Foundation allowed the award to stand because the decision "was made in good faith." Hence, the criteria for selecting winners can be changed and are already considerably divergent from the specifications of Alfred Nobel's will. However, Charles Janeway could not be included among the 2011 laureates because he had succumbed to cancer eight years earlier, although no one questioned his seminal contributions (12). The posthumous awarding of the Nobel Prize to Ralph Steinman provides another precedent to show that the rules can be changed.

SCIENCE BEFORE THE NOBEL PRIZE

The Scientific Revolution began in the 17th century and was almost 200 years old by the time the Nobel Prize was instituted. It is worth noting that such landmark discoveries as the laws of thermodynamics and electromagnetics, optics, the theory of evolution, and the atomic nature of matter were already established before the end of the 19th century. Although Nobel may have intended to accelerate the pace of science in certain fields by creating a prize, it is difficult to make the argument that the Nobel Prize is essential as an incentive for scientific progress. In fact, there is widespread consensus that although the prize recognizes outstanding contributions to the field, most recipients have already received plenty of recognition by the time that they are selected. As George Bernard Shaw once observed, the Nobel Prize "is a life-belt thrown to a swimmer who has already reached the shore in safety." At his wife's insistence, he accepted the Nobel Prize in Literature but donated the money to support the first translation of August Strindberg's plays from Swedish to English (13).

The Nobel Prize was created in the optimistic spirit that characterized the fin de siècle transition from the 19th to the 20th century, when Europe was at the zenith of its global power and science was widely regarded as a social good. By the end of the 19th century, science and technology had produced the germ theory of disease, railroads, the internal combustion engine, and the telegraph, which had revolutionized human health, travel, and communication, respectively. The decade of the 1890s had seen the first effective therapeutics against infectious disease in the form of serum therapy, which would lead Emil von Behring to receive the first Nobel in Physiology or Medicine in 1901. The First World War had not yet shattered innocent optimism about science by introducing aerial bombing, chemical weapons, and unrestricted submarine warfare. The Second World War would bring unimagined horrors, including genocide founded on scientific racism using tools of science in the form of gas chambers. In the mid-20th century, the world would learn firsthand of the dual-use nature of science with the creation of nuclear and biological weapons. Ambivalence regarding the contribution of science to the social good was already apparent in the early 20th century when the 1918 Nobel Prize was awarded to Fritz Haber for the discovery of nitrogen fixation, which allowed the synthesis of both fertilizers and explosives. That prize was also controversial given that Haber was widely credited for the development of chemical warfare, a source of family shame that is speculated to have led to the suicides of his first wife and son.

BENEFITS AND LIABILITIES OF THE NOBEL PRIZE

The effects of the Nobel Prize may be considered from the vantage points of individual scientists, the scientific enterprise, and the greater society. At the level of individuals, there is little question that Nobel Prizes can be good for recipients, for a prize brings money and instant recognition. Granted, the loss of privacy can be an irritation. Nobel laureate Richard Feynman observed: "The Nobel Prize . . . is a pain in the neck. It cuts off lots of things I would like to do in a sort of easy way, like a normal human being. I didn't see that the publicity would be so terrible" (14).

The Nobel Prize is good for the scientific reputations of winners and boosts the frequency with which their publications are cited by others (15). They may also find it easier to publish their subsequent scientific work, since papers with a Nobel laureate as a prominent author are more likely to be accepted (16). The prestige can free recipients of certain responsibilities, allowing them to devote more time to their work. The prize may even carry some health benefits, as Nobel laureates tend to live longer than those who are nominated but not selected (17). Alternatively, one might conclude that not receiving the prize has a detrimental effect. The realization that one's work is not of Nobel caliber may discourage some scientists from continuing their careers. As Hidde Ploegh has noted, "The obsession with awards . . . can ultimately devalue the passion and ingenuity of so many who will never share that limelight" (18). The consequences of not receiving the prize may be particularly acute for those who feel their research is Nobel-worthy. Some suspect that the chemist Gilbert Lewis, who first described covalent bonds, may have committed suicide after receiving 35 Nobel nominations but never the prize itself (19).

Receiving the Nobel Prize can have adverse effects on the career trajectory of a recipient. Analysis of other awards received by Nobel laureates shows an increasing rate before the Nobel Prize with a reduced trend afterwards, suggesting that the prize represents a pinnacle (20). The mathematician Richard Hamming suggested that recipients find it difficult to work on smaller problems once they have received the enhanced recognition that comes with the prize (21). There is also the disturbing condition of "Nobel Prize disease" or "Nobelitis," in which laureates feel entitled to express foolish opinions on subjects outside their expertise (22) (note that this definition of Nobelitis differs from our definition in chapter 32). Some have championed eccentric causes such as megadose vitamin C (Linus Pauling), telepathy (Brian Josephson), and the production of electromagnetic signals by bacterial DNA (Luc Montagnier) (23). Others have gone so far as to squander their reputations and become academic pariahs, including James Watson, who expressed controversial views on race, gender, religion, and sexuality; the semiconductor innovator William Shockley, who advocated eugenics; and the PCR inventor Kary Mullis, who happily shared his eccentric views on just about everything, including AIDS, climate change, the big bang, and astrology (24–26). Unfortunately, the behavior of Nobel laureates has not always been noble (27).

The Nobel Prizes are arguably good for the public image of science because they stimulate interest in science and provide an opportunity to celebrate human achievement. However, it is conceivable that competition for a Nobel Prize perverts the economics of science. The Nobel Prize may subliminally or consciously encourage scientists to focus on the disciplines specified in Alfred Nobel's will—physics, chemistry, medicine, and physiology—to the detriment of others, like mathematics and much of biology. The Nobel Prize neglects interdisciplinary research, which is increasingly important for the solution of complex problems (28). There is also concern that fields may stagnate once they have been recognized by a Nobel Prize, since scientists conclude that its major problems have been solved. In this regard, it is interesting that most Nobel Prizes in immunology have been awarded in connection with antibody research, but after the 1984 and 1987 prizes for hybridoma technology and the generation of diversity research, the topic entered the doldrums as many surmised (incorrectly) that there were no major challenges left in the field (29). Hence, it is conceivable that the Nobel Prize, with its prestige and global visibility, could distort the trajectories of certain fields. Researchers challenging the prion hypothesis noted that it became increasingly difficult for them to obtain funding once Stanley Prusiner had been awarded the

Nobel Prize in 1997 (30). Awarding the prize for a particular discovery can create a false sense that that there is nothing major remaining to be done, divert scientists to different pastures, and slow further progress in a research area. The frequency of such distortions is unknown, but their potential occurrence is of concern if it means that important areas of study are ignored or abandoned prematurely.

Perhaps the greatest damage that the Nobel Prize does to science is the distortion of its history. Once a prize is given, a new narrative of discovery is immediately created, irrespective of the facts, which quickly takes hold in the minds of scientists and the public. Although the majority of Nobel laureates are richly deserving of recognition, the contributions of others are diminished, and with time the true story may be lost unless historians carefully document the contributions of others. Unfortunately, most scientific fields lack historians, and the details of discovery are seldom written down unless some scientist takes it upon him or herself to do so. As generations pass, much of the historical knowledge in a field is only maintained in an oral tradition, if it is maintained at all. As an example, we consider the way that the path to solving the structure of DNA was distorted after the 1962 Nobel Prize was awarded to James Watson and Francis Crick (Box 21.1). We focus on the DNA story because there is so much information available about it, but do not doubt that similar stories exist for many other Nobel Prizes.

Box 21.1 The Nobel Prize distorts the history of DNA

In 1962, the Nobel Prize was awarded to Francis Crick, James Watson, and Maurice Wilkins "for their discoveries concerning the molecular structure of nucleic acids and its significance for information transfer in living material." This misrepresented the history of the discovery of DNA's central importance in biology and buried the contributions of many investigators. The molecular structure of nucleic acids was already known by the time the laureates began to work on the problem, including the knowledge that DNA contained equimolar ratios of adenine to thymine and cytosine to guanine, which had been painstakingly shown by Erwin Chargaff (72). The role of DNA in information transfer in living material had already been demonstrated in 1944 by a research team at Rockefeller University led by Oswald Avery (73). The principal contribution of Watson and Crick was the proposal of a double-helical structure for DNA that accounted for prior observations including Chargaff's rules and implicitly suggested a mechanism for replication. However, Rosalind Franklin had carried out innovative X-ray

(Continued)

Box 21.1 (Continued)

diffraction studies that anticipated DNA's helical structure. It is noteworthy that Watson and Crick's paper (74) was published in the same issue of *Nature* as lesser-known papers by Franklin and Wilkins, which reported the diffraction pattern data (75, 76). These data had been examined by Watson and Crick without Franklin's knowledge and were critical for their model. Franklin's findings were preceded by those of another woman crystallographer, Florence O. Bell, who in 1939 published that "the (DNA) fibre axis corresponds to that of a close succession of flat or flattish nucleotides standing out perpendicularly to the long axis of the molecule to form a relatively rigid structure, strongly optically negative, and showing double refraction" (77), a description that proved to be uncannily accurate. Also missing from the main story is the critical role of Jerry Donahue in steering Watson and Crick to the correct base structures capable of pairing by hydrogen bonding (78). By 1962, Avery and Franklin were dead, and the Nobel Committee awarded the prize to Crick, Watson, and Wilkins, excluding Chargaff, Bell, and Avery's coauthors of the landmark 1944 paper. Today, the Nobel Prize has so strongly associated DNA with Watson and Crick that many people think that they discovered the molecule, while Chargaff's rules and Avery's landmark contribution are less well-known except to students of history.

For society, the Nobel Prize serves the role of highlighting a scientific accomplishment that is recognized by fellow scientists as important. The prestige of harboring a Nobel laureate can enhance an institution's ability to raise funds. The number of Nobel laureates from specific countries has been used as a measure of scientific prowess and national pride. However, countries such as Brazil, which possesses a large and accomplished scientific establishment, have struggled with the absence of Nobel Prizes among their research institutions. In this regard, some have strenuously argued that a posthumous Nobel Prize should be awarded to Carlos Chagas, a Brazilian scientist credited with the discovery of American trypanosomes (31, 32). The Nobel Prize has arguably distorted the public view of science by portraying advances as a series of contributions by brilliant individuals rather than by the interactive and interdependent scientific community that exists in reality.

NOBEL PRIZE CONTROVERSIES

A review of the literature on the Nobel Prize reveals an institution fraught with politics and controversy, ranging from the choice of recipients to whether the prize should exist at all (1, 33, 34). These controversies provide useful insights into problems with the structure of the prize as it currently stands. Ever since the awarding of the first Nobel Prizes in 1901, the decision to award some individuals the Nobel Prize for discoveries to which others have contributed has been a recurring topic of contention. Emil von Behring received the inaugural prize in Medicine for the discovery and application of antitoxins (serum therapy), but his coauthor Shibasaburo Kitasato was omitted. Some suspected nationalism or racism behind this decision, but others have suggested that the committee may have simply been attempting to adhere strictly to the conditions of Nobel's will, which stipulated a single individual for each prize (35). Similarly, the 1901 prize in Physics was awarded to Wilhelm Roentgen for the discovery of X rays, and Philipp Lenard was left off despite the fact that he and Roentgen had been jointly honored with prior prizes for the discovery (36). In subsequent years, the prize has continued to be subject to intense criticism, both because of those selected and because of those omitted (Table 21.1).

As we discussed in chapter 9, the 1952 prize for physiology or medicine was awarded solely to Selman Waksman for the discovery of streptomycin, even though it was actually his graduate student Albert Schatz who found the antibiotic while

TABLE 21.1 Some controversial Nobel Prizes in the sciences[a]

Year	Discovery	Awardee	Controversy	Reference(s)
1901	Antibodies as antitoxins	Emil Adolf von Behring	Shibasaburo Kitasato's contributions not acknowledged	34
1901	Discovery of X rays	Wilhelm Roentgen	Philipp Lenard not recognized despite the fact that both had shared other prizes	35
1914	Caloric reaction	Robert Bárány	Exclusion of Julius Eduard Hitzig and Josef Breuer	83
1918	Synthesis of ammonia	Fritz Haber	Haber also developed gas warfare	84
1923	Insulin	Frederick Banting and John Macleod	Exclusion of C. H. Best and J. B. Collip	85
1923	Electron charge	Robert Millikan	Exclusion of Harvey Fletcher; allegation of data manipulation	86–88
1926	Carcinogenesis caused by *Spiroptera*	Johannes Fibiger	Discovery found to be erroneous	89
1930	Raman effect	C. V. Raman	Exclusion of G. S. Landsberg and L. I. Mandelstam	90

(Continued)

TABLE 21.1 (Continued)

Year	Discovery	Awardee	Controversy	Reference(s)
1944	Nuclear fission	Otto Hahn	Exclusion of Lise Meitner and Fritz Strassmann	91
1949	Prefrontal lobotomy	Egas Moniz	Procedure has arguably done more harm than good	33, 92
1952	Streptomycin	Selman Waksman	Exclusion of Albert Schatz	37, 41, 93
1957	Parity violation	Tsung-Dao Lee and Chen Ning Yang	Exclusion of Chien-Shiung Wu	94
1961	Carbon assimilation in plants	Melvin Calvin	Exclusion of Andrew Benson and James Bassham	95
1962	Structure of DNA	James Watson, Francis Crick, and Maurice Wilkins	Exclusion of Erwin Chargaff	96
1964	Lasers	Charles Townes, Nicolay Basov, and Aleksandr Prokhorov	Exclusion of Theodore Maiman	97
1968	Genetic code	Robert Holley, Har Gobind Khorana, and Marshall Nirenberg	Exclusion of Heinrich Matthaei	98
1973	Ethology	Konrad Lorenz	Lorenz accused of supporting Nazi ideas	99, 100
1974	Cell biology	Albert Claude, Christian de Duve, and George Palade	Exclusion of Keith Porter	101
1975	Tumor viruses	David Baltimore, Renato Dulbecco, and Howard Temin	Exclusion of Satoshi Mizutani	100
1978	Cosmic microwave background radiation	Arno Penzias and Robert Wilson	Exclusion of Ralph Alpher	102
1979	Computer-assisted tomography	Godfrey Hounsfield and Allan Cormack	Exclusion of William H. Oldendorf	103

TABLE 21.1 (Continued)

Year	Discovery	Awardee	Controversy	Reference(s)
1980	Determination of DNA sequence	Paul Berg, Walter Gilbert, and Frederick Sanger	Exclusion of Allan Maxam	104
1982	Theory of critical phenomena in connection with phase transitions	Kenneth Wilson	Exclusion of Leo P. Kadanoff and Michael E. Fisher	86
1983	Nucleogenesis	Willy Fowler and Subrahmanyan Chandrasekhar	Exclusion of Fred Hoyle	105
1984	Discovery of pulsars	Antony Hewish and Martin Ryle	Exclusion of Jocelyn Bell Burnell	104, 106
1986	Growth factors	Stanley Cohen and Rita Levi-Montalcini	Exclusion of Viktor Hamburger	107
1989	Oncogenes	J. Michael Bishop and Harold Varmus	Dominique Stehelin protested that he was not included	108
1993	Gene splicing	Philip Sharp and Richard Roberts	Exclusion of Louise Chow and Tom Broker	109
1997	Prions	Stanley Prusiner	Allegations that prize was premature and reduced funding for competitors	30, 110
1998	Nitric oxide	Robert Furchgott, Louis Ignarro, and Ferid Murad	Exclusion of Salvador Moncada and John Hibbs	111–113
1999	Signal hypothesis	Günter Blobel	Exclusion of David Sabatini, Bernhard Dobberstein, and others	114
2000	Neurotransmitters	Arvid Carlsson, Paul Greengard, and Eric Kandel	Exclusion of Oleh Hornykiewicz	115
2001	Catalytic asymmetry synthesis	K. Barry Sharpless, Ryōji Noyori, and William Knowles	Exclusion of Henry Kagan	116, 117
2002	Solar neutrinos	Raymond Davis, Jr. and Masatoshi Koshiba	Exclusion of John Bahcall	118
2003	Magnetic resonance imaging	Paul Lauterbur and Peter Mansfield	Exclusion of Raymond Damadian	43–45

(Continued)

TABLE 21.1 (Continued)

Year	Discovery	Awardee	Controversy	Reference(s)
2004	Discovery of ubiquitin-mediated protein degradation	Aaron Ciechanover, Avram Hershko, and Irwin Rose	Exclusion of Alexander Varshavsky	119
2006	RNA interference	Andrew Fire and Craig Mello	Exclusion of Victor Ambros, Gary Ruvkun, and David Baulcombe	120, 121
2008	HIV	Luc Montagnier and Françoise Barré-Sinoussi	Exclusion of Robert Gallo	122, 123
2008	Association of human papillomavirus with cancer	Harald zur Hausen	Exclusion of Nubia Muñoz	123
2008	Green fluorescent protein	Osamu Shimomura, Martin Chalfie, and Roger Tsien	Exclusion of Douglas Prasher	124
2008	Broken symmetry	Makoto Kobayashi, Toshihide Maskawa, and Yoichiro Namba	Exclusion of Nicola Cabibbo and Giovanni Jona-Lasinio	125
2009	Charge-coupled device	Willard Boyle and George Smith	Exclusion of Eugene Gordon and Michael Tompsett	126
2010	Graphene	Andre Geim and Konstantin Novoselov	Other contributors not acknowledged, including Walter de Heer and Philip Kim	127
2011	Innate immunity	Ralph Steinman, Jules Hoffmann, and Bruce Beutler	Numerous contributors not acknowledged, including Ruslan Medzhitov, Shizuo Akira, and Bruno Lemaitre	128
2013	Higgs boson	Peter Higgs and François Englert	Many scientists excluded, including Gerald Guralnik, Carl Richard Hagen, Tom Kibble, and scientists at CERN	129

TABLE 21.1 (Continued)

Year	Discovery	Awardee	Controversy	Reference(s)
2014	Blue light-emitting diode	Isamu Akasaki, Hiroshi Amano, and Shuji Nakamura	Oleg Losev, Nick Holonyak, Gertrude Neumark, and Herbert Paul Maruska excluded	130, 131
2017	Gravitational waves	Rainer Weiss, Kip Thorne, and Barry Barish	Thousands of scientists at LIGO (Laser Interferometer Gravitational-Wave Observatory) excluded	132
2020	CRISPR/Cas9	Emmanuelle Charpentier and Jennifer Doudna	Virginijus Šikšnys, Francisco Mojica, Feng Zhang, Ruud Jansen, Eugene Koonin, and Yochizumi Ishino excluded; controversial article by Eric Lander neglected Charpentier and Doudna	133, 134

[a] We have listed instances for which we have identified literature documenting the specific controversy. This list is likely to underestimate the number of controversial prizes.

working alone in a basement laboratory. The omission of Schatz remains highly controversial today and is aggravated by Waksman's refusal to credit Schatz as a codiscoverer (37–39). Adding to the controversy, the Nobel Committee admitted that they had never heard of Schatz, raising questions as to the thoroughness of their deliberations (39, 40). Some errors may result from the tendency of the Nobel Committee to consult discreetly with senior scientists in individual fields and thus receive relatively narrow advice from individuals who have some interest in the outcome (41). Similar criticisms of inadequate investigation haunted the exclusion of student assistant Charles Best from the Nobel Prize for the discovery of insulin a generation earlier. Instead, the 1923 prize was shared by Frederick Banting and his department chair, John Macleod, who claimed credit even though Banting and Best had performed the critical experiment while Macleod was away on vacation (42). More recently, the exclusion of Raymond Damadian from the 2003 prize in Physiology or Medicine remains unexplained and highly controversial (43, 44). Damadian proposed the concept of magnetic resonance imaging, published a paper in *Science* on tumor detection by magnetic resonance, and applied for a patent several years before either of the laureates Paul Lauterbur and Peter Mansfield had published their first papers on the topic. Damadian's disappointment led him to take out full-page ads in the *New York Times, Washington Post,* and *Los Angeles Times* (45).

In addition to scientists who did not share in the Nobel Prize for breakthroughs to which they contributed, there is a long list of those who were never recognized by the Nobel Committee for what seem like worthy discoveries (46). Prominent among these are Carlos Juan Finlay, who showed that yellow fever is transmitted by mosquitoes (47); Dmitri Mendeleev, creator of the periodic table; Josiah Willard Gibbs, who created the field of chemical thermodynamics; Ernest Starling, who formulated Starling's law of the heart and made many other contributions to physiology; inventors Nikola Tesla and Thomas Edison; Fritz Schaudinn and Erich Hoffmann, who discovered the agent of syphilis; Oswald Avery, who showed that DNA is responsible for bacterial transformation (Box 21.1); Lisa Meitner, for nuclear fission, whose exclusion from the 1944 prize may have reflected a combination of ignorance, misogyny, and anti-Semitism (48); Edwin Hubble, a pioneer in extragalactic astronomy; J. Robert Oppenheimer, the father of the atomic bomb, who first predicted the existence of black holes; Albert Sabin and Jonas Salk, who developed polio vaccines (see Box 20.1); Jocelyn Bell Burnell, for the discovery of pulsars (Box 21.2); and many, many others. Mendeleev came close to receiving the prize in 1906 and 1907, but his candidacy was undermined by the Swedish laureate Svante Arrhenius, who harbored a personal grudge against him (1). It is rumored that the 1912 Physics prize was to be awarded to Tesla and Edison, but the plan was scuttled because each refused to share the prize with the other. Avery's slight has been attributed to the skepticism of influential Swedish chemist Einar Hammarsten, as well as to Avery's reluctance to engage in self-promotion (49, 50). Hubble missed out because astronomy was not recognized as a branch of physics until the year of his death (51). Nevertheless, we note that many of those who have been passed up for the Nobel Prize have received other awards and recognition, and most scientists who make great contributions receive the respect of their peers irrespective of whether they are Nobel laureates. The fact that we mention these scientists by name is a testament to their immortality in science.

It is evident from the cited examples that the criticisms associated with the selection of winners principally fall into two categories: exclusion of individuals who are considered worthy by their peers, and allegations of inadequate research or objectivity on the part of the Nobel Committee in making its decisions. We note that these problems are not unique to the Nobel Prize. In 2013, the Canadian microbiologist Michael Houghton turned down the $100,000 Gairdner Award because he felt that his collaborators Qui-Lim Choo and George Kuo should have been recognized as well for the discovery of hepatitis C virus. Nevertheless,

Box 21.2 More fun than a Nobel Prize

Jocelyn Bell Burnell fell in love with astronomy from an early age. Her interests led her to enroll in the Physics Department at the University of Glasgow, where she found that she was the only woman in a class of 50 students. Undaunted, she graduated with honors and entered a graduate program in radio astronomy at Cambridge University. Battling imposter syndrome and the fear that she would be thrown out of Cambridge, she began a thesis project under the direction of Professor Antony Hewish. Her assignment was to help build a radio telescope and use it to scan for extremely distant celestial bodies called quasars. While poring over miles of chart data, she discovered an unusual string of pulses one and a third seconds apart. At first her advisor attributed it to interference or an artifact produced because she had set up the telescope incorrectly. Eventually another similar signal was detected, and Jocelyn Bell Burnell, at the age of 24, had discovered the first *pulsars*, rotating neutron stars formed after the collapse of supernovas. This monumental discovery would be recognized by a Nobel Prize in 1974, but the prize was awarded to her mentor, Hewish, and to Martin Ryle, head of the radio astronomy group, and not to her. Many have regarded this as an egregious oversight that epitomizes sexism in science (79). The astronomer Fred Hoyle vigorously protested the omission of Bell Burnell (80), and some attributed Hoyle's own omission from the Physics Nobel Prize eight years later to his intemperate response (81). Others cite it as yet another

Jocelyn Bell Burnell in 1967 and 2009. Photo credit: (left) Roger W. Haworth, under license CC BY-SA 2.0, (right) Astronomical Institute, Academy of Sciences of the Czech Republic, under license CC BY-SA 3.0.

(*Continued*)

Box 21.2 (Continued)

instance of senior scientists claiming credit for the discoveries of their trainees. But Jocelyn Bell Burnell has never complained. In fact, she jokes that she is better off: "If you get a Nobel Prize, you have this fantastic week and then nobody gives you anything else. If you don't get a Nobel Prize you get everything that moves. Almost every year there's been some sort of party because I've got another award. That's much more fun" (82). In 2018, she was awarded the Special Breakthrough Prize in Fundamental Physics, worth U.S. $3 million. She donated the entire amount to create scholarships for women, refugees, and students from historically underrepresented groups who are interested in becoming physics researchers.

in 2020, he accepted the Nobel Prize for the discovery of hepatitis C, sharing it with Harvey Alter and Charles Rice. In citing this episode, we do not imply any criticism for Houghton for his decisions; we only regret that the Gairdner and Nobel Prize committees were not more inclusive. The current policy of the Nobel Committee to seal its records for 50 years means that few if any contenders alive at the time the prize is awarded will ever have the opportunity to understand the decision-making process. The opacity of the process, the absence of remedy when errors are made, and the imperiousness of the decisions are additional criticisms on top of the structural problems cited above. The nomination process has also come under criticism because of national chauvinism in some fields, which results in the preferential nomination of chemists from the United States (52).

Given the high visibility of the Nobel Prize, the selection committee has a tremendous responsibility in assuring that the prize is given to the most meritorious scientists, since these individuals often serve as role models for younger scientists. Consequently, when the prize is given to those who exclude junior colleagues, to those who claim a large share of the credit through self-promotion, or to those who distort history to claim credit, the prize does harm by glorifying less-than-ideal role models (42).

The Nobel Prize exemplifies the "great man" theory of history, which posits that history is shaped by special individuals who transform the world through their unique insight and other qualities. This viewpoint was promulgated by Thomas Carlyle, who wrote that "The history of the world is but the biography of great men." However, we counter that luck and serendipity have perhaps even greater roles in shaping the history of science than the actions of single individuals (see chapter 14) and note that no prize, outside of lotteries, is given for sheer luck. Modern historians recognize that the "great man" theory of history is an incomplete view of how the world truly works. Although we agree that certain breakthroughs, such as Newtonian physics, can be credited to specific individuals, we concur with Thomas Kuhn that scientific revolutions can truly occur only after seminal concepts are accepted through the contributions of many individuals (53) (see chapter 11). People are shaped by the society in which they live, just as the actions of individuals can transform society. In this sense, the Nobel Prize often distorts the history of science and medicine, creating a new narrative. This a major concern, since history gives us an understanding of how progress is made (see chapter 9).

Perhaps the greatest objection is that the Nobel Prize reinforces a flawed reward system in science in which the winner takes all and the contributions of the many are neglected by disproportionate attention to the contributions of a few. Although there are always going to be some scientists who are principally driven by a desire for personal glory and wealth, there are many others for whom the joy of discovery, the pleasure of working with colleagues and trainees, and the opportunity to serve society are the primary rewards of science. Perhaps Richard Feynman said it best:

> I'm appreciated for the work that I did. And for people who appreciate it, and I notice that other physicists use my work, I don't need anything else. I don't think there's any sense to anything else. I don't see that it makes any point that someone in the Swedish Academy decides that this work is noble enough to receive a prize. I've already got the prize. The prize is the pleasure of finding the thing out. The kick in the discovery, the observation that other people use it. Those are the real things. The honors are unreal to me. I don't believe in honors. (54)

ALTERNATIVES

Few are happy with the current state of the Nobel Prize. The overwhelming majority of articles that comment on the Nobel Prize express some dissatisfaction and/or focus on controversies such as those described above. There continue to be regular and frequent calls for reform, but thus far these have fallen on deaf ears. In 2022, the astronomer Martin Rees assailed the limitation to three winners, writing in *Time* magazine that "the Nobel Committee's refusal to make an award to more than three people has led to manifest injustices, and given a misleading impression of how science actually advances" (55). Ed Yong, writing in *The Atlantic*, noted that the Nobel Prizes "distort the nature of the scientific enterprise, rewrite its history, and overlook many of its most important contributors" (56). Others have suggested that the Nobel Prize be expanded to recognize additional areas of science (57). In this regard, the narrow focus on physics, chemistry, medicine, and physiology has been criticized for creating a two-tiered system in which some fields get much publicity while others that are critically important to humanity, such as climate science, do not get commensurate attention (58). In fact, it is difficult to find defenders of the Nobel Prize in its current form. Those content with the status quo possibly include living Nobel laureates, the Nobel Foundation, and the nation of Sweden, which takes great pride in awarding the premier prize of science. To this list might be added a select group of scientists who are optimistic about their prospects for a prize in the near future.

In 1975, an editorial in *Nature* argued for the abolition of the Nobel Prize on the grounds that (i) there are not enough spots to go around, and consequently any prize results in inequity; (ii) the reward of only a few fields of study creates a sense that some fields are more important than others; and (iii) the prizes confer an unreasonable amount of credit on the recipients (33). We do not think that the Nobel Prize should be abolished, since its benefit to science in highlighting scientific accomplishment probably outweighs the liabilities associated with the problems described above. Nevertheless, it is worthwhile to consider alternatives to the current prize mechanism.

The dilemmas and controversies increasingly posed by the Nobel Prizes has led some to propose that groups rather than individuals share in the prize (59). In this regard, the completion of the Human Genome Project has been suggested to merit a Nobel Prize, given the enormous technological achievement and the benefits that have accrued from the information learned. The absence of a prize awarded for the genome achievement might reflect the inability of the current mechanism to recognize more than three individuals and to appropriately acknowledge all who contributed (60–62). In recent years, the Nobel Committee has continued to stumble into more and more criticism as it insists on limiting the prize to three individuals, as evident from criticisms involving prizes awarded for the Higgs boson (63), gravitational waves (64), CRISPR (clustered regularly interspaced short palindromic repeat) technology (65), the antimalarial drug artemisinin (66), and ancient DNA sequencing (67).

CONCLUSIONS

Over the past century, the Nobel Prize has established itself as the most visible celebration of scientific achievement. The prize is an institution that by its very existence promotes science, and there is no denying that the overwhelming majority of Nobel laureates are highly deserving of the honor. Alfred Nobel, the Nobel Foundation, and the nation of Sweden should have the gratitude of the scientific community for hosting, maintaining, and managing the prize. However, our analysis and the many others that we have cited also suggest that the Nobel Prize has become a recurring source of controversy. Perhaps it is time to reconsider how it is awarded. In our view, the major problem inherent in the Nobel Prize is the limitation of the prize to three individuals. This is not a new criticism. However, as time passes, the problem grows worse because most science is now done collaboratively. With the exception of a few unusual contributions, it is difficult and potentially unfair to limit credit for a specific scientific advancement to three individuals. Marvin Gozum (68) has argued that the current system of awarding the prize to individuals is obsolete and that the prize should be given prospectively for achieving a specific goal, but it is difficult to see how this change in emphasis would avoid controversy.

A better solution to the problem of recognizing a few scientists while unfairly excluding others might be to eliminate the stipulation that a Nobel Prize can be awarded to no more than three individuals, such that even large groups can be recipients. The requirement that honorees must be living at the time of the award might also be reconsidered, as it seems unfair to exclude individuals who made equivalent contributions merely because they were unfortunate to die prematurely. Recognizing groups instead of individuals would shift the emphasis away from individuals to human achievements in science and make the prize a celebration in which all of humanity can partake. In this regard, it is noteworthy that the Nobel Peace Prize has been given to organizations such as Amnesty International (1977), United Nations Peacekeeping Forces (1988), and Médecins sans Frontières (1999), and even to supranational states as the European Community (2012). In 1971, the American Association for the Advancement of Science awarded the Rumford Premium to research teams rather than individuals "in an effort to reflect the way in which much scientific research is conducted today" (68, 69). Furthermore, the Principe de Asturias Prize has already moved to recognize the reality that scientific advancements are often made by collaborations and recognizes "individuals, institutions, or groups of individuals or institutions" (70). In 2013, the Asturias Prize for Technical and Scientific Research was given to Peter Higgs, François Englert, and CERN (Conseil Européen pour la Recherche Nucléaire, or the European Organization for Nuclear Research), while in the same year the Nobel Committee limited recognition to Higgs and Englert, leaving out the entire CERN organization. Clearly, without CERN and the thousands of engineers and physicists who toiled to make the Large Hadron Collider work, there would be no Higgs boson and no Nobel Prize for this discovery. Similarly, the 2017 Asturias Prize was awarded to Rainer Weiss, Kip Thorne, Barry Barish, and LIGO (Laser Interferometer Gravitational-Wave Observatory). Interestingly, the Asturias Prize criteria have also previously specified that a nominee should be "an outstanding role model," a criterion that would have eliminated some prior controversial Nobel Prize winners from consideration.

We are cognizant that the criticisms of the Nobel Prize can also be applied to other awards in science. However, the Nobel Prize stands alone in prominence, whether or not this is justified, and consequently its disproportionately high value makes the problems in credit recognition much greater and more acute (71). Given that the winner-takes-all economic system of science may be responsible for many current problems, including scientific misconduct, shifting recognition from individuals to fields could have a beneficial effect on the state of science. Recognizing scientific achievements while more fairly allocating the credit might promote more cooperation between scientists within fields and avoid the potential personal rivalries and liabilities that can accompany competition in the quest for a Nobel Prize. Although some may argue that changing the focus from individuals to groups would run counter to Alfred Nobel's will, which aimed to recognize individuals, we note that the committee has already altered some of the will's original stipulation that the prize should be awarded to a single individual who is responsible for advancements made in the preceding year. Recognizing scientific groups has the theoretical shortcoming that some might conclude that all the major problems in an area have been solved, which could discourage bright minds from entering the field. We would counter that fields recognized by a prize could use the opportunity to recruit scientists to take on new questions, and some fields might even gather multiple Nobels, an accomplishment that would recognize a particularly productive group of scientists. Furthermore, in cases in which prizes are awarded to large groups, the award money that currently goes to individuals could be used instead to support young scientists in the field. Using award funds to support promising scientists in training would ensure that the money is used directly to support science and thus directly benefit the scientific enterprise. A Nobel Prize that more fairly allocates credit would do justice to all who contribute to scientific advances, avoid most controversies now plaguing the prize, and potentially improve the conduct and pursuit of science. For now, we can only imagine how such a reformed Nobel Prize system might work (Box 21.3).

Box 21.3 An alternative Nobel Prize scheme

Imagine that it is early 2013, and the Nobel Committee has received several nominations for the 2013 Nobel Prize in Physics. One nomination stands out, asking for recognition of the discovery of the Higgs boson by CERN scientists in 2012. The committee reviews the scientific contributions of the many people who contributed and decides to award the Nobel for "completion of the standard model" to the particle physics community. The announcement is made in early October, and the Nobel Foundation asks any scientist who contributed to the completion of the standard model to request a prize certificate that will be mailed for a nominal fee. The organization receives thousands of requests, including from engineers who built the Large Hadron Collider and physicists ranging from graduate students to retired professors. The diplomas are illustrated with the Nobel medal and include the name of the contributor. The prize money and processing fees are used to create new prizes in nascent fields or to create fellowships for individuals in under-resourced regions. Humanity celebrates another landmark scientific achievement without controversy. In subsequent years, prizes are awarded for the discovery of gravitational waves, human genomes, CRISPR, immunotherapy, etc., using the same mechanism without controversy. Science, humanity, and all who contributed are winners, and there are no losers.

22 Rejected Science

Rejection is in the fabric of what we do. We send our papers, carefully crafted to consider every angle and interpretation of our hard won data, and "Slap!" we're squashed like vermin.

Mole (1)

Dear Editors,
Thank you for the rejection of our paper. As you know, we receive a great many rejections, and unfortunately it is not possible for us to accept all of them. Your rejection was carefully reviewed by three experts in our laboratory, and based on their opinions, we find that it is not possible for us to accept your rejection. By this we do not imply any lack of esteem for you or your journal, and we hope that you will not hesitate to reject our papers in the future.

Mole (2)

Roughly two-thirds of the manuscripts submitted to a typical scientific journal are rejected (3), about the same proportion of time that an excellent professional baseball player fails to get a hit. Such selectivity is arguably essential for journals

Thinking about Science: Good Science, Bad Science, and How to Make It Better, First Edition.
Ferric C. Fang and Arturo Casadevall.
© 2024 American Society for Microbiology.

to retain their attractiveness, which is based in part on selectivity and impact factor (see chapter 26). Here we face a paradox in the scientific publishing process in which journal selectivity is equated to quality and authors are incentivized to submit their papers to the very journals that are most likely to reject them. Nevertheless, while this rejection rate may not approach the exceedingly high rejection rates encountered in National Institutes of Health (NIH) study sections (4, 5), each rejected manuscript represents hundreds of hours of work relegated, at least temporarily, to the rubbish bin.

There are two types of rejections: rejection by editors without review and rejections by editors after peer review by experts in the field. Editorial rejection, also known as "desk rejection," is rapid but very frustrating for authors for it usually comes without a detailed critique. Editorial rejection was relatively rare prior to the 1980s, as most journals published the majority of their submissions or obtained reviews for submitted manuscripts. However, as impact factor mania took hold (see chapter 26), many authors began to submit their papers to higher-impact journals that became flooded with submissions and were forced to institute a triage system to rapidly reject papers thought unlikely to be accepted. Editorial rejection is justified by journals on the grounds that it hastens the process by returning papers to the authors for rapid submission elsewhere. Editorial rejection can result when an editor spots a fatal flaw or may result from other factors such as poor English grammar, insufficient novelty, or content outside the scope of the journal (6). Interestingly, the day of the week for article submissions may also be a factor. An analysis of papers editorially rejected by rheumatology journals found that weekend submissions were far more likely to be rejected than those made on other days of the week, with Wednesday submissions having the lowest rejection rate (7). Although an explanation for this quotidian effect was uncertain, it suggested an unexpected source of bias that might be influencing editorial decisions.

From the author's viewpoint, an editorial rejection is a waste of time and effort. Since submission to a journal can take hours, the effort involved is substantial. Editorial rejections generally result in submission to the next-lower-tier journal, where the process may repeat itself, to produce a cascade of wasted time and effort coupled with increasing frustration. In contrast, rejection after review generally provides a rationale for the decision that can help authors to improve their paper for resubmission to the same journal or elsewhere. Although desk rejection always carries a risk for the journal in turning down an important manuscript, studies of editorially rejected papers indicate that most triaged papers are published in journals with lower impact factors, and their rates of citation by other articles correspond to the average for their journal (8). However, there is little evidence that editors can accurately predict the future citation rates for any given article (9).

Although most rejections are quietly and graciously accepted as an inevitable part of the process, we know as editors that occasionally rejections are appealed, and rarely there is even a vehement response. Rejected authors generally aim their anger at journals and editors, often without reflection. One author, after being informed that his manuscript was outside the scope of the journal, announced to one of this book's authors that he would never again submit an article to our journal. Another notified the journal that the reviews of his manuscript had been sent to the NIH to provide his study section with proof that reviewers in his field were hopelessly incompetent. In light of such intemperate responses, perhaps it is appropriate to reflect on the importance of rejection in the scientific process.

Let us first freely acknowledge (from ample personal experience) that rejection is painful. We agree with the old saying, "Honest criticism is hard to take, particularly from a relative, a friend, an acquaintance or a stranger" (attributed to Franklin P. Jones). However, we would assert that rejection is central to science. Science is a community endeavor in which experts attempt to achieve consensus regarding the present state of understanding in their field. This consensus is constantly under reevaluation. Although scientific knowledge may be tentative and provisional, it is not (and should not be) a trivial matter to change the scientific status quo. The more sensational or unexpected the discovery, the greater the burden of proof demanded by others. As James Randi famously observed, "If I told you that I keep a goat in the backyard ... and if you happened to have a man nearby, you might ask him to look over my garden fence. ... But what would you do if I said 'I keep a unicorn in my backyard'?" (10). Once a scientist makes a discovery, the task of amassing evidence to convince reviewers and skeptical competitors begins. Although it is not necessary to convince every last holdout (e.g., Peter Duesberg or Michael Behe on human immunodeficiency virus (HIV) and evolution, respectively) for new information to be incorporated into the corpus of scientific understanding, it is essential to convince a critical mass of workers in the field. Otherwise, the work will lack impact, whether valid or not.

As described in chapter 1, the scientific method evolved from the ancient Greek traditions of mathematical logic and rhetoric (11, 12). The Greeks valued the derivation of a logical conclusion from a succession of rational steps and revered the individual who could persuasively argue a point in public. Accordingly, scientists who have made a new discovery must systematically support their conclusions and then proceed to convince a skeptical community of the veracity of their claims. It is the duty of one's fellow scientists to challenge and critically scrutinize each piece of new information before accepting it. The process of questioning, demanding multiple lines of evidence and reproducibility, and testing the predictive power of new ideas makes our knowledge more secure. This is what makes science uniquely powerful as a way of understanding the natural world.

IS PEER REVIEW CENSORSHIP?

Given the unpleasantness of having one's work rejected (1), as well as a desire for more-rapid communication of scientific findings, some scientists have expressed nostalgia for the good old days when nearly any submitted manuscript was accepted for publication, and some have even compared peer review to censorship (13, 14). After all, neither Isaac Newton nor Charles Darwin had to submit to the indignity of peer review prior to publication! Let us then explore the latitude provided to authors in scientific manuscripts and attempt to distinguish the processes of peer review and censorship. In dissecting these issues, we hope to provide authors with tools for approaching the comments and criticisms that inevitably follow peer review. Furthermore, we hope that delineating the differences between peer review and censorship will encourage flexibility in authors, reviewers, and editors when dealing with controversial and speculative viewpoints.

To approach this question, let us first consider the historical relationship of science to other disciplines. At the outset of the Scientific Revolution, the major struggle of science was with religion. The ordeals of Galileo Galilei provide a case in point. Although Galileo is a scientific luminary, in his time he encountered problems with peer review by the Inquisition. Objections from church-appointed reviewers were not merely dogmatic. Significant questions were raised with regard to the heliocentric theory, including the fact that it could not explain the absence of stellar parallax, as briefly mentioned in chapter 9. The problem of parallax reflected the fact that, according to the heliocentric theory, the angle to a star should change with the time of year as the Earth goes around the Sun. This was a legitimate scientific criticism that would not be resolved until the 19th century, when technological advances allowed Friedrich Bessel to make the first demonstration of stellar parallax. Galileo was in fact not the first astronomer to run afoul of church censors, as Vatican Decree XXI had already declared that "this whole chapter can be deleted because it admittedly deals with the truth of the earth's motion," in reference to Copernicus' *De Revolutionibus* (15). Although associating the Inquisition and contemporary scientific peer review may seem extreme, a case can be made that the Inquisition represented a review by Galileo's learned peers. Despite the scientific criticisms of the heliocentric theory, Galileo was initially defiant and recanted only when shown the instruments of torture. In our experience, authors of scientific papers nowadays are generally happy to make revisions to get their papers accepted, and encouragement from torture devices is hardly ever needed anymore. Hence, things do appear to have changed for the better in the area of scientific publishing.

Censorship is defined by the dictionary as "examination in order to suppress or delete anything considered objectionable" (16). The word originates from the Roman censors, magistrates charged with both taking the census (for tax purposes) and maintaining public morality, or *regimen morum*. Peer review has been more specifically defined as "the evaluation of scientific research findings for competence, significance and originality by qualified experts" (17, 18). Peer review of manuscripts as it presently exists is taken for granted, but its history is much more recent than that of censorship. Although the peer review of scientific manuscripts dates back to the Royal Society of Edinburgh in 1731, peer review was irregularly performed by most journals until the latter half of the 20th century. While some journals, like the *British Medical Journal*, routinely sent all manuscripts to outside experts for an opinion prior to publication, *Science* and *JAMA* did not employ outside reviewers until 1940, relying only on editors' assessments for publication decisions, and the *Lancet* did not implement external peer review until 1976 (17, 19, 20). *Nature* did not require external peer review until 1973, with prior editors sometimes accepting manuscripts after only internal review by the journal (21). It is worthwhile to note that outside peer review was not always well accepted by authors in earlier times. Albert Einstein was incensed that the editor of the *Physical Review* sent his paper for outside review and was quoted (21) as having written to the editor that "[I] had not authorized you to show [our manuscript] to specialists before it is printed. I see no reason to address the—in any case erroneous—comments of your anonymous expert. On the basis of this incident I prefer to publish the paper elsewhere."

One of the problems with instituting outside peer review was that there were very few copies of a manuscript available, since duplicating relied on having the entire text copied anew or using carbon copies generated on typewriters, a technology that flourished until the 1970s. The critical technological advance of the photocopier in 1959 greatly facilitated the dissemination of manuscripts to multiple reviewers, and the recent development of the Internet, which allows manuscript files to be sent anywhere on the globe in seconds, has further enhanced the process. The information revolution has also increased the number of potential reviewers, since experts can be easily found in online databases such as PubMed or from journal internal records.

Peer review became essential because new incentives for publication dramatically increased the number of research papers (PubMed lists more than one and a half million articles published during the past year alone). Peer review allows journals to select the best papers for publication and helps busy scientists to prioritize the scientific literature while providing some quality control. There is some evidence that peer reviews improve the quality of publications, although much more work needs to be done in this area. An analysis of more than 27,000 papers for statistical content revealed an improvement after peer review (22). Adding statistical reviewers improves the quality of data reporting, while referring authors to guidelines appears to have no effect (23). In the clinical literature, there is evidence that peer review improves the reporting of clinical trial data (24). In addition, the interventions of editors can improve the quality of reporting on clinical trial data (25).

The stakes for having one's manuscript published in the relatively short list of selective and elite journals have become high, as decisions for hiring, promotion, and funding have become heavily reliant on publication record. One of the fascinating aspects of the sociology of science is that scientists prefer to publish in journals that present the greatest hurdles, which translate into scientific prestige. Whether based upon impact factor, reputation, or expertise, etc., the venue chosen for publication can have a significant impact on the visibility of a study and the fortunes of the authors. Hence, the most desirable venues for scientific publication are those in which articles are rigorously peer-reviewed and editors routinely reject manuscripts on the basis of priority, an imprecise term that is meant to convey importance, preference, suitability, and interest to the readership. The prestige of a journal has become a surrogate measure for the quality of the work itself.

The current system persists despite abundant evidence of imperfections in the peer review process (26, 27). Most scientists would agree that peer review improves manuscripts and prevents some errors in publication (28). However, although there is widespread consensus among scientists that peer review is a good thing, there are remarkably little data to show that the system works as intended (29–31). In fact, studies of peer review have identified numerous problems, including confirmatory bias, bias against negative results, favoritism for established investigators in a given field, address bias, gender bias, and ideological bias (28, 29, 32, 33). Reviewers are much more likely to accept a paper from a well-known scientist, such as a Nobel laureate, than when the scientists are less known. This was shown in a study that found that reviewers were more likely to agree to review a paper authored by a Nobel laureate when asked, and recommended rejection only 23% of the time when the laureate was the only author listed, compared to 65% of the time when a relatively unknown investigator was the only author listed (34).

Smith wrote that peer review is "slow, expensive, ineffective, something of a lottery, prone to bias and abuse, and hopeless at spotting errors and fraud" (31). Chance has been shown to play an important role in determining the outcome of peer review (35), and agreement between reviewers is disconcertingly low (27). Bauer has noted that as a field matures, "knowledge monopolies" and "research cartels," which fiercely protect their domains, suppress minority opinions, and curtail publication and funding of unorthodox viewpoints, are established (36). In response, experienced authors learn to negotiate reviewer hurdles by embracing conservatism and avoiding speculation, although some have complained that this response has the effect of "dumbing down" the scientific literature (37). Journals continue to experiment with alternative peer review models to remedy perceived shortcomings: PNAS has multiple tracks for manuscript submission, *Nature* undertook a brief trial of open review, and the elimination of reviewer anonymity has been discussed extensively (30). A prospective randomized study of blinding reviewers to author names or the origin of the manuscript revealed no effect on the detection of errors (38). However, blinding reviewers to author names can affect the tenor of the review since reviewers tend to be more favorable about a study when they know the identity of the authors (39).

The journal *PLoS One* reviews manuscripts for methodological soundness but not for perceived significance to the field, a judgment that is left to the readers. *PLoS One* also provides readers with the option of rating the papers and appending comments (40). The *EMBO Journal* has created a transparent editorial process in which communications to and from the editors along with the text of reviews are available to readers (41). Many other journals have endorsed what is being called "open peer review," in which both reviews and the identity of peer reviewers are published along with papers, although attitudes regarding open peer review vary greatly among researchers (42). Some fear that this increases the burden on reviewers, making it more difficult for editors to recruit qualified reviewers, and concerns about compromised reviewer candor have also been raised, particularly when unequal power relationships exist between authors and reviewers (43). In 2022, the journal *eLife* announced that it would no longer use reviewer assessments to decide whether a paper would be accepted or rejected. Rather, any paper approved by an editor for in-depth review will be provisionally accepted and posted as a preprint that is available for review by any interested readers as well as by invited reviewers (44). The change has stimulated vigorous debate (45). Although each of these models has potential advantages, no model that is clearly superior to the current system has yet emerged.

Returning to the questions of censorship, it is self-evident how foibles in peer review can create a major problem with scientific acceptance, for peer reviewers are the major gatekeepers for the printed word (32). Proponents of HIV denial or intelligent design like to compare scientific peer review to censorship (46–48). But the truth is that the scientific community has provided ample opportunity for these ideas to be publicly aired, arguably more than they deserve, and ultimately rejected. That is not censorship. Misrepresenting these discredited ideas as victims of censorship risks minimizing the true threats of scientific censorship, such as when government officials delete politically sensitive remarks by scientific agency heads (49) and surgeons general (50), alter reports by government scientists (51), or prohibit the publication of sensitive data (52).

Publishing in peer-reviewed journals remains the major mechanism for the dissemination of scientific knowledge. In recent years, preprints have emerged as a new mechanism for the dissemination of scientific results, but with an important caveat that preprints are made available before they have been peer-reviewed (Box 22.1). The peer review of scientific manuscripts is clearly distinct from these examples of censorship. However, if reviewers prevent authors from any discussion of controversial or speculative viewpoints or if editors are overzealous in screening manuscripts for perceived newsworthiness or consistency with prevailing dogma, there is a danger of blurring the distinction between peer review and censorship. If a reviewer obstructs the publication of a manuscript because it competes with or questions his or her own work, there is an ethical dimension as well. As journal

---

Box 22.1 Preprints and rejected science

A preprint is a complete version of a research paper that is shared publicly before it has been peer-reviewed and published. Preprints were long used in physics to communicate research results and obtain feedback but were only widely adopted in the biological sciences in the late 2010s (59). The use of preprints in the United States and United Kingdom got a tremendous boost once the NIH and Wellcome Trust allowed researchers to cite them in grant applications. Citing a preprint is preferable to merely inserting preliminary data in a grant application because a citation takes no space and allows the information to be fully developed in a research paper. For time-sensitive information, preprints also offer the opportunity for researchers to get their message out without the considerable delays imposed by the peer review process. This has proven to be invaluable during the COVID-19 pandemic, which has solidified the value of preprints for the scientific community (60, 61). Preprints also provide an opportunity to stake priority claims in competitive situations and allow researchers to make their results public even if their paper has been rejected, although this must be balanced against the potential for other researchers to "scoop" preprint authors by rushing their competing papers into publication (62). This concern may be mitigated by the recognition of preprints in priority claims and the "scoop protection" policies increasingly offered by journals (63).

editors, we are often privy to a kind of grammatical "courtship ritual" as authors attempt to maneuver their views past the intellectual hurdles imposed by reviewers. The analogy to a courtship ritual is fitting if one considers that a successful ritual results in the birth of a scientific paper. The typical struggle involves disagreements over significance, with major battles centering on words like "indicates," "suggests," "demonstrates," "is consistent with," "establishes," and "proposes." The complexity of the English language, with its 500,000-plus words, provides a rich resource for compromise. However, the effort spent in linguistic negotiations raises the questions of whether such effort is necessary and might even represent a subtle form of censorship. Reviewers should not try to rewrite papers to fit their own biases. It is one thing to insist that conclusions are supported by evidence. However, some latitude is appropriately given to authors for extrapolation and even speculation. Even more importantly, excessive influence by reviewers can stifle legitimate scientific debate and encourage conformity (26).

The overwhelming majority of rejected papers are eventually published in other journals (Box 22.2). However, many authors of rejected papers fail to address the

---

Box 22.2 The fate of rejected papers

Authors of rejected papers can take solace in the knowledge that most papers that are initially rejected by a journal are subsequently published in other journals. In 2000, the *Annals of Internal Medicine* reported that 69% of the papers that it had rejected were eventually published in other journals after an average interval of 18 months. This delay presumably reflects the time required to identify another journal, obtain peer review, make revisions, and complete processing of the paper by the journal (64). Although 31% of rejected papers were not found among subsequent publications, this does not necessarily mean that the data were not published, since researchers may rewrite papers and modify titles after initial rejection. In the intervening two decades, subsequent studies have confirmed that most rejected papers are eventually published. In 2007, the *American Journal of Neuroradiology* reported that 56% of rejected manuscripts were eventually published in 115 different journals after 16 months (65). Interestingly, a follow-up analysis of those publications reported that they had lower citation rates than papers accepted by the journal, suggesting that the original editorial decision was predictive of impact (66). However, this interpretation must be made with caution, as journal reputation may affect citation rates,

(Continued)

Box 22.2 (Continued)

independent of article quality. In recent years, similar findings have been reported by *Clinical Otolaryngology* (67), *Anesthesia* (68), and *Intensive Care Medicine* (69), which reported publication success of 55%, 67%, and 44% for papers rejected from their journals, respectively. We acknowledge that these reports are all from clinical journals, but our personal experience leads us to believe that similar outcomes would be found for basic science papers. The eventual publication success rate for initially rejected papers should encourage authors in most cases to take their rejected manuscripts, make every effort to improve them, submit them to other journals, and persist until the work is published. While pursuing publication, authors also have the option of making their data public through the preprint mechanism (Box 22.1). Data in notebooks or unpublished manuscripts are not useful to the scientific community. Once work has been completed, it behooves researchers to ensure that the knowledge gained is made available to the community, where it may eventually prove to be useful in unexpected ways.

---

original criticisms, probably thinking that these are specific to the journal that rejected them. In this regard, a study showed that only about a third of the substantive criticisms were addressed, with the majority of changes being in the abstract and title (53). This represents many missed opportunities by authors to improve their work and raises questions about the effectiveness of peer review as a mechanism to improve paper quality.

In summary, peer review is very different from censorship, but editors need to be careful to maintain the distinction. A respect for the wisdom of age requires us to give Galileo the final word here: "Long experience has taught me this about the status of mankind with regard to matters requiring thought: the less people know and understand about them, the more positively they attempt to argue concerning them, while on the other hand to know and understand a multitude of things renders men cautious in passing judgment upon anything new" (54).

Returning to the subject of rejection, it is human nature to set the bar lower for our own data than for someone else's, hence the aggravation of jumping through various hoops set by reviewers before a manuscript can be published. Moreover, reviewers are human—mistakes are made. The journal *Nature* still expresses regret over forcing Hans Adolf Krebs to publish his discovery of the tricarboxylic acid cycle somewhere else (55). An entire website has been devoted to the rejections endured by Nobel laureates (56), although we would hasten to add that lots of poor-quality work gets rejected too. The appropriate response to reviewers, though not always the first one that comes to mind, is to patiently address critiques whether they seem well informed or not. And let us not forget that authors are human. In our personal experience, specious criticism does not sting nearly as much as critiques that are right on target. Those extra experiments insisted upon by reviewers often turn out to provide valuable corroboration and occasionally even spare an author from committing embarrassing mistakes permanently into print.

The most common reasons for rejection of a paper are scope, the lack of a hypothesis (see chapters 2 and 3), insufficient novelty, low priority or impact on the field (see chapters 8 and 26), lack of real-life relevance, improper study design, inadequate sample size or statistical analysis, suboptimal data collection, poor writing or presentation, a failure to adequately support the conclusions, or insufficient rigor (see chapter 6). Many of these deficiencies can be corrected, and editorial rejection often simply means that the journal is not a good fit for the paper. There is a learning curve to performing research, and disappointing assessments by editors and reviewers should be taken as opportunities for improvement, not only for the paper in hand but also for future research and papers yet to be written.

Failure will always be a part of science (Box 22.3), and rejection will always be tough to take. But until the skeptics are convinced, the author's job is not done. Reviewers make mistakes, to be sure, but they are trying to do an essential, difficult, and generally thankless job. Most of the time, the manuscript

---

Box 22.3 The importance of failure in science

The history of science records its successes, but the lab notebooks of scientists record countless failures. As the historian Ann-Sophie Barwich has observed: "Science fails. It seems to fail at a high rate and with regularity. Experiments go wrong, measurements do not deliver the anticipated results, probes are contaminated, models are misleadingly simplistic or not representative, and some inappropriately applied techniques lead to false positives. One may wonder why science is so successful despite such prevalent failures" (70). The physicist and Nobel laureate C. V. Raman is said to have remarked, "If I never fail, how will I ever learn?" (71). Oswald Avery, one of the earliest molecular biologists, who discovered that DNA is responsible for genetic inheritance, frequently remarked, "Disappointment is my daily bread. ... I thrive on it" (72).

But the knowledge that all scientists experience disappointment and failure doesn't make it any less painful (73, 74). The challenge for scientists in training is to learn to become motivated rather than deflated by failure. Students who are beaten down by failure may find themselves spending less and less

(Continued)

Box 22.3 (Continued)

Oswald Avery in 1937, thriving on disappointment. Reproduced with permission of the Rockefeller Archive Center.

time in the lab and daydreaming about their upcoming vacation. Those who are destined to become successful scientists are energized by experiments that don't work or ideas that don't pan out, responding with new resolve, new ideas, and an irrational optimism that things will surely go better the next time.

review system works. As Winston Churchill once observed about democracy, peer review "is the worst [system] … except all the others that have been tried." Authors of rejected manuscripts can take solace in the fact that the overwhelming majority of rejected papers are eventually published somewhere (Box 22.2). How can we keep the system working? Reviewers can strive to provide reviews that they themselves would be willing to receive (57). This may be particularly challenging when one's own work has been recently rejected (58), but the Golden Rule remains a good principle in reviewing, as in other aspects of life. Even when a decision is made to reject a manuscript, reviews should be respectful, constructive, and reasonable, focusing on issues that are truly substantive. Authors, for their part, should carefully consider critiques before firing back injudiciously. For the rebuttal accompanying a revised manuscript, the author should take the time to respond point by point to each concern. Reviewers' critical comments and suggestions for experiments may be disputed but should not be ignored. When possible, it is often the best course of action for authors to provide additional data that resolve uncertainty and satisfy reviewers' concerns. A collegial but rigorous engagement between reviewer and author is at the very heart of science itself.

23 Unfunded Science

Despite being widely used by the scientific community, very few efforts have been made to turn the scientific method on peer review itself. … The evidence suggests there may be significant problems in the current way that research decisions are being made.

Susan Guthrie, RAND Europe (1)

Find two scientists together, and chances are good they are complaining about grants. The research community has been in a sustained funding crisis, exacerbated by an earlier period of growth that created new funding commitments and recruited additional scientists to the workforce. For many years, resources in the system have been insufficient to support demands for research funds, and scientists have had to devote enormous time and effort into competing over the limited funds available. Robert Siliciano, a prominent virologist, testified to a U.S. congressional committee that 60% of his time was dedicated to seeking research funding (2). There are simply not enough resources for the number of scientific mouths to feed.

Today, the cost of doing science has increased so much that it is largely a government-supported enterprise. Cutting-edge physics experiments require construction of highly sophisticated and costly instruments such as the Large Hadron Collider (LHC), Laser Interferometer Gravitational-Wave Observatory (LIGO), National Ignition Facility, and the Hubble and Webb telescopes. Each

Thinking about Science: Good Science, Bad Science, and How to Make It Better, First Edition.
Ferric C. Fang and Arturo Casadevall.
© 2024 American Society for Microbiology.

of these instruments and facilities requires billions of dollars to build and operate with budgets outside the means of individuals and private foundations; hence they are reliant on governmental support. Much of the physics, chemistry, and biology basic science research in the United States is supported by the National Science Foundation (NSF), which had a budget of more than U.S. $10 billion in 2022. Similarly, the primary source of biomedical research funding in the United States is the National Institutes of Health (NIH), with an annual budget of more than U.S. $40 billion in 2022. Despite this seemingly generous support, both the physical and biomedical sciences are woefully underfunded relative to the number of meritorious project proposals.

The NIH-supported research enterprise consists of two groups: intramural researchers, housed in NIH facilities, and extramural investigators, who are mostly housed in universities, medical schools, institutes, and industry. The ratio of funds spent on the intramural and extramural programs is roughly 1 to 10. In both cases, the allocation of funds is made according to peer review, but the NIH uses very different mechanisms for assessing investigators in these programs. Intramural investigators are usually evaluated through retrospective peer review, where their recent accomplishments are used to make funding decisions, a mechanism similar to that used by the Howard Hughes Medical Institute. In contrast, funding allocations to the extramural program, which comprises the overwhelming majority of the NIH budget, is allocated by a mechanism of prospective peer review in which scientists must write grant proposals detailing future work that are reviewed and criticized by a panel of experts known as a study section. The difference in funding mechanisms used by the intramural and extramural programs is significant because it shows that there is already some flexibility in the approach used by the NIH to distribute its research dollars. In this chapter, we will focus primarily on the prospective peer review mechanism used to allocate funds to extramural investigators, but many other agencies employ a similar approach. The fundamentals of NIH extramural peer review have not changed in more than a half-century. The process involves writing a proposal that is reviewed by a panel of "peers" and assigned a priority score that is converted to a percentile ranking. The NIH then funds proposals depending on the amount of money available, with the "payline" being that percentile ranking up to which funding is possible. At the time the system was designed, paylines exceeded 50% of the grant applications received. However, more-recent decades have witnessed a precipitous drop in the proportion of grants that are funded. Today's paylines and success rates have hovered at historically low levels, as low as 10% in some institutes. Despite a drastic reduction in the likelihood of funding success, the essential features of NIH peer review and funding allocation have not changed.

Would a successful business organize its research and development department so that employees spend more than half their hours writing detailed 5-year plans, and then provide resources for only a small fraction of them, leaving the rest to languish? Would an army provide weapons for only one-fifth of its soldiers and ask the rest to prepare written applications describing what they would do if they had them? Of course not. Yet that has essentially been the status of the scientific enterprise. After adjustment for inflation, the NIH budget and national investment in research and development have been in a prolonged state of stagnation (3, 4). A greater emphasis on centrally defined research priorities in an era of declining budgets has had a particularly harsh impact on individual investigator-initiated research, the traditional engine of scientific progress. R01 grants awarded between 2000 and 2007 exhibited a precipitous 46% decline and have not recovered (5). The American research establishment has been facing the most prolonged funding crisis in its history. After a doubling in funding at the beginning of the 21st century, the budget of the NIH was flat from 2003 to 2015, translating into a 25% reduction in actual buying power after taking inflation and the increasing costs of research into account (6).

The ongoing funding imbalance has caused lasting harm to the nation's scientific enterprise, undermining both productivity and innovation and discouraging new scientists from entering the workforce. The crisis has come at a particularly inopportune time, as research has a crucially important role to play in addressing the world's most pressing challenges (7). For some scientists, their very jobs are at stake. This is because salary support for many American scientists is more dependent on grant revenues than in other countries (8). Additional casualties of the funding crisis are more difficult to measure but are nevertheless real: deteriorating morale and a perceptible decline in scientific collegiality and cooperation. As David Sarnoff once observed, "Competition brings out the best in products and the worst in people" (9).

PEER REVIEW REFORM

As grant success rates began to tank, the NIH undertook an initiative to reform its peer review process. Called "Enhancing Peer Review," the program was initiated in 2007 and subsequently modified in response to feedback from numerous scientists. Reform of the current peer review process was motivated by a concern about the enormous administrative burden of the review process (10). Other justifications included complaints about review quality, very low funding rates among new investigators, and the declining NIH budget. The fact that none of these concerns was new (11, 12) served as a warning that solutions would not come easily.

The stated goals of NIH peer review reforms were to reduce the administrative burden associated with the grant process, enhance review quality, and increase support for new or early-stage investigators. The central elements of the program included (i) new funding targets for early-stage investigators (within 10 years of completing a terminal research degree or medical residency), (ii) shortened applications (a reduction from 25 to 12 pages), (iii) a new 1- to 9-point scoring system with separate scores for individual criteria (impact, investigators, innovation, feasibility, and environment), (iv) limitation of grant applications to two submissions, and (v) incentives for long-term reviewers.

IMPACT OF NIH REFORMS

At the time, we commended the NIH for their efforts, as some of the improvements were long overdue (13). For example, abbreviating the length of grant applications reduced the workload for applicants and reviewers and lessened the emphasis on experimental minutiae. In fact, compelling arguments can be made for even shorter applications, particularly for established productive investigators (14), and many non-NIH grant applications are now less than 12 pages. The renewed emphasis on new and early-stage investigators was welcome news for scientists early in their careers, who had watched the average age of R01 recipients rise steadily. These individuals represent the future of science and warrant special consideration. However, the benefits of other changes were less clear. A 9-point scale decompressed the scoring system, but few reviewers make use of the entire range. Scoring itself remained a subjective process, and the new scale failed to mitigate the problems inherent in selecting the most meritorious projects when resources are so limited (15). The quality of reviews deteriorated as reviewers provided terse and sometimes cryptic statements to support their decisions, making revisions more difficult.

THE PROBLEMS WITH PEER REVIEW

For a system that determines the fate of scientific proposals, peer review is remarkably unscientific. Analyses have concluded that the NIH peer review system is statistically weak, imprecise, and prone to bias (16, 17). Problems with the review of applications to U.S. federal agencies have been apparent since the 1970s. A study of 150 proposals submitted to the NSF found that the likelihood that a grant would be funded was dependent largely on chance and reviewer assignment (18). More recently, an experimental study where 43 reviewers were asked to review 25 NIH applications using the same methods used by the agency found no agreement in how reviewers translated their opinions into numerical scores (19). In fact, the authors concluded that it "appeared that the outcome of the grant review depended more on the reviewer to whom the grant was assigned than the research proposed in the grant" (19). At its extremes, the error and variability in the review process become almost laughable. One of our colleagues witnessed an application to receive a perfect score of 1.0 when it was submitted as part of a program project application, but the identical application was unscored as a stand-alone R01 proposal. Perhaps the most remarkable aspect of the entire grant review process is that scientific agencies continue to distribute funds with little or no evidence that their system of evaluation can identify the best and most important science. Despite the billions of dollars spent by U.S. science agencies to fund science since World War II, very little investigation has been performed to examine the predictive accuracy of study section peer review. With more than a half-century of study section assessments on record, it would be interesting to know the frequency with which major scientific discoveries were recognized and anticipated by study sections. For example, what fraction of applications scored above or below the 10th percentile has been associated with major recognized scientific discoveries during the past 50 years? Similarly, what percentage of important scientific discoveries that were initially reviewed as proposals was rejected? Putting a stronger scientific foundation and accountability into peer review would enhance confidence in the system and facilitate evidence-driven improvements (20).

What is the desired product of scientific research? This question does not have a simple answer, but one measurable outcome is the generation of primary research publications, which are in turn cited by other publications. Remarkably, NIH study sections are unable to accurately predict which grant applications are likely to exhibit the highest publication productivity. Although an analysis of more than 130,000 NIH-funded grant applications suggested a correlation between percentile scores and productivity (21), those findings contrasted with earlier studies showing poor predictive power for grant application peer review. Consequently, we reanalyzed the subset of the data for the grants awarded scores in the 20th percentile or better and found that the predictive ability of peer review was scarcely superior to what would be achieved by random chance and that differences in the median productivity exhibited by grants with high or low scores within this range were trivial (22). Our results corroborated earlier studies of more than 400 competing renewal R01 applications at the National Institute of General Medical Sciences (23) and 1,492 R01 applications at the National Heart, Lung, and Blood Institute (24). Hence, available evidence makes a powerful case that the primary mechanism for biomedical research funding allocation in the United States is inadequate for prioritizing which applications to fund. The aforementioned analyses were preceded by studies suggesting that the NIH peer review process lacks statistical rigor. Only two or three reviewers in a typical study section carefully read an individual grant application and provide comments, and this reviewer sample size is too small to provide an acceptable level of precision (17). This criticism is not unique to the NIH, as studies from many countries have identified problems with the precision of grant peer review. In Canada, Nancy Mayo and her colleagues found that the use of only 2 primary reviewers results in considerable randomness in funding decisions that could be improved by involving an entire 11-member review panel in the assessment of each application (25). Nicholas Graves and colleagues examined variability in scores for the National Health and Medical Research Council of Australia and concluded that 59% of funded grants could miss funding simply based on random variability in scoring (26). An analysis of applications to the Australian Research Council found interrater reliability for reviews to be poor (27), and researchers in Finland did not find that the reliability of grant peer review is improved by panel discussions (28). A French study observed that individual reviewers do not even tend to exhibit agreement on the weighting of criteria used for the grant review process (29).

A central weakness in the current system may be that experts are being asked to confidently predict the future of a scientific project, an inherently uncertain proposition. In this regard, the University of Pennsylvania psychologist Philip Tetlock showed that experts not only fare poorly in attempting to predict the future but also overrate their own abilities to do so (30). Another question is whether publication productivity is even the best metric on which to judge scientific success. Are study sections able to recognize potentially transformative research? Probably not, because intense competition for funding encourages both reviewers and applicants to be more cautious. The very structure of the NIH peer review system may encourage conformity and discourage innovation (31) of the type that could lead to scientific revolutions (see chapter 11) (32). As Nobel laureate Roger Kornberg once observed, "In the present climate especially, the funding decisions are ultra-conservative. If the work that you propose to do isn't virtually certain of success, then it won't be funded. And of course, the kind of work that we would most like to see take place, which is groundbreaking and innovative, lies at the other extreme" (33). The NIH recognized this problem and created the Transformative Research Award Program, but of course, this failed to solve the problem that transformative breakthroughs are often only evident as such after the fact (34) (chapter 11) (Box 23.1).

There is also the critically important issue of bias, which comes from many factors (Table 23.1). Sources of potential bias in peer review include cronyism and preference or disfavor for particular research areas, institutions, individual

---

Box 23.1 Transformative research that almost wasn't

Katalin Karikó has been justly celebrated for her seminal contributions to mRNA vaccine development, now credited with saving millions of lives during the COVID-19 pandemic. Her recent honors include the Vilcek Prize, the Princess of Asturias Award, a Breakthrough Prize, and the Lasker-DeBakey Clinical Medical Research Award. Many expect the Nobel Prize to be next. But Karikó repeatedly failed to obtain research grants to pursue her mRNA research in the United States, which ultimately contributed to her decision to join a biotech company in Germany (120). Were it not for her determination in the face of rejection and disappointment, the world might be in a very different place today. What other transformative research might we be missing because of our flawed funding strategy (121)?

Katalin Karikó. Photo credit: Szegedi Tudományegyetem, under license CC BY-SA 4.0.

---

scientists, gender, or professional status. Reviewer bias can potentially have a major effect on the course of science and the career success of individual applicants. One meta-analysis of peer review studies found evidence of gender bias, such that women were approximately 7% less likely to obtain funding than men (35). Studies focusing specifically on the NIH have found comparable success in men and women submitting new R01 applications but lower success rates for women submitting renewal applications (36). There is also a continuing concern about racial bias in NIH peer review outcomes. Despite a number of initiatives taken in response to a study showing that Black applicants were significantly less likely to be awarded NIH funding after controlling for educational background,

TABLE 23.1 Potential sources of bias in grant application peer review

Bias	Reference(s)
Race	37–40
Gender	35, 36, 110, 111
Institution/institutional size	41, 112
Geographic location	113
Investigator experience	114
Diverse research teams	115
Research field	12
Interdisciplinary research	116
Novelty	117, 118
Conflicts of interest	12, 119

country of origin, training, previous awards, publication record, and employer characteristics (37, 38), as yet there is no evidence that the racial gap in funding success has improved (39). In fact, a 2021 study again reinforced the notion that African American grant applicants experience bias in the review process that reduces their chance of being funded (40). NIH peer reviewers tend to give better scores to applications closer to their area of expertise, and several studies have suggested that reviewers are influenced by direct or indirect personal relationships with an applicant (12). Although the NIH also continues to tinker with the scoring criteria for grant applications in the hope of reducing institutional bias, its own internal analyses have found that about two-thirds of research funding goes to just 10% of the applicant institutions (41).

The influence of grant reviewers in determining the fate of an application is directly proportional to the payline (i.e., the lower the payline, the greater the reviewers' influence). This is highly germane to the current system, for it makes single individuals disproportionately powerful in their ability to influence the outcome of peer review. When generous paylines are available, applicants are likely to succeed even if there are scientific disagreements among applicants and reviewers. However, with shrinking paylines, a negative assessment by a single individual is often sufficient to derail a proposal. In this environment, a few individuals can profoundly influence the direction of research in an entire field. Reviewers are typically appointed for four-year terms, allowing them to influence their fields for extended periods of time. A Bayesian hierarchical statistical model applied to 18,959 R01 proposals scored by 14,041 reviewers found substantial evidence of reviewer bias that was estimated to impact approximately 25% of funding decisions (42). Theodore E. Day performed a computer simulation of peer review and found that very small amounts of bias could skew funding rates (42). "Targeting" research on the basis of program priorities can exacerbate the problem of bias and perversely lead to missed opportunities in basic research. The history of science is filled with stories of landmark discoveries by scientists who were looking for something else entirely (see chapter 14)—a third of anticancer drugs have been found by serendipity rather than by targeted cancer drug discovery research (43). Yet funding agencies continue to attempt to target research funding to perceived priority areas, while support for undirected investigator-initiated projects has declined sharply (5).

Both applicants and reviewers have adapted to the funding crisis in ways that may be counterproductive to science. Applicants have responded by writing more grant applications, which takes time away from their research. As most applications are not funded, this largely represents futile effort. As previously alluded to, some scientists estimate that half or more of their professional time is spent in seeking funding (2). Pressures to obtain funding can incentivize applicants to engage in questionable research practices and unethical behavior, including making unsubstantiated claims, withholding crucial information, and requesting funding for work already completed (44). On the other side, reviewers are asked to decide between seemingly equally meritorious applications and may respond by prioritizing them on the basis of "grantsmanship" (45), the ability to put together a polished application that checks off all the boxes, which makes a grant application easier to read but has never been shown to correlate with research productivity or innovation. One of the most controversial aspects of NIH grant policy was a decision to limit applicants to two submissions of a research proposal (46, 47). Under this policy, at a time when paylines were as low as 6%, many projects deemed meritorious by study sections were not only rejected but prohibited from resubmission for 37 months. With the rapid pace of science, this led to the death of many perfectly good ideas. Although this policy was eventually rescinded to allow applicants to resubmit their projects as new grants (48), substantial damage was done.

Peer review is used in both the ranking of grant applications and the evaluation of scientific papers. However, there are significant differences in how peer review of grant applications and papers operates. For grant applications, reviewers are chosen by an administrator who may or may not have in-depth knowledge of the relevant field, and review panels do not necessarily include the expertise necessary to review all proposals. For papers, reviewers are chosen by an editor who usually has expertise in the subject matter and can select reviewers with specific expertise in the subject area. Hence, a major difference between study section and manuscript peer review is that the latter is more likely to achieve a match between subject matter and expertise. Accordingly, grant review is a more capricious process than manuscript review, and a single rogue reviewer can sink an application by assigning low scores without even needing to provide a convincing rationale for those scores. Publication decisions are made by editors, who can directly discuss areas of disagreement with authors and overrule single negative reviews at their discretion. Furthermore, authors have the option to appeal rejection decisions or submit their work to another journal. In contrast, there is no process for negotiation with scientific review administrators and little or no alternative to NIH funding. Another major difference is that the negative consequences of peer review differ for manuscript and grant applications, since the former usually find another publishing venue (see chapter 22), whereas a denied grant application means that the proposed work cannot be done. Therefore, peer review of grant applications is of much greater importance for science than peer review of scientific manuscripts.

A critical aspect of the current crisis is that success rates for grant applications have fallen by more than two-thirds since the 1960s (49), and yet the system for funding allocation has essentially remained the same. A 2015 survey of researchers submitting proposals to the National Aeronautics and Space Administration (NASA), the NIH, and the NSF showed that even highly productive researchers faced a 50% likelihood of not obtaining funding in the current cycle, resulting in the defunding of one-eighth of active programs following three such cycles (50). The authors of this survey estimated that at current funding rates, 78% of applicants would be unable to obtain federal funding for their research. This raises two obvious questions: (i) why has the system remained the same, and (ii) why do scientists persist in this low-yield activity? Although we are not privy to discussions and decisions that have occurred among government leaders, it seems likely that the system has remained the same in the hope that national funding allocations will improve and because of the inertia involved in changing a mechanism that has seemed to work relatively well for decades. As to why scientists persist in trying, the literature on the psychology of gambling behavior may provide some clues. People feeling desperate about their prospects will purchase lottery tickets as a surrogate for hope (51). Desperation is certainly prevalent in today's scientific community (52). Entrapment in a system due to a previous investment of time and resources is also commonly invoked as an explanation for gambling (53), and many scientists have difficulty envisaging an alternative career path. In fact, current trends in science demand so much specialization (54) that most scientists are unable to shift into fields where funding may be more plentiful. Intelligence and a high level of executive function, as seen in most scientists, are correlated with susceptibility to maladaptive decision-making and the "gambler's fallacy" (55). Risk-taking behavior may even have a neurological basis. Optimism has been described as a *sine qua non* for scientists (56), and irrational optimism correlates with reduced tracking of estimation errors by the right inferior prefrontal gyrus of the cerebral cortex (57). Finally, there is unfortunately an obvious fact that not all scientists persist in seeking funding. Many leave academia or abandon research entirely.

We identify the following critical issues as persistent problems in governmental grant applications:

1. **A persistent imbalance between resources and applicants.** Problems with peer review are more evident during times of scarce funding. There is no escape from the relentless math of a growing number of applications chasing a shrinking number of grant dollars. When available funds do not increase significantly, diversion of funds to early-stage investigators can make things worse by heightening competition among senior investigators and reducing available resources for scientists during what should otherwise be their period of peak productivity (58). In addition to their direct scientific contributions, senior investigators represent an invaluable source of wisdom, guidance, and inspiration for younger scientists. Furthermore, as new investigators become established (a transition that occurs rapidly in science), the shift in funding priorities that allowed their early success will soon disappear and leave them to compete with other senior investigators for a shrinking pool of resources. This raises the uncomfortable question of whether we should be training more scientists when there is already a shortage of support for those already trained. Clearly, there is a need for recruiting new investigators to maintain the pool of scientists and to provide fresh outlooks on scientific problems. However, the strategy for ensuring the success of new investigators needs to be considered carefully. Short-term increases in support for trainees and new investigators will only exacerbate the dearth of grant support and job opportunities as these trainees progress in their careers, unless the total level of support for research is substantially increased.

2. **Disincentives for novelty.** Reviewer biases favor topics well understood and appreciated by the study section and disfavor less conventional ideas or understudied topics. This leads to greater homogeneity in science. Applicants learn to write conservative proposals to avoid creating targets for reviewers. The playful curiosity and open-ended thinking that characterize the best science (33, 59) have become increasingly rare as scientists are driven by funding anxiety to propose safe, conservative, short-term projects. (Of course, conservatism is no guarantee of success in the peer review arena; reviewers may then characterize such a proposal as "unambitious" instead of "risky.") There is a basic distinction between the peer review of manuscripts, in which the work has been done, and the peer review of grants, in which the work is being proposed. The difference is as fundamental as reviewing a movie versus forecasting the weather. Scientists have a limited ability

to predict *a priori* which experimental paths will be most fruitful. Therefore, grant reviews at best involve probability and uncertainty. If only projects that are certain to succeed are given support, then it becomes a virtual certainty that many worthwhile projects will fail to receive support. One could argue that most projects in which one can predict success with certainty are, by their very nature, unlikely to be the type of highly innovative science that leads to major breakthroughs, since certainty in prediction means that one is operating within existing boundaries of knowledge. Study sections normally base their decisions on consensus and thus function within the sphere of what Thomas Kuhn called "normal science" (see chapter 11) (32), which discourages innovative deviations from established paradigms. The drive toward conformity ignores the essential role of serendipity in science (60) (see chapter 14)—unexpected results are often the most exciting and fruitful ones. The pressure to eschew innovation will not be alleviated by a few Transformative Research, Eureka, New Innovator, or Pioneer awards. What is actually needed is not separate funding earmarks for high-risk research but rather sufficient breathing room to accommodate a few high-risk ideas as an appropriate feature of any research project.

3. **Variable reviewer expertise.** Another important difference between peer review of manuscripts and grants is the expertise of the reviewers. Journal editors can consult any scientist in the world about a manuscript, but an NIH grant review is essentially limited to the expertise of the 20 to 30 or so scientists in the room, who may each be outstanding in their field but cannot hope to encompass the entire range of applications considered. Some of the most active scientists can rarely participate in study sections because they are submitting their own applications virtually every grant cycle. Grants are typically reviewed by at least some individuals with substantially less expertise in a given subject than the applicant, and such reviews tend to focus on critiquing "grantsmanship" rather than the science itself. The grant application, originally intended only as a tool to facilitate the distribution of research funds, has thereby acquired an undeserved status as an object of obsession. When there is insufficient funding available for the number of quality proposals submitted, funding decisions become increasingly capricious. Decisions to deny funding must be justified, so critiques become packed with unhelpful and generic demands for more preliminary data, expected results, anticipated pitfalls, alternative approaches, and timelines, etc. Such criticisms may not be the true reason for denying funding, but they force applicants and reviewers to prolong the dance of revision and re-review. There has always been a disconnection between the science that is described in grants and that which is performed in reality, but today the chasm can be absurdly wide.

4. **Disparities in funding among investigators.** What is the optimal amount of funding for a single investigator? This is a fundamental question for which there

is no answer. Clearly, research is very difficult, if not impossible, in the absence of funding. At the other extreme, one can imagine that efficiency declines as groups become very large and the efforts of a single investigator become diffused. This problem might be studied using available economic tools and the results used to optimize the allocation of funds, but this issue, like peer review, should be the subject of rigorous scientific analysis. Study sections should perhaps consider not only the productivity of an applicant but the productivity relative to dollars awarded, with the caveat that the number of publications alone should not be the sole parameter considered (61). In fact, it might be argued that new tools should be developed to help study sections objectively gauge the productivity of investigators and the potential impact of proposed research. While the awarding of grants on the basis of productivity seems reasonable in principle, it should be recognized that this creates a positive feedback loop that can aggravate the already highly inhomogeneous distribution of research funding. *Nature* reported that 200 scientists had 6 or more grants from the NIH in 2007 and that just 19 researchers accounted for 165 research grants totaling $160 million (62). Expensive "big science" technology-driven projects and large clinical trials are important contributors to this trend. The success of a fortunate few prominent scientists in the current funding environment cannot conceal the funding woes of the majority, in a mirroring of the widened gap between the haves and have-nots in the economy at large. The NIH peer review reform initiative considered a variety of suggestions to address the maldistribution of grants, such as requiring principal investigators to spend at least 20% effort on each grant, but failed to arrive at a consensus. At the very least, it would seem important to study the relationship between funding amount and productivity for research laboratories to determine the optimal amount of funding to maximize efficiency. A study conducted by the National Institute of General Medical Sciences found that laboratory productivity tends to plateau at annual funding levels over $750,000, consistent with the idea that moderation in lab size, as in most things, may be best (63).

5. **Lengthy review process.** With regard to federal agency grant applications, investigators must wait months to receive a critique of their application. Such delays may represent purposeful inefficiency, so that revised applications cannot be quickly resubmitted. Just as manuscript reviews are performed in a few weeks and immediately made available to authors, there is no reason that grant critiques cannot be provided to applicants within a similar time frame.

6. **Administrative burden.** In recent years, the scarcity of research funding has been paralleled by burgeoning administrative burdens. It was estimated that 42% of the total faculty research time is consumed by administrative activities required for compliance with research grants, such as progress reports,

accounting, animal protocols, human protocols, and select agent regulations, etc. (64). An Australian study documented that applicants for funding spent 38 days working on grant applications even after an administrative "streamlining" process, which in aggregate represented 614 wasted years of scientist time (65). The hours spent filling out grant applications accomplish nothing for society or science and represent a colossal waste of time for highly trained individuals. Although scientists owe society full accountability on these important issues, the enormous energies spent on paperwork introduce friction into the scientific process that contributes to inefficiency and lower productivity. A better system would review only the science and then obtain evidence for regulatory compliance for fundable applications.

7. **Wasted time and human power.** Withholding funding will not make investigators more productive, nor will it be easy for unfunded investigators to come up with the additional preliminary data needed to make an amended proposal more persuasive. Once an investigator struggles through a few years of inadequate funding, it becomes less and less likely that their research program will ever return to a competitive level. The system as it exists is exceedingly wasteful in that we are not allowing many highly trained and competent scientists to work to their full potential, and some are unable to work at all. The increased diversion of scientists' efforts into grant procurement rather than research, along with the temptation to pursue funds earmarked by Congress for specific programs rather than continuing the natural lines of investigation initiated by the individual investigator, results in reduced productivity. In response to a news feature in *Science* entitled "U.S. Output Flattens and NSF Wonders Why" (66), John Moore replied, "The number of papers that are written is diminishing because scientists are able to spend less time writing papers! Instead, we spend ever-more time on … writing, rewriting, and re-rewriting grant applications as the NIH's pay line drops to catastrophically low levels" (67). The funding shortage continues to undermine recruitment of the best and brightest young minds to research. The costs will be difficult to measure—it is hard to quantify the discoveries not made and the great scientists who never were.

THE CASE FOR A MODIFIED LOTTERY FOR FUNDING GRANT APPLICATIONS

With funding remaining scarce and the ability of peer review to distinguish among meritorious grants questionable, we have suggested that the NIH distribute research funding through a modified lottery system (68). Systematic studies have shown that NIH grant peer review fails in its primary goal of stratifying meritorious applications when it comes to predicting the primary research outcome of citation metrics. In fact, we found that scientists cannot predict the subsequent productivity of funded proposals for the top 20% of ranked proposals (22). But despite such data, the CSR (NIH Center for Scientific Review) has continued to defend its methods (69). As already discussed, attempted reforms in NIH peer review have failed to address the inherent unfairness of the system (13). In addition, the NIH spends a lot of money on grant peer review. The annual budget of the CSR exceeds $100 million, which pays for more than 24,000 scientists reviewing approximately 75,000 applications and attending approximately 2,500 panel meetings (69). The costs are not only economic. Writing and reviewing grants are extremely time-consuming and divert the efforts of scientists away from doing science itself. Specifically, the NIH is asking scientists who perform peer review to perform the impossible, i.e., to discriminate among the best proposals, an endeavor that results in arbitrary decisions, causes psychological stress for both reviewers and applicants, and doesn't even necessarily support the most important science. Recognizing the flaws in the current grant funding process, some scientists have suggested alternative approaches that would represent a radical departure from the present peer review system. Johan Bollen has suggested having scientists vote on who deserves funding (70). Michele Pagano has recommended basing funding for established scientists on track record and a one-page summary of their plans (14). In fact, this approach has some empirical support, as prior publication productivity has been shown to correlate with future productivity of R01 grant recipients (71), but would need an accommodation for young investigators to succeed, as they have not yet established a track record. John Ioannidis has proposed several options ranging from awarding small amounts of funding to all applicants to assigning grants randomly or basing awards on an applicant's publication record (72). We have proposed that the NIH adopt a hybrid approach based on a modified lottery system (68, 73, 74) (Fig. 23.1).

The debate over the optimal strategy for allocating funds for scientific research has interesting parallels with the decisions involved in making financial investments. In 1973, the economist Burton Malkiel published his now-classic book, *A Random Walk down Wall Street* (75). Malkiel argued that investors cannot consistently outperform stock market averages and, therefore, a passive investment strategy can be just as effective as an active one. In fact, very few professional investors consistently outperform the market. A study called "Does Past Performance Matter?" by S&P Dow Jones found that only 2 out of 2,862 funds were able to remain in the top quarter over five successive years, worse than might be predicted by random chance alone—"If all of the managers of these mutual funds hadn't bothered to try to pick stocks at all—if they had merely flipped coins—they would, as a group,

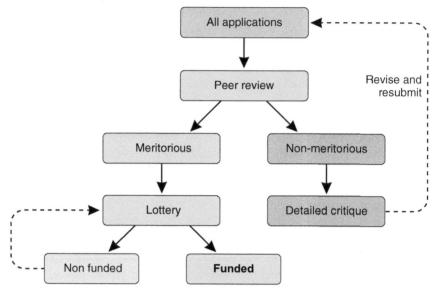

FIGURE 23.1 *Proposed scheme for a modified funding lottery. In stage 1, applications are peer-reviewed to determine those that are meritorious or non-meritorious on the basis of conventional peer review. Non-meritorious applications may be revised and resubmitted. In stage 2, meritorious applications are randomized by computer and funding is awarded to as many applications as funds permit on the basis of randomly generated priority scores. Percentages or total grants placed in the meritorious and non-meritorious categories could be adjusted by the amount of funding available. Similarly, the number of times that a meritorious grant can be reentered in the lottery could be decided administratively.*

probably have produced better numbers" (76). Even Warren Buffett has instructed in his will to "Put 10% in short-term government bonds and 90% in a very low-cost index fund. ... I believe the long-term results from this policy will be superior to those attained by most investors—whether pension funds, institutions, or individuals—who employ high-fee managers" (77). In 2007, the statistician Nassim Nicholas Taleb published the acclaimed book *The Black Swan* (78), which argued that the most influential events were both highly improbable and unpredictable. According to Taleb, investors should not attempt to predict such events but instead should construct a system that is sufficiently robust to withstand negative events and maximize the opportunity to benefit from positive ones. Applied to science, this suggests that it may be futile for reviewers to attempt to predict which grant applications will produce unanticipated transformational discoveries. In this regard, our review of revolutionary science in chapter 11 suggests that historical scientific revolutions lack a common structure, with transformative discoveries emerging from puzzle-solving, serendipity, inspiration, or a convergence of disparate observations (34). Consequently, a random strategy that distributes funding as broadly as possible may maximize the likelihood that such discoveries will occur. Taleb underscores the limits of human knowledge and cautions against relying on the authority of experts, emphasizing that explanations for phenomena are often possible only with hindsight, whereas people consistently fail in their attempts to accurately predict the future.

Four European economists have raised the question "Given incomplete knowledge of the market, is a random strategy as good as a targeted one?" (79, 80). A computer simulation was performed using data from British, Italian, German, and American stock indices. The authors compared four different conventional investment strategies with a random approach. Over the long run, each strategy performed similarly, but the random strategy turned out to be the least volatile, i.e., the least risky strategy with little compromise in performance. Given that assigning funds for investment or research allocation each involves a wager on future success with incomplete information, these lessons from the world of finance have relevance to science funding. Among the advantages of index funds are that randomization of the investment process can reduce "herding behavior" and financial "bubbles" (which raises the question of whether we are heading for precision medicine and genome editing "bubbles"—but that is a discussion for another time). An indexed strategy for picking stocks reduces the administrative costs associated with fund management, just as a modified lottery system for grant allocation could reduce the administrative costs of review.

As we consider reform proposals for grant peer review, it is important to state some basic principles that we believe are likely to be accepted by most scientists. First, we recognize that there are qualitative and quantitative differences among research proposals. Clearly, not all scientific projects are equally meritorious. We currently rely on the assessment of experts in the form of peer review to determine those differences. An ideal system would be a meritocracy that identified and funded the best science, but the available evidence suggests that the current process fails in this regard, and the goal might in fact be impossible. Second, we argue that some form of peer review will be required for funding allocation. Although we have catalogued many problems with the current peer review system, it is essential to have grant proposals evaluated by panels of scientists who have expertise in the area. Although experts may not be able to discriminate between meritorious proposals, they are still generally able to weed out proposals that are simply infeasible, are badly conceived, or fail to sufficiently advance science. Third, scarce research funds should be distributed in a fair and transparent manner. While fairness is likely to be partly in the eye of the beholder, there are mechanisms that are generally acknowledged to be fair. Specifically, there is a need to neutralize biases in funding decisions. Otherwise, the enormous power of reviewers at a time of unfavorable paylines will distort the course of science in certain fields. In this regard, there is evidence for increasing inefficiency in the translation of basic discovery into medical goods (81, 82). Although the causes for this phenomenon are undoubtedly complex, any bias in funding decisions affects the type of research done, which in turn influences potential downstream benefits for society. Should the review process favor new investigators? A case can certainly be made for the importance of providing support to new investigators, as they represent the future of science (83). This should not be taken to suggest that older investigators are unimportant. In fact, higher publication productivity has been seen for competing renewals than for new grants, and for projects directed by senior investigators (84), recognizing that established investigators have significant advantages relative to new investigators with regard to experience, prior productivity, reputation in the field, and laboratories that are already established and productive. In a world of plentiful research funds, new investigators can compete successfully for funding with established laboratories. However, in times of funding scarcity, differences between established and new investigators can become magnified to favor established investigators over new ones. Established investigators benefit from the so-called "Matthew effect," whereby those with resources and prestige are more likely to receive further rewards (85). Consequently, steps should be taken to improve the opportunities for new investigators as a matter of science planning policy. A modified lottery system could immediately benefit young investigators by creating a more level playing field.

Given overwhelming evidence that the current process of grant selection is neither fair nor efficient, we instead suggest a two-stage system in which (i) meritorious applications are identified by peer review and (ii) funding decisions are made on the basis of a computer-generated lottery (Fig. 23.1). The size of the meritorious pool could be adjusted according to the payline. For example, if the payline is 10%, then the size of the meritorious pool might be expected to include the top 20 to 30% of applications identified by peer review. This would eliminate or at least mitigate certain negative aspects of the current system, particularly bias (Table 23.1). Critiques would be issued only for grants that are considered non-meritorious, eliminating the need for face-to-face study section meetings to argue over rankings, which would bring about immediate cost savings. Remote review would allow more reviewers with relevant expertise to participate in the process, and greater numbers of reviewers would improve precision. Funding would be awarded to as many computer-selected meritorious applications as the research budget allows. Applications that are not chosen would become eligible for the next drawing in four months, but individual researchers would be permitted to enter only one application per drawing, which would reduce the need to revise currently meritorious applications that are not funded and free scientists to do more research instead of rewriting grant applications. New investigators could compete in a separate lottery with a higher payline to ensure that a specific portion of funding is dedicated to this group or could be given increased representation in the regular lottery to improve their chances of funding. Although the proposed system could bring some cost savings, we emphasize that the primary advantage of a modified lottery would be to make the system fairer by eliminating sources of bias (Box 23.2). The proposed system should improve research workforce diversity, as any woman or applicant from a historically underrepresented group (HUG) who submits a meritorious application will have an equal chance of being awarded funding. There would also be benefits for research institutions. A modified lottery would allow research institutions to make more-reliable financial forecasts, since the likelihood of future funding could be estimated from the percentage of their investigators whose applications qualify for the lottery. In the current system, administrators must deal with greater uncertainty, as funding decisions can be highly unpredictable. Furthermore, we note that program officers could still use

---

Box 23.2 Randomization as a mechanism for bias reduction

Several studies have shown that the likelihood of success or failure in grant review is dependent on random factors, such as reviewer assignment (19). Adding the effects of bias (Table 23.1) introduces additional unpredictability. The use of randomization in the modified lottery scheme (Fig. 23.1) can eliminate bias. Our analysis has shown that scientists cannot reliably discriminate between highly meritorious proposals to predict their eventual productivity (22). In our proposed scheme, we acknowledge that bias can still enter the process in the selection of meritorious and non-meritorious proposals, but this selection is not where ultimate funding decisions are made. The critical step where funding decisions are currently made is during the ranking of proposals, since the order of priority determines funding. Unfortunately, ranking is prone to both conscious and unconscious biases on the part of reviewers that can mean the difference between success and failure based on nonscientific criteria such as gender and race. In other words, the current system of funding allocation by ranking is already a lottery, but one that suffers from bias due to the lack of randomization.

selective pay mechanisms to fund individuals who consistently make the lottery but fail to receive funding or in the unlikely instance that important fields become underfunded due to the vagaries of luck.

The proposed system would treat new and competing renewal applications in the same manner. Historically, competing applications have enjoyed higher success rates than new applications, for reasons including that these applications are from established investigators with a track record of productivity. However, we find no compelling reason to justify supporting established programs over new programs.

Although we recognize that some scientists will cringe at the thought of allocating funds by lottery, the available evidence suggests that the system is already in essence a lottery without the benefits of being random (22). Furthermore, we note that lotteries are already used by society to make difficult decisions. Historically, a lottery was used in the draft for service in the armed forces. Today, lotteries are used to select students for charter schools (86), to determine the order of selection in the National Basketball Association draft, to issue green cards for permanent residency, and even to allocate scarce medical resources (87). Modified lotteries have been advocated as the fairest way in which to allocate vaccines and organs for transplantation (88, 89). If lotteries could be used to select those who served in Vietnam, they can certainly be used to choose proposals for funding. We are not the only ones to arrive at this idea (90).

The institution of a funding lottery would have many immediate advantages. First, it would maintain an important role for peer review at the front end, to decide which applications are technically sound enough to merit inclusion in the lottery. Second, it would convert the current system, with its biases and arbitrariness, into a more transparent process. Third, it would lessen the blow of grant rejection since it is easier to rationalize bad luck than to feel that one failed to make the cut due to a lack of merit. Fourth, it would relieve reviewers from having to stratify the top applications, since it is increasingly obvious that this is not possible. Fifth, meritorious but unfunded proposals could continue to have a shot at receiving funding in the future instead of being relegated to the dustbin. Sixth, it would be less expensive to administer, and some of the funds currently used for the futile exercise of ranking proposals could be devoted instead to supporting actual scientific research. Seventh, it should decrease cronyism and bias against women, members of HUGs, and new investigators. Eighth, it would give administrators in research institutions a greater capacity to make financial projections based on the percentage of their investigators who qualify for the lottery. Ninth, the system would be less noisy, would be fairer, and might promote new areas of investigation by removing favoritism for established fields that are better represented in review panels. Tenth, the realization that many meritorious projects remain unfunded might promote more-serious efforts to improve research funding and study alternative approaches to peer review. In fact, the success rate of the lottery would provide a clear number for society and politicians to understand the degree to which meritorious research proposals remain unfunded, and this would hopefully lead to an increased budgetary allocation for research and development. Under the current system, the underfunding of science is hidden by the fallacious mantra that the worthiest science continues to be funded, which provides an excuse for inaction. An NSF report found that 68% of applications were rated as meritorious but only a third were funded (91).

INCREASING INTEREST IN LOTTERIES FOR FUNDING ALLOCATION

Since our proposal for a modified lottery was first suggested in the *Wall Street Journal* in 2014 (73) and then presented in more detailed form in 2016 (68), there have been hopeful signs of movement toward the use of lotteries for funding allocation. The realization that the current system is broken has increased interest in alternative approaches and specific consideration of a modified lottery (92–95). Modified or partial lotteries have been endorsed by others (96–99), and in 2013, the Health Research Council of New Zealand became the first major funder to use a lottery mechanism to distribute research funds. A survey of investigators reported that 63% favored using a lottery to fund exploratory grants, with only 25% opposed (100). Opinions regarding lotteries for other types of grants were more evenly divided, with those who had succeeded in obtaining funding by lottery more favorably disposed to this approach (100). The Swiss National Science Foundation has instituted a system of drawing lots to discriminate among equally meritorious proposals (101). Similarly, Germany is piloting the use of randomization for the selection of grant proposals (102, 103). Beginning in 2023, the British Academy will trial a partial randomized allocation system (104). The Novo Nordisk Foundation has announced that it will be using a modified lottery to select recipients of its Project and Exploratory Interdisciplinary Synergy grants (105). Clearly, a randomization mechanism would be a huge change from the historic practice of having scientists rank proposals. As with any change, finding its advantages and disadvantages will take time, and we suggest that major agencies interested in using this mechanism pilot its use and obtain prospective data that can be rigorously analyzed to make better decisions. Similar suggestions and recommendations have been made by others (99).

CONCLUDING REMARKS

The biologist E. O. Wilson compared scientists to prospectors searching for gold (106): "In the 17th, 18th and 19th centuries, making scientific discoveries was like picking nuggets off the ground." But prospecting today is more challenging. The rewards are still great, but the big finds are more elusive. Targeted initiatives would direct all scientists to look for new lodes in the same place, while "transformative research" initiatives aim to fund only those who strike it rich. Neither strategy is optimal. Society must accept that science, as Ioannidis astutely observed, is an inherently "low-yield endeavor" (107). However, this low-yield endeavor has consistently improved the lot of humanity since the Scientific Revolution of the 17th century and remains humanity's best bet for finding solutions to deal with such challenges as climate change, pandemics and disease, a faltering green revolution, and the need for new energy sources (108, 109). To continue to reap the maximal benefits of scientific exploration, researchers must be encouraged to search as far and wide as possible, leaving no stone unturned, even though only some will be successful in their quests. As Taleb has written, "The reason markets work is because they allow people to be lucky, thanks to aggressive trial and error, not by giving rewards or incentives for skill" (78). We must provide scientists with an opportunity to get lucky.

Peer review remains a cornerstone of science and a true leveler of the playing field on which applicants compete. We recognize that it is imperative for the quality and fairness of peer review to be optimized during times of resource scarcity. However, the problem of inadequate resources cannot be compensated for by changes in the mechanism by which available funds are allocated. In fact, peer review cannot work effectively when funding is so limited (15). The reinvigoration of science in the United States and other countries with strong scientific communities will require bold action that considers the needs of the entire research workforce and not only the scientific elite. The foremost priority is to restore a balance between funding and applicants that returns paylines to reasonable levels. As the former NIH director Elias Zerhouni acknowledged, "Peer review doesn't need to be as stringently quality-focused when there is a lot of money" (6). The current system for awarding research grants is error-prone, vulnerable to bias, wasteful, and discourages innovation. America must put its scientists back to work making discoveries instead of endlessly writing grants.

24 Retracted Science

A man who has committed a mistake, and doesn't correct it, is committing another mistake.

Attributed to Confucius

The most extreme remedy for a seriously flawed scientific publication is retraction from the literature. It is important to clarify what this means. Retracted papers are not physically removed from journals. Rather, retraction means that a publication is designated as invalid. The term "retraction" is admittedly imprecise and somewhat problematic since retracted publications remain accessible in the literature and may continue to be read and cited. A retraction is announced by means of a retraction notice that typically includes the cause for retraction. Prior to the 21st century, when most journals were printed on paper, a retraction notice was published in a later volume of the journal, and such notices were not easily linked to the original paper. Today, most journals use electronic publishing formats that allow a retraction notice to be seen whenever a retracted article is accessed or identified in a searchable database such as PubMed. Retracted articles are also marked in a manner as to indicate that they have been retracted. However, the system of labeling retracted articles is not uniform, and not all retracted papers are clearly marked as such. This chapter will consider the reasons for retraction and what they can tell us.

Trends in retracted publications provide an indicator of the failure rate of the scientific process. Retracted publications represent wasted resources and can

Thinking about Science: Good Science, Bad Science, and How to Make It Better, First Edition.
Ferric C. Fang and Arturo Casadevall.
© 2024 American Society for Microbiology.

erode public confidence in science, which can in turn translate into cynicism and reduced support. There is a consensus that retracted publications are the proverbial "tip of the iceberg" (1), signifying a much larger body of poor-quality scientific work that has made its way into the published literature. If that is the case, then a significant proportion of the scientific literature is wrong. Although scientists have always comforted themselves with the thought that science is self-correcting (chapter 18), the immediacy and rapidity with which knowledge disseminates today means that incorrect information can have a profound impact before any corrective process takes place. For example, a fraudulent *Lancet* paper that claimed an association between the measles, mumps, and rubella vaccine and autism was not retracted until 12 years later (2), raising false alarm among the public regarding vaccination that has regrettably persisted. It has also become all too apparent that misinformation can be as effectively disseminated as valid information via modern technology (3–5). Thus, even retracted papers, like the fraudulent measles vaccine study, continue to be cited and used as evidence to buttress false claims (6).

Although concerns about the relationship between pressure to publish and research fraud are not new (7), the sharp rise of retracted papers is a relatively recent phenomenon (8). Retractions were rare before the 1980s but rose exponentially in the early years of the 21st century. The accelerating rate of retractions was an early indicator of serious trouble in the scientific enterprise. Even though retractions for misconduct constitute a tiny fraction of all research publications, they risk discrediting science as a whole. Retractions in areas of great public interest, such as medicine and global warming, are particularly dangerous because they undermine confidence in scientifically grounded policy recommendations. Hence, bad science is bad for society.

In 2010, the journalists Ivan Oransky and Adam Marcus launched a blog called *Retraction Watch*, which is devoted to the examination of retracted articles "as a window into the scientific process" (9); sadly, they have had no trouble finding material. Over the past decade, *Retraction Watch* has become an important institution in science, which maintains a detailed record of retractions. The fact that it is run by investigative journalists is reflected in its interest in the deeper causes of retraction, and the site provides a repository of information regarding events, causes, and background information associated with retractions.

Most retracted papers fall into one of two general categories: those retracted because of scientific misconduct and those retracted because of errors in methodology, conclusions, or approach. One might refer to these as "dishonest" and "honest" retractions, respectively. Although each has a different cause, and the consequences to authors are very different, their immediate effects are the same in that they undermine the credibility of science. An early study of the retraction problem analyzed the cause of retraction for 788 retracted papers and found that error and fraud were responsible for 545 (69%) and 197 (25%) cases, respectively, while the cause was unknown in 46 (5.8%) cases (10). Although the implication was that most papers are retracted for honest errors, subsequent work that we performed in collaboration with Grant Steen showed that most retractions are actually due to misconduct, and this will be discussed in more detail later in this chapter.

HONEST RETRACTIONS

Honest retractions occur from errors of methodology, conclusions, or approach. We call them "honest" because no malicious intent is involved. Reducing honest retractions will require implementing reforms that reduce the number of errors, and we will discuss errors in more depth in chapter 25, along with some suggestions for prevention in chapter 31.

DISHONEST RETRACTIONS

Dishonest retractions occur when the deliberate manipulation or fabrication of results is discovered, prompting a retraction of the published work. Reducing dishonest retractions might be achieved by either increasing the penalties or reducing the incentives for misconduct. Severe penalties ranging from the loss of reputation to criminal charges of fraud are already in place, raising doubts about whether additional penalties would be effective. The adoption of uniform standards for reporting retractions and perhaps the institution of a centralized database for documenting scientific misconduct might help to ensure that individuals guilty of misconduct are recognized. Scientists are conscious of reputation, and unflattering notoriety remains a potent disincentive for cheating. Nevertheless, there is a limited role for sanctions. Although it is important for misconduct to have consequences, unequivocal instances of misconduct are uncommon and represent only a portion of undesirable behavior. Therefore, we would focus efforts on reducing the incentives for dishonest actions by scientists.

REASONS FOR RETRACTION

Our interest in the problem of retractions began when one of us (F.F.) was the editor in chief and the other (A.C.) was an editor at the journal *Infection and Immunity*, which is published by the American Society for Microbiology (ASM) and has a focus at the intersection of microbiology and immunology. Of more than 28,000 articles in its first 40 years of existence, *Infection and Immunity* issued only 15 retractions.

Six of these were issued in a single year (2011) and arose from a single laboratory (11–16). Eight articles retracted by *Infection and Immunity* were found to contain digital figures that had been inappropriately manipulated (11–18) (see chapter 18 for details of the Mori case). Six of the others were retracted by the authors after they determined their previously reported findings to be unreliable: two were unable to confirm their original results (19, 20), one discovered that a cDNA library was actually obtained from another organism (21), and three found a critical reagent to be impure (22–24). The remaining article was retracted due to extensive plagiarism (25). This is a reasonably representative sample of the reasons for manuscript retraction discussed in guidelines from the Committee on Publication Ethics (COPE) (26, 27).

Although journals have an important role to play, they do not have primary responsibility for investigating possible scientific misconduct. That responsibility rests with an author's institution (28, 29) and, if funding from the U.S. Department of Health and Human Services is involved, the Office of Research Integrity. Nevertheless, if an editor has concerns about the validity of data in a submitted manuscript, the editor has the prerogative to request that authors provide their raw data for review. If misconduct is suspected, the journal can contact the institution and recommend an inquiry. Once an institution has determined that misconduct involving research publications has occurred, journals are obligated to consider retraction of the work. In a case involving repeated instances of digital-figure manipulation that resulted in six retracted *Infection and Immunity* articles, another journal initially raised the question of misconduct, and the author's institution performed a thorough investigation before informing *Infection and Immunity* of its concerns. After receiving this notification, *Infection and Immunity* performed an independent review of the evidence, requested a response from the author(s), and then reached a decision to retract the articles in question after consultation with multiple editors and members of the ASM Publications Board.

A COPE survey of Medline retractions from 1988 to 2004 found 40% of retractions of articles to be attributed to honest error or nonreplicable findings, 28% to research misconduct, 17% to redundant publication, and 15% to other or unstated reasons (27). Research misconduct is classified as falsification or fabrication, with falsification defined as the manipulation of materials, processes, or data to misrepresent results and fabrication defined as reporting the results of experiments that were not actually performed (30). Plagiarism refers to the misrepresentation of another's ideas or words as one's own and includes self-plagiarism, sometimes referred to as redundant publication. While some have criticized the term "self-plagiarism" on semantic grounds (31), it has nevertheless proven to be a useful way to describe the practices of publishing the same article in more than one journal or recycling large sections of text in more than one article.

It is important that retracted publications remain available in the literature even though they are declared to be invalid. Retracted articles provide a record of scientific information that was once viewed as valid but is no longer considered to be so. However, several studies show that retracted publications can continue to be cited after retraction. For example, a 2022 study of 3,000 retracted and 3,000 control papers found that retraction only reduced citation rates by about 60% (32). Why do retracted papers continue to be cited? Sometimes they may be cited as an example of a flawed study, but in other cases, the papers are cited because other researchers are unaware that they have been retracted (33). This suggests the need for additional technical improvements to highlight retracted literature and a need for scientists to take greater care in reading the papers they are citing. Efforts are underway to address this problem, which will require coordination among publishers and enhanced education for researchers (34).

ETHICAL GUIDELINES AND RETRACTION POLICY

A 2004 survey found that many scientific journals lacked formal retraction policies (35). However, the journals published by ASM, including *Infection and Immunity*, have long had specific guidelines for ethical conduct and retractions, which are detailed in the Instructions to Authors (36). These guidelines define plagiarism, as well as the fabrication, manipulation, or falsification of data. In addition, the ASM guidelines distinguish between retractions, which are reserved for major errors or misconduct that call the conclusions of an article into question, and errata or authors' corrections, which rectify minor errors. The issue of manipulation of computer-generated images is specifically addressed, with image processing acceptable only if applied to all parts of an image (see chapter 17).

Either publishers or authors may initiate a retraction (26, 37), and most journals today are willing to retract flawed articles with or without authors' consent (38). Authors are consulted regarding the wording of a retraction notice, but final decisions are at the discretion of the editor, who should take care to ensure that the notice accurately represents the reasons for retraction (39). The bloggers at *Retraction Watch* have successfully advocated for transparency in retraction notices (40). We concur with the COPE guidelines that notices should state who is issuing the retraction and the reason for the retraction in order to distinguish misconduct from error. The goal in writing a retraction notice is to be clear, accurate, and fair, with fairness applying to both the authors and journal readership. Beyond this basic information, we are reminded of William Galston's observation that some things must be shrouded "for the same reason that middle-aged people should be clothed" (41). A limitation is that retractions associated with ethics investigations often occur before the investigation is complete. Consequently, retraction notices may not have complete information on misconduct. Additional information may be obtained by cross-referencing authors with findings from *Retraction Watch* and the U.S. Office of Research Integrity.

As a reader once commented to us, "there is no statute of limitation on retractions." In 1955, Homer Jacobson published an article called "Information, Reproduction and the Origin of Life" in the journal *American Scientist* (42). Fifty-two years later, after learning that creationists were citing his article as evidence for the divine origin of life, he decided to retract the article (43). Similarly, in 1920 the *New York Times* published an editorial mocking the aerospace pioneer Robert Goddard for suggesting that a rocket could function in the vacuum of space (44), stating that Goddard "seems to lack the knowledge ladled out daily in high schools." The newspaper later retracted their article on July 17, 1969, following the successful launch of Apollo 11.

CAN RETRACTED ARTICLES BE REPUBLISHED?

In theory, a retracted article may be revised and republished, with removal of any erroneous, falsified, fabricated, or plagiarized content. This approach can be particularly useful when extensive corrections of erroneous data are involved (45, 46). In practice, however, authors of a retracted article may find republication to be a challenge. If misconduct has taken place, the authors may be subject to sanctions from the journal, which prohibit resubmission within a specified time frame. Misconduct compromises the trust between author and editor, and in such cases, authors may find it awkward to later approach the same journal to request consideration of a previously retracted article. In addition, the passage of time may have reduced the significance of the reported findings such that the article is no longer assigned high priority by the journal. Nevertheless, there are instances in which a retracted article has been corrected and republished by the same or another journal (25, 47–51). Scientists, it would seem, also believe in redemption.

JOURNALS DIFFER IN RETRACTION FREQUENCY

To determine whether journals differ in frequency of retracted articles and whether there is a relationship between retraction frequency and journal impact factor, we carried out a PubMed search for retracted articles among 17 journals ranging in impact factor between 2.00 to 53.484. We defined a "retraction index" for each journal as the number of retractions in the time interval from 2001 to 2010, multiplied by 1,000, and divided by the number of published articles with

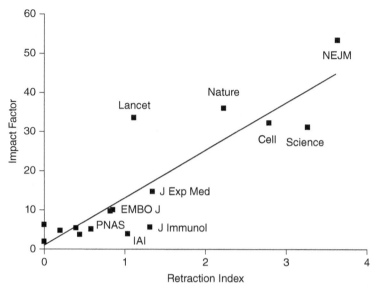

FIGURE 24.1 *Correlation between impact factor and retraction index. The 2010 journal impact factor (142) was plotted against the retraction index as a measure of the frequency of retracted articles from 2001 to 2010 (see text for details). Journals analyzed were* Cell, EMBO Journal, FEMS Microbiology Letters, Infection and Immunity, Journal of Bacteriology, Journal of Biological Chemistry, Journal of Experimental Medicine, Journal of Immunology, Journal of Infectious Diseases, Journal of Virology, Lancet, Microbial Pathogenesis, Molecular Microbiology, Nature, New England Journal of Medicine, PNAS, *and* Science.

abstracts. A plot of the journal retraction index versus the impact factor revealed a surprisingly robust correlation between the journal retraction index and its impact factor ($P < 0.0001$ by Spearman rank correlation) (Fig. 24.1). Although correlation does not imply causality, this preliminary investigation suggested that the probability that an article published in a higher-impact journal will be retracted may be higher than that for an article published in a lower-impact journal. Others have also noted that articles in journals with higher impact factors are more likely to be retracted (52, 53).

The correlation between a journal's retraction index and its impact factor suggests that there may be systemic aspects of the scientific publication process that can affect the likelihood of retraction. When considering various explanations, it is important to note that the economics and sociology of the current scientific enterprise dictate that publication in high-impact journals can confer a disproportionate benefit to authors relative to publication of the same material in a journal with a lower impact factor. For example, publication in journals with high impact factors can be associated with improved job opportunities, grant success, peer recognition, and honorific rewards, despite widespread acknowledgment that impact factor is a flawed measure of scientific quality and importance (see also chapter 26) (54–59). Hence, one possibility is that rates of fraud and scientific misconduct are higher in papers submitted and accepted to higher-impact journals. In this regard, the disproportionally high payoff associated with publishing in higher-impact journals could encourage risk-taking behavior by authors in study design, data presentation, data analysis, and interpretation that subsequently leads to the retraction of the work. Another possibility is that the desire of high-impact journals for clear and definitive reports may encourage authors to manipulate their data to meet this expectation. In contrast to the crisp, orderly results of a typical manuscript in a high-impact journal, the reality of everyday science is often a messy affair littered with nonreproducible experiments, outlier data points, unexplained results, and observations that fail to fit into a neat story. In such situations, desperate authors may be enticed to take shortcuts, withhold data from the review process, overinterpret results, manipulate images, and engage in behavior ranging from questionable practices to outright fraud (60). Alternatively, publications in high-impact journals have increased visibility and may accordingly attract greater scrutiny that results in the discovery of problems eventually leading to retraction.

Monica M. Bradford, executive editor of the journal *Science*, was quoted in a *New York Times* article as suggesting that the extra attention high-impact journals receive might be part of the reason for their higher rate of retraction—"Papers making the most dramatic advances will be subject to the most scrutiny" (61). This proposition was tested in a follow-up study that we carried out with Grant Stein to analyze the time between publication and retraction; however, we found no correlation between journal impact factor and time to retraction (62). If the sole explanation for the higher retraction rate in high-impact journals was increased attention, one might have expected problematic papers published in high-impact journals to be retracted more quickly. Nevertheless, our findings did not conclusively disprove Bradford's explanation. In fact, it is possible that multiple factors contribute to the correlation between retraction index and impact factor. Whatever the explanation, the phenomenon appears deserving of further study. The relationship between retraction index and impact factor is yet another reason to be wary of judging papers on the basis of a journal's reputation or impact factor (chapter 26).

IMPACT OF RESEARCH MISCONDUCT

Science must try to be self-correcting, and retractions provide a critically important function by rectifying the scientific record. However, the system is far from perfect. As we have already noted, it is likely that only a small percentage of scientific misconduct results in retraction. Sensational new claims attract scrutiny and are more likely to be refuted by subsequent research (63–75). However, reports based on falsified or fabricated data may be more difficult to detect if the conclusions happen to be true. Retractions often do not occur for years after publication (76–79), which is perhaps understandable given the time required for other researchers to attempt to replicate results and for institutions to perform thorough investigations (80), but this means that erroneous information remains in circulation for prolonged periods before correction (81). Moreover, it is disheartening that retracted articles continue to be cited, sometimes for decades afterward (82–88).

It is not difficult to surmise the underlying causes of research misconduct. As we discussed in chapter 18, misconduct represents the dark side of the hypercompetitive environment of contemporary science, with its emphasis on funding, numbers of publications, and impact factor (89). With such potent incentives for cheating, it is not surprising that some scientists succumb to temptation. Funding agencies and journals provide regulations and disincentives for misconduct, but these may be inadequate if the incentives are too great and even counterproductive if the penalties are excessively harsh. Another response to misconduct has been to increase formal ethics instruction for research trainees. While this effort may be worthwhile, there is little evidence of its effectiveness (90). When a prominent article is retracted, a common refrain is, "Why didn't the reviewers catch that?" In fact, many would-be retractions are caught during the review process. However, without access to raw data, it is unrealistic to expect that even careful and highly motivated reviewers can detect all instances of falsification or fabrication.

Plagiarism is a more complex matter, as it is based upon a modern concept of intellectual property that dates back only to 18th-century Europe (91). The rise of the Internet has facilitated plagiarism, but technology has also arisen to facilitate the detection of plagiarism or redundant publication (92–95). Some have suggested that plagiarism is a culturally relative concept, which is less likely to be regarded as an unethical practice by some scientists in non-Western countries or those belonging to a younger generation (96–102). However, we do not share this view. Scientists must be explorers, and it is best if they do not precisely follow the wagon ruts left by their predecessors but instead strike out on their own paths, using their own words. Most journals strictly prohibit plagiarism or self-plagiarism.

At a personal level, findings of research misconduct are likely to result in the end of a scientific career. Our analysis of publications and grant funding success for scientists found to have committed misconduct revealed a 92% decline in publication output and significantly reduced funding after censure by the Office of Research Integrity (103). The message to scientists is simple: *Don't do it.*

IMPACT OF RETRACTION ON THE SCIENTIFIC ENTERPRISE

Apart from the negative effects of retractions for authors and the loss of money and effort, a highly visible paper that is retracted can have potentially large effects on science. Analysis of the citation network of a single paper published in 2012 in *Nature* and retracted in 2014 shows that high-profile publications can develop extensive citation networks before they are retracted (Fig. 24.2) (104). This particular paper was cited 187 and 1,626 times in 2014 and 2015, respectively, demonstrating both continuing citation after retraction and rapid contamination of the citation network by an invalid publication (103). The effects of erroneous evidence on the conduct of science is unmeasurable, but retracted papers that have been cited are likely to influence subsequent work. Erroneous or fraudulent data are by definition not reproducible, and it is reasonable to surmise that this contributes to the reproducibility crisis (chapter 7).

A COMPREHENSIVE REVIEW OF RETRACTIONS IN THE BIOMEDICAL LITERATURE

As the number and frequency of retracted publications are indicators of the health of the scientific enterprise and because retracted articles represent unequivocal evidence of project failure, irrespective of the cause, in 2012 we and our colleague Grant Steen performed a comprehensive review of retracted articles indexed in the PubMed database. At the time, the PubMed database listed more than 25 million articles, primarily relating to biomedical research, that had been published since the 1940s. Retracted articles were classified according to whether the cause of retraction was documented fraud (data falsification or fabrication), suspected fraud, plagiarism, duplicate publication, error, unknown, or other reasons (e.g., journal error, authorship dispute).

We identified 2,047 retracted articles, with the earliest retracted article published in 1973 and retracted in 1977. To understand the reasons for retraction, we also consulted reports from the NIH Office of Research Integrity and other published resources (105, 106), in addition to the retraction announcements in scientific journals. Use of these additional sources of information resulted in the reclassification of 118 of 742 (15.9%) retractions in an earlier study (10) from error to fraud. For example, a retraction announcement in *Biochemical and Biophysical Research Communications* reported that "results were derived from experiments that were found to have flaws in methodological execution and data analysis," giving the impression of error (107). However, an investigation of this article conducted by Harvard University and reported to the Office of Research Integrity indicated that "many instances of data fabrication and falsification were

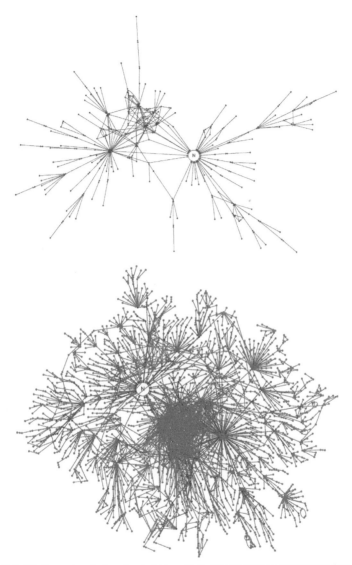

FIGURE 24.2 *Citation network of a retracted paper. Each dot represents a publication that cited the retracted paper. Although the effects of the information in the retracted paper on subsequent publications are unknown, the network density provides an indication of the influence that a single paper can have on its field. If the cited information is incorrect, whether due to error or misconduct, the impact on other research is likely to be detrimental. Figure was published in reference 104 and this article is distributed under the terms of the CC BY 4.0 International License (http://creativecommons.org/licenses/by/4.0/), CC0 1.0 (http://creativecommons.org/publicdomain/zero/1.0/) applies to the data made available in this article, unless otherwise stated.*

found" (108). In another example, a retraction notice published by the authors of a manuscript in the *Journal of Cell Biology* stated that "in follow-up experiments . . . we have shown that the lack of FOXO1a expression reported in figure 1 is not correct" (109). However, a subsequent report from the Office of Research Integrity stated that the first author committed "research misconduct by knowingly and intentionally falsely reporting . . . that FOXO1a was not expressed . . . by selecting a specific FOXO1a immunoblot to show the desired result" (110). In contrast to earlier studies, we found that most retracted articles were retracted because of some form of misconduct, with only 21.3% retracted because of error. The most common reason for retraction was fraud or suspected fraud (43.4%), with additional articles retracted because of duplicate publication (14.2%) or plagiarism (9.8%). Miscellaneous reasons or unknown causes accounted for the remainder. Thus, for articles in which the reason for retraction is known, three-quarters were retracted because of misconduct or suspected misconduct and only one-quarter were retracted for error.

A marked recent rise in the frequency of retraction was confirmed (8, 86), but was not uniform among the various causes of retraction. A discernible rise in retractions because of fraud or error was first evident in the 1990s, with a subsequent dramatic rise in retractions attributable to fraud occurring during the subsequent decade. A more modest increase in retractions due to error was observed, and increasing retractions for plagiarism or duplicate publication were a recent phenomenon, seen only since 2005. The recent increase in retractions for fraud could not be attributed solely to an increase in the number of research publications, as retractions for fraud or suspected fraud as a percentage of total articles increased nearly 10-fold since 1975.

The relationship between journal impact factor and retraction rate that we observed earlier was also confirmed using this much larger database of journals and retracted papers, but the relationship depended on the cause of retraction. Journal impact factor showed a highly significant correlation with retractions because of fraud or error but not with those because of plagiarism or duplicate publication. Moreover, retractions for fraud or error and those for plagiarism or duplicate publication were encountered in distinct subsets of journals, with differences in impact factor and limited overlap. Articles retracted for fraud took a considerably longer time to retract than those retracted for other reasons, perhaps reflecting the time required to investigate an allegation of fraud. A few authors were responsible for multiple retractions, and nearly all articles retracted by authors with 10 or more retractions were retracted because of fraud. Some retracted articles exhibited a rapid and sustained decline in citations after retraction, but others continued to be cited.

In addition to confirming a recent rise in the incidence of retractions, this study provided several additional insights. Perhaps the most significant finding was that most retracted articles result from misconduct, and nearly half of retractions were for fraud or suspected fraud. In addition to analyzing a larger sample that encompassed all retractions in the biomedical research literature at the time, our study differed from previous analyses in the use of alternative sources of information, including reports from the Office of Research Integrity, *Retraction Watch*, news media, and other public records. The U.S. Office of Research Integrity was formed in 1992 and is charged with oversight of misconduct allegations involving research sponsored by the U.S. Department of Health and Human Services. The consideration of secondary sources led to changes in the perceived cause of retraction in 158 instances, leading us to conclude that for many retractions, the retraction notice is insufficient to ascertain the true cause of a retraction. The flood of retraction has continued. Although some have suggested that the rate of retractions may be plateauing (111), this conclusion may be premature in view of the length of time it takes for journals to retract articles. Data from the *Retraction Watch* database show that the percentage of papers in science and engineering rose steadily between 2000 and 2020 (112) (Fig. 24.3). It is sobering to note that the PubMed database has increased in size by about 30% in the 10 years since our study, but the number of retracted articles has risen by more than 400%.

We were disappointed to find that not all articles suspected of fraud were retracted. The *Lancet* and *British Medical Journal* expressed serious reservations about

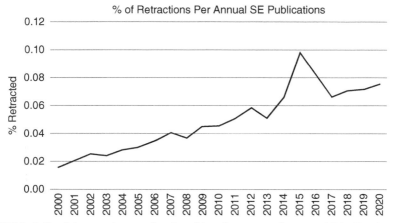

FIGURE 24.3 *Trends in retractions as a function of time. Percentage of papers published in science and engineering that were retracted are based on data from the Retraction Watch Database and National Science Foundation. Graph is reproduced from reference 112 and reprinted here with permission.*

the validity of the Indo-Mediterranean Diet Heart Study after the primary author was unable to present original records to document ethics review and informed consent (113, 114), yet the original articles had not been retracted (115, 116). Several articles authored by Mark Spector when he was working in the laboratory of Efraim Racker remained in the literature (117, 118), despite documentation that Spector committed data fabrication (119). R. K. Chandra was found to have committed fraud in the performance of clinical trials, but only a single article was retracted (120), even though considerable evidence was obtained to suggest that other publications were also fraudulent (28). Therefore, we conclude that the number of articles retracted for fraud represents an underestimate of the actual number of fraudulent articles in the literature.

Although some retraction notices are specific and detailed, many are uninformative or opaque. In 119 instances, no information regarding the reason for retraction was provided by the journal. Notices are often written by the authors of the retracted article themselves (121), who may be understandably reluctant to implicate themselves in misconduct. Furthermore, investigation of suspected misconduct is a lengthy process, and retraction notices are frequently made before the full results of investigations are available. Among 285 investigations concluded by the Office of Research Integrity from 2001 to 2010, the length of investigation averaged 20 months and ranged up to more than 9 years (110). Policies regarding retraction announcements vary widely among journals, and some, such as the *Journal of Biological Chemistry*, routinely declined to provide any explanation for retraction (the journal has since changed its policies). These factors have contributed to the systematic underestimation of the role of misconduct and the overestimation of the role of error in retractions (10, 122), and speak to the need for uniform standards regarding retraction notices (27). The analysis of authors with large numbers of retractions showed us that fraudulent articles can go undetected for many years. In fact, such cases are frequently revealed only fortuitously when exposed by an attentive reviewer or whistleblower (123).

Since our publication in 2012, other groups have confirmed the results. Analysis of retractions listed in PubMed from January 1, 2013, to December 31, 2016, identified 1,082 retractions (2.5 per 10,000 entries), for which misconduct was identified in 65% (124). Similar results were found in single-country studies involving Turkey (125) and Brazil (126). In addition, a new form of misconduct has emerged since our study, which is an increasingly frequent cause for retraction: fake peer review (Box 24.1). Among 697 papers retracted from Iran from 2001 to 2016, fake peer review was the third most common cause after duplication and plagiarism, accounting for 21% of all retractions (127). The emergence of new forms of misconduct illustrates the need for continued vigilance.

Box 24.1 Fake peer review

The early 2010s witnessed the emergence of a new form of misconduct, fake peer reviews (143). As noted in chapter 22, the sequential development of the photocopy machine and the Internet greatly increased the number of scientists who could review a manuscript. However, this accessibility was also a source of vulnerability. Many journals invite authors to suggest potential reviewers of their work. This can be particularly helpful when editors are handling papers outside their usual areas of expertise. In 2011, a journal noticed a number of suggested reviewers with email addresses unassociated with a known institution. Moreover, these individuals rapidly returned favorable reviews that contained numerous grammatical errors. When the editors contacted the reviewers using their institutional email addresses, it became clear that someone else was spoofing their identities to provide fake reviews. Unfortunately, this may signify a new trend in scientific misconduct (105).

ARE RETRACTIONS BECOMING MORE FREQUENT?

Overall, manuscript retraction appears to be occurring more frequently, although it is uncertain whether this is a result of increasing misconduct or simply increasing detection due to enhanced vigilance. In one of the earliest studies examining this issue, Grant Steen reviewed 742 retracted articles and found that the number of retracted articles has risen approximately 10-fold over the past decade, with the greatest increase among those retracted due to misconduct (8). Although errors certainly account for a substantial proportion of retracted articles (122), Steen argued that many retractions are a consequence of deliberate attempts by an author to deceive (128). Most scientists feel that research misconduct is uncommon. However, a meta-analysis of survey data reported that 2% of scientists report having committed serious research misconduct at least once, and one-third admit to having engaged in questionable research practices (60). Given the stigma associated with retractions and the challenges in detecting misconduct, it is likely that retractions represent only the tip of the iceberg (129).

Whether or not retractions have stabilized (111, 112, 130) (Fig. 24.3) apparent trends in retractions due to misconduct must be interpreted cautiously since findings of misconduct require institutional investigations that can last for years. Furthermore, retraction notices do not always accurately report misconduct, especially if the retraction precedes the conclusions of a related misconduct investigation. These caveats notwithstanding, if the rate of total and misconduct-related retractions is slowing, this would be a hopeful sign that increased efforts to establish standards and educate trainees about figure preparation and publication ethics might be having some effect.

CONCLUDING COMMENTS

The analysis of retractions underscores the importance of vigilance by reviewers, editors, and readers and investigations by institutions, government agencies, and journalists in identifying, documenting, and mitigating research misconduct (131). But while the rising rate of retractions suggests a need for increased attention to ethics in scientific training, this alone is unlikely to be successful in curbing poor research practices. The rise in the rate of retractions raises a red flag concerning the health of the scientific enterprise itself (132). Although articles retracted because of fraud represent a very small percentage of the scientific literature, it is important to recognize that (i) only a fraction of fraudulent articles are retracted; (ii) there are other more common sources of unreliability in the literature (60, 133–135); (iii) misconduct risks damaging the credibility of science; and (iv) fraud may be a sign of underlying counterproductive incentives that influence scientists (136, 137). A better understanding of retracted publications can inform efforts to reduce misconduct and error (see chapter 25) in science.

Given that most scientific work is publicly funded and that retractions for misconduct undermine science and its impact on society, the surge of retractions suggests a need to reevaluate the incentives driving this phenomenon. Elsewhere in this volume we argue that increased retractions and ethical breaches may result, at least in part, from the incentive system of science, which is based on a winner-takes-all economics that confers disproportionate rewards to winners in the form of grants, jobs, and prizes at a time of research funding scarcity (132, 137, 138). In chapter 31, we propose a set of reforms to strengthen the scientific enterprise, ranging from improved training of scientists to the identification of mechanisms to provide more-consistent funding for science (132, 137). Solutions to address the specific problem of retractions may include the increased use of checklists by authors and reviewers; improved training in logic, probability, and statistics; an enhanced focus on ethics; the formation of a centralized database of scientific misconduct; the establishment of uniform guidelines for retractions and retraction notices; and the development of novel reward systems for science (Table 24.1) (132). Dedicated national agencies, such as the U.S. Office for Research Integrity, can play an invaluable role in supporting and overseeing institutional investigations of alleged misconduct. An ongoing dialogue is warranted to find effective measures to improve the quality and reliability of the scientific literature.

TABLE 24.1 Retraction problems and suggested solutions

Problem	Reform	Suggested solutions
Honest retractions	Methodological	Embracing philosophy with formal training in logic, epistemology, and metaphysics
		Rigorous training in probability and statistics
		Increased use of checklists
Dishonest retractions	Cultural	Development of new reward systems with an emphasis on quality and a recognition of team science
		Reconsideration of the priority rule
		Establishment of a centralized database of scientific misconduct
		Enhanced focus on ethics

The large number of retractions during the first two decades of the 21st century represents a disturbing trend. Although correction of the scientific record is laudable *per se*, erroneous or fraudulent research can cause enormous harm, diverting other scientists to unproductive lines of investigation, leading to the unfair distribution of scientific resources, and, in the worst cases, even resulting in inappropriate medical treatment of patients (139, 140). Furthermore, retractions can erode public confidence in science. Any retraction represents a tremendous waste of scientific resources that are often supported with public funding, and the retraction of published work can undermine the faith of the public in science and their willingness to provide continued support. The corrosive impact of retracted science is disproportionate to the relatively small number of retracted articles. The scientific process is heavily dependent on trust. To the extent that misconduct erodes scientists' confidence in the literature and in each other, it seriously damages science itself. As Herbert Arst has noted, "All honest scientists are victims of scientists who commit misconduct" (141). And yet retractions also have tremendous value in signifying that, at least sometimes, science can correct its mistakes.

25 Erroneous Science

Delay is preferable to error.

Thomas Jefferson, in a letter to George Washington, 1792 (1)

Error is inherent to science (2). While error is not to be feared, it must be anticipated and prevented whenever possible. However, true science is inseparable from error. In the 1993 decision of Daubert v. Merrell Dow Pharmaceuticals, the U.S. Supreme Court used the existence of a measurable error rate as one of their criteria to distinguish legitimate from junk science (3) (see chapter 16). This chapter will discuss the problem of error in science.

"Error" is defined by the dictionary as "an unintentional deviation from truth or accuracy" (4). High school students learn that scientific errors can be random or systematic. Random errors can occur with any measurement and tend to occur in both directions around the true value, such that an increasing number of measurements can improve accuracy. Systematic errors tend to occur in one direction and cannot be compensated by additional measurements. One can add to this another category called "human error," which refers to the well-documented propensity of human beings to commit procedural errors such as forgetting to add a reagent or control. Error is a complex topic, and error analysis is an entire field of study rooted in formal mathematics. This chapter is not intended as a primer on error analysis. Rather, we wish to introduce the notion that error is part of science by discussing examples of errors that occur in research and highlighting how the discovery of error can allow scientists to do better science.

Thinking about Science: Good Science, Bad Science, and How to Make It Better, First Edition.
Ferric C. Fang and Arturo Casadevall.
© 2024 American Society for Microbiology.

ERROR IN THE BIOMEDICAL SCIENCES

There has been relatively little scholarly work on the nature of error in the biomedical sciences. As a follow-up to our 2012 study of retractions (chapter 24), we analyzed the nature of errors resulting in the retraction of papers in the biomedical literature. A published retraction notice usually contains an explanation of why an article was retracted, although the amount of detail provided can vary widely. Both misconduct-related and error-related retractions are heterogeneous, as there are multiple types of misconduct and error.

The database used for our study included more than 2,000 English language articles identified as retracted in the PubMed database (5). The study of error-related retractions focused on 423 articles that were not associated with scientific misconduct. To understand the causes of error, retraction notices were reviewed and the causes of retraction were classified into eight categories: irreproducibility, laboratory error, analytical error, contamination, control issues, programming problems, control problems, or other. Laboratory errors included subcategories of unique errors, contamination, DNA-related errors, and control problems. Analytical errors included programming errors. The "other" category included cases in which retraction was triggered by inadvertent duplicate publication on the part of the publisher.

Error-related retractions comprised approximately 1/5 of the retracted articles but only 1/500 of the more than 20 million journal articles indexed as of that date. The articles retracted for error were found to have three general causes: lab error (236, 55.8%), analytical error (80, 18.9%), or lack of reproducibility (68, 16.1%) (Fig. 25.1 and Table 25.1). Other miscellaneous causes of retraction included publisher errors leading to duplicate publication, priority claim errors, and editor-initiated retractions due to manuscript errors. Error-related retractions originated from 31 countries, with a distribution similar to that previously observed for retractions resulting from misconduct or suspected misconduct (5). Although PubMed entries are not indexed by country of origin, this distribution generally reflected the countries where most research is performed.

Over half of all error-related retractions were attributed to lab error. Although most instances of lab error were unique circumstances, we were able to subdivide lab errors into unique errors (54%), contamination (31%), DNA sequencing and cloning errors (13%), and problems with controls (2%). Unique errors were those that were specifically associated with the study in question and did not fall into the other categories. Lab errors sometimes resulted in multiple retractions when the error was propagated across different studies. For example, an error in an assay to measure the concentration of the drug fospropofol resulted in six retractions (6). Contamination was the single largest cause of retraction due to lab error, accounting for half of all retractions. Isolated episodes of contamination sometimes resulted in multiple retractions when common reagents were used

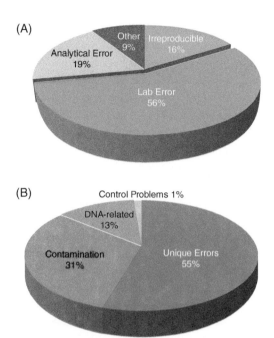

FIGURE 25.1 *Major causes of error. (A) Distribution of categories of error leading to retraction. (B) Distribution of subcategories of lab error.*

TABLE 25.1 Categories of errors before and after 2000[a]

Category	Retractions			
	Pre-2000 (n)	Post-2000 (n)	Total [n (%)]	P value[b]
Irreproducibility	38	30	68 (16.1)	0.0017
Laboratory error	97	139	236 (55.8)	0.2293
Unique	43	85	128 (54.2)	0.0119
Contamination	44	30	74 (31.3)	0.0002
DNA-related	8	22	30 (12.7)	0.1118
Control	2	2	4 (1.7)	0.6411
Analytical error	20	60	80 (18.9)	0.0071
Other	8	31	39 (9.2)	0.0155
Total	163	260	423	

[a] Table adapted from data published in reference 6.
[b] P values calculated by 2-tailed Fisher's exact test using raw numbers of retracted papers.

in multiple experiments. For example, an instance of contamination in which a reagent inadvertently contained cobra toxin or phospholipase resulted in the retraction of 13 articles (7), providing an example of the vulnerability of studies that rely on a single-source reagent.

We found only six notices attributing the cause for retraction to contamination of cell lines or the use of inappropriate cells. This was surprising because the problem of cross contamination of cell lines has been known for decades and remains a major concern in cell biology research despite several calls to action to address this pervasive problem (8–10). Unfortunately, the biological research community has not mounted an adequate response (11). As early as 1981, a compilation of dozens of articles tainted by cell line contamination was published (12). The problem has continued to be amply documented in the literature: a study of 550 leukemia cell lines found that 15% were contaminated with other cells (13), and another study found widespread contamination of tumor cell lines (14). Amanda Capes-Davis and her colleagues (15) compiled a list of more than 350 contaminated cell lines and stressed the need for validation of cell lines. More recently, a study using genetic cross-filing identified contamination in six commonly used human adenoid cystic carcinoma cell lines, including HeLa cells and cells from nonhuman species (16). The prevalence of contamination among other types of commonly used cell lines has been estimated to be as high as 20% (17). The paucity of retractions attributed to contaminated cell lines suggests that the literature contains many unretracted but potentially erroneous studies. Fields that rely heavily on the use of cell lines may be particularly affected by this problem, including cancer research, in which lack of reproducibility has been documented to be a major concern (18).

Analytical errors comprised the next-largest category of retracted results, involving 19% of causes identified in error-related retraction notices. Unlike contamination, which is easily defined, analytical errors comprised a heterogeneous group that included errors in computer program coding, data transfer, data forms, spreadsheets, statistical analysis, calculations, crystallographic analysis, and histological interpretation. As with contamination of common reagents, analytical errors can result in multiple retractions for articles that rely on the same erroneous database or computer program. An error in an in-house program used for analyzing crystallographic data that inadvertently reversed two columns of data resulted in inverted electron density maps and led to the retraction of five articles (19). It is noteworthy that the same program had been used for other articles in preparation (19); had the error not been caught in time, additional retractions would likely have resulted. Incorrect clinical data entry into case study forms led to seven retractions (20). Programming and computer-related errors were blamed for twelve retractions. In recent years, there has been increasing concern about the proliferation of commercial modeling programs that are used without a full understanding of the limitations of the source code and the assumptions made by users (21). It is, therefore, possible that programming errors will make a higher contribution to error in the future. Today, computer programs are used in everything from electronic laboratory notebooks to statistical calculations to image analysis. Computers greatly increase the capacity of investigators to analyze data but also introduce vulnerabilities that may be inapparent to researchers. Surveys have shown that scientists tend to trust software obtained from other sources and fail to validate software using known data sets (21). This suggests the possibility that software errors in commonly used programs could lead to a propagation of errors in different labs.

Retractions due to an inability to replicate published results accounted for 16% of error-related retractions (Table 25.1). The irreproducibility category included all notices in which the authors withdrew an article and cited an inability to reproduce prior results without providing an explanation for the problem. Clearly, this category must have included some studies that were retracted due to laboratory and analytical errors, but the authors did not provide sufficient information in the retraction notice to make those assignments. The irreproducibility of the scientific literature has become a topic of increasing interest and concern, even in the general press (22–24). Recently, the Nobelist Frances Arnold received kudos from colleagues after she quickly retracted a paper describing a novel enzymatic synthesis of beta-lactams (25). This was a finding of great potential importance since beta-lactams are widely used antibiotics. However, when Arnold's lab was unable to reproduce the published work because of missing notes from a student, she announced the retraction on Twitter: "It is painful to admit, but important to do so. I apologize to all. I was a bit busy when this was submitted and did not do my job well."

Numerous explanations have been suggested for the poor reproducibility of scientific studies, including inadequate characterization of key resources such as antibodies, model organisms, knockdown reagents, constructs, and cell lines (26); insufficiently detailed methods (27); publication bias favoring positive results (28); and random variations interpreted as significant results (29). The latter problem is particularly relevant to the biological sciences, which comprise most of the articles indexed in the PubMed database. An application of Bayesian statistical tests suggests that the commonly accepted criterion for statistical significance ($P < 0.05$) is not sufficiently stringent (30), which could lead to false-positive associations that are not reproducible (Box 25.1). The use and misuse of the P value and the flawed association of statistical significance with scientific veracity led some statisticians to propose in 2019 that the use of P values and statistical tests be demoted from its current prominence in biomedical sciences and replaced with a new system that weighs all sources of evidence (31).

Concern about reproducibility of scientific studies has led to the emergence of entities that offer to reproduce research results for a fee (32). This is particularly important for companies that license intellectual property from universities, given the low rate of reproducibility of findings from academic laboratories (see chapter 7) (18, 33). The pressure to publish has been blamed for reduced quality of scientific work. Brown and Ramaswamy analyzed the quality of protein structures and found an inverse correlation between quality and the impact factor of the journal in which the work was published, with the "worst offenders" regarding the publication of

Box 25.1 Types of error

Errors in hypothesis testing are classified into two types. A type I error is a false positive, or rejection of the null hypothesis when it is actually true. A type II error is a false negative, or the acceptance of the null hypothesis when it is actually false. Statistical significance and P values refer only to type I errors, i.e., to distinguish a false-positive from a true-negative result.

Type I versus type II error.

erroneous structural data being the high-impact general science journals (34). This suggests that publishing incentives may perversely encourage haste and error.

We attempted to discern temporal trends by comparing the causes for error before and after the year 2000 (Table 25.1). This year was chosen arbitrarily because it represents a time roughly corresponding to the widespread introduction of several new technologies in the biological sciences, including next-generation DNA sequencing, mass spectrometry, and RNA interference. The number of retractions caused by unique laboratory error events increased from 27.0 to 32.7% ($P = 0.0019$), a trend that may reflect the increasing complexity of laboratory techniques used in biological research. Comparison of the pre-2000 and post-2000 retracted literature revealed no significant decrease in the prevalence of DNA sequencing-related errors despite improvements in the technologies available for DNA sequencing. Nevertheless, there were interesting findings from the comparison of retractions before or after

2000. First, the percentage of retractions that admitted irreproducibility without providing an explanation was reduced from 23.3 to 11.5% ($P = 0.0017$). Although we do not know the explanation for this trend, it is possible that the cost of retraction in terms of reputation and prestige has led investigators to provide more information to support their actions. Second, the number of retractions attributed to contamination was significantly reduced from 27 to 11.5% ($P = 0.0002$). In this instance, the explanation may be improved analytical techniques that provide more information with regard to sample purity. Furthermore, the widespread use of kits for carrying out molecular and biochemical techniques could be associated with a reduction in inadvertent contamination of reagents generated by individual investigators. However, just because a reagent originates from a commercial source does not guarantee its purity, and an increased reliance on commercial reagents raises the possibility that contamination at the source could simultaneously impact many laboratories. Third, there was a significant increase in retractions attributed to analytical error, rising from 12.2 to 23.1% ($P = 0.01$). Although it may be too early to identify the causes for this trend, it is possible that studies that generate large amounts of numerical data make data manipulation errors more likely (35).

We acknowledge several limitations to our study. First, by necessity, we relied on information in retraction notices to describe the source of error accurately. Such information has been shown to be potentially misleading, such as when misconduct is disguised by invoking error or irreproducibility (5). Second, the information in retraction notices represents the authors' version of events, and such notices do not generally receive peer review. Third, the number of retractions studied represents a small fraction of all problematic articles in the literature. Our experience with the paucity of retractions caused by cell contamination despite widespread recognition of the problem suggests that there may be biases in the type of studies that are retracted, such that in some fields, the retraction option is not used as frequently as it should be. Fourth, the information quality in retraction notices is highly variable, ranging from detailed explanations to terse and uninformative statements. Fifth, there is a striking lack of uniformity among journals in the quality of retraction notices (36). However, these limitations also suggest areas where the literature can be improved. For example, establishing criteria for retraction notices and for standardizing the information in notices could significantly improve this aspect of the scientific literature.

Although the number of problematic articles in the literature cannot be precisely ascertained, it is almost certain that retraction notices represent a small fraction of the erroneous literature. Some articles widely regarded as erroneous remain in the literature despite published concerns and, in some instances, calls for their retraction (some prominent examples are listed in Table 25.2). The paucity of retractions for contaminated cell lines suggests that problematic articles in some

TABLE 25.2 **Examples of unretracted articles containing significant errors**

Authors	Journal (year)	Claim	Errata	Description of error
Traver (58)	*Proceedings of the Entomological Society of Washington* (1951)	The author reported a mite infestation of her own scalp that was refractory to treatment and undetectable to others.	None	The author is now felt to have suffered from delusional parasitosis (58).
Steinschneider (59)	*Pediatrics* (1972)	Multiple cases of SIDS (sudden infant death syndrome) were reported in a single family.	None	The children's mother was convicted of murder (60).
Davenas et al. (61)	*Nature* (1988)	Serial dilutions of anti-IgE that eliminated the presence of any anti-IgE molecules were reported to remain capable of stimulating basophil degranulation.	None	A team of observers visited the authors' laboratory and pronounced the results to be a "delusion" (62). An independent group subsequently failed to reproduce the original findings (63), although their conclusions were not accepted by the original authors (64).
Fleischmann and Pons (65)	*Journal of Electroanalytical Chemistry* (1989)	Nuclear fusion was reported from the electrolysis of deuterium on the surface of a palladium electrode at room temperature ("cold fusion").	Erratum (66)	Potential sources of error were identified (67), and an independent group of physicists monitoring experiments in Pons's laboratory was unable to detect evidence of fusion (68).
Bagenal et al. (69)	*Lancet* (1990)	Patients treated at a complementary therapy center for breast cancer exhibited higher mortality than those receiving conventional care.	None	A statistical analysis subsequently assessed the differences between cases and controls to be "so small that no conclusion could be made" (70).

TABLE 25.2 (Continued)

Authors	Journal (year)	Claim	Errata	Description of error
Chow et al. (38)	*Nature* (1993)	Combinations of mutations conferring resistance to three antiretroviral drugs were incompatible with HIV replication.	Erratum (40), acknowledging that additional unrecognized mutations were present	Other groups reported that multiresistant mutant HIV is still able to replicate (39, 71).
Bellgrau et al. (72)	*Nature* (1995)	Expression of CD95L (FasL) was reported to prevent rejection of mismatched transplanted tissue.	Erratum (containing a minor correction only)	Other investigators were unable to reproduce the findings in other experimental systems (73–75). The author of an accompanying commentary retracted his commentary (76).
Gugliotti et al. (77)	*Science* (2004)	RNA was reported to form hexagonal palladium nanoparticles.	None	A former collaborator concluded that the hexagonal crystals were a solvent artifact (78, 79).
Labandeira-Rey et al. (80)	*Science* (2007)	Panton-Valentine leukocidin (PVL) was reported to be required for staphylococcal virulence.	None	Virulence phenotypes previously attributed to PVL were found to be due to a mutation in the *agr* P2 promoter (81).
Wyatt et al. (82)	*Science* (2010)	Nonribosomal peptides (aureusimines) were reported to be required for staphylococcal virulence.	Erratum (83) acknowledging second site mutation	Virulence phenotypes previously attributed to aureusimines were found to be due to a *saeS* mutation (84).
Wolfe-Simon et al. (41)	*Science* (2011)	The bacterium GFAJ-1 was reported to substitute arsenate for phosphate in nucleic acids.	Editor's note (44)	Eight critical technical comments were published, followed by two articles showing that GFAJ-1 DNA does not contain arsenate (42, 43).
Regan et al. (85)	*Journal of the National Cancer Institute* (2012)	CYP2D6 genotyping did not predict responsiveness of breast cancer patients to tamoxifen.	None	Deviation from Hardy-Weinberg equilibrium suggests genotyping errors (85).

scientific fields may be ignored without being retracted. The mechanism of issuing errata or retracting articles is a useful tool to correct isolated errors, but errors that systematically affect entire fields (like cell contamination) require other remedies, such as the development of new standards moving forward.

The percentage of scientific articles requiring correction ranges from 0.5 to over 3%, depending on the field (37). Many corrections are minor, do not undermine the central conclusions of an article, and may be simply noted in errata (publisher's corrections) or corrigenda (authors' corrections). The incidence of errata and corrigenda has been stable over time, whereas the incidence of error-related retractions has been rising over the past three decades (Fig. 25.2). This suggests that errata/corrigenda and error-related retractions are independent and largely unrelated phenomena. However, we have occasionally observed correction notices that report major problems with publications. For example, a 1993 report in *Nature* that combination antiviral chemotherapy halted HIV replication (38) was later found to be erroneous (39), but a correction was issued instead of a retraction (40) (Table 25.2). In 2011, a bacterium was reported to incorporate arsenic rather than phosphorus into its DNA (41). This sensational finding was subsequently shown to be erroneous (42, 43), and the journal issued an editor's note (44), yet the original article has not been retracted. In other instances, we note that authors of erroneous articles have responded to criticism in a prompt and transparent manner (45), and we laud their actions as an example of the (potentially) self-correcting nature of science.

In summary, our systematic study of error-related retractions suggested that a few broad categories of error are responsible for most error-related retractions. These categories suggest general areas in which scientists are most likely to encounter problems during the conduct of research. Some retraction notices provide details regarding how errors were discovered, which, in turn, suggest ways in which

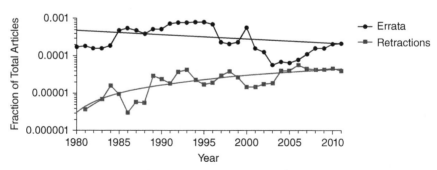

FIGURE 25.2 *Errata and error-related retractions over time. Data from PubMed from 1980 to 2011, inclusive. Decline in errata is nonsignificant (P = 0.07); increase in error-related retractions is significant (P<0.0001).*

such problems may be avoided in the future. Many encountered problems could have been avoided by an experimental design that was more robust, incorporated more controls, or included independent secondary methods to verify results. For example, the ubiquity of contamination, with its potential for unraveling multiple studies, suggests the need for greater attention to the characterization of reagents, cell lines, vectors, etc. Errors involving the inadvertent use of incorrect cell lines might be reduced by the generation of validated bar-coded cell lines that can be easily verified before carrying out experimental work, a step that has been formally proposed by an international panel of scientists (11). We have suggested the use of checklists as a mechanism for reducing error (46), and the journal *Nature* has introduced a pre-submission checklist in an effort to reduce errors in manuscripts. We conclude that the analysis of error in the retracted literature has the potential to improve science by informing the best practices for research. Given the high personal, financial, and societal costs associated with erroneous scientific literature, the implementation of research practices to reduce error has the potential to significantly improve the scientific enterprise. On the basis of our analysis of the retracted literature, we suggest the adoption of specific measures to reduce experimental error (Table 25.3). We also note recommendations by others on how to reduce error, ranging from increased statistical training to the development of incentives to reduce error and changes in laboratory culture (2).

Finally, and perhaps most important, our analysis revealed major problems in the mechanisms used to correct the scientific literature. These problems range from inadequate information in retraction notices to the continued presence of publications known to be erroneous in the literature and the use of errata to report major flaws in articles that should instead be retracted. Both the scientific community and society are dependent on the integrity and veracity of the scientific literature, which is now being questioned in the general media (22–24). This raises concern that future public support for the scientific enterprise could be eroded and scientific findings of major societal importance might not be heeded. We are hopeful that our findings, together with our earlier report that most retractions are due to misconduct (5), will stimulate discussion to develop standards for dealing with error in the scientific literature and actions to improve its integrity. Failure to do so would itself constitute a serious error.

ERRORS IN THE PHYSICAL SCIENCES

Although much of the discussion of error thus far has been based on our studies in the biomedical sciences, we note that erroneous science is found in all scientific disciplines. A Web-based proficiency testing survey of 110 analytical chemists identified 230 causes of error, of which the most frequent were sample preparation

TABLE 25.3 Examples of common errors and suggested remedies

Category	Type	Example[a]	PMID[b]	Suggested remedy
Irre- producibility		*"Thus, our own present findings, along with those from other labo- ratories, contradicts major findings from our previous report."*	20134478	Increase redundancy in experi- mental design, more replications, aim for more-robust statistics
Lab error	Data entry	*"Incorrect data were found to have been included on the case report forms and subsequently in the databases of some studies."*	15504385	Double entry or the creation of duplicate databases (86)
		Data pertaining to a dopamine receptor D4 polymorphism, rs4646984, were mistakenly pasted into a column containing serotonin transporter gene (5-HTTLPR) data.	20050156	
	Contamination	*"The cell line ACC3 which was used in the study . . . was reportedly found cross contaminated with HeLa."*	22359742	Cell line authentication and good cell line practice (10, 11)
	Selection	*"The mice that had been ordered for the experiments described in our published article were, in fact, mistakenly mice that were doubly deficient in . . . receptor."*	21209288	Double-check critical reagents, preferably by a second individual
	DNA-related	*"A mistake was made during the sequencing of the smk-1 allele from the . . . strain that was used in our studies."*	20584918	Validation of results by independ- ent methods and incorporating more redundancy into experi- mental design
		Methodological problem that resulted in a misidentification of a pseudogene as a novel mutation.	21135394	
		"Resequencing of the cDNA clone obtained in the Y2H screen resolved a particularly GC-rich region upstream of the calcyon start codon that had been misread before and indicated that the calcyon coding sequence is out of frame with the GAL4 activation domain."	17170272	

Data analysis	Statistical analysis	These errors relate to the numbers of families included and several erroneous values in the tables. This failure led, among others, to an extremely high correlation between the male subjects' mates and subjects' parents in the facial proportion jaw width/face width.	19129144	Statistical consultation and independent assessment of data-base integrity
		"Errors in the statistical calculations and interpretation of the analyses presented in that article."	16983065	
	Computation	*"Due to a decimal place error for body weight… we have had to re-analyze our data… upon re-analysis with the corrected value we have found our conclusions to change in a way that does not warrant publication."*	17193703	Double-check critical calculations, preferably by a second individual
		"Dilution errors were made in calculating the numbers of bacilli present in the lungs."	7594697	
		"We rechecked the original data and found that the equation used to correct for differences in weights between groups of animals was incorrect."	3882286	
Programming	Coding error	As a result of a bug in the Perl script used to compare estimated trees with true trees, the clade confidence measures were sometimes associated with the incorrect clades.	17658946	Validate program outputs with known databases and results
Controls		*"A change in international convention of testing occurred between the time that the majority of controls were investigated and the time that the majority of patients were investigated. We incorrectly assumed these tests to be entirely equivalent."*	9303950	Identification of correct control group

[a] Quotations taken from the retraction notice.
[b] PubMed identification number of the retraction notice.

(16%), equipment failure (13%), human error (13%), and calibration mistakes (10%) (47). The category of "human error" was large and ranged from inattention to poor decisions. For disciplines that are highly reliant on instruments to make observations, such as physics, the state of the instrument can be a major source of error, as seen in two widely publicized incidents. In 1989, a pulsar created by a 1987 supernova explosion was first thought to be spinning at the rate of 2,000 times per second, but the result was ultimately attributed to a television camera glitch (48). More recently, in 2012, neutrinos were reported to travel faster than light, but this claim was ultimately retracted and blamed on a loose cable (49). Erroneous conclusions can also result from failing to consider all the variables that can affect measurement. In 2015, an early report of gravity wave detection was subsequently found to be in error because the investigators had not taken cosmic dust into account (50). As astronomy has increasingly relied on costly space explorations, instances of embarrassing errors have repeatedly crippled missions, including software glitches and the highly embarrassing crash of the Mars Climate Orbiter because the spacecraft team was using English units of measurement and the navigation team at the Jet Propulsion Laboratory was using metric (51). As the physical sciences have become increasingly sophisticated, their instruments and research infrastructure are dependent on computers, and any error can produce a catastrophic failure. Such sensitive machines are also vulnerable to errors caused by unanticipated effects. The Large Hadron Collider (Box 21.3), which identified the Higgs boson, is so large that it can be distorted by minute gravitational differences induced by the rotation of the moon; consequently, the position of the moon must be taken into account for its operation (52).

ERRONEOUS SCIENCE IN PERSPECTIVE

Errors are inherent to the scientific process and will continue to occur whenever humans perform research. Nevertheless, many errors can be prevented. Fields differ greatly in the amount of error that they will tolerate. Particle physics, with its reliance on complex instruments backed by strong theoretical predictions, has adopted the so-called five-sigma standard before findings are accepted. Five-sigma refers to a statistical criterion of a P value of 3×10^{-7}, or approximately 1 chance in 3.5 million that one rejects the null hypothesis in error (53). In contrast, the biomedical sciences often accept a P value of 0.05, which implies a 5% error in rejecting the null hypothesis. One of us (A.C.) experienced these variations in field error tolerance during his training. During his graduate work in physical biochemistry, acceptable error was usually limited to a few percent, but he was told during his postdoctoral work in immunology not to believe anything less than "a 10-fold effect." This experience reflected the fact that physical chemistry, with its close relation to physics, relied heavily on instrumental measurements that provided results grounded in a well-established theoretical framework, while immunology studied complex *in vivo* processes that were poorly understood or controllable. Hence, immunologists attempted to ensure the veracity of their observations by focusing on large effects. This should not be taken to mean that smaller biological effects are unimportant—for example, a doubling of blood sugar levels is enough to make a person diabetic, and a doubling of blood pressure is enough to cause a stroke. The difference between blood sugar or blood pressure and immune response parameters is that the first two are tightly controlled, whereas the latter can exhibit greater variability. Clearly, a scientist's perspective on error depends on the scientific field in which one works.

SCIENTIFIC ERROR AS PROGRESS

It is naïve to pretend that programming errors that lead rockets to crash is anything less than a disaster for science and all those involved. However, when the detection of error invalidates a result, we learn that the finding is wrong, which can signify progress, particularly when this allows researchers to avoid such errors in the future. On the other hand, some scientific errors such as the biological concept of race have resulted in lasting harm to humanity for generations long after the original notions have been disproven (Box 25.2). Learning from errors can make science more resilient, just as airplane crash investigations over time have made plane travel extremely safe. When major errors surface in a field, scientists have the

---

Box 25.2 The worst error in science?

Errors within science can be corrected by subsequent scientific work, but scientific errors that permeate popular culture can cause lasting harm. It has been argued that the greatest error in science was the classification of humanity into different races (55) (see also chapter 16), a conclusion that evolved into scientific racism and has led to some of the greatest crimes in human history, including slavery, genocide, and the marginalization of entire peoples. Today, we know that individuals within the species *Homo sapiens* share 99.9% of their DNA and there is no molecular or genetic basis

(Continued)

Box 25.2 (Continued)

for categorizing humans into separate races. Fortunately, scientists have moved to correct this error in recent decades, and close to 90% of anthropologists now reject the concept as flawed (56). One might argue that elements of scientific racism were present before the Scientific Revolution, as evidenced by the treatment of native peoples after the European discovery of the Americas, and that the embrace of this concept by 19th-century science was the incorporation of an immoral idea into science rather the other way around. Viewed from that perspective, the invalidation of scientific racism by late-20th-century science can be considered as science correcting a societal error rather than a scientific one. However, scientific racism unfortunately remains rooted in certain segments of society, and the historical harm done to millions by racism cannot be undone. The cautionary tale of race science and its consequences reminds scientists to carefully consider the societal implications of their work.

opportunity to learn and modify their work to prevent future errors. In this regard, honesty in admitting error and publicizing its causes are acts that should prompt praise and admiration. Retracting papers after honest errors are identified corrects the scientific record and increases the trustworthiness of one's work (54). A frequent refrain is that science is self-correcting, but this first requires the recognition of errors to correct.

26 Impacted Science

The use of journal impacts in evaluating individuals has its inherent dangers. In an ideal world, evaluators would read each article and make personal judgments.

Eugene Garfield, inventor of the impact factor (1)

Future historians and sociologists looking back on our time may be bewildered by the preoccupation of scientists with the journal impact factor (JIF), a measure of the frequency with which articles in a journal have been cited during the previous two years, divided by the number of published articles. The JIF was conceived by Eugene Garfield in 1955 to help librarians identify the most influential journals based on the number of citations, so that these journals could be included in the Science Citation Index (1). The first ranking of journals by impact factor was published in 1972 (1). Over time, the JIF became regarded as a surrogate measure of journal prestige and, by extension, desirability. Today, the value of a scientific publication is increasingly judged by the IF of the journal in which it is published, which in turn influences the ability of scientists to be appointed, promoted, and successfully funded. This has created a new variable in the prestige economy (chapter 19) that has distorted the scientific enterprise and created many negative consequences. Volumes have been written on the misuse of the JIF in judging publications, scientists, and scientific work (2–5), and we will not revisit those

Thinking about Science: Good Science, Bad Science, and How to Make It Better, First Edition.
Ferric C. Fang and Arturo Casadevall.
© 2024 American Society for Microbiology.

arguments here, but there has been relatively little discussion of the effect of JIF mania on scientific progress. This chapter examines this fascinating phenomenon in the sociology of science.

A QUESTION OF VALUES

When scientists prioritize publishing in journals with high JIFs, they are making a value statement. For a discipline centered on reason, logic, method, and process, this is a perverse choice, since the value of a publication is its information content, which is independent of where the article is published. In fact, the use of JIF for policy and decision-making in academia has been shown to be fallacious on both inductive and deductive grounds (6). Since the information content of a manuscript does not change depending on where it is published, then the preference for publishing in venues with higher JIF can be viewed as a social construct, i.e., "a concept that exists not in objective reality, but as a result of human interaction and because humans agree that it exists" (7). The focus on JIF as a surrogate measure of quality and importance is a relatively recent phenomenon in science, as we can recall a time when journal prestige and JIF did not exert the powerful influence they do today (Box 26.1). The increasing popularity of the JIF has occurred during a time of growing efforts to develop rankings of people, goods, and services, which could be an example of societal trends reflecting back on science (Box 26.2).

Box 26.1 How to choose a journal? Then and now

Sometime in the early 1980s, one of us (A.C.) asked his Ph.D. advisor, Loren Day at New York University, how to choose a journal to which to submit a scientific manuscript. Day said, "It's easy," and picked up some of the paper journal issues that were lying around the laboratory. At the time, the Day Laboratory received paper copies of *PNAS*, *Biochemistry*, and the *Journal of Biological Chemistry*. Electronic article searches were a specialized activity carried out by librarians after a formal request, and it was common for laboratories to subscribe to the journals that were most important for their research. Day opened an issue of *Biochemistry* and pointed to the page where the editors were listed, all of whom were very prominent scientists in the field. He explained that you picked a journal based on the type of content it published and for its editors, since editors were the gatekeepers to publication; publishing an article in a journal with a distinguished editorial board consisting of prominent working scientists gave it credibility and prestige. Two decades later, this method of choosing journals had largely disappeared, as many scientists were choosing journals based largely on the JIF. This meant that many scientists now preferred to publish in journals with the highest IF irrespective of the quality of the editorial board. In fact, many of the most selective journals with high JIF employed professional editors who were not active scientists to serve in a gatekeeper role. Although other variables such as open access, whether the publisher is a society or a for-profit organization, scope, and page limitations are also factors in choosing a journal, the JIF can be an overriding consideration for many scientists. Consequently, papers are frequently first submitted to journals with the highest IF and then cascaded down a disheartening series of desk rejections and resubmissions until a home for publication is found. Since submitting a manuscript is a complex process that can take hours, this approach consumes considerable time and effort.

Box 26.2 The cult of numerology in science

Science is not immune to social trends. It is noteworthy that impact factor mania arose during a general trend that began in the late 20th century to quantify the strengths and weaknesses of goods and services, often in the form of rankings (105). This resulted in numerical rankings for colleges and universities, businesses, movies, cars, kitchen appliances, and practically anything that could be measured, a craze that was facilitated by the development of the Internet, where rankings became a popular form of clickbait. Viewed in this context, it is perhaps unsurprising that science would also embrace rankings, as science reflects the values of the society in which it operates. The popularity of rankings is not necessarily a bad thing, since lists may help consumers make better choices. For example, rating refrigerators on the basis of energy efficiency is helpful to consumers who want to understand the long-term costs of using the appliance. The JIF fits nicely with this trend since it is a simple value that is easily understood. However, scientific papers are not refrigerators, and the value of a complex scientific work is difficult to express as a simple number. Moreover, journals publish papers of varying impact (52), just as stores sell more than one type of refrigerator. Judging a paper by the journal in which it is published may be as fraught as choosing a refrigerator based on the store that is selling it.

CAUSES OF IMPACT FACTOR MANIA

It is not immediately obvious why scientists as a group would embrace the impact factor of a journal as an indicator of the importance of a publication and, by extension, the quality of individual scientists and their work. It has been suggested that the seemingly irrational focus directed on the impact factor amounts to "impact factor mania" (8). Is this diagnosis accurate? "Mania" is defined as "an excessively intense enthusiasm, interest, or desire; a craze" (9). Evidence of an excessively intense enthusiasm and desire is apparent in the fixation of many scientists on publishing their work in journals with the highest possible impact factor. Evidence of a craze is perhaps more elusive, but after considering that the definition of "craze" is "to cause to become mentally deranged or obsessed; make insane," we note some trends consistent with the definition. Although the term "mentally deranged" is more applicable to individuals than to a field, some manifestations of field behavior are consistent with insanity. Obsession is evident in the behavior of certain scientists to shop for high-impact journals (10). Therefore, we conclude that according to this definition, the life sciences are arguably in a state of mania regarding the JIF. The analogy of this behavior to disease has been made by others, who have referred to the same phenomenon as "journal mania" (3) and "impactitis" (11). We agree with portraying the problem as a medical disease to emphasize the harmful consequences of judging a paper based on where it is published rather than on the value of its content, and in the hope that the condition might someday be curable. Although IF mania seems irrational, in a way the behavior is actually rational for an individual scientist because it confers disproportionate benefits to those who succeed in placing their work in high-IF journals.

In keeping with the disease metaphor, we have titled this chapter "Impacted Science." In clinical medicine, impaction is a severe condition resulting from chronic constipation that is very painful and causes great distress. Details regarding the therapeutic process of disimpaction are best left to medical textbooks, but suffice to say that this procedure was often assigned to junior physicians, and the patient feels much better afterwards. As physician-scientists, we embrace the view that to successfully treat a disease one must identify its causes. Impact factor mania cannot be a condition inherent to science given its recent emergence, but rather must reflect conditions in the contemporary scientific ecosystem that have allowed it to arise and persist. We also subscribe to an economic view of human behavior in which choices are often made in response to incentives. Accordingly, there must be compelling incentives for scientists to engage in the unscientific behavior of linking the quality of science with publication venue. There are aspects of the current scientific enterprise that can at least render the behavior of impact factor mania understandable, if not completely logical.

Hyperspecialization of science

As science has succeeded as an enterprise, it has become ever more specialized (chapter 10) (12). The increasing specialization of science has made it increasingly difficult for scientists working in different fields to understand each other's work. Relying on publication in highly selective journals as a surrogate measure of quality provides a convenient, if intellectually lazy, alternative to attempting to read and understand a paper outside one's own specialty. At least one author has attributed impact factor mania to laziness on the part of senior scientists and bureaucrats (3).

Paucity of objective measures of the importance of scientific work

There are presently more than 25,000 scientific journals and well over one million new articles published each year. There is a pressing need to prioritize this massive amount of information. The reliance of scientists and granting agencies on the impact factor as a measure of scientific quality is rooted in the need for quantitative measures of importance in science (chapter 8) (13). In this context, the impact factor of the journal where work is published is used a surrogate marker of importance, despite the fact that citation frequency for a journal does not predict citation frequency for individual papers (2, 5, 14). Although we have proposed a scheme to assess the importance of scientific work (13) (chapter 8), our approach remains subjective and does not readily lend itself to quantitative analysis. Hence, impact factor mania is driven by the absence of other readily available quantitative criteria to assess the importance of scientific articles.

Use in faculty hiring and promotion

A major contributor to impact factor mania has been the incorporation of JIF into the criteria used for hiring and promotion (15). Young scientists are told that they need papers in high-impact journals to get a job and be promoted. Some universities have even developed formal criteria for promotion based on the JIF. For example, promotion criteria at the University of Bern once specified that papers submitted for promotion consideration need to be in the upper third of the journals in the relevant discipline, as ranked by JIF (16) However, a study of 1,525 publications submitted by 64 candidates for promotion at this institution revealed a poor correlation between the JIF and the relative citation ratio (16), which accounts for differences in the sizes of fields and their impact on citation metrics (Box 26.3), demonstrating the unfairness of using JIF for promotion decisions.

Hypercompetition for funding and jobs

In the United States, success rates for grant applications are at historic lows, and the imbalance between applicants and academic positions for scientists has reached

Box 26.3 Other numbers used for the measurement (and mismeasurement) of science

Although the impact factor is the best-known number that is used and misused to rate science, it has some competition. For example, the *Eigenfactor* has been developed to measure not only citations but also the quality of the journals that cite a work (106). A journal's Eigenfactor therefore reflects not only the number of citations of its articles but also the structure of the citation network. Google Scholar is a bibliometric website that provides a running tally of citations accrued by scientific papers and scientists. Google Scholar also computes the *h-index*, developed by Jorge Hirsch (107), as an author-level measure of bibliometric output that reduces the number of publications and their citations to a single number for each scientist. Simply put, the *h*-index is the highest number of papers that have been cited at least the same number of times (for instance, a scientist with 30 papers each cited 30 or more times would have an *h*-index of 30). Although the *h*-index has significant advantages over judging authors by the impact factor of the journals where they publish their work, it has other disadvantages, including giving full credit for partial contributions to publications, failing to correct for different citation rates in different fields, and disadvantaging investigators early in their careers who necessarily have lower indices than their more senior colleagues (108). More-recent efforts have led to the development of *standardized author-level citation metrics* that adjust for authorship position and field (109). Concerns with impact factor as a measure of science have led to efforts to create improved publication-level metrics, such as the *relative citation ratio* (110). This metric quantifies citations relative to other articles in the same field. Although each of these numbers measures something, we believe that no single number can truly measure the totality of a scientist and their contributions. All performance metrics must be used with caution and in full recognition of their limitations. As the developers of the Eigenfactor acknowledged, the gold standard for scholarly assessment remains "reading the scholar's publications and talking to experts about their work" (106).

a crisis point (17–19) (see chapters 20 and 23). Review panels are routinely asked to discriminate between grant applications that seem equally meritorious, and search or promotion committees likewise must decide which of many highly qualified candidates are most deserving of hire or professional advancement. In this environment, quantitative bibliometric tools seem to offer an objective way to measure researcher performance (20).

Benefits to selected journals

High-impact journals produce a desirable commodity when a paper is accepted for publication. These journals create a scarcity of publication slots by rejecting the vast majority of submissions, resulting in a monopoly that restricts access and corners the market for excellent articles. This in turn creates a sense of exclusivity that encourages more submissions, as scientists equate low acceptance rates with greater merit (21) (see chapter 22). Ironically, high-impact journals are rewarded for rejecting manuscripts. A central error made by many scientists is in assuming that exclusivity is an indicator of exceptional quality, an assumption that is not always justified (22). Nevertheless, the impact factor of a journal is a better predictor of subsequent citation frequency than an indicator of article quality (23). *Nature*, *Science*, and *Cell* collectively account for 24% of the 2,100 most cited articles across all scientific fields (24). The COVID-19 pandemic was accompanied by a major increase in the JIF of infectious disease journals resulting from the blockbuster effect of a small number of very highly cited papers that were not necessarily research related (25). In fact, there is no evidence that this blockbuster effect was related to research quality, but the resulting swings in JIF have upended the traditional ranking of infectious disease journals (25).

Benefits to scientists

Publication in prestigious journals has a disproportionately high payoff that translates into a greater likelihood of academic success. This has been referred to as acceptance to the "Golden Club," with rewards in the form of jobs, grants, and visibility (26). There is also a tendency for publication in highly selective journals to beget more success, perhaps in part because greater visibility attracts capable and ambitious trainees and also because editors and reviewers may apply more-liberal standards in assessing submissions from authors who publish regularly in prestigious journals (27). This helps to perpetuate the cycle. The tendency for the rich to get richer and the poor to get poorer was designated the "Matthew effect" by the sociologist Robert K. Merton (28), in reference to the biblical verse Matthew 13:12—"For to everyone who has, more will be given, and he will have abundance; but from him who does not have, even what he has will be taken away." The economic consequences of the Matthew effect (chapter 19) explain why prominent scientists have an interest in the perpetuation of impact factor mania.

National endorsements

Some nations have developed schemes to rate the productivity of their scientists, depending on the impact factor of the journals in which their papers are published. Brazil has established a "Qualis" scale based on the average impact factor of their publications, which is used to grade students and faculty (29). China has offered monetary rewards to editors who increase the impact factors of their journals (30), and China, Turkey, and South Korea provide cash bonuses to scientists who publish in journals with high impact factors (31). Spain instituted the use of bibliometric indicators such as the JIF for the evaluation of scientific research in 1989, which was followed by a doubling of scientific output from 1991 to 1998 compared to 1982–1990 despite flat or shrinking research funding (32). Whether this association is causal is uncertain, and we note that those time periods differ regarding both the extent of personal computer use and the local political environment, with the decade of the 1990s seeing greater integration of Spain into the European community. One is also reminded of Goodhart's law, as discussed in chapter 19 ("When a measure becomes a target, it ceases to be a good measure"). Nevertheless, such associations may encourage other nations to follow suit.

Prestige by association

Despite the well-known admonition not to judge a book by its cover, the cachet of the most highly selective journals is readily transferred to its contents. Although one may resent the disproportionate influence of the most prestigious journals, one cannot deny that many important and high-quality research articles are published there, and articles are judged by the company they keep. However, the high impact factors of selective journals result from a subset of articles that are extremely highly cited; the rest benefit from reflected glory (Fig. 26.1). Of course, there are also many articles published in less prestigious venues that are of equal or greater quality, and these may be unfairly neglected.

IMPACT IS NOT IMPORTANCE

Scientific impact and importance (chapter 8) are not the same. One dictionary defines "impact" as "a powerful effect that something, especially something new, has on a situation" while "important" is defined as "necessary or of great value" (33). These definitions distinguish impact as something occurring recently, whereas importance has the potential to become timeless. The related term "impacted" means "blocked" (34), and we note that the current obsession with impact may be blocking scientific progress. Apart from these very different definitions, impact and importance differ in our ability to measure them. Impact is measurable by its effect on the present. The focus of scientists on impact is evident for the increasing use of this word in journal titles (35). For example, impact may be measured by media attention or a flurry of citations immediately following publication. However, the fact that impact is measurable does not mean that

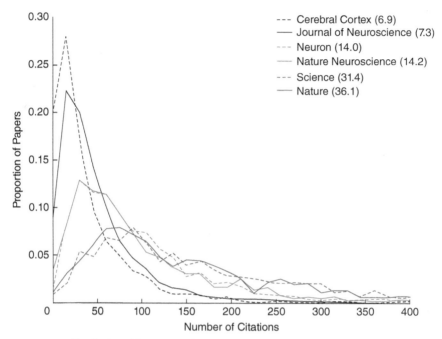

FIGURE 26.1 *Distribution of the number of citations for neuroscience articles in six major journals, 2000–2007. Journal impact factors are shown in the legend. The distributions show substantial heterogeneity and overlap, indicating that the journal in which an article is published is not highly predictive of the number of citations of individual articles. Reprinted with permission from (52) Kravitz DJ, Baker CI. 2011. Front Comput Neurosci 5:55. doi: 10.3389/fncom.2011.00055.*
© 2011 Kravitz and Baker, under license CC BY-NC.

high-impact papers are necessarily important in a more fundamental and lasting sense. For example, papers describing common experimental methods or review articles that compile information tend to be more highly cited than those describing fundamental conceptual advances (36). A survey of biomedical scientists found that highly cited papers were more likely to describe evolutionary advances than surprising and revolutionary findings (37). Hence, citation rates tend to capture only one type of importance. Attention in the popular press signifies short-term recognition, but this impact is transient and fleeting unless the work is truly important. In contrast, importance may not be initially appreciated, and science lacks precise quantitative tools with which to measure importance (chapter 8) (13). While some scientific findings have both high impact and importance, such as the description of the structure of DNA by James Watson and Francis Crick (38), which has been cited more than 17,000 times since publication, Gregor Mendel's seminal studies that were foundational for the field of genetics (39) were neglected for the next 35 years and cited fewer than half a dozen times during the period before their rediscovery by Hugo de Vries and Carl Correns (40). Focusing on impact belies the fact that scientists are more interested in importance. There are also examples of high-impact papers that are not important, such as the widely publicized paper describing a bacterium purported to incorporate arsenic instead of phosphorus into its DNA (41). Although this work has been cited more than 500 times as of 2023, the central finding is now known to be false (42, 43) (see chapter 25), providing an example of high impact but low importance. Sadly, many papers in the literature are seldom if ever cited (44, 45) and thus lack both impact and importance. Alternative measures of impact, such as "Altmetrics," which are based on online mentions of a work, do not solve this problem and have not been found to necessarily correlate with scientific quality (46). If impact is not equivalent to importance, then there is a danger that IF mania could distort the course of science by diverting scientists to research areas that seem most likely to be cited rather than allowing them to pursue their natural intellectual interests and curiosity, which could lead to unsuspected and important findings. This concern is compounded by the difficulty in assessing the importance of a scientific discovery when it is first made (13, 37, 47) (chapter 8).

PROBLEMS WITH IMPACT FACTOR MANIA

We would argue that the current impact factor mania, whether applied to individual researchers, publications, or grants, is seriously misguided and exerts an increasingly detrimental influence on the scientific enterprise. We would particularly emphasize the following concerns.

1. **Distortions of the scientific enterprise.** The greatest distortion caused by impact factor mania is the conflation of paper quality with publication venue rather than the actual content of the paper. This encourages the branding of science and scientists with journals in which work is published. The likelihood of obtaining funding, academic promotion, selection for awards, and election to honorific societies becomes dependent, at least in part, on publication venue. This distorted value system has become self-sustaining, with the editors of high-impact journals commanding far great power and influence than is healthy for the scientific enterprise. Papers are more likely to be cited in high-impact journals if they are in large, established fields (2). Thus, an excessive emphasis on bibliometrics may perversely steer scientists away from understudied areas of research. Perhaps its most pernicious effect is its potential influence on how scientists work and what they choose to study. If impact is not importance and we reward impact through the JIF mechanism, then we run the risk that impact factor mania will influence the types of questions that scientists pursue while

more important work is neglected or not done. One can imagine that in today's highly competitive environment, investigators focusing on important questions that are not viewed as trendy or high-impact may not survive (48). Unfortunately, there is no way to measure the consequences of work that is not done.

2. **Decline of certain disciplines.** Since the size of a field potentially determines the maximal number of citations, it stands to reason that investigators publishing papers in smaller fields will be cited less and journals specializing in such areas will have lower JIFs. This can have various detrimental effects. For example, it may be more difficult to recruit investigators to such fields, perpetuating their small size. Smaller fields also have fewer journals, which may result in more journal self-citations, a practice that is frowned upon by some bibliometric organizations. For example, the journal *Zootaxa* was temporarily suspended from Journal Citation Reports (Clarivate) because of its disproportionate number of self-citations, despite a lack of evidence that the citations were improper (49). Once a journal loses its JIF, it will attract fewer submissions, which in turn means fewer high-quality papers, which can trigger its decline.

3. **The full impact of a scientific discovery may not be apparent for many years.** The journal impact factor is calculated by dividing the number of total citations over the number of articles published in the previous two years. However, as we have discussed in earlier chapters, the impact of truly important and novel findings often takes more than two years to be fully realized (4). Thus, the central premise of measuring the importance of scientific journals or papers based on impact factor is fundamentally flawed. An unanticipated consequence of the way the impact factor is calculated is that the most novel and innovative research may be less attractive to some journals because such work, by its very nature, will have its major influence during a time period that does not contribute to the calculation. The history of science is replete with anecdotes of individual scientists doing fundamentally important work that was not appreciated at the time, and there are no features that can reliably distinguish breakthrough papers at the time of their publication (47). Journals that reject manuscripts because they cannot immediately appreciate the importance of the work may be turning down the next penicillin, PCR, or tricarboxylic acid cycle (50). Problems with shortsighted peer review are not limited to journals; an analysis at the National Institute of General Medical Sciences (NIGMS) has concluded that current grant review procedures have an error rate of approximately 30% and result in many meritorious applications going unfunded (51).

4. **Inappropriateness of JIF as a predictor of quality.** The use of journal IF to measure the quality and impact of individual papers is invalid from a statistical standpoint. The high IF of selective journals results from their ability to attract a few papers that are very highly cited (2, 5, 52). In fact, replacing the mean

citation values used to calculate the JIF with the median number of citations reduced the numerical ranking of journals by 30 to 50% (53). Since the JIF is an average value, comparing the citation impact for two publications in a given journal is no better than flipping a coin (54). Publication venue is a poor predictor of the number of times that an individual article will be cited. Thus, for most authors, the benefits of publishing in high-IF journals result more from their association with other papers in the same journal that happen to be highly cited than from the worthiness of their own content. In other words, publishing in a high-IF journal may be easier than producing highly cited work (55). Adding to the inappropriateness of using JIF as an indication of paper quality is the fact that some journals routinely manipulate journal content and acceptances to increase their JIF (56), which can be viewed as a form of scientific misconduct (57).

5. **Access to high-impact journals is restricted.** Today, the gateway to publishing in the highest-IF journals is determined by a small cadre of professional editors and the reviewers whom they select. Although these individuals generally do a superb job in selecting papers that will be highly cited, there can be little confidence that they are selecting the most important science. In fact, the high retraction indices associated with high-IF journals suggest that not all their decisions are wise (chapter 24). Since the choices made by editors and reviewers have a tremendous impact on the careers of authors whose papers are selected or not selected, and the papers published by these journals can have an enormous influence on the work of other scientists, a handful of elite journals are exerting a disproportionate effect on who does science, what work they do, and how it is valued. This in turn encourages researchers to work in areas favored by the high-IF journals and their editors.

6. **Emphasis on high impact means that many meritorious studies will not be funded or published.** Asking scientists to conduct only high-impact research creates a strong bias that discourages high-risk research and reduces the likelihood of unexpected breakthrough discoveries. We suggest an analogy between scientists and foraging predators, which employ random Brownian movement when prey is abundant and use more-complex yet still random Lévy flights when searching for sparser prey (58, 59). Scientists in search of new discoveries cannot hope to investigate the full scope of nature if they must limit their searches to areas that a consensus of other scientists judge to be important. As Vannevar Bush observed more than 75 years ago, "Basic research is performed without thought of practical ends. . . . Many of the most important discoveries have come as a result of experiments undertaken with very different purposes in mind. . . . It is certain that important and highly useful discoveries will result from some fraction of the undertakings in basic science, but the results of any one particular investigation cannot be predicted with accuracy" (60).

7. **Journal impact factor and individual article citation rate are poorly correlated.** The majority of a journal's impact factor is determined by a minority of its papers that are very highly cited (14). It is well-known that the impact factor does not necessarily predict the citation prospects of other papers published in the same journal (2, 5, 24, 61), and the relationship between impact factor and citation rate may be weakening (62). Furthermore, when the impact of science is analyzed by other criteria, there is at best a very weak positive correlation with the impact factor of the journal in which the work was published, whereas journal rank by impact factor correlates inversely with the reliability of research findings (5, 63–65).

8. **Citation rate is an imperfect indicator of scientific quality and importance.** Citation rate is highly dependent on factors such as field size and name recognition (55, 66). Studies suggest that journal impact factor may correlate poorly with the true value of research to a field (67, 68). Review articles and descriptions of new methods tend to be disproportionately cited (69, 70). Moreover, an emphasis on citation rate as a measure of impact perversely discourages research in neglected fields that are deserving of greater study.

9. **Delays in the communication of scientific findings.** The disproportionate rewards associated with publication in high-impact journals create compelling incentives for investigators to have their work published in such journals. Authors typically cascade their articles through multiple submissions as they shop for the highest-IF journal that will publish their work. Since the time required for the peer review of each submission can take months, this delays the communication of scientific findings and slows the progress of science. Such effort can consume considerable time and resources since investigators often respond to reviewers by performing additional experiments in an effort to convince journals to accept their papers. The multiple submissions also consume reviewers' and editors' time and delay the public disclosure of scientific knowledge. Publication delay slows the pace of science and can directly affect society when the manuscript contains information important for drug and vaccine development, public health, or medical care. For an investigator, the time spent in identifying a high-impact journal can result in a loss in citations (10). Collectively, these effects translate into major inefficiencies in the dissemination of scientific information.

10. **Creation of perverse incentives.** The pressure for publication in high-impact journals may contribute to reduced reliability of the scientific literature. Articles retracted due to data falsification or fabrication are disproportionately found among high-impact journals (63, 64), and errors also appear to be more frequently encountered (64, 65). The most selective journals demand clean stories and immaculate data, which seldom match the reality of laboratory investigation,

where experimental work can produce messy results. Hence, some investigators may be tempted to cut corners or manipulate data in order to enjoy the disproportionate rewards associated with publishing in prestigious journals. Impact factor mania also pressures journals to raise their impact factors, which can lead to practices that range from gaming the system to outright editorial misconduct. Some journals have been reported to pressure authors to include more citations to their own journal to artificially increase the journal's impact factor. In 2013, a scheme was uncovered in which three journals conspired to cite each other's papers in a mutual effort to increase their impact factors, a conspiracy that was blamed in part on the use of journal impact factor by the Brazilian funding agency to judge the quality of articles (71).

11. **New opportunities for misconduct.** The perverse incentives associated with impact factor mania have created new opportunities for misconduct as editors, journals, and reviewers try to game the system to increase their JIF. This has led to such unethical practices as "citation stacking," in which editors ask authors to cite articles in their journals to increase their citation rates, and "citation cartels," in which authors and journals agree to cite one another to raise their impact factors (72). Citation-related misconduct can also occur at the level of reviewers. An analysis of 69,000 reviews found that 0.8% of reviewers asked authors to cite their own publications in a manner that was felt to be suspicious (73). Although this practice is not always inappropriate, egregious cases have been encountered, such as a reviewer for the journal *Bioinformatics* who requested an average of 35 additional citations in each review, mostly to their own publications (74).

12. **Impact factor mania as a tragedy of the commons.** Impact factor mania persists because it is useful to certain scientists, certain journals, and the bureaucracy of science, particularly in certain nations. These interests synergize to maintain the malady despite its obvious flaws and detrimental consequences for science. Benefits to individuals and special interest groups do not translate into an overall benefit for the scientific enterprise. In keeping with the winner-takes-all economics of science (chapter 19) (75), impact factor mania benefits a few, creates many losers, and distorts the process of science, yet can be understood as rational behavior by individual scientists because of the large rewards accrued by those who succeed. In 1968, Garrett Hardin authored an essay where he used the phrase "tragedy of the commons" to describe in economic terms a situation in which individuals carry out behavior that is rational and in their self-interest but detrimental to the community (76). In this regard, impact factor mania exemplifies a tragedy of the commons in the midst of the scientific enterprise because the behaviors associated with IF mania benefit

certain individual scientists to the detriment of the community (8, 77). Impact factor mania will continue until the scientific community makes a concerted effort to break this destructive pattern of behavior.

CALLS FOR REFORM

Scientists' unhealthy obsession with impact and impact factors has been widely criticized (2–5, 14, 21, 69, 78–85). Yet many feel trapped into accepting the current value system when submitting their work for publication or judging the work of others. Some efforts to counter impact factor mania have been noteworthy.

DORA

The American Society for Cellular Biology, in concert with journal editors, scientific institutions, and prominent scientists, organized the San Francisco Declaration on Research Assessment (DORA) in December 2012 to combat the use of the journal impact factor to assess the work of individual scientists (86). Some prominent scientists in the biomedical research community spoke out in support of DORA. Nobel Prize winner Harold Varmus advocated the redesign of the curriculum vitae (CV) to emphasize contributions instead of specific publications (87). Varmus decried a "flawed values system" in science and lamented the fact that "researchers feel they will win funding only if they publish in top journals" (87). Although the DORA initiative has not been as influential as was initially hoped, it has made a difference in certain institutions. The University of Bern discontinued the practice of using JIF in promotion and replaced it with the relative citation ratio (Box 26.3) (16). An assessment of the impact of DORA five years after the declaration found that it had enhanced awareness of the problems with JIF and led several funding organizations to deemphasize its use (88). Given that impact factor mania is a cultural-sociological phenomenon and that cultural changes take time, perhaps it is still too early to assess the full impact of this movement.

Boycotting high-impact journals

In 2013, Nobelist and former *eLife* editor in chief Randy Schekman criticized the monopoly of what he calls "luxury journals" in an editorial published in the British newspaper *The Guardian*. Schekman vowed henceforth not to publish in *Nature*, *Cell*, and *Science*, stating that the disproportionate rewards associated with publishing in those journals distorts science in a manner akin to the effects of large bonuses on the banking industry (86). Although such efforts are well-intentioned, we remain skeptical that boycotting the impact factor or the "luxury journals" will be effective, because the economics of current science dictate that scientists who succeed in publishing in such journals will accrue disproportionate rewards. This continues to be an irresistible attraction. Even if the journal impact factor were to disappear tomorrow, the prestige associated with certain journals would persist, and authors would continue to try to publish there. Most scientists do not actually know the impact factors of individual journals—only that publication in such journals is highly sought after and respected. For instance, it is not widely known that *Science* is actually ranked only 27th among all journals in impact factor (2021 JIF), lower than journals such as *CA—A Cancer Journal for Clinicians* and *World Psychiatry*. Similarly, we fear that boycotts of specific prestigious journals would hurt the trainees of those laboratories by depriving them of high-visibility venues for their work. In lieu of *Science, Nature,* or *Cell,* Schekman has recommended that authors submit their best papers to *eLife*. Upon its founding, the managing executive editor described *eLife* as "a very selective journal" (89), but in 2022 the journal announced that it would no longer use reviewer assessments to decide whether a paper would be accepted or rejected, and the impact of this editorial policy change on submission rates remains to be seen.

The Leiden Manifesto

The Leiden Manifesto for research metrics was proposed in 2015 and named after the city in The Netherlands where it was presented (35). This framework was developed by experts on bibliometrics and consists of 10 principles for use in data evaluation. Like DORA, it cautions against the assessment of investigators and their output in simplistic numerical rankings, given problems with bibliometric information and what it measures, but goes further in codifying principles that could be considered a guideline for best practices in this developing field (35).

WHAT INDIVIDUAL SCIENTISTS CAN DO

Despite widespread recognition that the impact factor is often misused, the misuse continues and is likely to continue until the scientific community acts to reduce its value to those who benefit from its use. Young scientists may feel trapped in a situation that is no fault of their own and powerless to do anything about. As long as a critical mass of scientists continues to submit their best work to highly selective journals, impact factor mania, or its equivalent, is likely to persist. Nevertheless, scientists are not powerless to confront this social construct.

Thus far the many criticisms of the IF seem to have had little effect on its inappropriate use, creating a sense of resignation to the flawed value system that pervades science today. Most scientists feel helpless to fight the system and simply play the game of trying to get their work into a journal with the highest IF. Others express concern that their fields, which have historically published their work in respected but low-IF, society-supported journals, may go extinct if forced to compete on the basis of JIF in academic settings (77). The JIF poses particularly strong headwinds for young investigators as they seek to establish their reputations (90). However, there are some encouraging signs that JIF mania may have begun to abate. A group of scientists and organizations have signed on to the DORA initiative (see above) in the hope of discouraging the use of JIF in hiring and promotion decisions (91). A change in the format of the CV to emphasize accomplishments and deemphasize papers has been advocated; such an approach is reflected in the new requirements for biosketches used in grant applications to the National Institutes of Health. Some eminent scientists have called upon researchers to place less emphasis on IF and more emphasis on quality and service when selecting an appropriate journal to publish their work (86, 92).

Although JIF mania is likely to continue its stranglehold on the biomedical sciences for the foreseeable future, we are hopeful that this is a temporary aberration in the long trajectory of the history of science. Below we suggest ten actions that could help science shake off this malady.

1. **Reform review criteria for funding and promotion.** We strongly advocate the assessment of research quality by peers in an individual's field rather than using simplistic measures based on the quantity of papers and prestige of the publication venue. Academics should resist attempts by institutions to use JIF in promotion or tenure decisions. When evaluating the performance of scientists, a focus should be on scientific contributions rather than publications in prestigious journals. Academic administrators should be educated that impact factor is an inadequate measure of individual achievement (93) and should be informed of the DORA principles. Individual scientists may sign up to support the DORA initiative at http://am.ascb.org/dora/ (94).

2. **Consider the use of diverse metrics.** The listing of journal IF in CVs is a new phenomenon that presumably reflects a desire by the individual to elevate the importance of a published work by demonstrating that it was published in a frequently cited journal. This rationale is flawed, since the journal IF is an average of article citation frequency, which provides no indication of how frequently the specific article will be cited in the future. This practice should be discouraged, and such information should be disregarded.

   If metrics are to be used for judging individual scientists and their projects, review committees and administrators should consider a diverse range of parameters. This is one of the messages in the Leiden Manifesto (35). Although we do not suggest that the quality of scientists' work can be reduced to a single number, we do note that alternative citation metrics, such as the number of citations, h-index, relative citation ratio, and adjusted standardized citation metrics (Box 26.3), are probably better measures of scientific impact than the JIF. In addition, evaluations that employ diverse and adjusted metrics are likely to provide more insight than assessments based on single simple criteria.

3. **Increase interdisciplinary interactions.** The specialization of science has made it increasingly difficult for scientists to understand research outside their own field (12). However, there is no substitute for actually reading and understanding a scientific article. This requires that scientists who are asked to judge the productivity of another scientist working in a different field must acquire a working familiarity with that field. Increasing opportunities for interaction between researchers from different fields in training programs and at seminars and meetings will help to improve the quality of research assessment as well as stimulate transdisciplinary or multidisciplinary research.

4. **Encourage elite journals to become less exclusive.** Letters of rejection often state that "we regret that we receive many more meritorious submissions than we can publish." Why shouldn't elite journals expand to accommodate all meritorious articles? Artificial restrictions on journal size serve to perpetuate the current wasteful system that requires authors to cascade serial submissions from one journal to another.

5. **Address current imbalances in research funding and the scientific workforce.** As long as there is an unreasonable level of competition for grants and jobs, methods to compare scientists with one another are going to be utilized. Scientists must work with policymakers to alleviate the current shortages of research funding and job opportunities that have created the current crisis.

6. **Return to essential scientific values.** Ultimately, the only cure for impact factor mania must come from scientists themselves. If scientists fail to curb their current impact factor mania, they will pass onto their trainees a distorted value system that rewards the acquisition of publications in exclusive journals rather than the acquisition of knowledge and one that promotes an obsession with individual career success over service to society. Moreover, the current insistence on funding only high-impact projects is skewing the focus of research efforts and increasing the likelihood that important avenues of investigation will be overlooked. Scientists must return to essential scientific values that place an emphasis on research quality and reproducibility, the advancement of knowledge, and service to society over the accumulation of publications in prestigious journals.

7. **Diversify journal club selections.** Journal clubs are popular in research institutions and are generally considered essential for the training of young scientists. A typical journal club format involves the review, discussion, and critique of

selected scientific papers to inform participants about new developments in science. Journal clubs are often dominated by papers from a small number of high-IF journals, based on the assumption that such papers are more likely to represent cutting-edge science. Although there is no question that many papers published in high-IF journals are outstanding, the persistent selection of such papers for journal clubs perpetuates the misperception that JIF equals quality. Discussing interesting articles from more-specialized society journals in a journal club can counteract this impression and might help to improve journal club discussions, which too often degenerate into discussions of why a particular paper was published by such a high-impact journal.

8. **Do not judge papers based on the reputation of the journal.** In scientific conversations, one often hears scientists justifying findings by saying that the work was published in a high-IF journal. This lazy and all-too-common practice contributes to IF mania by replacing critical assessment with prestige by association. Justifying scientific quality based on publication venue makes little sense given the wide range of citation impact of individual papers published by high-IF journals. Science should be judged by the quality and interest of the data and their reproducibility.

9. **Discuss the misuse of the IF in ethics courses.** All scientists should be aware of the pervasiveness of IF mania, how the IF is calculated, how the IF influences scientist behavior, and the ways in which some journals attempt to game the system and elevate their IF (95). These issues can be incorporated into the graduate curriculum and discussed in seminars on publication ethics for established scientists, postdoctoral fellows, and research staff. This information would allow scientists to understand the limitations of this parameter for any use other than comparing journal citation averages.

10. **Cite the most appropriate sources in scientific papers.** By definition, papers in high-IF journals are cited, on average, more frequently. Authors may preferentially cite papers from high-IF journals due to their greater visibility. Some evidence to support this notion has been obtained from a study in which investigators independently reviewed research submitted to a single meeting and rated it according to quality and newsworthiness (23). The subsequent citation performance of the work showed only a very weak correlation with the independent scientific assessment, whereas the IF of the journal was the strongest predictor of subsequent citation by other papers. If this is a general phenomenon, then the positive feedback loop between IF and citation may represent another example of the Matthew effect in science (28), in which papers in high-IF journals are more frequently cited than comparable papers in low-IF journals for reasons unrelated to their quality and importance (96). Authors may also preferentially cite papers from high-IF journals to give the

appearance of virtue by association. Such problems can be avoided if authors take care to cite the most appropriate original source of a statement irrespective of the journal in which it is published. With today's emphasis on citation productivity, it is more important than ever for authors to be as complete and accurate as possible in referencing the scientific literature.

CONCLUSIONS

Today, there is a widely held notion that investigators must publish in high-IF journals to obtain grants, jobs, or promotions. In this economic framework, a journal's IF correlates with quality, and the mean IF of papers in a particular journal is taken to imply that all papers published by that journal are of similar quality. Since IF is proportional to the frequency of citations and inversely proportional to the number of papers published, journals are rewarded for being highly selective, excluding articles that are not anticipated to be highly cited in the near future. This exclusivity creates an artificial scarcity that is conflated with quality. In turn, there is tremendous pressure on researchers of all ranks to publish in high-IF journals, as acceptance by such journals is considered indicative of high-quality work (in many cases, by other scientists who have not even read the paper), which brings disproportionately high rewards to the author. This reward system can create incentives for behaviors that are not conducive to good science, including secrecy, haste, misconduct, and error (65, 97, 98). Consistent with this notion are the positive correlations between the proportion of papers retracted from a journal and its impact factor, which we called the retraction index (64), and the proportion of retractions that result from misconduct (65) (chapter 24). Even when there is no obvious misconduct, there is a concern that many studies in highly cited journals are not reproducible (99) (chapter 7). Although lack of reproducibility can have various causes, ranging from inadequate controls to misconduct, the demand by high-IF journals for clean stories with a clear message may tempt some investigators to selectively report their data, thereby reducing the likelihood that the work can be reproduced. The possibility that a focus on impact over importance is distorting the course of science should be of tremendous concern to all scientists, even those who benefit from the status quo. All of this may be having an effect on the overall efficiency of science, as historical trends in research expenditures and measurable outputs in the biomedical sciences suggest that the frequency of revolutionary breakthroughs may be declining (100).

IF mania continues despite broad condemnation because it is useful to certain elite investigators, journals, and funding organizations (8). As long as resources and positions remain scarce, the perverse competitive cycle driven by IF mania will continue despite the overall damage that it causes to the scientific enterprise. However, there are encouraging signs that the JIF craze has crested and that the mania may be abating. Utrecht University in The Netherlands has abandoned the use of JIF in evaluating published work during hiring and promotion decisions (101), but as a sign of how entrenched the practice is, the decision was criticized by some scientists (102). Other universities may be following suit, and a pan-European agreement supported by the European Union was announced in 2022 to curtail the use of JIF in academic evaluations (103). There is also bibliometric evidence for a declining influence of the JIF. An analysis of citation trends in the 20th century found that the relationship between journal IF and citations increased until 1990 but subsequently declined. Furthermore, the proportion of highly cited papers outside the top journals is increasing (62, 104). If these trends continue, we expect to see more diversity in highly cited sources, which would be good for science. However, even if the importance of the JIF is in decline, an adverse impact on the course of science from decades of IF mania will continue for years to come. A renewed effort is needed to return science to an emphasis on rigor, reproducibility, and responsibility while encouraging scientific curiosity in all its forms. Together we can disimpact science. Science will feel much better afterwards.

27 Risky Science

Research is a way of taking calculated risks to bring about incalculable consequences.

Celia Green (1)

Science and technology have created both unprecedented opportunities and risks for researchers and the societies in which they operate. As Isaac Asimov observed, "The dangers that face the world can, every one of them, be traced back to science. ... The salvations that may save the world will, every one of them, be traced back to science" (2). Managing this risk is a never-ending challenge, and when to limit the scope of research in order to mitigate risk poses a difficult dilemma. Since much scientific research occurs at the boundary of the known and unknown, scientists must continually face risks that are incompletely understood. Here we consider risks to both experimenters and society.

PERSONAL RISKS

The history of science contains many examples where the pursuit of knowledge placed the researcher at risk because of dangers inadequately appreciated at the time. Although scientists can die from laboratory accidents, often involving known risks and lapses in safety precautions (Box 27.1), here we emphasize personal risks from research in which the degree of danger was not known at the time the work

Thinking about Science: Good Science, Bad Science, and How to Make It Better, First Edition.
Ferric C. Fang and Arturo Casadevall.
© 2024 American Society for Microbiology.

Box 27.1 Biosafety versus biosecurity

The terms "biosafety" and "biosecurity" are often used interchangeably, and while related in the sense that both aim to prevent harm from dangerous biological agents, they differ in meaning. The Merriam-Webster online dictionary defines "biosafety" as "safety with respect to the effects of biological research on humans and the environment" and "biosecurity" as "security from exposure to harmful biological agents" (59). In practice, biosafety relates to protocols and procedures used in the research environment to ensure the safety of laboratory personnel. In contrast, biosecurity is more expansive and includes protecting the safety of society, agriculture, and material from biological agents. To distinguish the terms further, consider that wearing eye protection in the laboratory is part of biosafety, while safeguarding dangerous microbes from terrorists would fall under biosecurity. Some aspects of biosafety and biosecurity clearly overlap. For example, good biosafety practices to prevent infection in the laboratory with a contagious agent are also a form of biosecurity since they safeguard society from a spillover infectious disease event.

---

was performed. Table 27.1 lists just some of the scientists believed to have died from an occupational exposure during their research. Some, such as those pursuing new elements, died from exposure to new gases generated in the course of their research, such as fluorine (3). Others, such as Jesse Lazear and Clara Maass, died during human experimentation to investigate the route of yellow fever transmission, while Daniel Carrión died following self-inoculation of *Bartonella*-infected tissue (Box 27.2). Numerous casualties from radiation poisoning occurred in the early days after the discovery of radioactivity and X rays, of which Clarence Dally and Marie Curie are well-known victims. There are numerous well-documented episodes of scientists becoming infected with pathogenic microbes acquired in laboratory accidents. An authoritative review in 1979 reported several thousand episodes involving bacteria, fungi, parasites, and viruses (4), and many additional cases have occurred since then (5).

SOCIETAL RISKS

Although science has provided tremendous benefits to society, the application of science and technology toward military objectives has produced a tremendous increase in the potential destructiveness of war, as exemplified by the development

TABLE 27.1 A few of the scientists believed to have been injured or killed from research-related exposures

Scientist	Year	Exposure	Outcome	Reference(s)
Georg Wilhelm Richmann	1753	Lightning	Electrocution	64
Carl Wilhelm Scheele	1786	Heavy metals	Renal failure, death	65
Ascanio Sobrero	1846	Nitroglycerin explosion	Facial scarring	66
Jerome Nickles and Paulin Louyet	1869 and 1880	Fluorine	Poisoning	3
Daniel Carrión	1885	*Bartonella bacilliformis*	Fatal Oroya fever	60
Vera Bogdanovskaia-Popova	1897	Chemical explosion	Death	67
Jesse Lazear and Clara Maass	1900 and 1901	Mosquito bite	Yellow fever	68
Clarence Dally	1904	X rays	Skin cancer	69
Elizabeth Fleischman-Ascheim	1905	X rays	Cancer	70
Alexander Bogdanov	1928	Blood transfusion	Transfusion reaction	71
Marie Curie	1934	Radiation	Aplastic anemia	72
Harry Daghlian and Louis Slotin	1945 and 1946	Nuclear accident	Radiation sickness	73, 74
Andrei Zheleznyakov	1987	Novichok	Exposure sequelae	75
Karen Wetterhahn	1997	Dimethylmercury	Mercury poisoning	76

Box 27.2 Daniel Carrión, medical martyr

In 1885, when Daniel Alcides Carrión García was a medical student in Lima, Peru, he became aware of an unproven hypothesis that a local skin condition called *verruga Peruana* (i.e., Peruvian warts) was linked to an acute illness called "Oroya fever." He asked for permission to experiment on himself to test this hypothesis, but his teachers refused. Undaunted, Carrión coaxed a physician friend, Evaristo Chavez, to help him inoculate his arms with material taken

(*Continued*)

Box 27.2 (Continued)

from a teenage boy with *verruga Peruana*. Approximately one month later, Carrión developed fever and hemolytic anemia characteristic of Oroya fever and ultimately succumbed to the disease. Chavez was tried as an accessory to murder but acquitted, and Carrión was declared a national hero and medical martyr (60). Today we recognize that *verruga Peruana* and Oroya fever are different clinical manifestations of infection with the hemotropic bacterium *Bartonella bacilliformis*. Although there are many examples of self-experimentation in history, the ethical considerations remain controversial (61–63).

Daniel Alcides Carrión García (1857–1885).

of nuclear weapons. However, other than the exceptional circumstances when scientific research was used to develop new weapons, such as gas warfare and nuclear weapons, most of the science and technology that has increased the lethality of war was not developed specifically for use in warfare *per se*. Consequently, the societal risks from research should be distinguished from the human penchant to use any available technology as a tool for war. One event that highlights the societal risk from medical research is the 1977 reappearance of the H1N1 influenza virus, which had disappeared from circulation in human populations since 1957 but was maintained in laboratories. In 1977, H1N1 influenza reemerged in human populations and subsequently became endemic (6). The close identity between laboratory strains and the 1977 epidemic strain heightened suspicion of a lab leak, as the influenza virus ordinarily undergoes genetic drift each season (6). In 2020, the initial detection of the severe acute respiratory syndrome coronavirus 2 (SARS-CoV-2) in Wuhan, China, which resulted in the COVID-19 pandemic, raised similar fears about a possible laboratory escape event (7). However, establishing the origin of SARS-CoV-2 has proven difficult because of insufficient information about its natural reservoir. At the time of this writing, the preponderance of evidence and opinion in the scientific community is that the SARS-CoV-2 emergence resulted from zoonotic transmission from a natural reservoir, possibly with involvement of an intermediate host (8). This will be further discussed in chapter 30.

EXAMPLES OF POTENTIALLY RISKY RESEARCH

Large Hadron Collider (LHC)

The LHC was built by CERN (Conseil Européen pour la Recherche Nucléaire, or the European Organization for Nuclear Research) for the sole purpose of studying high-energy particle physics; this led to confirmation of the existence of the Higgs boson. Before the instrument was used, some suggested the possibility that high-energy collisions could unleash doomsday events through the creation of hypothetical particles known as *strangelets*, which can be viewed as microscopic black holes. In response, CERN commissioned studies, which concluded that such events were highly improbable (9), and the machine began operations in 2008. To date there has been no evidence to support the existence of such threats, even as higher energies have been explored. It is noteworthy that reassurances on safety are based on probability calculations of known and assumed parameters, as reflected in the final paragraph of the CERN report:

> Before closing this introduction, we wish to make a general remark: All estimates concerning production probabilities and subsequent properties of various objects at the LHC necessarily involve certain theoretical assumptions. Some, for example the invariance of physical laws under space and time translations, are so general that they do not need to be explicitly stated. Others are based on extrapolations of known properties of hadronic systems. They will be explained in the following sections whenever they are used. Here we want only to emphasize that no estimates are absolutely assumption-free. (9)

Statistical reasoning in risk analysis has been criticized on the grounds that error or uncertainty in the parameters used might result in a misleading assessment of the likelihood of catastrophic events (10). Perhaps the greatest reassurance on the safety of the LHC comes from nature itself, since cosmic rays with energies much higher than anything produced by the LHC reach the Earth, yet the universe remains intact.

Transgenic species, gene drives, and synthetic biology

These techniques straddle the space between basic science and technology, with some making such rapid progress that they are already being applied in industry. For example, transgenic animals and plants are already part of agriculture and pharmaceutical production. Transgenes are also used to create novel life forms, such as the popular GloFish, which express fluorescent proteins to create fish with artificial colors. Gene drives and synthetic biology are earlier in their development, but each has the potential to produce unintended consequences, thus falling under the rubric of risky science. Whereas gene drives are a form of genetic engineering that allow a deleterious gene to spread rapidly through a population (11), synthetic biology is a field that encompasses all of molecular biology with principles incorporated from engineering, chemistry, and physics (12), with the potential to generate new components, systems, or life forms that do not exist in the natural world. The construction of transgenic species has already proven tremendously useful to humanity in the form of pest-resistant crops and drug-producing microorganisms, etc., while both gene drives and synthetic biology offer tremendous power to alter the biosphere. The concerns with transgenic species, gene drives, and synthetic biology all stem from uncertainty about possible unforeseen outcomes in applying such powerful technologies. For example, the introduction of transgenic corn and cotton into agriculture has been followed by escape of the genetically engineered genes into wild related species by hybridization (13, 14). Although humans have been genetically modifying other species for thousands of years through selective breeding, the notion of giving a species new properties by directly modifying its DNA rapidly and permanently with transgenes, gene drives, or synthetic biology raises a new set of biosafety and ethical issues that can trigger strong opposition. In recent years, a movement has emerged against genetically modified organisms (GMOs), focusing primarily on edible plants, leading to bans in certain countries and the mandatory labeling of foodstuffs regarding their GMO content. Although such movements are often well-intentioned, there is concern that they could prevent the development of much-needed new crops that are more productive, less resistant to pests and diseases, and more likely to thrive in warming climates. A case in point is the development of golden rice, so named because it is rice expressing beta-carotene, which is a precursor of vitamin A. Vitamin A deficiency is widespread in many poor countries where rice is a staple diet and is associated with high childhood mortality. Substitution of regular rice with golden rice could treat the vitamin deficiency without disrupting the diet and farming practices in those regions, and there is a widespread consensus that it is safe. However, the introduction of golden rice has elicited significant opposition from groups opposed to GMO foods, and its use remains limited (15).

Artificial intelligence (AI)

Humanity is currently living through a period of tremendous changes brought on by the information revolution (16). Computer programs able to perform functions that ordinarily require human intelligence, such as decision-making, perception, translation, and conversation, are included under the umbrella of AI. Such programs may be tremendously useful in all aspects of human endeavor by complementing human intelligence with the speed, power, and deep memory of computer systems. Nevertheless, there are concerns. In the short term, there are concerns about loss of privacy, work, and security associated with ever more powerful programs. In the long term, there are concerns that such programs can exceed the capacity of human brains and, once unleashed, could be difficult to control. As AI programs become more complex and linked to enormous databases, there is an increasing possibility of unpredictable negative consequences.

Nanotechnology

This is a branch of science and inquiry that deals with very small structures (less than 100 nanometers, or about one-thousandth the width of a human hair). The central idea in nanotechnology research is the creation of novel minute structures with new properties that can perform new functions. One example of nanotechnology is the creation of buckminsterfullerene C_{60}, also known as fullerenes or buckyballs. When fullerenes are arranged into a cylinder, they form carbon nanotubes. There is intense interest in these materials for numerous applications in medicine, construction, and electrical conductivity (nanoelectronics). Given the combinatorial versatility of carbon bonds and the many possible types of geometric structure, the number and types of fullerene and carbon nanotube-like compounds are astronomical, and each has different capabilities, properties, and toxicities. The promise of carbon nanomaterials is that they could trigger revolutionary changes in cancer therapy and drug delivery (17). However, as with any new technology, there are concerns about how these materials interact with the biosphere. Here again, their novelty means we have little or no precedent for calculating risk. There is already evidence that occupational exposure to nanofibers can trigger inflammatory changes and tissue remodeling resembling the responses to asbestos (18), which is a potent carcinogen.

THE GAIN-OF-FUNCTION RESEARCH CONTROVERSY

Gain-of-function (GOF) research refers to the alteration of a microorganism to confer a property that may make it more dangerous, such as acquiring the ability to be transmitted to mammals. Unfortunately, the term "GOF" is a poor descriptor for the research, which has caused considerable confusion since it literally describes many different properties that have little to do with biosafety or biosecurity. For example, cloning a gene into a bacterium to produce a protein such as insulin literally gives it a new function, but such a modification would not be of concern and is not considered GOF research. In current parlance, GOF research usually refers to modifications that enhance the characteristics of a microorganism, often a virus, to increase its threat potential to humans or animals. This might best be illustrated with concrete examples. In 2012, two research groups reported the adaptation of the H5N1 bird flu virus to permit mammalian transmissibility and the molecular changes responsible for this new property (19, 20). This was a major finding, because H5N1 was known to be an avian influenza virus that occasionally caused human infections with high mortality but had not shown a propensity for human contagiousness or spread. Hence, adaptation of the virus to allow mammalian transmission immediately raised concerns about biosecurity and biosafety (Box 27.1). In the area of biosecurity, there was concern that knowledge about key amino acids responsible for mammalian transmission could lead a bad actor to engineer the natural H5N1 virus for use as an agent of bioterrorism or biological warfare. In the area of biosafety, there was concern that the creation of such viruses in the laboratory could lead to inadvertent human infections with the potential to escape into the community. Proponents of GOF-type experiments argued that this line of work is essential to identify basic mechanisms of pathogenesis that will inform new treatments and vaccines. Opponents of GOF-type experiments argue that there is no historical evidence to support the usefulness of this information and that the risks outweigh the benefits. The debate on GOF-type experiments of this nature is ongoing, with no evidence that the arguments advanced by one side are found persuasive by the other.

Early in the 2000s, the debate over GOF-type experiments was focused more on biosecurity and the concern that information generated by such experiments could be used to make new weapons for bioterrorism and biological warfare. Biosecurity concerns were heightened by the mailing of anthrax spores to several politicians and media organizations shortly after the terrorist attacks of September 11, 2001. The mailing of anthrax spores illustrated how disruptive a biological attack could be, resulting in the prolonged closure of government buildings and mail facilities. Such concerns were heightened by publications in the early 2000s describing the ease by which manipulation of pathogens could increase their virulence. For example, in 2002, the insertion of the interleukin-4 gene into ectromelia virus was found to neutralize vaccine immunity (21), which suggested an approach for turning smallpox virus into a biological weapon that could no longer be protected against by vaccines (22, 23). In the following year, a paper describing how poxvirus complement evasion proteins could be used to bypass innate humoral immunity (24) further elevated concerns about the intentional manipulation of pathogens to enhance their virulence. Adding still more fuel to the fire, a paper was published describing a mathematical model that analyzed the milk supply in the United States and identified vulnerabilities that might allow purposeful contamination with botulinum toxin to result in widespread disruption (25).

In response to these concerns, the United States government convened the National Science Advisory Board for Biosecurity (NSABB), on which one of us (A.C.) served from 2005 to 2014. The NSABB worked for many years to develop the concept of dual-use research of concern (DURC), which aimed to define and isolate the small number of research activities for which there was a high likelihood that the information generated could be misused for malevolent purposes (26). In 2012, the NSABB became embroiled in the GOF controversy, when the U.S. government asked for prepublication advice regarding the GOF experiments in which H5N1 virus was adapted for mammalian transmissibility. The NSABB recommended revision of the manuscripts to reduce the likelihood that the research could be used by bioterrorists (27), but the papers were subsequently published without major modifications (28). Perhaps the most significant outcome of that controversy was to sensitize the scientific community to the potential risk entailed by certain types of research, but there was no consensus on whether such research should be prohibited, with proponents and opponents each arguing for the benefits and risks of such work, respectively. Furthermore, the question of whether scientific data should be redacted during publication to reduce the chance of malevolent use was left unanswered. In 2013, the *Journal of Infectious Diseases* reported a new variant of botulinum toxin that was not neutralized by available antibody antidotes (29) but redacted the protein sequence to minimize its potential for misuse (30). Redaction of sensitive data prior to publication raises a variety of ethical and legal issues that remain unsettled (31).

In the second decade of the 21st century, with no further biological attacks since the mailing of anthrax spores in 2001, concerns about risky research gradually shifted from biosecurity to concerns about biosafety and the risks for society from inadvertent release of dangerous pathogens. Remarkably, the FBI concluded that a government scientist who didn't feel that enough was being done to safeguard the community from bioterrorism was responsible for mailing the anthrax-laden letters (32, 33). More recently, the catalysts for increased security were a series of highly publicized incidents highlighting potential societal risks from well-intentioned biological research. In 2014, smallpox virus was discovered in an old freezer at the National Institutes of Health campus in Bethesda, Maryland, despite assurances that such virus samples had been destroyed (34). In other unrelated incidents, the

Centers for Disease Control and Prevention inadvertently mailed a sample of the deadly H5N1 influenza virus instead of another influenza virus sample, and there was concern that laboratory personnel may have inadvertently been exposed to anthrax spores in a research laboratory (35). In addition, attention was shifting from conventional biological weapons such as anthrax spores to the concept of pathogens with pandemic potential (PPP), which were primarily viruses with a demonstrated history, or potential, for causing pandemics. PPPs included poxviruses, influenza viruses, and coronaviruses, among others. Again, the battle lines were divided between those most concerned about risks (36, 37) and those who focused on potential benefits, including one of us (A.C.) (38, 39). The assessment of the risks involved varied greatly among the participants in the debate, ranging from large (40) to minuscule (41). In this environment, the government mandated pauses for GOF-type experiments in influenza virus research (42), which eventually resumed with some limitations and additional oversight provided by the NSABB (43).

Suffice it to say that nothing had been definitively resolved regarding so-called GOF experiments when SARS-CoV-2 unleashed a pandemic in 2019. It is ironic that much of the debate during the prior two decades was about influenza virus variants such as H5N1 avian influenza, but the culprit this time turned out to be a coronavirus, a class of viruses that was underestimated in its pathogenic potential prior to the outbreak of SARS in 2003. The COVID-19 pandemic has validated and humbled the views of both camps in the GOF debate (44). For opponents of GOF experiments with PPP, the COVID-19 pandemic has vividly illustrated the risks associated with the inadvertent or deliberate release of such agents into human populations, while also showing that earlier minimization of risks by proponents of GOF experiments was inappropriate, given how unprepared the world was to handle a pandemic (44). Allegations of inadequate biosafety precautions at the Wuhan Institute of Virology, disagreement about whether scientists at the institute were performing GOF research, and a lockdown by the Chinese government on information relating to the origins of SARS-CoV-2 have only increased concerns (45, 46). On the other hand, prior knowledge about coronaviruses gained from decades of painstaking research resulted in the rapid development and deployment of diagnostics, vaccines, antibody-based therapies, and small-molecule antiviral agents, reinforcing the importance of basic research in the pathogenesis and immunity of viral infections for preparation against future outbreaks. However, the question of whether GOF-type experiments are needed to investigate and prepare for threats from nature remains unresolved.

At the heart of the dispute on GOF-type experiments is uncertainty about the relative value and risks of such experiments. Since the controversy involves questions of values and judgment, it is difficult to see how a consensus that addresses all concerns can be achieved. Given this impasse, Michael Imperiale and one of us

(A.C.) have proposed a different approach, which focuses on first deciding what needs to be known about a particular pathogenic microbe and then considering the various experimental approaches that could provide the necessary information (47). In their synthesis, GOF-type experiments might be justified by the importance of the question asked and whether the proposed experiments are able to meet specific criteria, and then they should only be done with the greatest safety precautions. These authors and others have also called for the establishment of a national mechanism to help investigators to navigate questions of methodology, publication, and the need for GOF-type experiments (48). It is envisioned that such a review board would be modeled on the highly successful Recombinant DNA Advisory Committee of the National Institutes of Health (49) that guided research during the molecular biology revolution to minimize risks while still allowing the development of numerous new technologies and therapies.

ROGUE SCIENCE

Rogue science refers to scientific work that occurs outside of norms of society and the consensus of the scientific community. For example, in 2012, a businessman arranged to dump 100 tons of iron ore dust in the Pacific Ocean off the Canadian coast to determine whether it would foster the growth of plankton and trap atmospheric carbon (50). This experiment was performed in violation of a United Nations moratorium on ocean fertilization experiments and raised concern about the possible promotion of toxic algal blooms. In another instance, in 2018, the Chinese scientist He Jiankui announced to the world that he had used CRISPR (clustered regularly interspaced short palindromic repeats) to perform genome editing in human embryos, which was in violation of guidelines and norms against the application of this technology in humans (51). Fortunately, such examples of rogue science are relatively rare. However, when they occur, they carry significant risk to experimental participants and the environment and could also trigger oppressive regulations and undermine public trust in science. Minimizing risk through careful experimentation, applying the precautionary principle of "better safe than sorry," following regulations, consulting colleagues, and conducting research in an open environment are essential to maintaining public trust in science.

MINIMIZING AND MITIGATING RISK

Today, a multilayered structure is in place to minimize risk. In the United States, students learn about biosafety and the ethical conduct of research in courses that form an integral part of their training. Institutions also provide training for specialized areas of research that involve handling of human blood samples, animal research, and working with radioactive materials. At the institutional level, there are Institutional Biosafety Committees (IBCs), which were originally established to supervise recombinant DNA work but subsequently evolved as bodies to review and approve all research carrying potential danger, such as working with pathogens or dangerous chemicals including carcinogens. All human research must be reviewed and approved by Institutional Review Boards (IRBs), which also oversee the performance of the work through regular progress reports. Grant applications for the funding of science routinely require assurances that the work will comply with biosafety standards, and research involving pathogens with pandemic potential, select agents, or gain-of-function experiments is subject to additional layers of review. However, currently existing protocols focus on known risks, and both science and scientists remain vulnerable to unknown risks and to individuals who defy the norms. Self-regulation by the scientific community can be an important component of risk mitigation. In the early 1970s, advances in molecular biology allowed the expression of foreign genes in microbes. This led to the concern that genetic manipulation might result in dangerous new organisms. In 1975, a group of scientists, lawyers, and ethicists met at the Asilomar Conference Center in California to consider the risks and benefits of recombinant DNA technology. Applying the precautionary principle, they drafted a document, which articulated risks and benefits, and provided guidelines and recommendations for future work, including the need for containment of modified organisms (52). Similar efforts have been applied to human gene editing and synthetic biology.

Despite such measures, recent setbacks in international efforts to investigate the origins of the COVID-19 pandemic (53, 54) have underscored the inadequacy of our present institutions and norms designed to mitigate the risks of science. Although sober assessments of potential risks and benefits sound like a good basis for decision-making in principle, the benefits of research are often difficult to know at early stages (see chapter 8), and the estimates of risk are strongly influenced by whether one has a fundamentally optimistic or pessimistic attitude (55). As seen in the debate over risks and benefits associated with GOF research, scientists do not necessarily agree where to draw the line. It has further been argued that the societal risks of research are widely distributed, and decisions regarding risky science should not be left to scientists alone but must be informed by social and political considerations, as well as by scientific and technical assessments of risks and benefits (56, 57). New multidisciplinary mechanisms of oversight with procedural transparency, broad international support, and enforcement authority are needed. If we fail to strengthen current processes to oversee and limit scientific risk, the consequences may include both excessive restrictions on beneficial research and adverse scientific impacts on human society and the environment (58). That is a risk the scientific community cannot afford to take.

28 Authoritarian Science

> *A new scientific truth does not triumph by convincing its op-*
> *ponents and making them see the light, but rather because its*
> *opponents eventually die, and a new generation grows up that is*
> *familiar with it.*

Max Planck (1)

In this chapter, we visit the effects of authority in science. Specifically, we will explore how the authority that comes with the prestige of being an accomplished scientist is used and misused in the development of science. As is our custom, we begin with definitions. Dictionary.com defines "authoritarian" as "favoring complete obedience or subjection to authority as opposed to individual freedom" (2). Other definitions of the word stress that this term is often used in reference to the political organization of states. Such definitions and usages do not quite fit our needs, since science, while somewhat hierarchical, does not operate in an environment that demands complete obedience or subjection to authority. However, science might be viewed as authoritarian in that it values and relies on the authority of those who have made major contributions, possess special expertise, and/or are held in high esteem by their fellow scientists.

Western science in the Late Middle Ages and early modern era certainly met the definition of authoritarian, in that thought and inquiry were channeled into the received wisdom of antiquity, whereby scientific phenomena were defined

Thinking about Science: Good Science, Bad Science, and How to Make It Better, First Edition.
Ferric C. Fang and Arturo Casadevall.
© 2024 American Society for Microbiology.

by Aristotelian physics, and astronomy was explained by the geocentric theory and the Ptolemaic system. In this regard, modern science dates its stirrings to the struggles to overturn authority by new theories that accommodated new evidence, such as the struggle between Galileo Galilei and the Catholic Church. Later, with the advent of the Copernican and Newtonian revolutions, science embraced the notion that all knowledge is provisional, thus rejecting authoritarian trappings and accepting that progress in answering any one question inevitably leads to more questions. Hence, contemporary science is not authoritarian, but its authority in the modern world means that scientists may have disproportionate influence in certain matters within and outside their expertise. It is important to consider how that authority may be used for the benefit or harm of humanity.

Perhaps the best example of authoritarian science in the modern world is provided by Trofim Lysenko, who dominated Soviet biology and genetics for much of the mid-20th century. Lysenko rejected Mendelian genetics in favor of his own views of heredity, which combined obsolete Lamarckian concepts with his own logic and beliefs. These beliefs were then imposed onto Soviet science, and those who questioned them ran the risk of being labeled political opponents. Dissenting scientists were removed from their positions, and some were imprisoned or executed. Lysenko's rise and influence paralleled that of Stalin, and Lysenko's views about biology and inheritance were endorsed by the state because they fit within state policies. Mendelian genetics, which was making great progress in the rest of the world, was condemned in the Soviet Union as counterrevolutionary (3). Crop failures resulting from policies based on Lysenkoism resulted in mass starvation and cost millions of lives. The case of Lysenko provides an example of pseudoscience (chapter 16) becoming authoritarian science, enforced with the tools of the state, including the secret police. The consequences were devastating for Russian science and the Soviet people (4).

AUTHORITY IN SCIENCE

Apart from the case of Lysenko, whose authority on scientific questions came from the power of the state, authority in science is more frequently the result of scientific discovery and achievement. In general, this authority is provided by the respect of peers. However, some scientists can transcend their scientific reputations to achieve more widespread fame, and public acclaim endows them with an aura of authority even in areas far outside their expertise. Public fame and acclaim can translate into extraordinary influence in other realms. A good example of this phenomenon is the famous 1939 letter by Albert Einstein to President Roosevelt, alerting him of the potential of using the enormous energy generated by nuclear fission in the design of atomic weapons (5), an event that catalyzed the United States to begin the Manhattan Project. In this situation, Einstein was not an expert in nuclear fission but was well-known to the public from his contributions to the popularized theory of relativity. The letter had been written by Leo Szilárd, in consultation with other physicists who were experts in nuclear fission and aware of its potential and dangers. Yet only Einstein signed the letter. Although the other scientists were more knowledgeable on the specific topic, the decision was made to have Einstein send the letter, as his name would carry more weight. In other words, Einstein's popularity and recognized accomplishments gave him greater authority to convey a message to the President of the United States than true authorities in the field.

Whereas the Einstein letter to Roosevelt is an example of using authority in science to promote a field (nuclear fission) with enormous consequences for the outcome of the Second World War and the postwar world, scientific authority does not always promote science. A well-known example is the influence of Lord Kelvin on the disciplines of geology and biology in the late 19th century. Lord Kelvin was one of the best-known scientists of his time, who achieved fame for formulating the first and second laws of thermodynamics. When Darwin's *On the Origin of Species* was published in 1859, Lord Kelvin argued against the mechanism of evolution on the grounds that insufficient time had elapsed for speciation to occur, based on his own estimate of the age of the earth, which relied on extrapolating time from the assumed rate of cooling of a large molten rock. Unfortunately, this estimate did not consider radioactivity, which had not yet been discovered. Lord Kelvin estimated the age of the earth to be 100 million years, which turned out to be wrong by a factor of 45. This episode provides an example where the authority conferred upon a scientist by remarkable achievements in one field enabled him to provide opinions in another field that were erroneous.

Another well-known episode of contrarian authority was the opposition of Rudolf Virchow and Max Joseph von Pettenkofer to the germ theory of disease in the late 19th century. Both Virchow and von Pettenkofer were eminent scientists known to the public, who wielded great influence in their time. Virchow is credited with greatly advancing the cellular and pathologic basis of disease, and von Pettenkofer was a hygienist who helped to create the field of public health. Virchow believed that infectious diseases were a consequence of cellular degeneration followed by subsequent opportunistic infection (6), while von Pettenkofer also disputed the role of microbial contagion, but instead attributed illness to putrescent substances in soil and the resulting toxic vapors called "miasmas" (7). The opposition of Virchow and von Pettenkofer to the germ theory promoted by Koch and Pasteur was more akin to competing theories of disease causation and hence not directly analogous to the polemics of Lord Kelvin in evolution and geology. Nevertheless, the common thread is that the authority wielded by these prominent scientists delayed the acceptance of the germ theory.

Virchow also believed that epidemics were a result of social conditions, such as poverty, poor diet, and hazardous working conditions. Drawing upon his observations during a typhus epidemic in mid-19th-century Germany, Virchow advocated for government intervention to improve living conditions (8). Similarly, von Pettenkofer advocated improvements in nutrition and wastewater disposal (9). Unfortunately, his opposition to germ theory as the explanation for cholera had disastrous consequences for the city of Hamburg, where thousands of deaths occurred because the civic leaders accepted his views and neglected to clean up the water supply (7). So strong were his convictions that in 1892 von Pettenkofer publicly ingested live *Vibrio cholerae* bacteria after first swallowing bicarbonate to neutralize his stomach acid, to disprove that these bacteria caused cholera. It remains somewhat of a mystery to this day why he became only mildly ill, and he continued to insist to his dying day that cholera was not caused by bacteria (10). Although they were wrong in dismissing the primary role of microbes in epidemics, Virchow and von Pettenkofer were not completely in error. Today, we recognize the importance of host resistance, social conditions, and sanitation as major determinants of public health.

NOBEL DISEASE

Given that the Nobel Prize is considered to be the epitome of scientific achievement and recognition (chapter 21), Nobel laureates enjoy a privileged position in the scientific hierarchy of recognition, credit, and authority. Regrettably, a few laureates have deviated from their fields of expertise to espouse opinions in unrelated fields. This phenomenon has been called "Nobel disease" (11) or "Nobelitis" (12) by others (as noted previously, the definition of "Nobelitis" in this context is different from that we propose in chapter 32). A partial list of Nobel laureates who have strayed from their fields of expertise is provided in Table 28.1. Some of the adopted causes are relatively benign, such as Linus Pauling pushing the benefits of vitamin C based on his personal experience and beliefs. However, even the relatively benign focus on vitamin C had significant costs. Because of Pauling's prominence and popularity, significant clinical work was required to test and debunk his claims for vitamin C, which diverted resources from other problems. More distressingly, other Nobel laureates, such as Philipp Lenard, Julius Wagner-Jauregg, Johannes Stark, William Shockley, and James Watson, have promoted unscientific ideas on race and lent their authority to support deplorable racist theories (13–16). Hence, Nobel disease can carry

TABLE 28.1 Some Nobel laureates who have strayed from their areas of expertise

Nobel laureate	Accomplishment	Non-expert advocacy	Reference(s)
Philipp Lenard (1905)	Cathode ray tube	Nazi Aryan policies	25
Alexis Carrel (1912)	Perfusion pump	Eugenics, Nazi racial theories	26
Johannes Stark (1919)	Doppler effect	Nazi Aryan policies	25
Julius Wagner-Jauregg (1927)	Hyperthermia for syphilis	Racial genetics	15
Wolfgang Pauli (1945)	Exclusion principle	Mental-physical synchronicity	27, 28
Linus Pauling (1954, 1962)	Chemical bond theory	Prevention of the common cold by vitamin C	29
William Shockley (1956)	Discovery of transistor	Racial genetics	16
James Watson (1962)	DNA structure	Race and intelligence	14
Julian Schwinger (1965)	Quantum electrodynamics	Cold fusion	11
Ivar Giaever (1973)	Electron tunneling in superconductors	Climate change denialism	11
Brian Josephson (1973)	Quantum tunneling	Homeopathy, cold fusion	30
Nikolaas Tinbergen (1973)	Bird behavior	Causes of autism	11
Arthur Schawlow (1981)	Lasers	Facilitated communication for autism	11
Kary Mullis (1993)	Polymerase chain reaction	AIDS and climate change denialism	31
Richard Smalley (1996)	Fullerenes	Evolution denialism	11
Louis Ignarro (1998)	Nitric oxide	Herbal supplements	32
Luc Montagnier (2008)	HIV discovery	Homeopathy, origins of severe acute respiratory syndrome coronavirus 2 (SARS-CoV-2)	33, 34

significant societal costs. Such cautionary tales may serve as a warning for future laureates who believe that prestigious prizes confer intellectual infallibility and show that scientific authority can be co-opted for political and nonscientific purposes. Even the most accomplished scientists can hold erroneous beliefs when they deviate from the methods and principles of science.

FIELD AUTHORITY

Apart from these well-known instances where scientific authority has influenced politics, every scientist is aware of other scientists in their own fields who hold strong opinions on one subject or another and are able to sway the opinion about scientific ideas and the direction of research. In chapter 10, we considered how most scientists work within fields (17), which constitute the sociological units of science organized around a specific epistemic pursuit. Scientific fields have leaders who influence thought and behavior within the field, sometimes persisting through their followers even after they have departed the scene. For example, even within well-established fields founded on predictive and well-validated theories, such as quantum mechanics, there can be a wide diversity of opinions on fundamental problems that trace their origins to disagreements between such foundational figures in physics as Albert Einstein and Niels Bohr (18). Nevertheless, in his autobiography, the great German physicist Max Planck (Fig. 28.1) observed (see the epigraph that begins this chapter) that eminent scientific authorities may not change their minds when they are wrong, but their succession by younger generations of scientists allows old dogmas to be overturned. Paul Samuelson summarized this as "Science makes progress funeral by funeral." The MIT economist Pierre Azoulay has recently presented data to support this concept, showing that the premature deaths of eminent scientists are followed by a decline in publications by their collaborators and a surge in contributions by other scientists. This

FIGURE 28.1 *Max Planck (1858–1947), c. 1930. In addition to his memorable quote regarding scientific progress and his pioneering contributions to quantum mechanics, Planck had admirable moral character. In contrast to some other German physicists, he opposed anti-Semitism and the ousting of Jewish scientists from academia (35). Photograph by Transocean.*

dynamic may play an important role in the continuing development of scientific fields into new directions (19). The so-called "Planck's principle" was noted with interest by Thomas Kuhn in his seminal work on scientific revolutions (chapter 11) (20).

AVOIDING AUTHORITARIAN FOIBLES

The lesson from the examples we have provided here is that scientists must remain open to new ideas and take care not to stray from the epistemic principles of science. Had Virchow and von Pettenkofer accepted the germ theory of disease, they might have made more progress in their own fields of study. With incorporation of the germ theory, Virchow could have gleaned more insight into microbe-mediated cellular and tissue damage and elucidated mechanisms by which damaged tissues promote microbial invasion and replication. Similarly, von Pettenkofer, who was already an expert on sanitation, could have more effectively improved municipal water supplies to reduce the transmission of waterborne diseases. For the Nobelists with Nobel disease, danger can arise from the inappropriate use of authority conferred for achievements in specific areas to promote beliefs without a scientific basis.

The bottom line is that science rightly values expertise, but not authority. As no one can be an expert at everything, when considering important questions, it best to seek advice from as many experts as possible. We close with the wisdom of ancient experts:

Socrates: "I neither know nor think I know" (21).

Confucius: "To know that we know what we know, and that we do not know what we do not know, that is true knowledge" (22).

Leonardo da Vinci: "Whoever in discussion adduces authority uses not intellect but rather memory" (23).

Galileo: "In the sciences the authority of thousands of opinions is not worth as much as one tiny spark of reason in an individual" (24).

29 Deplorable Science

> This means that to entrust to science—or to deliberate control according to scientific principles—more than scientific method can achieve may have deplorable effects.
>
> Friedrich August von Hayek (1)

> Science does not have a moral dimension. It is like a knife. If you give it to a surgeon or a murderer, each will use it differently.
>
> Wernher von Braun (2)

> If you want to do bad things, science is the most powerful way to do them.
>
> Richard Dawkins (3)

In earlier chapters we have discussed a variety of desirable adjectives applied to science, including elegant, rigorous, important, and others. However, in this chapter, we must explore the darkest corners of this remarkable human endeavor, deplorable science.

Once again, we begin with definitions. The Merriam-Webster online dictionary defines "deplorable" as "deserving censure or contempt," while Dictionary.com defines it as "causing or being a subject for grief or regret; lamentable" and "causing or being a subject for censure, reproach, or disapproval; wretched; very bad" (4).

Thinking about Science: Good Science, Bad Science, and How to Make It Better, First Edition.
Ferric C. Fang and Arturo Casadevall.
© 2024 American Society for Microbiology.

These definitions are themselves complex, and some are dependent on the social mores of the time. However, we are interested in definitions that transcend time and historical trends, those that are not dependent on the extant social system. Hence, we will focus on definitions in which deplorable actions cause grief and actions in which science has caused human harm. We define deplorable science as scientific inquiry involving direct and deliberate human harm. This definition will permit us to navigate the complex intersection between deplorable and unethical, terms that are related but not synonymous (all deplorable science is unethical, but not all unethical science is deplorable). We recognize that this working definition will exclude science that some readers may consider deplorable, such as use of animals in research, use of embryonic tissues obtained in abortions, etc., but we seek to draw a clear *cordon sanitaire* around scientific activities that we hope all can agree are unacceptable. Furthermore, we distinguish deplorable science from science applied to war in an effort to gain a military advantage (Box 29.1).

Defining deplorable science as requiring or involving deliberate human harm takes us initially into the biological sciences. We will exclude self-experimentation

---

Box 29.1 Should a chapter on deplorable science include a quote from Wernher von Braun?

We began this chapter with a series of quotes, including one from von Braun, observing that science, like a knife, can be wielded for good or evil. We included the quote with some hesitation because of the moral ambiguity associated with its author. Wernher von Braun was a brilliant engineer who made major contributions to rocketry and was considered a hero in the United States in the 1950s and 1960s for helping to launch the space program that ultimately delivered a man to the Moon in 1969. However, those accomplishments followed a dark past when he worked for Nazi Germany and developed the infamous V-2 rocket, the world's first ballistic missile that was fired indiscriminately at London, where it killed thousands of civilians during World War II. Furthermore, his rocket research and production relied on slave labor, which is a crime against humanity (27). Details of von Braun's work for Nazi Germany came out only after his death because they were kept secret by the U.S. government. We remain uncomfortable with von Braun but use his quote because it captures the duality of science, and we wonder whether he was thinking of his own career as a creator of weapons of mass destruction and of vehicles for exploration.

Saturn V rocket. Designed by a team led by Wernher von Braun, the Saturn V is here shown at the launch of the Apollo 11 mission at 9:32 a.m. EDT on July 16, 1969 (28). Courtesy of NASA.

or medical research that meets the criterion of deliberate harm that is carried out with subject informed consent, performed in the open and within a regulatory framework for ethical human research. For example, deliberate infection of men with *Neisseria gonorrhoeae* carried out in human volunteers who have given their consent and are studied in experiments supervised by Institutional Review Boards is not deplorable science, even though some of the infected individuals will develop disease in the form of urethritis, which is treated afterwards with antibiotics (5, 6). More recently, human volunteers have been deliberately infected with severe acute respiratory syndrome coronavirus 2 (SARS-CoV-2) to learn about the pathogenesis of COVID-19 through a controlled infection (7). Such an experiment was considered ethical by some, since the treatments available at the time were primarily antibody based, with small-molecule antiviral therapies available in case the subjects developed severe COVID-19, although concerns about unknown long-term effects were raised (8). Experimental human research is sometimes the only means available to obtain important medical information, and sometimes such experimentation carries risk of harm to the subjects involved. In the case of

N. gonorrhoeae, there is no animal model to study gonorrhea, and deliberate human infection provides a means to study the course of the disease and test vaccines.

Here we consider five illustrative examples of medical research that meet these conditions, while sadly noting that there is no shortage of historical examples of human experimentation with pathogenic microbes that fall within our definition of deplorable science (9). In 1966, Henry Beecher described twenty-two unethical or questionably ethical medical studies (10). We focus on five studies for which there is a considerable body of literature to analyze their causes and failures.

Tuskegee study. The infamous Tuskegee study was a prospective observational study of the natural course of syphilis in an African-American population between 1932 and 1972 (11). The study recruited individuals with latent syphilis and observed them but did not offer them therapy. At the time the study began, there were already some therapies for syphilis in the form of salvarsan and arsenic-based compounds. Although these were toxic and not very effective, they were considered standard therapy at the time. Over the course of the 40-year study, penicillin therapy was developed, which is highly effective and continues to be used today. By 1972, the year the study became public, 28 participants had died from syphilis, 40 spouses had been infected, and 19 children had congenital infections (11). The deplorable aspect of the study was that patients known to have syphilis were untreated despite the availability of effective agents, and that decision led directly to human harm.

Guatemala syphilis experiments. A U.S.-funded study carried out from 1946 to 1948 deliberately infected uninformed individuals with syphilis to determine whether prophylactic administration of penicillin could prevent infection (12). Serological monitoring continued until 1953. These experiments involved the deliberate infection of several hundred men who were not informed of their infections and did not give consent. Deliberate human infection with a microbe known to be capable of causing serious or life-threatening disease should have been anticipated to result in human harm, which makes these experiments deplorable.

German and Japanese experimentation on prisoners. During World War II, cruel, inhumane, and deadly experiments that resulted in immense suffering in study participants were carried out by German and Japanese physicians and scientists on prisoners, in research ranging from the effects of chemical weapons to hypothermia to human infections with agents of germ warfare such as *Yersinia pestis.* Such research often, but not always, was performed in the pursuit of information that would inform some aspect of the war effort, providing particularly dastardly examples of deplorable science (13).

Untreated cervical neoplasia in New Zealand and India. At least two studies have been reported in which women with cervical neoplasia were not treated to evaluate the natural progression of this condition (14). One study occurred in New Zealand (15) and the other in India (16). In the New Zealand study, a physician decided in the 1960s not to treat young women with cervical dysplasia since he did not believe this to be a premalignant lesion, despite a scientific consensus at the time that these lesions could progress to cervical cancer (15, 17). The Indian study similarly followed women with cervical dysplasia for more than a decade without offering treatment (16). In both the New Zealand and Indian studies, some women who did not receive therapy for cervical neoplasia died of cervical cancer.

Willowbrook hepatitis study. The Willowbrook State School was a home for children with developmental abnormalities that existed in Staten Island, New York City, and operated from 1953 to 1987. Conditions in the facility were very substandard, and approximately 90% of all children developed infectious hepatitis, now known to be caused by hepatitis A virus. To study the natural course of hepatitis, investigators deliberately infected children with hepatitis A virus through the administration of a stool extract (10, 18). Many of these children suffered the symptoms of acute hepatitis including jaundice, vomiting, and anorexia, which must have been particularly distressful for them given their impaired ability to communicate as a result of their developmental disabilities. Although most of the children would have become infected naturally, the action of the investigators made hepatitis a certainty, and this action brings the studies into the deplorable domain.

The five examples cited above have the common denominator that humans were hurt by the research and not informed of the risks of their participation nor asked for their consent. Not coincidentally, each example involved a population that was vulnerable in some respect. While each case meets our definition of deplorable science, we do not mean to imply that they are comparable in degree. Certainly, the sin of commission in torturing and killing prisoners for dubious medical research is not the same as infecting children with a disease that most would have acquired anyway, is usually transient with complete recovery, and for which medical care was provided. In this regard, we note that one of the investigators in the Willowbrook study wrote a defense of the project, where he stated that informed consent was obtained from the parents of the children and emphasized their efforts to provide care for highly disabled children living in terrible conditions (19). Hence, while we do not condone any of these studies, we would not create false equivalences between them.

When evaluating deplorable science, one must also wrestle with the question of whether the study produced useful information (Box 29.2). Before considering this thorny issue, we state up front that we reject the Machiavellian dictum that the end justifies the means. Furthermore, we are also concerned that any attempt to glean useful information from deplorable science might be used to justify it and thus tread carefully when evaluating any gain in knowledge that came at the price of deliberate human suffering. Of the five studies considered here, the Willowbrook study differs from the others in having produced important medical information, since it led to the discovery of two forms of hepatitis, now known to be caused by the hepatitis A (HAV) and B (HBV) viruses, and helped to establish

---

Box 29.2 Is it ethical to use findings from unethical research?

This is a difficult question that has been extensively discussed for Nazi experimentation but continues to arise, such as regarding Chinese research involving recipients of organs transplanted from executed prisoners (29). Ethical arguments in favor of using such data cite the potential benefits of such information to others. Furthermore, ethical standards in science have evolved over time, and applying contemporary standards to earlier research can be fraught in less extreme cases (30, 31). Arguments against use include the unreliability of unethical research and the implicit condoning of such research, with a lack of respect and concern shown for the victims. Might this encourage future researchers to seek forgiveness rather than ask permission (32)? These concerns are magnified when the unethical research practices are ongoing, as a willingness to use the research findings may result in further harm to others (33). The ethical challenges and real-life implications of these questions are evident in the case of the unethical gene editing of human embryos in a misguided attempt to make them HIV resistant (34) (chapter 27). The children are clearly innocents, yet the information from this unethical experiment could conceivably help them and others in whom gene editing is being considered. The responsible scientist was imprisoned, as his actions were judged to be not only unethical but also criminal.

How should society and science deal with such information? The Declaration of Helsinki flatly states that unethical research findings should not be published (24). In a survey of journal editors, most respondents indicated that they would reject articles that describe ethically uncertain research (35). On the other hand, a majority of physicians, bioethicists, and scientists indicated in another survey that they would use data from unethical research if they thought that it would save a life or improve quality of life for an individual or society as a whole (36). The ethical code of the American Medical Association states: "In the rare instances when ethically tainted data have been validated by rigorous scientific analysis, are the only data of such nature available, and human lives would certainly be lost without the knowledge obtained from the data, it may be permissible to use or publish findings from unethical experiments" (37).

Kristine Moe has argued that any use of such data must acknowledge "the incomprehensible horror that produced them," but that it would be wrong to allow "the inhumanity of the experiments to blind us to the possibility that some good may be salvaged from the ashes" (38). However, Arthur Schafer has responded that this should only apply when "the value of the data acquired is of great value to mankind," and in most cases of unethical research, "the morally appropriate policy ... would seem to be denunciation of the method of acquisition together with a refusal to use the data in any way" (39). The benefits to science and society from maintaining strict ethical standards for research are great. Accordingly, the bar for using findings from unethical research should be high.

the benefits of gamma globulin preparations to prevent these diseases. However, any consideration of the value of information gathered in deplorable science experiments must also consider whether the means used were necessary for the knowledge gained. In this respect, we note that the Willowbrook study was conducted over two decades and involved both observational and interventional components. One might well argue that the contributions of the study could have been obtained from observation alone and did not require the deliberate infection of children with developmental abnormalities. David Rothman has argued that neither the Tuskegee nor Willowbrook infection experiments produced much useful science (20), although in the case of Willowbrook, this has been disputed by one of the investigators (19). Certainly, the availability of alternative means of inquiry to produce the same information without causing human harm would negate any justification for the work done.

Learning about deplorable science is an important aspect of ethical education in both medicine and the biomedical sciences. This in turn requires knowledge about the history of medicine and science and the incorporation of this information into ethical training courses (21). The examples considered above occurred in Japan, Germany, India, New Zealand, Guatemala, and the United States, and all occurred in recent historical times. Hence, we cannot ascribe them to different ethical considerations in remote times or associate them with specific countries or cultures. This also means that all scientists must remain alert to prevent such research from happening again.

In fact, one of us (F.F.) served on a study section in which a prominent research group requested funding for a study involving the deliberate infection of human subjects in a low-income country with a bacterium that sometimes causes long-term neurological complications. The researchers were attempting to determine the characteristics of various bacterial strains that are responsible for their different clinical manifestations. Although the project had been approved by an Institutional Review Board, and the participants were to be informed of potential consequences of the deliberate infection, the review panel declined funding because they regarded the studies as too risky to justify the potential harm to participants. In particular, the panel was concerned that the compensation of participants might be viewed as coercive, the researchers had declined to take responsibility for providing medical care that might be required as a result of the study, and the information gained would be unlikely to directly benefit the study population. This illustrates that the threat of deplorable science continues to be with us, and the existence of Institutional Review Boards and informed consent, while important, are not a guarantee of safety.

We note that the Tuskegee, Willowbrook, and New Zealand studies each came to public attention following exposés in the general media. This is important because the existence, conduct, and results of these studies were already known to members of the medical community, and some of the results had even been reported at meetings. The Willowbrook study was example number 16 in the Beecher article published in the *New England Journal of Medicine* (10). In each of these studies, some colleagues objected, but objections within the structure of the medical profession were not sufficient to stop or modify the research. Hence, safeguards both within and outside of the profession are needed. In fact, some practitioners of deplorable research have been celebrated in their own times, such as the University of Pennsylvania dermatologist Albert Kligman, who developed pioneering treatments for acne and fungal skin infections. Regrettably, many of Kligman's breakthroughs were achieved in research on prison inmates, mostly Black, from the 1950s through the 1970s (22). As a result, these men, many of whom were illiterate, were exposed to dangerous substances including chemicals, pathogens, and carcinogens. Some had lasting adverse health consequences from the exposures. In his lifetime, Kligman was esteemed as a giant in his field, and his name was given to awards, lectureships, and laboratories. These have now been removed, as the field has belatedly begun to address the troubling ethics of his clinical studies (23).

Deplorable science can ultimately lead to research reform. Nazi human experimentation exposed during the 1947 Nuremberg trials led to the 1964 Declaration of Helsinki, which established international ethical principles guiding research in human subjects (24). The Tuskegee and Willowbrook studies became public scandals in the 1970s and led the U.S. Congress to create a commission in 1974. Several years later, that commission produced the Belmont Report, which recommended principles for human research that are codified today (25). In New Zealand, news about the untreated neoplasia study led to enhanced protections for patients in human research, as well as educational reforms (15, 17).

Thinking about past deplorable science with the particular goal of preventing such research in the future, we are struck by how all of this could have been prevented by practicing the principle of non-maleficence, which is derived from the medical maxim *Primum non nocere*, or "First, do no harm." Both acts of commission and acts of omission can lead to harm. Acts of commission, such as those that occurred in the Guatemala syphilis study and Willowbrook hepatitis study, would have been precluded by following the dictum of doing no harm. Similarly, acts of omission such as occurred in the Tuskegee syphilis or New Zealand untreated neoplasia studies could lead to patient harm since the consensus at the time was that treatment was beneficial. Thus, not providing treatment caused direct harm to patients. Had the investigators, regulators, or funders of these studies asked whether patients with syphilis or cervical dysplasia needed treatment, they would have obtained an affirmative response by standards of their time, which could have resulted in study cessation or modification. Had they asked themselves whether they would be happy for a family member to participate in such research, would the research still have been done? Going forward, the best protection against future deplorable science is to create a multitiered defense system to prevent such research, as we have suggested for improving the quality of science in other respects (26). This would include ethical training for all investigators, close adherence to principles of ethical experimentation in the design and implementation of scientific research, institutional review of all human research, transparency in study design, and frequent review of ongoing projects that includes outside experts not associated with the institution or project. Deploring unethical science is not enough—it must be prevented.

IV FUTURE SCIENCE

30 Plague Science

Fear and panic could destroy the city as much as the plague itself. ... Quacks preyed on the poor with their neverfail miracle drugs. ... People lied to each other—and to themselves.

Daniel Defoe, *A Journal of the Plague Year* (1722) (1)

Humanity has faced plagues since time immemorial, but only since the end of the 19th century has science helped to confront the terror of epidemics and mitigate their effects. The rise of the germ theory during the decades of 1850 to 1880 posited a causal association between microbes and disease, providing humanity with a new approach to infectious diseases that led to improvements in sanitation, vaccines, serum therapy, and, eventually, small-molecule antimicrobial agents. As the biological sciences progressed, the tools available to confront plagues increased considerably. The pace of progress is evident when considering three great pandemics that have occurred since the rise of the germ theory of disease: 1918 influenza, human immunodeficiency virus (HIV), and severe acute respiratory syndrome coronavirus 2 (SARS-CoV-2) (Table 30.1). When one considers that humanity has been defenseless against infectious diseases throughout most of its history, the progress made against plagues is a breathtaking testament to the benefits that science can bring to our species. As we write this chapter on "plague science," the world continues to recover from COVID-19, a pandemic of historic proportions that infected more than 675 million people and claimed more than

Thinking about Science: Good Science, Bad Science, and How to Make It Better, First Edition.
Ferric C. Fang and Arturo Casadevall.
© 2024 American Society for Microbiology.

TABLE 30.1 A comparison of pandemic responses

Time to:	Pandemic		
	1918 influenza	HIV	SARS-CoV-2
Recognition	Months	Years	Weeks
Agent discovery	14 years	3 years	Weeks
Diagnostic test	Decades	5 years	Weeks
First antiviral	Decades	6 years[a]	4 months[b]
Vaccine	Decades	None to date	1 year

[a] Azidothymidine (AZT).
[b] Remdesivir.

6.8 million lives (the actual excess death toll is closer to 20 million). Despite an unprecedented international effort, which resulted in the development of multiple effective vaccines and the administration of more than 13 billion vaccine doses, the emergence of new viral variants has frustrated predictions by public health experts and slowed economic recovery. Although we have not yet seen the last of SARS-CoV-2, we have seen enough to comment on the present with the hindsight of the recent past.

COVID-19 RESPONSE SUCCESSES

Amidst the COVID-19 calamity, one might argue that science is one of the few aspects of the human response that has worked relatively well. To many scientists, the pandemic has provided a vivid reminder of the importance of investment in basic science, public health, and a robust scientific workforce. Within weeks of the announcement of a new severe respiratory illness in China, biomedical scientists had identified the cause as a new coronavirus and developed rapid and accurate diagnostic tests. In short order, academic and pharmaceutical scientists began to test existing drugs to identify some that had antiviral activity, and despite some blunders, as will be described below for hydroxychloroquine, it was rapidly determined that remdesivir, a drug originally intended for Ebola virus, was active against SARS-CoV-2. Antibody-based therapies were developed, first by taking a page out of history and deploying convalescent plasma, and then through the development of monoclonal antibodies. Probably the greatest impact on the course of the pandemic came from the development of effective vaccines (Fig. 30.1). Of these, the mRNA-based vaccines were developed within a single year and proved highly effective in reducing mortality. The remarkably rapid manufacture and deployment of COVID-19 vaccines, including some based on novel technologies, is clearly one of the great success stories of the pandemic response. Nevertheless, in considering

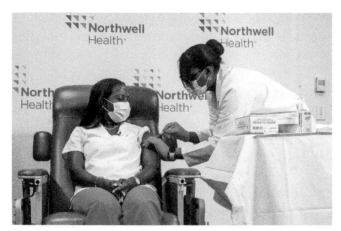

FIGURE 30.1 *First COVID-19 vaccine recipient in the United States. Sandra Lindsay, a nurse from Queens, New York, was the first COVID-19 vaccine recipient in the United States. Lindsay is currently the Vice President for Public Health Advocacy at Northwell Health. She received the Presidential Medal of Freedom for her contributions to public health and in recognition of being the first individual to receive the COVID-19 vaccine in the United States (85).* Image courtesy of Northwell Health.

the "warp speed" with which these vaccines were brought forward to the public, it is important to remember that the conceptual and technological advances that made COVID-19 vaccines possible took place over the preceding decade or more, and viewed in this light, the full developmental timeline of the COVID-19 vaccines is not an aberration in comparison to those of other novel vaccines (2, 3). If not for the basic and in some senses revolutionary vaccine research that took place over many years before SARS-CoV-2 was first recognized, a vaccine might still be a distant dream.

COVID-19 RESPONSE FOIBLES

Despite the many advances in preventing and treating COVID-19, there were also missteps as the world scrambled to respond to a deadly new pathogen. Sadly, the initial spirit of international cooperation has not been sustained, hampering efforts to provide equitable vaccine distribution or ascertain pandemic origins (4, 5). It has been humbling for the United States to lead large, high-income countries in per capita deaths from COVID-19 (6), even with its wealth and scientific expertise. We are all too aware of the needless illnesses and deaths that resulted from misguided political leadership, inadequate preparation, delayed responses, fragile supply chains, health disparities, and vaccine hesitancy (7). But we will not dwell on these issues here. Rather, we would like to review the COVID-19 pandemic through the prism of the three R's of research integrity: rigor, reproducibility, and responsibility. These form the fundamental pillars of the foundation of science. It is appropriate that we devote more attention to the foibles than to the successes so that we can learn from the mistakes and missed opportunities. What could have been done better? What needs to improve?

RIGOR

In chapter 6, we attempted to define rigorous science. We proposed a framework that includes redundant experimental design, sound statistical analysis, recognition of error, avoidance of logical fallacies, and intellectual honesty (8). Has COVID-19 research been rigorous? Certainly, much of the work that brought us vaccines and new therapies has been. But unfortunately, other instances fell far short of rigor. Perhaps the most prominent example early in the pandemic related to the use of hydroxychloroquine for the treatment of COVID-19.

Chloroquine and hydroxychloroquine, drugs used for the treatment of malaria and rheumatic diseases, were suggested as possible therapies for SARS in 2003, but were not formally studied in a clinical setting. Shortly after the COVID-19 pandemic began to take off in China, rumors of a possible benefit of these agents began to appear in social media. On February 19, 2020, a letter was published in an obscure Asian journal, which mentioned a press briefing reporting that chloroquine improved clinical outcomes compared to control treatment in more than 100 patients with COVID-19 in China (9). No data were provided, however, and the paper initially attracted little notice. However, on March 4, 2020, an article from a prominent research group in France led by Didier Raoult appeared in the *International Journal of Antimicrobial Agents*, reviewing the *in vitro* antiviral activity of chloroquine against coronaviruses and referring to the reportedly favorable experience in China (10).

On March 13, two cryptocurrency investors, Greg Rigano and James Todaro, posted a non-peer-reviewed paper online, touting chloroquine as "an effective treatment for COVID-19" (11). The authors claimed some prestigious institutional affiliations, which were subsequently disavowed, but nevertheless their paper was tweeted by Elon Musk to over 40 million followers, and one of the authors was interviewed by Tucker Carlson of Fox News, who publicized the claims to millions of viewers. On March 16, a preprint from Raoult and his group reported the use of hydroxychloroquine with or without azithromycin in 26 patients, who were compared with 16 controls receiving standard care (12). Viral clearance as measured by PCR was reported to be more rapid in recipients of hydroxychloroquine and much more rapid in those who also received azithromycin. The paper was accepted for publication soon afterward.

Red flags about this paper were raised almost immediately, but not before hydroxychloroquine was being touted by President Donald Trump as a "game changer" that would be fast-tracked by the U.S. Food and Drug Administration (FDA) for approval. On March 28, the FDA issued Emergency Use Authorization for hydroxychloroquine and chloroquine in patients with severe COVID-19. Among the concerns regarding the French study were the small sample size, the lack of randomization (which resulted in poorly matched study and control groups), the use of different diagnostic assays, and the failure to account for six patients who had been initially enrolled in the study (13). It was also noted that the editor in chief of the journal publishing the article was one of the authors, and the review process took less than 24 hours. The same research group subsequently published a paper reporting that hydroxychloroquine and azithromycin were 92% effective in more than 1,000 patients with early COVID-19 (14). However, the follow-up study did not include a control group. Raoult was quoted as saying that randomized controlled trials are unnecessary and unethical in deadly infectious diseases and appeal only to statisticians "who have never seen a patient" (15). Meanwhile, he initiated a lawsuit against Elisabeth Bik, one of the scientists who had criticized his original paper on hydroxychloroquine.

We now know from numerous subsequent clinical studies involving thousands of patients that hydroxychloroquine, with or without azithromycin, is *not* beneficial for patients at any stage of COVID-19, nor for the prevention of infection, and may even be associated with a higher risk of death (16). With very few exceptions, the results are highly consistent. The FDA Emergency Use Authorization was withdrawn on June 15, 2020. Yet hydroxychloroquine was given to countless patients on the basis of non-rigorous science and continued to be given long after the withdrawal of FDA approval. We'll have a bit more to say about hydroxychloroquine when we discuss reproducibility.

As if to prove the adage that "history repeats itself—the first time as tragedy, the second as farce" (17), no sooner had enthusiasm for hydroxychloroquine begun to wane when a new unproven treatment began to gain rapidly in popularity—ivermectin. Ivermectin is a macrocyclic lactone used to treat parasitic infections in humans and other animals. Early in the pandemic, Australian scientists reported that ivermectin could inhibit the replication of SARS-CoV-2 *in vitro* (18). It was quickly pointed out that the concentration of ivermectin required to inhibit viral replication greatly exceeded concentrations achievable from normal human dosing of the drug, but unfortunately this warning went unheeded.

Soon there were reports that ivermectin could prevent SARS-CoV-2 infection or reduce disease progression in patients with mild to moderate COVID-19 (19). A systematic review found that most clinical studies of ivermectin failed to meet predefined eligibility criteria and suffered from imprecision and a high risk of bias (20). The authors of the review concluded that it is uncertain whether ivermectin is beneficial in COVID-19. But this message was drowned out by ringing endorsements from groups like America's Frontline Doctors, a group of physicians closely aligned with right-wing political organizations, which had originally advocated the use of hydroxychloroquine to treat COVID-19. A related group calling themselves the Front Line COVID-19 Critical Care Alliance, or FLCCC, led by a critical care physician in Wisconsin named Pierre Kory, lent their voices to these efforts. Kory was quoted as saying, "My dream is that every household has ivermectin in their cupboard. And you take it upon development of the first symptom of anything approximating a viral symptom. ... Even if it's not COVID, it's safe to take it and it's probably effective against that virus" (21).

During that latter part of the summer of 2020, an explosive rise in ivermectin prescriptions was observed in the United States. High doses of ivermectin can produce a variety of adverse effects including gastrointestinal symptoms, seizures, respiratory failure, and coma. Poison centers found themselves swamped with calls about ivermectin overdoses. The FDA even issued a warning not to take ivermectin for COVID-19, noting that some people were even taking ivermectin preparations intended for the deworming of horses. Off-label ivermectin use continued to be popular on social media even though careful reviewers found serious flaws in the purported evidence for its benefits in COVID-19. An independent analysis failed to find a single clinical trial demonstrating a benefit from ivermectin in COVID-19 that did not contain "either obvious signs of fabrication or errors so critical that they invalidated the study," including the same patient data being used to represent multiple subjects, non-random patient selection, numbers unlikely to occur naturally, incorrectly calculated percentages, and an inability of local health organizations to corroborate that the studies took place (22, 23).

Were these aberrations? Extreme cases perhaps, but systematic analyses suggest that deficiencies in the rigor of COVID-19-related research have been commonplace. A study of 686 COVID-19 clinical research articles found shorter time to publication and lower methodological quality than for other articles in the same journals (24), suggesting a lowering of the usual standards for publication in the time of plague.

Another issue that cannot be ignored is the important role of social media in disseminating and amplifying misinformation. A commentary has discussed an urgent need for improved understanding and stewardship of global collective behavior, cautioning that it has become easy to connect and share information through social media, but "in contexts where decisions depend upon accurate information, such processes can undermine collective intelligence and promote dangerous behavior" (25).

In some cases, scientists themselves have contributed to the spread of misinformation (15, 26). Science does not defer to authority, and open dissent and rigorous debate are considered essential for the critical assessment of new ideas (chapters 1, 22, 28). However, the COVID-19 pandemic has revealed a downside to the public debate of scientific questions. Nearly every public health advisory, whether about face masks, lockdowns, school closures, or vaccines, has been endlessly scrutinized and debated in the press and in social media, often with the best of intentions. Unfortunately, this has exacerbated existing societal divisions, fueled the politicization of science, and eroded confidence in scientists, public officials, and institutions. A lack of trust in government institutions has undermined public health efforts and provides an ominous warning for responses to future pandemics. A country-level analysis demonstrated an inverse correlation between confidence in institutions and COVID-19 burden (27), and national resiliency to COVID-19 has positively correlated with trust (28). As the epidemiologist Jay Kaufman observed, "Science alone can't heal a sick society. ... Science is a social process. ... To restore faith in science, there must be faith in social institutions" (29). Our social divisions are literally killing us.

REPRODUCIBILITY

In chapter 7, we discussed reproducibility in science and the fundamental expectation that an experimental result can be confirmed by others. We explained how there are many reasons why experimental results may be difficult to replicate (30), and studies like the Reproducibility Project (chapter 7) have found that only a minority of landmark findings from experimental psychology could be replicated (31).

One unusual but important reason for a lack of reproducibility is scientific fraud. Obviously, fabricated or falsified data will be difficult to replicate because such data are not based on true experimental observation. As detailed in chapter 24, our 2012 analysis of more than 2,000 retracted articles indexed by PubMed found that most retractions resulted from misconduct, including fraud, duplicate publication, and plagiarism (32). According to the blog *Retraction Watch*, more than 300 articles relating to COVID-19 have been retracted at the time of this writing. This is a modest number in view of the more than 350,000 PubMed-indexed articles published on COVID-19, but it is nevertheless instructive to look at a few of the most high-profile retractions. Two of the retracted articles were published in the most prestigious and selective clinical journals in the world, the *Lancet* and the *New England Journal of Medicine*.

The *Lancet* paper, published in May 2020, purported to describe a multinational registry analysis of 96,032 patients hospitalized with COVID-19 and found that hydroxychloroquine and chloroquine were associated with decreased in-hospital survival (33). The *New England Journal of Medicine* paper, published the following month, purported to analyze 8,910 patients hospitalized with COVID-19 and found no association between the use of angiotensin-converting enzyme inhibitors or angiotensin receptor blockers and in-hospital mortality (34). Note that both of these findings have been subsequently confirmed by other legitimate studies, so what is in dispute is not the bottom-line findings of these articles. The problem is that the data appear to have been made up out of thin air. The retracted articles claimed to draw from an enormous registry including patients from 671 hospitals on six continents. One would expect such a monumental effort to involve hundreds if not thousands of contributors. However, suspicions quickly arose when no one could identify even a single hospital that had contributed to this registry, and major discrepancies were noted between the number of reported cases for some regions and data from independent sources (35).

Although both papers were retracted within weeks of publication, significant damage was done. Clinical trials had been suspended and international guidelines had been revised on the basis of the published findings. Editors and reviewers were apparently seduced by the reputation of the first author (a respected Harvard professor) and the power of big data. The pressures of the pandemic may have allowed the papers to slip through the peer review process with less than the usual scrutiny. Sapan Desai, the founder of Surgisphere, the company that claimed to have assembled the registry, turned out to be a surgeon with multiple prior malpractice claims and little experience in data analytics. An investigation of his company identified just a handful of employees, several of whom had no background in science or data analysis, and the company was occupying rented office space. Amit Patel, a coauthor on the studies, turned out to be Desai's brother-in-law. The first author, Mandeep Mehra, admitted to not having seen any primary data (36). Although Desai has not admitted any wrongdoing, the Surgisphere papers are now believed to have been a hoax based on fabricated data. The editor in chief of the *Lancet* concluded that the study was a "monumental fraud" (37). Nevertheless, the papers continued to be cited despite their retraction; each has been cited more than 1,000 times, and many citing articles appeared to be unaware that the papers had been retracted.

Further evidence of peer review failure can be seen in two retracted articles by a psychologist named Harald Walach. Walach published articles claiming that COVID-19 vaccines were responsible for two deaths for every three infections prevented and that face mask use results in hypercapnia in children (38, 39). Although the papers were strongly criticized for their nonsensical findings and quickly retracted by the journals *Vaccines* and *JAMA Pediatrics*, they were widely circulated and continued to be cited by anti-vaccine and anti-face mask advocates.

A more common challenge for reproducibility in clinical research is heterogeneity. The concept of evidence-based medicine was introduced in 1992 to shift clinical decision-making from a process dependent on "intuition, unsystematic clinical experience and pathophysiologic rationale" to one that incorporated evidence from clinical research (40). Although originally conceived as an integration of these different sources of knowledge, over time the hierarchy of evidence has evolved to strongly favor the results of randomized clinical trials over observational data, with pathophysiological principles and expert opinion given lowest priority. However, person-to-person variability poses an important limitation on the value of clinical trial data. In a heterogeneous population, a treatment may only benefit a subset of people, while others experience no benefit or even adverse effects. A failure to recognize heterogeneity can obscure the benefits of an effective treatment. It is recognized that to yield reproducible results, randomized clinical trials must be performed under conditions in which the subjects are as homogeneous as possible with regard to demographics, clinical features, and outcomes. However, too much homogeneity in study populations can reduce their relevance to more variable real-world situations, while heterogeneity can reduce the reproducibility of clinical trials. Precision or personalized medicine may offer a solution if subsets of patients can be identified on the basis of clinical criteria, genetics, or biomarkers that permit the targeting of therapeutic interventions to those who are likely to benefit. Heterogeneity has proven to be a challenge, no more so than in COVID-19, in which clinical outcomes are extremely variable and dynamic over time.

It has been recognized since the early days of the pandemic in Wuhan, China, that some patients infected with SARS-CoV-2 can remain asymptomatic, and those with symptoms often recover spontaneously after a mild illness. Only a minority progress to more severe illness, which may be complicated by respiratory failure, hyperinflammation, or thromboembolic events (41). Clinical outcomes have been found to vary widely depending on age, sex, comorbidities, race or ethnicity, hospital, country, viral strain, and immune status. This creates enormous challenges for the assessment of the efficacy of therapeutic interventions. Viral load declines from the time of initial presentation in virtually all COVID-19 patients (42), so the potential clinical impact of an intervention is critically dependent on timing. Failure to adequately control for heterogeneity in study populations appears to be common in the COVID-19 literature, and an analysis of randomized trials included in World Health Organization (WHO)-sponsored network meta-analyses found that most trials included patients with a range of clinical severity that may have influenced outcomes but could not be adequately assessed from the data provided (43). Such variance can cause effective interventions to appear ineffective, or vice versa.

One example of this is the use of convalescent plasma in patients with COVID-19. It is now firmly established that an important determinant of clinical outcomes in COVID-19 is the presence of neutralizing antibodies directed against the spike protein, in particular, to the receptor-binding domain of the spike protein. Neutralizing antibody titers are predictive of vaccine- or infection-induced immunity to subsequent infection. The administration of neutralizing monoclonal antibodies is protective in patients who are at high risk of disease progression. The timing of antibody responses is also important; patients with delayed neutralizing antibody responses are at risk for more-severe illness even though they eventually develop high antibody titers.

Thus, it seemed highly predictable that the administration of convalescent plasma from patients who have recovered from COVID-19 should be beneficial to patients with early infections. Indeed, an epidemiologic study revealed an inverse correlation between convalescent plasma use and mortality in the United States, estimating that its deployment saved about 100,000 lives in the first year of the pandemic (44). However, demonstrating a benefit using the randomized controlled trial mechanism proved to be unexpectedly challenging, with clinical trials, observational studies, and even meta-analyses yielding different conclusions (45–49). Some of the negative results almost certainly reflect late use in critically ill patients and units of insufficient titer, but for others the outcomes are difficult to reconcile (50). Patient heterogeneity is likely to be a major reason for the discrepant results. Although contemporary analytical methods attempt to control for clinical variables, viral load, viral variants, timing of plasma administration, illness severity, donor and recipient antibody titers, other treatment interventions, and various potential outcomes, controlling for each of these variables is extraordinarily challenging. The conflicting results regarding convalescent plasma in COVID-19 have revealed a vulnerability in our approach to evaluating treatment interventions and its reliance on randomized trials that presume patient uniformity.

Some critics of evidence-based medicine, like Mark Tonelli at the University of Washington, have advocated a more expansive and transdisciplinary approach to clinical decision-making that incorporates pathophysiological knowledge, provider experience, and patient preferences along with clinical trial data (51). In addition, the precision or personalized medicine movement emphasizes the unique characteristics of a patient rather than the generalizations that can be made about a large cohort of patients. Such an approach might be better suited to answer questions such as which COVID-19 patients are most likely to benefit from convalescent plasma.

Other weaknesses of evidence-based medicine have been revealed in considering the efficacy of face masks in preventing SARS-CoV-2 infection. More than three years after the beginning of the pandemic, the clinical trial evidence base for face masks remains woefully inadequate (52) and lags far behind the evidence from laboratory research, natural experiments, real-life data, and observational studies (53).

RESPONSIBILITY

Serious questions regarding scientific responsibility have arisen during the pandemic. Although SARS-CoV-2 was initially assumed to have arisen as a zoonotic pathogen making a species jump from bats to humans, possibly via an intermediate host (54), subsequent attention was focused on the possibility of transmission from a lab leak involving scientists at the Wuhan Institute of Virology, where coronavirus research is performed (55). Speculations were made about whether so-called gain-of-function research (see chapter 27) to identify viral adaptations that facilitate human infection might have inadvertently led to the emergence of SARS-CoV-2 (56). At this time, the weight of scientific evidence favors a natural origin (57–63). Nevertheless, the lack of transparency on the part of the Chinese government and the politicization of the issue have raised uncomfortable questions about whether scientists are adequately considering and mitigating the public health risks of research with dangerous pathogens (see chapter 27). As Dr. Tedros Ghebreyesus, the Director-General of the WHO, has noted, "Lab accidents happen" (64), and it is incumbent upon scientists to be cognizant of these hazards and to take every possible precaution to minimize the risk to the public.

These concerns must be carefully weighed against the known and potential benefits of virological research, which are essential for understanding viral pathogens and how to combat them. The rapid recognition of SARS-CoV-2 and the development of accurate diagnostics and effective therapeutics and vaccines were only possible because of previous decades of research on coronaviruses. It would be a tragic overreaction to fears of a lab leak to shackle further virological research so that the world is powerless to deal with future pandemic threats (65). As discussed in chapter 27, the scientific community addressed somewhat analogous concerns at the dawn of the recombinant DNA era in the 1970s, leading to a temporary moratorium on certain types of research until consensus guidelines were developed at the Asilomar Conference (66). Since that time, recombinant DNA technology has brought dramatic improvements in medicine, agriculture, and virtually all areas of biological science. Similarly, it will be imperative to find a way forward that balances the enormous societal benefits of virological research with the need to ensure that the risks of research with dangerous pathogens are carefully mitigated.

In his classic description of the Black Death in London, *A Journal of the Plague Year*, Daniel Defoe wrote, "It mattered not from whence it came" (1). But the origin of the pandemic does matter. Whether by zoonotic spillover or lab accident, we need to understand how the pandemic began in order to take appropriate measures to prevent this from happening again. Nevertheless, a unanimous consensus regarding SARS-CoV-2 origins is not required for society to take immediate actions to both reduce human-wildlife contact leading to spillover events and to reduce the risks of research with dangerous pathogens that could result in lab leaks. We cannot allow lingering uncertainty to distract us from these important tasks.

Another important responsibility of scientists is to place the interests of society above their own. But this is easier said than done, and self-interest may have contributed to delays in the acceptance that SARS-CoV-2 is airborne. It is now widely recognized that the predominant route of SARS-CoV-2 transmission is via the airborne transmission of respiratory droplets of a wide range of sizes, generated by routine activities such as breathing, speaking, and singing, as well as by coughing or sneezing. Although transmission is most likely at short distances, crowding and poorly ventilated indoor spaces can increase the likelihood of longer-range transmission. This has had major implications for the prevention of COVID-19 infection by distancing, face mask use (Fig. 30.2), adequate ventilation, and the avoidance of crowds.

Public statements by infection control specialists early in the pandemic emphasized that SARS-CoV-2 was spread by "droplets" rather than by airborne transmission, by which they meant the deposition of large, virus-laden droplets onto environmental surfaces and hands by coughing or sneezing, and the subsequent transfer of viruses by direct contact. It followed that transmission could be prevented by handwashing and environmental disinfection. We now know that this is untrue. Unfortunately, the erroneous belief in droplet spread was based on outdated concepts and misinterpretations of earlier research, which were quickly challenged by aerosol experts from fields such as physics, engineering, and chemistry (67). In June 2020, the journal *Clinical Infectious Diseases* published a letter from 239 aerosol experts calling for the medical community to address the possibility of airborne SARS-CoV-2 transmission (68). However, this suggestion was strenuously resisted by more than 300 members of the infection control community, who wrote a response insisting that COVID-19 spread resulted from "droplets and close contact" and accusing the aerosol scientists of sowing "confusion and fear" (69).

In chapter 10, we considered the issue of specialization in science and the emergence of scientific fields (70, 71). In particular, we have noted that although fields have benefits for scientists, they can also promote resistance to outside opinions and serve to sustain dogmas that impede scientific progress. Trish Greenhalgh at Oxford University has arrived at similar conclusions in her assessment of the droplet versus airborne transmission debate that delayed the institution of measures such as face mask use, distancing, ventilation, and the avoidance of crowds, and resulted in more infections and time and effort wasted on ineffective measures. Using a framework proposed by the French sociologist Pierre Bourdieu, she argued that infection control specialists rejected dissenting views from experts who primarily came from other fields in order to preserve

(A)

(B)

FIGURE 30.2 *Reduction of airborne virus transmission by face mask use. (A) Seattle police officers during the influenza pandemic, December 1918. From the National Archives. (B) New York City police officers during the COVID-19 pandemic, June 2020. Courtesy of edenpictures. Used under license CC BY-SA 2.0.*

the primacy of their own discipline regarding matters of infection control (72). In doing so, they appeared to disregard the precautionary principle and failed in their responsibility to place society's interests above their own. It took more than a year for the WHO and Centers for Disease Control and Prevention to finally acknowledge that SARS-CoV-2 transmission is predominantly airborne. This almost certainly delayed the implementation of an effective public health response to the pandemic.

CONCLUSIONS

Although we have focused on some examples in which science fell short of its ideals during the COVID-19 pandemic, in many ways the scientific community acquitted itself well. We have already mentioned the remarkable success of vaccine manufacturing efforts. Scientists threw themselves wholeheartedly into COVID-19 research, putting their other interests on hold. Collaborative and team science allowed the rapid execution of multinational trials, and data sharing, open access, and preprint servers facilitated the timely dissemination of information. Critical knowledge regarding presymptomatic spread and superspreading, nonpharmaceutical interventions, diagnostic testing, distinction between protective and detrimental immune responses, ventilatory management, thromboembolic complications, and viral variants was quickly and widely shared, allowing substantial improvements in clinical outcomes as the pandemic proceeded.

Nevertheless, both the strengths and weaknesses of the contemporary scientific enterprise have been on ample display. Scientists produced a massive number of publications, but wading through this literature to find the truth has not been easy. As Robert Peter Gale wryly observed, the pandemic would be quickly ended if SARS-CoV-2 could be conquered by publications, guidelines, and interminable meetings alone (73). Due to the urgency of the pandemic, publication standards were sometimes compromised (74). Limitations of randomized clinical trials as an exclusive source of knowledge to guide clinical decision-making were exposed. Important questions were raised about the potential dangers of scientific research on pathogenic microbes, while the origins of the pandemic are still debated.

Although vigorous debate is often regarded as a sign of a healthy scientific enterprise, open bickering among scientists about public health policies instead contributed to increasing societal polarization and undermining of public trust in science during the pandemic (75–77). Peter Sandman, an expert on risk communication, identified many mistakes made by officials in their messaging to the public, including a failure to communicate uncertainty and acknowledge error (78). Lockdowns and protective mandates were frequently met with backlash and hostility to public health officials, even as lockdowns and mandates prevented hospitals from becoming overwhelmed and lowered death rates (79). Declining trust in scientists has been accompanied by a loss of support for public health interventions and vaccination (80). Unfortunately, the loss of credibility and public resistance to expert guidance have proven persistent, even as pandemic waves have receded. Many states have passed legislation to weaken the influence of public health officials, leaving agencies even less prepared to deal with future pandemic threats than before (81). There are many things that should be undertaken to strengthen our public health infrastructure (82), but for now much of society seems to be moving in the opposite direction.

Lastly, social media has posed a formidable challenge by creating alternative sources of information and disinformation that compete with science in influencing policymaking and public opinion. COVID-19 has widened the partisan divide with regard to trust in medical scientists (83). Meanwhile, society has scarcely begun to address the threat posed by disinformation. With science as just one among many potential sources of information, the imperative of making science as rigorous, as reproducible, and as responsible as possible becomes clear. Otherwise, why should anyone believe scientists rather than other sources claiming to represent the "truth"? For science to retain its influence in society, efforts to make it more rigorous, reproducible, and responsible must be a ceaseless undertaking. Only scientists can carry the torch and continue to perform the high-quality research that will lead to better prevention and treatment, not only of COVID-19, but of future threats as well. As Albert Camus observed in *The Plague*, "What's natural is the microbe. All the rest—health, integrity, purity—is a product of the human will, of a vigilance that must never falter" (84). We can and must learn from our experience (Box 30.1) because it is a certainty that COVID-19 will not be the last pandemic.

Box 30.1 Ten Lessons from COVID-19 for Science

1. Pandemics are a test not only for science but for the partnership between science and public institutions.
2. Wealth and advanced technology do not necessarily translate into effective public health responses, particularly when there is a lack of public trust in institutions and deficient health infrastructure.
3. Scientific standards must be maintained even during public health emergencies.
4. Premature endorsement of ineffective therapies and delayed implementation of effective measures cost lives.
5. Public health messaging must not only be clear and transparent, but should acknowledge uncertainty, admit error, and explain underlying rationale.

(Continued)

Box 30.1 (Continued)

6. Coercive measures should be avoided whenever possible and employed for a limited time.
7. Scientists accustomed to vigorous debate must recognize the potential for open disagreement to lead to societal confusion, erosion of public trust, and worsening politicization of science.
8. National concerns cannot override the need for a sustained and cooperative international response.
9. Inequalities in health outcomes are aggravated in pandemics.
10. Future pandemic planning should include more robust supply chains and public health infrastructure.

31 Reforming Science

There are three basic flavors of incentive: economic, social and moral.

Steven D. Levitt and Stephen J. Dubner, *Freakonomics* (1)

All but foolish men know, that the only solid, though a far slower reformation, is what each begins and perfects on himself.

Thomas Carlyle (2)

There is no more delicate matter to take in hand, nor more dangerous to conduct, nor more doubtful in its success, than to set up as a leader in the introduction of changes. For he who innovates will have for enemies all those who are well off under the existing order of things, and only lukewarm supporters in those who may be better off under the new.

Niccolò Machiavelli (3)

By many measures, contemporary science has been a resounding success, particularly when one considers the impressive advances in technology and biomedicine. Yet this progress has come at a price, and one may still ask whether the scientific enterprise is healthy. A romantic ideal depicts scientists as intrepid and objective explorers in the relentless pursuit of truth (4). Although many, if not

Thinking about Science: Good Science, Bad Science, and How to Make It Better, First Edition.
Ferric C. Fang and Arturo Casadevall.
© 2024 American Society for Microbiology.

most, scientists are drawn into the business of science by curiosity about nature and a desire to improve the human condition through the accrual and application of knowledge, the everyday realities of science can be quite different. To be successful, today's scientists must often be self-promoting entrepreneurs whose work can also be driven not only by curiosity but by personal ambition, political concerns, and quests for funding. Individual scientists intensively compete with other scientists for resources (chapters 19 and 20), and many scientists are dependent on grant funding to provide part or all of their salaries (chapter 23). Although scientists would prefer to follow wherever their data may lead, research funding is often restricted to specific topics believed to represent the most urgent social priorities. Scientists are expected to demonstrate consistent productivity in terms of manuscripts (chapter 22), on which their future funding, promotion, and tenure depend. The greatest value is placed on journals with high "impact" (and commensurately high rejection rates, as an indication of selectivity) (chapter 26), and most of the glory goes to the first and last authors of a publication. In this environment, we see vulnerabilities and inefficiencies in the contemporary scientific enterprise. This penultimate chapter will briefly recap the most urgent concerns that we have discussed in earlier chapters and suggest some possible solutions.

Our assessment, bolstered by decades of work as professional scientists, leads us to conclude that the scientific enterprise is not as healthy as it could be, nor as it *needs* to be to effectively address the challenges facing humanity in the 21st century. The current hypercompetitive environment has created an insecure working environment for scientists; fostered poor scientific practices, including frank misconduct; and created widespread disillusionment throughout the scientific community, from trainees to senior investigators (chapters 17 to 21). Changes in scientific methods and culture are needed but will have a limited impact unless they are accompanied by fundamental structural reforms in the way that science is supported. This is because, as we have already discussed, many dysfunctional aspects of science are rational responses by scientists to incentives presented by the current system (chapters 19, 20, and 23). True reform will require altering these incentives by addressing fundamental structural aspects of the scientific enterprise that create the constraints under which science is performed. Since most science is supported by public funding, any structural reform will inevitably involve changes in the way governmental financial support is provided to scientists. As the allocation of public funds ultimately results from the political process, attempts to implement structural reforms will by necessity involve engagement with politicians and governmental agencies. Since countries differ in their political and scientific organization, structural reforms will differ from country to country. Although we focus on biomedical research in the United States, the system with which the authors are most familiar, we hope that readers in other countries will find many of the themes to be of universal relevance.

MAJOR PROBLEMS WITH BIOMEDICAL SCIENCE IN THE UNITED STATES

The primary problem of inadequate funding

At the root of most of the problems with American science today is a lack of sufficient resources to support the current enterprise (chapter 23). Grant paylines that commonly exceeded 50% in the 1960s are now below 20% in many disciplines. Overall success rates of NIH R01 research proposals, including both renewal and new applications submitted to the National Institutes of Health, have fallen by more than 50% since 1965 (Fig. 31.1). Once the ameliorating but only temporary effects of the 2009 ARRA (American Recovery and Reinvestment Act) stimulus funding came to an end, the full impact of the deficient federal investment in science began to be fully felt. Grant review panels were once again forced to decide between competing highly meritorious projects. While some competition is arguably good for science (chapter 20), excessive competition is demoralizing, destructive, and counterproductive. Funding agencies cannot continue to reject roughly four-fifths or more of grant applications without seriously damaging science. In the current climate, good ideas are going unsupported, opportunities are being squandered, and capable scientists are being lost. It may be tempting to demand an increased contribution of resources from researchers' institutions, but this is unlikely to be successful, at least in the near term, because institutional budgets have also taken a hit from depreciating investments and reduced revenues to state governments that find themselves unable to meet commitments made during better economic times. In fact, increasing evidence suggests that the indirect costs provided by federal grants are inadequate to meet the true institutional costs of doing research. A study from the University of Rochester found that the institution was required to

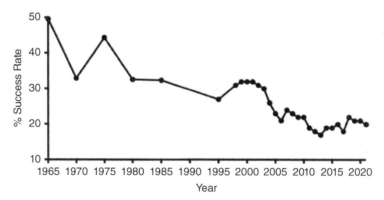

FIGURE 31.1 *Overall R01 grant success rates at the National Institutes of Health, 1965 to 2021. From reference 114 and the NIH Data Book (8).*

contribute 40 cents for every grant dollar generated by its new faculty hires in basic science, even though the investigators were highly successful, obtaining an average of $800,000 (1999-2006 U.S. dollars) in grant revenues per faculty member per year (5). Indirect cost rates thus create a perverse calculus in which the greater an investigator's success in obtaining grants, the greater the resulting burden on the institution.

An increasing emphasis on targeted research funding

The funding shortage has been exacerbated by a reduced emphasis on investigator-initiated projects (e.g., NIH R01-supported projects) in favor of targeted research and big science (4, 6). Investigator-initiated R01's declined by 15% as a proportion of the NIH research budget since 1997–1998 (7), while targeted research grants doubled over the same period (8). From these data we infer a worrisome trend in which a major funding agency is increasingly directing what scientists should work on rather than allowing scientists to follow their curiosity and identify the most important problems for study. Moreover, political influences on funding distort scientific priorities and neglect important areas of investigation. Targeted funding for currently recognized problems is politically more palatable than spending on what appear to be esoteric projects associated with basic research. While we recognize that public agencies that are mandated to improve the health of society must address immediate problems with targeted funding, it is important to realize that novel therapies and transformative discoveries are critically dependent on basic research. History has repeatedly shown that serendipity plays an important role in scientific progress, leading to such products as penicillin, Teflon, Viagra, and PCR, to name just a few (chapter 14).

The Manhattan Project and the war on AIDS are often touted as models for successful focused research, but these may represent the exception rather than the rule. The Manhattan Project was successful in part because the field of physics had advanced sufficiently to make the creation of a nuclear fission-based bomb feasible, and also perhaps because the physicists were lucky that their prototypes worked the first time (9) (chapter 13). In other words, the basic science on nuclear fission had already been done, so that the challenges in making the bomb were largely in the technical and engineering realms. Similarly, the war on AIDS was able to rapidly deliver effective antiviral agents because preceding decades of research on retroviruses and rational drug design had created a scaffolding on which to build a new program. It is noteworthy that the advances against AIDS relied on basic science investments made decades earlier at a time when there was little evidence that retroviruses were involved in human disease. For example, the highly successful protease inhibitors against human immunodeficiency virus (HIV), which did so much to lower mortality in the 1990s, relied on knowledge gathered in the 1970s that retroviral proteins are first expressed in a long peptide that must be cut by proteases for viral replication. Put another way, the curiosity of earlier scientists in studying viruses that caused tumors in animals combined with the prescient financial support of prior generations for basic research paid off with tremendous dividends once retroviruses became associated with human disease. Recently, we were reminded of the debt to the investments of prior generations when the pharmaceutical industry developed mRNA vaccines for COVID-19 in record time, relying on knowledge and technologies ranging from molecular biology to lipid chemistry that were decades in the making (10). Today, retroviruses are known to cause AIDS, leukemia, and various tumors, and drugs that target retroviruses are making a tremendous difference in human health. In contrast, no effective AIDS vaccine has yet been developed, possibly because the immunology is not sufficiently well understood, and the war on cancer has produced encouraging advances but also many disappointments. Hence, the success of targeted research may be a function of how much basic science is known about a problem.

Leaky pipelines

Although women receive nearly half of all doctoral degrees, they make up only about one-fifth of tenured science faculty in the United States (11). The "leaky pipeline" for women in science careers has many causes rooted in inequities and inequalities in science (chapter 15). Women with Ph.D. degrees and young children are 35% less likely than men with young children, and 28% less likely than women without children, to obtain tenure (11). The pipeline is also leaky for historically underrepresented groups (HUGs) in science (12–14), and evidence indicates that applicants from HUGs are less likely to receive research funding from the NIH, even after accounting for education, country of origin, training, employer, prior research awards, and publications (15). The leaky pipeline represents an enormous systematic loss of talent and diversity to science.

Increasing administrative burden

American scientists are increasingly burdened with regulations and administrative responsibilities that touch all aspects of research (16). Although some regulation is clearly necessary, there is a point at which the costs exceed the benefits. Other than publications and grant applications, much research-associated paperwork is designed to address concerns about animal welfare, patient safety, and the accountability of public funds (17). Although the scientific portion of an NIH research grant application has been mercifully shortened to 12 pages, the administrative portion is typically much longer; program project grants or training grants may reach hundreds of pages. Most research projects require annual progress reports, and ARRA-supported research required quarterly progress reports, even though in the research world, progress does not occur with sufficient regularity to justify a 3-month reporting interval. (Nevertheless, grateful recipients of ARRA funding gladly complied with this request.)

Regarding vertebrate animal experimentation, society has wisely insisted that vertebrate animals be treated humanely in research. However, there are some contradictions in the way that society deals with animal issues. For example, a typical animal euthanasia protocol for mice requires an extremely detailed description of the methods to avoid pain and suffering, yet a stroll into any hardware store reveals several methods for killing mice with extreme pain and suffering, including asphyxiation and/or cervical dislocation by mouse trap, terminal exhaustion on a glue pad, and poison. The contrast between the societal approaches to murine death in the lab and the private home creates the paradox that regulations in place to ensure laboratory animal welfare can slow research intended to help society, while citizens do not require any protocols to exterminate vermin. Is the life of a mouse living in the walls of a private home less valuable than that of a mouse in a laboratory cage? We make this analogy not to minimize the need for humane animal experimentation but rather to highlight societal contradictions in how it approaches animal welfare. (We hope the solution won't be to increase the regulatory requirements for exterminators or to ask citizens to fill out a protocol form before buying a mousetrap!) Paperwork requirements for clinical research are even more onerous. The Infectious Diseases Society of America has called attention to the excessive regulatory burden resulting from the Health Insurance Portability and Accountability Act and the "mission creep" of Institutional Review Boards (18, 19). No wonder that a graduate student wrote, "When I grow up … I fear I will become an administrator … rather than an investigator" (20).

Grant peer review

Review panels are able to accurately identify bad science but have a poor record of distinguishing highly innovative work or work that challenges existing dogma. Although reviewers can be counted on to identify the top 20 to 30% of grant applications, identifying the top 10% is impossible without a crystal ball or time machine. It is well documented that grant peer review is insufficiently precise to provide reliable rank ordering of applications (21) (chapter 23). Our own study revealed that grant reviewers cannot predict which top tier applications will be the most productive (22). Moreover, the present review system places an excessive emphasis on potential conflicts of interest, leading in many cases to the disqualification of reviewers best able to provide a knowledgeable review, with insufficient emphasis on competence, vision, and relevant expertise.

A workforce imbalance

One of the greatest concerns remains the imbalance among members of the research workforce and the funding currently available to sustain that workforce. In some respects, the current scientific workforce resembles a pyramid scheme with a small number of principal investigators presiding over an army of research scientists, postdocs, students, and technicians who have little autonomy and increasingly uncertain career prospects. Despite efforts to foster the career development of young and women investigators, senior researchers remain disproportionately men and increasingly senior. The demographer Brian Martinson has repeatedly raised concerns that the number of Ph.D.'s greatly outnumbers the number of available academic positions (23). He quotes Shirley Tilghman, former president of Princeton University and chair of the Biomedical Research Workforce Working Group, as saying: "Although the current system has succeeded in maximizing the amount of research performed ... it has also degraded the quality of graduate training and led to an overproduction of PhDs."

As one vivid example, Martinson has used the family tree of *Caenorhabditis elegans* researchers at WormBase to show how the popularity of this nematode model created a Malthusian crisis, whereby successful scientists working in this field trained more successful scientists focused on the same model, thereby creating a surplus of scientists competing for limited resources and an inevitable population collapse (24). The rapid intensification of competition for jobs and funding—3 orders of magnitude growth in the field over about 4 decades—is likely to have played a key role in the tragic Elizabeth Goodwin misconduct scandal at the University of Wisconsin. Goodwin's own students made the painful decision to turn her in for data fabrication in grant applications, derailing not only her research career but also those of her trainees (25). Martinson recalled the words of the preeminent demographer Nathan Keyfitz: "Only long afterwards did it occur to any of us, myself included, that in the stationary condition to which every system must ultimately converge, each scholar can have one student who will take up his work, that is, one successor during the course of his entire career—not one per year, but *one per 30 or more years.*"

Publish and still perish

The well-known "publish or perish" mentality can drive some scientists to write papers whether or not they have anything new and important to say. The result is an unwieldy scientific literature in which the most highly cited papers are concentrated in a small number of journals (26) while a large proportion of papers are cited infrequently or not at all (27). In 2010, Arif Jinha estimated that 50 million scholarly articles had been published since 1665, when the first modern journal appeared, with more than half published between 1985 and 2010 (28). Much of the recent scientific literature is repetitive, unimportant, poorly conceived or executed, and oversold; perhaps deservingly, much of it is ignored. However, the sheer number of papers generates an enormous burden on the peer review system (29). As the pool of qualified peer reviewers is inadequate to meet demand, the critical role of peer review in screening and correcting manuscripts prior to publication cannot properly function. These factors, as well as the difficulty publishing negative results (publication bias), serve to undermine the reliability of the scientific literature (30, 31).

Survival of the fittest

A Darwinian struggle for existence can produce multiple behaviors, including competition, aggression, reciprocity, altruism, cooperation, and sometimes, unfortunately, cheating. Each of these behaviors may be readily observed among contemporary scientists (chapters 17 and 18). Competition can be beneficial when it provides a motivation for resourcefulness and innovation. However, competition in excess can be a compelling incentive for scientific misconduct ranging from egregious fraud to more-subtle transgressions, such as selective reporting of results (32) (chapter 20). The need to distinguish one's work from that of competitors can induce scientists to exaggerate the importance of their findings or downplay potential caveats (33). While this may seem relatively innocuous in the short run, the cumulative effect can be the erosion of public confidence in the scientific method (34). Another casualty of the competitive environment is timely communication. Scientists often keep valuable information to themselves for fear of being "scooped" by competitors, and attempts to place work in the most highly selective journals may delay publication for months or years (chapter 26). The entire shape of the research effort is distorted as researchers scramble to conform their work to targeted funding opportunities and steer away from risky lines of inquiry or projects requiring a lengthy time investment. Unfortunately, in science, risk and reward often go hand in hand (35), so a focus on conservative projects means that opportunities to make revolutionary breakthroughs will be missed.

Winner takes all

The current scientific enterprise is a highly competitive environment with a winner-takes-all system in which the greatest rewards are given to individuals who excel in scientific discovery, publication prestige and quantity, and grant funding (chapter 19). Since its inception, science has operated with the priority rule, which means that when different scientists work on the same problem, the credit goes to the one who provides the answer first (36). Furthermore, success in science tends to beget more success, in a phenomenon that Robert Merton called the "Matthew effect" (37) (see also chapters 19, 23, and 26). The benefits of scientific success even appear to translate into better health, as recipients of Nobel Prizes live longer than those who are only nominated (38) (chapter 21). In any human endeavor, competition can be healthy when it promotes ingenuity and creativity and harnesses primal human energies. Thus, it has been argued that the priority rule in science is beneficial to society (36). However, a winner-takes-all system has inherent dangers, including a greater likelihood of cheating by participants. In this regard, scientific fraud is a form of cheating that can lead to publications (as well as retractions). Fraudulent publications that affect public perception and policy, such as the 1998 *Lancet* article that linked autism to the measles vaccine (39), can be destructive to society (chapter 18). Despite occasional high-profile instances of fraud that adversely affect public opinion, society has continued to hold an overwhelmingly positive view of science and scientists (40). Nevertheless, this trust is precious and must not be taken for granted, as illustrated by the rapid loss of trust in science and scientists that followed the incident known as "Climategate" (41), in which emails and computer files from climate scientists were hacked from a server at the University of East Anglia and released to the public.

Unfortunately, the political controversies associated with the COVID-19 pandemic, including debates about masking, vaccines, and the origins of severe acute respiratory syndrome coronavirus 2 (SARS-CoV-2), have taken a toll on the public view of science, with a significant decline in trust in medical scientists, particularly among those with self-described conservative political orientation (42, 43). Paradoxically, this lack of trust has occurred precisely at a time when the scientific establishment responded rapidly and effectively to the COVID-19 pandemic, producing rapid tests, vaccines, and antiviral medications in record time, reinforcing the notion that the benefits of scientific advances can only be fully realized in partnership with robust public support (see also chapter 30).

The priority rule

We have already discussed the priority rule and its effects on science in chapters 19, 20, and 21. Briefly, the priority rule means that credit and its associated rewards go to the first individual or group of individuals who announce a discovery irrespective of the number of scientists who have contributed to the solution of the problem. How did the priority rule become the dominant economic system in science for assigning rewards? Why do scientists accept this economic system? There are no simple answers to these questions, but the priority rule was in place long before science became a major human enterprise. For example, Isaac Newton and Gottfried Wilhelm Leibniz quarreled in the 17th century about priority in the invention of calculus. A list of scientific priority disputes is maintained by Wikipedia (44). In the early days of the Scientific Revolution, most inquiry was carried out by individuals with sufficient means and leisure to devote to thinking, experimentation, and the publication of books. In a system that began without prizes, salaries, or grant applications, priority may have been the only source of prestige among a small number of colleagues. However, the priority rule probably contributes to many of the current maladies in science. The effort to get there first and grab the largest share of the credit undoubtedly contributes to such practices as citation bias, secrecy, and the appropriation of others' ideas and data. The competitive environment created by the priority rule is vividly apparent in James Watson's inspection of Rosalind Franklin's X-ray diffraction data, as described in *The Double Helix* (45). A more recent example is provided by the dispute over priority in the discovery of HIV (46). The race to be first encourages risk-taking that can lead to scientific breakthroughs that benefit society while at the same time creating conditions that are detrimental to science. Hence, it may be worth reconsidering the priority rule and whether it is the best reward system for science.

Science as a team sport

Although the edifice of scientific understanding is sometimes envisaged as an accumulation of individual discoveries, in reality, science is a community effort comprising innumerable interdependent contributions (chapters 20 and 21). Credit is disproportionately awarded to principal investigators for what is truly the product of teamwork, and nearly all scientific contributions are heavily dependent on knowledge obtained earlier. As Newton famously remarked, he was able to see farther by standing on the shoulders of giants. In the spirit of an Amish barn-raising, a celebration of the collective achievement of science should subsume individual achievement.

THE CHALLENGE OF REFORMING SCIENCE

It can be argued that the Scientific Revolution has been the single greatest transformative event for humanity since the harnessing of fire. Science has cured disease, changed our perception of our place in the universe, unleashed the green revolution, taken us into space, and shrunk the world through rapid transportation and instant communication (chapter 11). However, science has also brought devastating weaponry and planetary degradation from natural resource extraction, pollution, and climate change. Regardless of one's stance on the benefits and liabilities of the Scientific Revolution, science remains humanity's best hope for solving some of its most vexing problems, from feeding a burgeoning population to finding alternative sources of energy to protecting our world against planet-killing meteorites. The year 2022 saw a breakthrough in fusion research (47) and witnessed a dramatic example of the power of science to protect our planet when NASA deflected the orbit of a small asteroid to test the technology of using high-energy impacts to deflect life-threatening meteorites (48). And yet, notwithstanding the importance of the scientific enterprise, only a tiny fraction of the human population is directly involved in scientific discovery. It is therefore crucial for society to nurture and sustain the fragile scientific enterprise and optimize its functioning to ensure the continuing survival and prosperity of mankind.

Since eclipsing theology as a framework for understanding the natural world and separating itself from philosophy as an intellectual discipline, science has reigned as the supreme arbiter for certain types of knowledge for nearly two centuries. The association of science with technology and national power has led to considerable public and governmental funding support. Looking back, the progress of science may appear inexorable, but there are increasing signs that this great human enterprise could benefit from introspection and retooling. Today, science finds itself increasingly besieged, and some of its disciplines are in outright crisis. Threats to science come from both within and outside of the scientific enterprise. External threats include increasing antiscientific attitudes expressed by the general public and politicians, skepticism about the scientific community's conclusions (the global warming debate is but one current example), inadequate funding, and increasing regulation. The discipline of virology is under attack over concerns about the origin of SARS-CoV-2 despite the weight of evidence indicating that the COVID-19 pandemic was most likely a zoonosis and the fact that virologically related research gave us vaccines and antiviral therapies that saved the lives of millions (49) (chapter 30). These external threats are exacerbated by inadequate efforts by scientists to educate and engage the public in a clear discussion of the benefits and limitations of scientific findings. Internal threats include dissatisfaction of scientists with many aspects of the business of science as it is performed today (including but certainly not limited to peer review and incessant pressure to obtain grants and publications) and the corrosive impact of research errors and misconduct, as reflected by an increasing number of retracted publications (chapters 24 and 25).

History teaches us that most, if not all, great human enterprises must undergo periodic cycles of self-examination and renewal to maintain their vigor. Examples of great reforms include Marius' revamping of the legion system that allowed the Roman Empire to survive for centuries, the abandonment of scholasticism during the early Enlightenment that ushered in the Scientific Revolution, and Abraham Flexner's creation of the modern medical school curriculum. Reforms are nearly always catalyzed by crisis and discontent, and perhaps we are approaching a time when fundamental reforms are needed for the scientific enterprise. However, history also tells us that reforms are usually bitterly resisted by the establishment, and any attempt at reforming science is likely to encounter strong headwinds.

Any movement to reform science must consider the problems and suggest solutions. In our view, there are changes that must engage societal and political processes (structural), whereas others can be made entirely within the scientific enterprise (methodological and cultural).

STRUCTURAL REFORMS

Renewed investments in science

In the 21st century, humanity is facing several potential existential crises in the form of rapid climate change, pandemics, and severe environmental degradation. With 8 billion people in need of food, shelter, and safety, our species faces monumental challenges. We believe that science can make a critically important contribution to navigating the near and distant horizon with its capacity to generate new technological solutions to major problems. However, for this to happen, we need a healthy and vibrant scientific enterprise, and that requires stable funding for science. The political dialogue over research funding in the United States and many other countries has become a recurring discussion about shrinking budgets and competing priorities. Since 1963, the federal investment in research and development as a percentage of gross domestic product (GDP) has fallen steadily (7). If the major problems facing humanity are to be addressed, then this trend must be reversed. To place this in some perspective, the 2011 federal investment in applied and basic research was equivalent to only about 1% of the estimated cost of the wars in Iraq and Afghanistan (50, 51). A 2010 report indicated that U.S. research output had become surpassed by Europe and Asia (52). The old cold war may be over, but a new period of great power competition has begun, and the threats facing the modern world are no less formidable than those that catalyzed governmental investment in science in the mid-20th century. As an independent group of economists concluded, the time is right for a major sustained government investment in infrastructure (53), and we submit that such an initiative must include the scientific enterprise. This will not only generate jobs in the short run, as a majority of grant expenditures are dedicated to personnel costs, but will lead to the discoveries that spawn new industries and create unimagined efficiencies in the long run. As the economists' report argued, "Labor costs will never be lower. Equipment costs will never be lower. The cost of capital will never be lower. Why wait?" (54). Scientific innovation is a valuable national resource that should not be squandered. Until increased resources are available, a diversion of more funding to untargeted funding mechanisms, such as investigator-initiated applications (R01s), could help to sustain the scientific enterprise during the present period of resource scarcity. Furthermore, devoting a portion of funding to institutions for salary support, rather than devoting all federal research support to individual projects, would provide greater stability to the system.

Balancing and renewing the scientific workforce

If a fundamental structural problem with science today is the inadequate financial support of the current scientific workforce, then an obvious potential solution is to reduce the size of the workforce (55). However, while this might reduce competition for funding in the short run, this action would be tremendously shortsighted, as experts from across the political spectrum agree that more scientists, not fewer, are needed to address society's many challenges and generate the innovative discoveries that will resuscitate the global economy (56–58). A publication from the National Academies of Sciences called *Rising Above the Gathering Storm* made a cogent argument to expand the pipeline of scientists, engineers, and mathematicians (59). The urgent need for more scientists documented in this report led to the National Math and Science Initiative (NMSI), a program designed to improve mathematics and science education and attract the best and brightest students to scientific careers. Notably, this was a bipartisan initiative, as the NMSI was initiated during the administration of President George W. Bush and was vigorously supported by the Obama administration, which added the "Educate to Innovate" program to enhance educational opportunities in science, technology, engineering, and mathematics.

It must be recognized that one of the greatest obstacles to recruiting the best and brightest students to scientific careers is the unhappiness of scientists working in the current environment. Anxiety over the future is at an all-time high, and there is concern that stopgap measures to set aside funds for new investigators only intensified competition for funds among senior scientists (60). This situation may have reached a crisis point in 2022 when it became clear that the majority of students in the biomedical sciences were opting out of doing more academic postdoctoral training, which is usually the gateway to academic research positions (61). If the poor morale of active scientists is not urgently addressed, all of the new initiatives will be for naught. It makes little sense to aggressively recruit bright young students to lifelong careers of struggle and uncertainty, especially when scientific training requires a tremendous investment in time and resources. Few scientists would go as far as Jonathan Katz, a Washington University in St. Louis physicist who wrote an essay entitled "Don't Become a Scientist!" (62), but many mentors nevertheless make their reservations known to their trainees, even if inadvertently. Recruitment becomes much more straightforward if trainees can envisage a clear path to career success. An NIH working group formed specifically to address the future biomedical research workforce found that an imbalance in supply and demand was the leading concern of the commenters providing input to the committee (63). Although the report acknowledged the specific needs of women and scientists from HUGs (14, 64), it did not make recommendations specifically aimed at increasing diversity. Solving the leaky pipeline will require not only initial recruitment efforts and investment, but the establishment of sustained support mechanisms and the institution of more flexible and family-friendly policies (65) (chapter 15). Less than 0.1% of the world's population is presently working as scientists or engineers (66), and only a fraction of this small percentage is involved in the generation of new knowledge. On this slender thread hangs society's future.

Recognizing the critical importance of basic research

We discussed the current emphasis on "translational" research in chapter 12 (7). While we acknowledge the importance of removing obstacles that impede the translation of basic discoveries into useful applications, we are concerned that an excessive focus on translation may eventually become a cautionary tale. Immediately following World War II, Vannevar Bush, President Harry Truman's science policy advisor, wrote, "It is wholly probable that progress in the treatment of ... refractory diseases will be made as the result of fundamental discoveries in subjects unrelated to those diseases, and perhaps entirely unexpected by the investigator ... [progress] results from discoveries in remote and unexpected fields. ... Government has provided over-all coordination and support; it has not dictated how the work should be done" (67). Elsewhere in the report, Bush astutely observed that "scientific progress on a broad front results from the free play of free intellects, working on subjects of their own choice, in the manner dictated by their curiosity" (67). In *Lives of a Cell*, Lewis Thomas wrote that "if I were a policy maker ... (I would) give high priority to a lot more basic research" (68). Sadly, the wisdom that basic research provides the essential raw material for practical applications appears to have less appeal to current policymakers. The dramatic decline in the success rate of grant applications seeking support for basic research needs to be urgently addressed (69).

Optimizing laboratory size

The efficiency of laboratories in functioning, exchanging ideas, and producing new information must be a function of the lab size, funding, and their composition. This raises the question: What is an optimal lab size? Remarkably, little or no scholarship has been done on this question. An interesting study by Jeremy Berg, the former head of the National Institute of General Medical Sciences, suggested diminishing returns once a laboratory has more than about $750,000 in direct costs (2006 U.S. dollars) (70). Other possible beneficial changes from optimizing laboratory size may be the creation of more principal investigator positions and a renewed emphasis on investigator-initiated projects.

Regulatory and review reform

An unfortunate aspect of regulations is that new requirements are added to older ones, but the paperwork burden never seems to be reduced. The aforementioned reduction in NIH grant length is a noteworthy exception. An effort to similarly streamline the administrative sections of grant applications and reporting requirements would be welcome, perhaps as part of a comprehensive effort to limit the regulatory burden on scientists to measures that actually serve a meaningful purpose. In addition, the mechanisms for grant peer review should be reexamined (71). Some scientists have advocated further reductions in the length of grant applications, particularly for established investigators (72), and emphasizing the scientist rather than the project (73), but any changes will result in winners and losers and are bound to be controversial unless the pot of money is expanded and overall levels of funding are restored to a reasonable level.

A scientific study of science

Despite the unquestioned success of science and the scientific method, it is remarkable how little we know about how to configure the scientific enterprise in an optimal manner to confront the problems facing humanity. For example, we do not know the answers to the following questions. How many scientists do you need to ensure a steady stream of innovation that will allow consistent economic growth? What is the optimal size of a research group? How well does peer review perform? What is the optimal duration of scientific training? What is the relationship between length of scientific training and subsequent success in science? What is the optimal award duration for a research grant to promote productivity without encouraging too much comfort and lassitude? Without answers to these questions, it is difficult to make the best choices as we struggle to restructure certain aspects of science. In fact, much of what we think we know in this realm is anecdotal and largely derived from individual experience. However, each of these questions could be the subject of rigorous study, and the answers of such studies could provide information to inform future decision-making.

METHODOLOGICAL REFORMS

Revising criteria for promotion

As the *Guinness Book of World Records* shows, the desire to be first, whether in science or in the consumption of hot dogs at one sitting, is human nature and, as such, unlikely to change. Scientists will continue to race to be the first to achieve a goal, and journals will continue to vie for original reports. Perhaps the best place for a methodological reform to replace the priority rule is at the level of academic promotion and in the awarding of symbolic rewards, such as scientific prizes (chapter 21). In the present system, promotion decisions typically depend upon performance review by other faculty members who have a limited understanding of an investigator's work, which increases reliance on surrogate measures of quality, such as grant dollars, bibliometric analysis, and journal impact factor (74) (chapter 26). A reform of the promotion process based on careful peer evaluations of scientific quality and the specific contributions of the authors might help to reduce the present emphasis on priority. Furthermore, increasing the value of collaborative publications when considering promotion could provide important incentives for greater cooperation between scientists.

Reembracing philosophy

Science traces its ancestry to natural philosophy, which in turn emerged from philosophy. Many early scientists were fully grounded in the philosophical fields of their time and contributed to both disciplines. René Descartes separated philosophy from theology while making seminal contributions to mathematics and physics. Leibniz discovered the calculus while at the same time proposing the metaphysical theory of monads. In this regard, it is worth remembering that Ph.D. degrees granted in the natural sciences are actually doctorates of philosophy. Unfortunately, scientific training today does not include significant instruction in philosophy despite the critical importance of the philosophical branches of logic, epistemology, and ethics to science. In fact, there are numerous instances in which philosophical thought has greatly influenced scientific discovery and vice versa. Albert Einstein credited the philosopher Immanuel Kant with inspiration that led to the theory of relativity, and Einstein's scientific contributions have in turn influenced philosophy (75).

Philosophy is currently regarded as including four branches: epistemology, logic, ethics, and metaphysics. We believe that the knowledge found in each of these branches can contribute significantly to improving the scientific enterprise. Formal training in logic and epistemology could reduce the number of errors by scientists (chapter 25). One common error in science is the attempt to make positive inferences from negative data (e.g., ruling out a mechanism or cause and effect from negative experimental data). Errors in logical thought can lead to dogma and affect the direction of entire fields of study. For example, the conclusion that humoral immunity has no role in protection against many intracellular pathogens was based on an inability to demonstrate the efficacy of antibodies with the methodology used (76). This conclusion evolved into dogma and greatly affected the direction of research, including the development of vaccines. However, this constituted a logical error because a possible protective role of antibodies could not be excluded by experiments that showed no protection. The original conclusion was eventually shown to be faulty when hybridoma technology allowed the generation of protective monoclonal antibodies. A stronger foundation in epistemology, supplemented by an awareness of philosophical issues involving language, might also avoid some of the problems created in science from the misuse of such terms as "descriptive" (77) (chapter 2) and "mechanistic" (78) (chapter 3) in the categorization of projects and papers. Ethics, or "moral philosophy," has already returned to scientific training as students are increasingly taught formal ethical principles in science, but much more could be done in this realm with the goal of reducing the number of dishonest retractions (chapter 18). Metaphysics, or the study of reality, is perhaps the most problematic branch of philosophy for scientists who regard it as a domain outside scientific pursuit. However, metaphysics is the antecedent of natural philosophy and is thus in the ancestral line of the scientific enterprise. Knowledge of metaphysical questions could help scientists to frame problems and perhaps approach the boundary without crossing it. At the Johns Hopkins School of Public Health, a new educational program named R3 for the three essential qualities of science, Rigor, Reproducibility, and Responsibility, was established in 2017, which aims to redress the need for formal epistemological and philosophical training in graduate education (Box 31.1).

Box 31.1 Training scientists as generalists

Today, graduate scientific training is highly specialized. This is an inevitable result of the success of science in spawning new fields that each represent a massive repository of human knowledge (chapter 10). The typical scientist today begins their science education by majoring in a specific scientific discipline and proceeding to graduate school, where they become immersed in a particular field. This mode of education is very successful at training individuals to be able to do deep research work but is notably different from the way that scientists trained in the early 20th century. The problem with such highly specialized training is that it channels individuals into narrow areas of research, which may not be conducive to addressing questions that arise in

(Continued)

Box 31.1 (Continued)

the interfaces between fields. When the psychologist Kevin Dunbar spent a year observing scientists in four molecular biology laboratories, he found that laboratory groups with a broader range of interests were better able to capitalize on unexpected findings. As described by David Epstein in his book *Range*: "Dunbar witnessed that … labs most likely to turn unexpected findings into new knowledge for humanity made a lot of analogies and made them from a variety of base domains. The labs in which scientists had more diverse professional backgrounds were the ones where more and more varied analogies were offered, and where breakthroughs were more reliably produced when the unexpected arose" (112).

Recognizing the need to train individuals with broader knowledge of science and philosophy, a new program was begun at Johns Hopkins School of Public Health known as R3 for the three R's of science: Rigor, Reproducibility, and Responsibility. The program, led by Dr. Gundula Bosch, is open to all students and provides courses on epistemology, communication, ethics, and causality (96, 113).

The three R's.

Enhanced training in probability and statistics

There is increasing recognition that improving the quality of scientific work will require more formal training in basic principles, including statistical theory (79). Scientific certainty is often defined in probabilistic terms. Knowledge of probability and statistics is essential for the design, execution, and interpretation of many scientific experiments. As the late statistician Stephen Lagakos observed, "Sure, you can lie with statistics … but it's a lot easier to lie without them" (80). Although most scientists have some knowledge of probability and statistics and can calculate P values using statistical software, the level of statistical expertise varies greatly among individuals. In fact, much deeper knowledge of the foundations of these disciplines is needed. For example, a 2010 paper in *Science* was retracted because the authors assumed that two correlated variables were independent while these variables were in fact additive and dependent (81).

Use of checklists

There is conclusive evidence that the use of checklists can reduce errors in human activities ranging from aviation to surgery (82). Science should be no exception. An increased use of checklists in the conduct of scientific experimentation and publication could conceivably reduce scientific errors and consequently improve productivity, reduce waste, and prevent retractions. Clearly, different types of checklists would be needed for different occasions. For example, a simple checklist can be constructed for a scientific result in which a stimulus appears to cause an effect (Table 31.1), which if followed would enhance conceptual rigor and reduce the likelihood of a false conclusion. In recent years, several journals have introduced checklists before submission to reduce the number of errors (83, 84). Although these are undoubtedly helpful in getting a paper submitted, we suggest that they would be even more helpful if consulted before the work is done since being aware of the requirements before work is completed could help ensure that it is done correctly. There is some encouraging evidence that the use of checklists has translated into improved publications in those journals that have introduced them (85, 86).

CULTURAL REFORMS

Reward system reform

The reward system in science is at the heart of many problems but is so ingrained that it will be difficult to reform. However, any attempt to reduce the incentives for cheating will have to address structural issues in the payoffs for scientists (chapters 19 and 21). In nature, winner-takes-all strategies have been associated

TABLE 31.1 Sample checklist for an observation in which a stimulus elicits an effect

☐ Does an increase or decrease in the magnitude of the stimulus translate into a commensurate effect? (e.g., a dose-response relationship)
☐ Does the effect ever occur without a stimulus? (e.g., false-positive rate)
☐ Does the stimulus ever fail to elicit an effect? (e.g., false-negative rate)
☐ What is the temporal relationship between stimulus and effect? (temporal causality)
☐ Is the effect reproducible? (reproducibility)
☐ Can the effect be measured by an independent technique? (validation through independent methodology)

with male evolutionary strategies that disproportionally reward risk-taking in the production of offspring (87). A reform to the payoff system might also have other benefits in addition to reducing the incentives for cheating. Given the increasing connectivity between fields and specialties of science, there is an increasing need for collaboration, yet a system of winner-takes-all is inherently unfair to collaborators. A different reward system could promote team science and thus promote the overall progress of the scientific enterprise (chapter 20). In this regard, we note that efforts are already ongoing to promote and recognize collaboration and team science that provide guidance for communication, sharing credit, and handling conflict (88).

Nobody here but us chickens

A culture that places a greater emphasis on the derivative and collaborative nature of scientific advances, along with a reduced emphasis on rewarding hypercompetitive behavior and the cult of the "rock star" investigator, would improve science. The evolutionary biologist David Sloan Wilson has recounted an attempt to improve the productivity of egg-laying hens by Purdue researcher William Muir (89). The approach of selecting the most productive individual hens from each group to breed the next generation was compared with selecting the most productive groups of hens. Unexpectedly, the latter approach was most successful because the individuals within successful groups had learned to function cooperatively, and the happier hens laid more eggs. Productivity plummeted when the star performers were grouped together, and all but three hens in this group were dead by the end of the experiment. After a lecture describing these results, a professor in the audience exclaimed, "That describes my department! I have names for those three chickens!" Unfortunately, many of us know similar chickens in our own departments.

A vision of a healthier scientific culture

Science functions best when scientists are motivated by the joy of discovery and a desire to improve society rather than by wealth, recognition, and professional standing. Despite current pressures, it is perhaps remarkable that many scientists continue to engage in selfless activities such as teaching and reviewing without compensation, decline to publish work that doesn't meet stringent standards for quality and importance, freely share reagents or knowledge without worrying about who gets the credit, and take genuine pleasure in supporting the efforts of other scientists. Such individuals should be recognized and emulated.

How to build a motivated research community

The systems biologist Uri Alon has written a thought-provoking essay on how to build a motivated research group (90). He concludes that individuals need to be matched with projects appropriate for their talents and passions and that they require both autonomy and connectedness with other members of the group. The research group is but a microcosm of the entire research community. A healthy scientific environment is one in which the freedom to do what one wants is complemented by support and stimulation from a community. This will enhance productivity and innovation.

IMPROVING THE QUALITY OF THE SCIENTIFIC LITERATURE

Finally, we return to the scientific literature, which has been a recurring focus of this book (especially chapters 2, 8, 9, 17, 18, 22, 26, and 30). The scientific literature is the critical system by which scientific findings are communicated and archived for subsequent reference and analysis. Hence, the reliability of the scientific literature is of the utmost importance to society. However, as we have amply documented, rising numbers of retracted articles (chapter 24), reproducibility problems (chapter 7), and inappropriately duplicated images (chapter 17) have increased concern that the scientific literature is unreliable (31, 91–93). Contributing factors may include competition (chapter 20), sloppiness, prioritization of impact over rigor (chapter 26), poor experimental design, inappropriate statistical analysis, and lax ethical standards (94, 95). Although the number of questionable publications represents a very small percentage of the total literature, even a few problematic publications can reduce the credibility of science. Hence, it is important to redouble efforts to improve the reliability of scientific publications. Critical systems are designed to be fail-safe. This does not mean that failures cannot occur, but rather that redundant and compensatory mechanisms are engineered into the system to detect and mitigate failures when they occur. Specific measures can reengineer the scientific literature so that it is better able to prevent and correct its failures.

Improving graduate and postgraduate training

Training is the foundation of all scientific endeavors. Contemporary graduate scientific training is designed to prepare trainees to perform deep investigation into a highly specialized area (96) but does not necessarily provide students with a broad scientific background. As discussed in chapter 10, students are taught in a guild-like environment by a mentor who may or may not have been trained in good scientific practices. Consequently, there is no guarantee that programs are consistently producing scientists who are adequately prepared to do good science. Improving postgraduate training to ensure that trainees are well versed in scientific rigor, statistical analysis, experimental design, and ethics (Box 31.1) can improve the quality of the scientific literature by improving the quality of the research itself.

Reducing errors in manuscript preparation

Roughly 1 of every 25 articles in the biomedical literature contains an inappropriately duplicated image (93) (chapter 17). The majority of incidents of inappropriate image duplication result from simple errors in figure assembly (97). However, a minority of these represent intentional efforts to mislead the reader, which constitutes scientific misconduct. Involving multiple individuals in figure preparation prior to manuscript submission may reduce the likelihood of error and also discourage intentional deception. Software has been developed that can aid in the detection of manipulated images. The *Journal of Clinical Investigation* began using Proofig (98) software to screen manuscripts and within a year tripled the number of problematic images detected, leading to the rejection of 13 papers after initial editorial acceptance but prior to publication (99). Increased vigilance on the part of reviewers and editors together aided by increasingly sophisticated software programs can reduce the incidence of problematic images.

Pre-submission criticism

Although peer review is intended to detect and correct errors prior to publication, the process involves only a small number of reviewers and is well-known to be imperfect (98). Critical input from a broader range of colleagues may lead to identification of weaknesses in a manuscript and allow authors to improve the quality of their published work. Pre-submission criticism may be informally obtained by asking others to read a manuscript before submission or by posting the manuscript on a preprint server and alerting colleagues in the field that the data are available in prepublication form (100). Both authors of this book have received pre-submission criticism of manuscripts posted as preprints that led to improvements. A more longstanding mechanism for obtaining pre-submission criticism is to present unpublished data at meetings and seminars.

Robust review and editorial procedures

After a manuscript is submitted for publication, the peer review and editorial processes are major checkpoints for quality improvement. Reviewers can identify errors, and training may improve their ability to detect problematic data. Journals can use software to identify plagiarism, image manipulation, or data anomalies (101–103). Dedicated statistical editors and reviewers can help to ensure that complex data sets are appropriately analyzed. Some journals have reacted to the epidemic of inappropriately duplicated images (see chapter 17) by requesting that copies of original blots and gels prior to cropping or image processing be submitted for inspection and archiving. For example, the *Journal of Clinical Investigation* requires authors to submit copies of the original Western blot images used for figure construction. Having access to the original data may discourage manipulation and can be used to address questions that may arise in figure presentation.

Post-publication criticism

The development of sites such as PubPeer allows readers to anonymously post critical feedback after a manuscript has been published (104). Such comments can alert the scientific community to potential problems concerning a published manuscript and allow the authors to respond. Some concerns may be easily addressed, while others may require correction or even retraction of an article. As it may be difficult to fully evaluate published results without access to the primary data, journals have a responsibility to respond to readers' concerns and to work with authors to resolve them. Historically, both journals and institutions have sometimes failed to live up to their obligations in addressing problematic articles and allegations of research misconduct (105–107) (chapter 18) (Fig. 31.2). Although post-publication review occurs relatively late in the process, it provides an important safeguard that allows even published findings to be corrected.

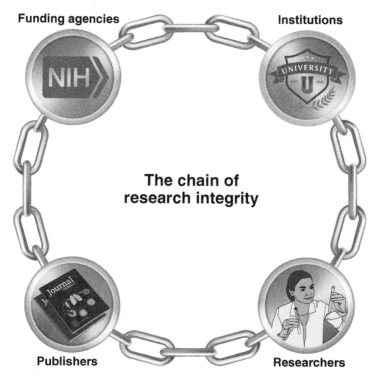

FIGURE 31.2 *The chain of research integrity. Publishers, funding agencies, institutions, and researchers each play essential roles in protecting the integrity of the research record (107).*

Increasing journal-based research

Journals tend to focus more on publishing scientific information than on analyzing their own performance and are often secretive about their publication practices. However, much of what we have learned about problems with the scientific literature has come from editors and journals willing to analyze and share their experiences. The experience of the journal *Molecular and Cellular Biology* in dealing with inappropriate image duplications (97) (chapter 17) suggests that the majority result from simple errors, although approximately 10% of these led to retractions, and that prepublication image screening may be more efficient than waiting to deal with problems after publication. Going forward, it will be important for more journals to examine their own experiences and share them with the scientific community to establish best practices and improve the entire publishing enterprise. In this regard, it is encouraging and noteworthy that journals such as the *Journal of Clinical Investigation* (99, 108) are publishing their experiences and that we are learning from them.

Fostering a culture of rigor

In recent decades, many life science researchers have learned to accept a culture of impact, which stresses publication in high-impact journals, flashy claims, and packaging of results into tidy stories. Today, a scientist who publishes incorrect articles in high-impact journals is more likely to enjoy a successful career than one who publishes careful and rigorous studies in lower-impact journals, provided that the publications of the former are not retracted. This misplaced value system creates perverse incentives for scientists to participate in a "tragedy of the commons" that is detrimental to science (109). The culture of impact must be replaced by a culture of rigor that emphasizes quality over quantity (chapters 6 and 26). A focus on experimental redundancy, error analysis, logic, appropriate use of statistics, and intellectual honesty can help make research more rigorous and likely to be true (110) (chapter 6). The publication of confirmatory or contradictory findings must also be encouraged to allow the scientific literature to provide a more accurate and comprehensive reflection of the body of scientific evidence (111) (chapter 7).

There is no single simple remedy for improving the reliability of the scientific literature. Scientific publishing is a complex ecosystem with multiple interacting components, each of which has a critical role to play in ensuring the integrity of the whole (107). Reengineering the system to incorporate multiple fail-safe features from data acquisition to post-publication review will better prevent, detect, and correct failures and result in a more reliable scientific literature. Science is a human endeavor and, as such, will never be perfect. Nevertheless, remarkable achievements have been made in science and technology, which remain humanity's greatest hope for the many challenges it currently faces. To obtain the full benefit of science, its literature must be reliable. For too long, science has relied on the mantra that it is self-correcting. Science can be self-correcting, but only through the concerted efforts of all scientists working at multiple levels. The steps outlined here provide a blueprint to begin this process.

CONCLUDING REMARKS

True reform will require addressing major structural, methodological, and cultural aspects of the scientific enterprise. For starters, this means a reduction in personal pressures on scientists, a greater institutional commitment to "hard" salary support, an emphasis on quality rather than quantity of publication, the fostering of a cooperative and collaborative culture, a reduced dependence on journal impact measures, and the development of more-stable and -sustainable sources of research funding. Additional structural changes will be required to enhance cooperation, allow risk-taking, reduce funding pressures, and provide more-flexible career pathways to prevent the ongoing loss of capable scientists along the pipeline (64). A society serious about confronting the real challenges of the future cannot afford to leave so many good scientists behind. The current global economic recession calls for intensive investment to renew the scientific infrastructure, which includes not only bricks, mortar, and equipment, but human resources as well. Nations that recognize this opportunity will be the ones that rule the future.

The question is not whether science is failing, but rather, whether the current scientific enterprise is as healthy as it should be. Our answer is that it is not failing when measured by a continuing string of discoveries, but neither is it healthy, as evidenced by the growing problem of retractions, inequities, and inequalities (chapters 15 and 24), as well as the many other maladies that we have discussed in this book. What we propose is nothing less than a comprehensive reform of scientific methodology and culture. However, the dialogue can begin by addressing specific problems in science, such as the problem of honest and dishonest retractions discussed in chapter 24.

Science has been so successful over the past three centuries that it should now be sufficiently secure to return to its philosophical roots. Embracing this central field of the humanities could also help scientists to be better communicators in their engagement with the public at large. Science can become methodologically more rigorous by adhering to the principles of epistemology and logic, which in turn could reduce the number of errors in scientific work (chapter 25). A better appreciation of ethics could also reduce the problem of dishonest retractions (chapter 24). An emphasis on probability and statistics in the graduate curriculum should be noncontroversial given its importance in experimental design and interpretation. The use of checklists is a simple pragmatic suggestion that could be expanded to incorporate principles of epistemology and logic. We call for a cultural change in which scientists rediscover what drew them to science in the first place. In the end, it is not the number of high-impact-factor papers, prizes, or grant dollars that matters most, but the joys of discovery and the innumerable contributions both large and small that one makes through contact with other scientists. For many of us, old habits may be too deeply entrenched to change, but we can start to foster a more cooperative scientific culture in our trainees. It would be naïve to believe that competition and personal ambition could or should be eliminated from science. Nevertheless, it is reasonable to ask whether the current scientific culture is allowing science to be as fruitful as it could be, particularly when the present system provides such potent incentives for behaviors that are detrimental to science and scientists. Only science can provide solutions to many of the most urgent needs of contemporary society. A conversation on how to reform science should begin now.

V AFTERWORD

32 Diseased Science

It is more important to know what sort of person has a disease than to know what sort of disease a person has.

Attributed to Hippocrates

Chapter 26 describes a widespread affliction of scientists known as *impact factor mania* (1), also referred to as *impactitis* (2), for which there appears to be no cure. Recognition of this condition has led us to consider whether additional unrecognized medical conditions may be unique or overrepresented among scientists.

Ahypothesemia. Characterized by the absence of a hypothesis; in its terminal stages, may manifest the symptom of HARKing (hypothesizing after the results are known) (3). Some scientists have hypothesized that this is a problem (4). See also chapter 2 and *hypothesosis*.

Amnesia originosa. An inability to recall the actual origin of an idea that one now regards as one's own. Afflicted individuals are able to present others' ideas as their own without guilt or attribution to the original source.

Appendiceal hypertrophy. A relatively new condition that first became manifest when journals began to allow supplementary data. Authors suffering from appendiceal hypertrophy stuff their papers with supplementary data irrespective of its relevance, perhaps hoping to induce data overload and reviewer fatigue. Reviewers, in particular those suffering from *experimentitis infinitum*

Thinking about Science: Good Science, Bad Science, and How to Make It Better, First Edition.
Ferric C. Fang and Arturo Casadevall.
© 2024 American Society for Microbiology.

(see below), may aggravate appendiceal hypertrophy by demanding additional information of uncertain value. Preventive measures include charging extra fees for supplementary data analogous to the taxes imposed on tobacco use.

Areproducibilia. The inability to obtain the same experimental result twice (5) (see also chapter 7). This is not necessarily a problem for individuals who publish irreproducible results and simply move on to leave other scientists to deal with the problems (6, 7). However, recurrent areproducibilia may impair scientific reputation, as subsequent work by the individual is not considered credible.

Borderline probability disorder. Afflicted individuals may dismiss the potential importance of results with $P = 0.06$ while unquestioningly accepting the importance of results with $P = 0.05$ (8, 9). See also *significosis*.

CNS depression. The feeling after one's paper has been rejected by *Cell*, *Nature*, and *Science* (1). The malady generally abates once the paper is published in a lower-tier journal.

Deja poo. The sensation that you have seen some results before—because someone is recycling the same crappy data (see chapters 17 and 18).

Dogmatitis. 1. Manifested by a courageous adherence to one's principles (benign). 2. Manifested by perversely clinging to disproven ideas (malignant).

Editorial dysfunction (**ED**). A condition experienced by authors in which prolonged periods of unresponsiveness to one's submitted manuscript are punctuated by brief intervals of false hope that finally terminate in rejection.

Experimentitis infinitum. A condition exhibited by reviewers who always demand more experiments irrespective of the amount of data already provided (10, 11). Also known as *status revisicus*.

Gelatophobia. The fear of getting scooped. Gelatophobia may lead to the premature emission of a manuscript to a journal before it is ready.

Gotchalism. A disease of reviewers who think they have spotted a fatal flaw in experimental design (12).

Honorrhea. An obsession with seeking or receiving awards. Tends to become chronic. See *Nobelitis*. There is no known cure, for such individuals can never be satisfied.

Hyperacute rejection. A condition in which the rejection email arrives in your inbox before the confirmation of submission (13) (see also chapter 22).

Hyperpromotosis. The recurrent overestimation of the importance of one's own findings and the zeal exhibited in broadcasting one's accomplishments are pathognomonic signs.

Hypothesosis. Characterized by an inability to recognize that not all research requires a hypothesis (4, 14) (see also chapters 2 and 3). May be complicated by the development of *ichthyophobia* (aversion to "fishing expeditions"). See *mechanitis*.

Impact factor mania. Also known as *impactitis* (2). A condition in which the perceived value of scientific work is based on the impact factor of the journal where the work is published rather than the content of the work itself (1). A highly contagious and debilitating condition for which there is no known cure, although effects may be mitigated by the DORA initiative (15).

Inflammatory vowel disease. Characterized by the recurrent excretion of irate letters to the editor.

Irritable brain syndrome (IBS). Common symptoms are alternating periods of flowing ideas and constipated thinking. May be complicated by bouts of cerebral flatulence.

Mechanitis. A condition exhibited by scientists who misuse the words "descriptive" and "mechanistic" while failing to recognize that careful description is essential to science and mechanisms are relative to the vantage point of the observer. The illness can be mitigated by reading chapters 2 and 3 of this book several times a day until symptoms subside. Prognosis is generally good, although relapses may be frequent.

Myiasis. A condition characterized by the repeated and excessive use of the word "my," as in *my* lab, *my* discovery, and *my* paper. The malady often coexists with *priorititis* (see below). The etymological relationship to a disease involving parasitic maggots is purely coincidental. Victims of myiasis fail to recognize that any scientific discovery reflects the contributions of many individuals. Myiasis may have serious long-term debilitating effects because it irritates colleagues and can lead to social isolation. Therapy is most effective if administered by scientists of higher rank.

Nobelitis. A rare but debilitating condition afflicting only the scientific elite (16) (see also chapter 21). May be manifested by auditory hallucinations involving telephone callers with Swedish accents. Seasonal incidence is frequently observed with rising anticipation in early fall followed by prolonged depression once the prizes are awarded and the afflicted individual has not been selected.

Obstinatus ani (OA). A condition characterized by stubbornness out of proportion to the available evidence. See also *dogmatitis*. OA has notably affected individuals in the fields of AIDS causation, climatology, and vaccine research (17). The diagnosis of OA can be made by asking an individual to state the evidence required to alter their stance and observing the (lack of) response. There is no known cure.

Obfuscous incommunicado (OI). A condition characterized by the inability of an individual to express themselves clearly. Afflicted individuals speak or write only in incomprehensible, jargon-laden prose. The *ennui* subtype is contagious and produces a sleep disorder of audiences. Potentially treatable through courses and workshops on scientific communication.

Obsessionis curriculum vitae (**OCV**). An unhealthy preoccupation with the length of one's resume. Variants include obsession with citation count and *h*-index.

PNAS envy. The sensation experienced when congratulating a colleague on their election to the National Academies of Sciences, publisher of the *Proceedings of the National Academy of Sciences*. Once affected individuals are elected to the academy, the condition may progress to *Nobelitis*.

Polyauthoritis. An emerging disease involving manuscripts in which the number of authors exceeds the number of data points.

Priorititis. A condition characterized by a need for an individual to make the case for his/her priority in a scientific discovery (18) (see also chapter 19). Priorititis is frequently associated with narcissism and may coexist with *myiasis* and *amnesia originosa*. If untreated, priorititis can lead to bitterness and social isolation.

Pseudohypoegotism. A condition characterized by insincere displays of humility. Afflicted individuals are known to exhibit recurrent humble-bragging, as in "I'd like to acknowledge the little people who really did all the work," "I am so humbled to receive this prestigious award," or "I felt so awkward receiving the prize from the King of Sweden because surely there are many more deserving scientists out there." Pseudohypoegotism is a generally benign condition with few consequences for science. However, pseudohypoegotism can be an irritant to chronically exposed colleagues.

Publicititis. A condition characterized by insatiable cravings for publicity and media recognition. Individuals with publicititis may badger institutions and journals to issue press releases for their work. Some authorities consider publicititis to be a variant of *hyperpromotosis*.

Retention deficit disorder. The inability to recall anything from the lecture you just heard or the article you just read.

Rigor purportis. Having the appearance of rigor (chapter 6) but requiring far less effort. Referred to as *rigor abortis* when any pretense of rigor is abandoned.

Significosis. Manifested by a failure to discern between biological and statistical significance (9). Individuals with significosis fail to realize that just because something is unlikely to have occurred by chance doesn't mean it's important (19). See also chapter 8 and *borderline probability disorder*.

Slime disease. Individuals with this condition are observed to explain any biological phenomenon in terms of biofilms.

If you recognize any of these symptoms, please see a (real) doctor immediately. You may be a scientist!

References

Chapter 1 – What Is Science?

1. **Nemeth E.** 2007. Logical empiricism and the history and sociology of science, p 278–302. *In* **Richardson A, Uebel T** (ed), *The Cambridge Companion to Logical Empiricism.* Cambridge University Press, Cambridge, United Kingdom. http://dx.doi:10.1017/CCOL0521791782.012.
2. **Merriam-Webster.** 2022. Merriam-Webster online dictionary. https://www.merriam-webster.com/dictionary/science.
3. **Sagan C.** 1990. Why we need to understand science. *Skeptical Inquirer* **14**:263–269.
4. **Science Council.** 2022. Our definition of science. https://sciencecouncil.org/about-science/our-definition-of-science/.
5. **Huxley TH.** 1876. *American Addresses, with a Lecture on the Study of Biology.* Macmillan and Co, London, United Kingdom.
6. **Wolpert L.** 1994. *The Unnatural Nature of Science.* Harvard University Press, Cambridge, MA.
7. **Cromer AH.** 1993. *Uncommon Sense: the Heretical Nature of Science.* Oxford University Press, Oxford, United Kingdom.
8. **Falagas ME, Zarkadoulia EA, Samonis G.** 2006. Arab science in the golden age (750-1258 C.E.) and today. *FASEB J* **20**:1581–1586. http://dx.doi.org/10.1096/fj.06-0803ufm.
9. **Bacon F.** 1620. *Novum Organum.* John Bill, London, United Kingdom.
10. **Descartes R.** 1637. *Discours de la Méthode Pour bien conduire sa raison, et chercher la vérité dans les sciences.* Jean Maire, Leiden, The Netherlands.
11. **Popper KR.** 1959. *The Logic of Scientific Discovery.* Hutchinson, London, United Kingdom.
12. **Kuhn TS.** 1962. *The Structure of Scientific Revolutions.* University of Chicago Press, Chicago, IL.
13. **Feyerabend P.** 1975. *Against Method: Outline of an Anarchistic Theory of Knowledge.* New Left Books, New York, NY.

Thinking about Science: Good Science, Bad Science, and How to Make It Better, First Edition.
Ferric C. Fang and Arturo Casadevall.
© 2024 American Society for Microbiology.

14. **Ziman JM**. 1968. *Public Knowledge: the Social Dimension of Science.* Cambridge University Press, Cambridge, United Kingdom.

15. **Resnik DB**. 2000. A pragmatic approach to the demarcation problem. *Stud Hist Philos Sci* **31**: 249–267. http://dx.doi.org/10.1016/S0039-3681(00)00004-2.

16. **Bak-Coleman JB, Alfano M, Barfuss W, Bergstrom CT, Centeno MA, Couzin ID, Donges JF, Galesic M, Gersick AS, Jacquet J, Kao AB, Moran RE, Romanczuk P, Rubenstein DI, Tombak KJ, Van Bavel JJ, Weber EU**. 2021. Stewardship of global collective behavior. *Proc Natl Acad Sci U S A* **118**:e2025764118. http://dx.doi.org/10.1073/pnas.2025764118.

17. **O'Brien TC, Palmer R, Albarracin D**. 2021. Misplaced trust: when trust in science fosters belief in pseudoscience and the benefits of critical evaluation. *J Exp Soc Psychol* **96**:104184. http://dx.doi.org/10.1016/j.jesp.2021.104184.

18. **Chavalarias D, Ioannidis JP**. 2010. Science mapping analysis characterizes 235 biases in biomedical research. *J Clin Epidemiol* **63**:1205–1215. http://dx.doi.org/10.1016/j.jclinepi.2009.12.011.

19. **Hume D**. 1739. *A Treatise of Human Nature.* John Noon, London, United Kingdom.

20. **Gould SJ**. 1997. Nonoverlapping magisteria. *Nat Hist* **106**:16–22, 60.

21. **Iaccarino M**. 2001. Science and ethics. As research and technology are changing society and the way we live, scientists can no longer claim that science is neutral but must consider the ethical and social aspects of their work. *EMBO Rep* **2**:747–750. http://dx.doi.org/10.1093/embo-reports/kve191.

22. **Bergstrom C, West J**. 2021. *Calling Bullshit: the Art of Skepticism in a Data-Driven World.* Penguin Random House, New York, NY.

23. **Wilson EO**. 2006. *The Creation: an Appeal to Save Life on Earth.* W.W. Norton, New York, NY.

24. **Stevenson A, Lindberg CA (ed)**. 2010. *New Oxford American Dictionary,* 3rd ed. Oxford University Press, Oxford, United Kingdom. http://dx.doi.org/10.1093/acref/9780195392883.001.0001.

25. **Wigner EP**. 1960. The unreasonable effectiveness of mathematics in the natural sciences. *Commun Pure Appl Math* **13**:1–14. http://dx.doi.org/10.1002/cpa.3160130102.

26. **Montgomery SL**. 2019. Why did modern science emerge in Europe? An essay in intellectual history. *Know* **3**:70–92. http://dx.doi.org/10.1086/701903.

27. **Poskett J**. 2017. Five millennia of Indian science. *Nature* **550**:332. http://dx.doi.org/10.1038/550332a.

28. **Blatch S**. 2013. Great achievements in science and technology in ancient Africa. *ASBMB Today* **12**:32–33.

29. **Rovelli C**. 11 July 2014. Science is not about certainty. *The New Republic.* https://newrepublic.com/article/118655/theoretical-phyisicist-explains-why-science-not-about-certainty.

30. **Feynman R**. 1964. Seeking new laws, p 149–173. *In Messenger Lectures on the Character of Physical Law.* MIT Press, Cambridge, MA.

Chapter 2 – Descriptive Science

1. **Albee E.** 1907. Descriptive and normative science. *Philos Rev* **16**:40–49. http://dx.doi.org/10.2307/2177577.

2. **Merriam-Webster**. 2022. Merriam-Webster online dictionary. https://www.merriam-webster.com/dictionary/descriptive.

3. **Casadevall A.** 2010. ASM launches mBio. *mBio* **1**:e00120-e10. http://dx.doi.org/10.1128/mBio.00120-10.

4. **Grimaldi DA, Engel MS**. 2008. Why descriptive science still matters. *Bioscience* **57**:646–647. http://dx.doi.org/10.1641/B570802.

5. **Grimes DA, Schulz KF**. 2002. Descriptive studies: what they can and cannot do. *Lancet* **359**: 145–149. http://dx.doi.org/10.1016/S0140-6736(02)07373-7.

6. **Darwin C.** 1985. Letter to A. R. Wallace 22 December 1857. *In* **Burkhardt F, Smith S** (ed), *The Correspondence of Charles Darwin.* Cambridge University Press, Cambridge, United Kingdom. https://www.darwinproject.ac.uk/letter/DCP-LETT-2192.xml.

7. **Marincola FM**. 2007. In support of descriptive studies; relevance to translational research. *J Transl Med* **5**:21. http://dx.doi.org/10.1186/1479-5876-5-21.

8. **Doyle AC**. 1986. *A Study in Scarlet*. Bantam, New York, NY.

9. **Aebersold R, Hood LE, Watts JD**. 2000. Equipping scientists for the new biology. *Nat Biotechnol* **18**:359. http://dx.doi.org/10.1038/74325.

10. **Bassler B, Bell S, Cowman A, Goldman B, Holden D, Miller V, Pugsley T, Simons B**. 2004. Editorial policy on genome-scale analyses. *Mol Microbiol* **52**:311–312. http://dx.doi.org/10.1111/j.1365-2958.2004.04124.x.

11. **Hallam A**. 1975. Alfred Wegener and the hypothesis of continental drift. *Sci Am* **232**:88–97. http://dx.doi.org/10.1038/scientificamerican0275-88.

12. **Gramling C**. 2 July 2021. A WWII submarine-hunting device helped prove the theory of plate tectonics. *Sci News*. https://www.sciencenews.org/article/fluxgate-magnetometer-submarine-plate-tectonics.

13. **Kell DB, Oliver SG**. 2004. Here is the evidence, now what is the hypothesis? The complementary roles of inductive and hypothesis-driven science in the post-genomic era. *BioEssays* **26**:99–105. http://dx.doi.org/10.1002/bies.10385.

14. **Papacosta P**. January 2005. Nobel prize for a "computer" named Henrietta Leavitt (1868-1921). *Status: a Report on Women in Astronomy*. https://aas.org/sites/default/files/2019-09/status-Jan05sm.pdf.

15. **Leavitt HS, Pickering EC**. 1912. Periods of 25 variable stars in the small Magellanic cloud. *Harv Coll Obs Circ* **173**:1–3.

16. **Hubble EP**. 1929. A spiral nebula as a stellar system, Messier 31. *Astrophys J* **69**:103–157. http://dx.doi.org/10.1086/143167.

17. **Wegener A**. 1915. *Die Entstehung der Kontinente und Ozeane*. Vieweg und Sohn, Braunschweig, Germany.

Chapter 3 – Mechanistic Science

1. **Hobbes T**. 1651. *Leviathan*. Cambridge University Press, Cambridge, United Kingdom.

2. **van Mil MHW, Postma PA, Boerwinkel J, Klaassen K, Waarlo AJ**. 2016. Molecular mechanistic reasoning: toward bridging the gap between the molecular and cellular levels in life science education. *Sci Educ* **100**:517–585. http://dx.doi.org/10.1002/sce.21215.

3. **Fischman DA**. 2003. The descriptive curse. *Scientist* **17**:18.

4. **Rajan TV**. 2009. Would Harvey, Sulston, and Darwin get funded today? *Scientist* **13**:12.

5. **Casadevall A, Fang FC**. 2008. Descriptive science. *Infect Immun* **76**:3835–3836. http://dx.doi.org/10.1128/IAI.00743-08.

6. **Allen GE**. 2005. Mechanism, vitalism and organicism in late nineteenth and twentieth-century biology: the importance of historical context. *Stud Hist Philos Biol Biomed Sci* **36**:261–283. http://dx.doi.org/10.1016/j.shpsc.2005.03.003.

7. **Craver CF, Darden L**. 2005. Mechanisms in biology. Introduction. *Stud Hist Philos Biol Biomed Sci* **36**:233–244. http://dx.doi.org/10.1016/j.shpsc.2005.03.001.

8. **Darden L, Craver CF**. 2002. Strategies in the interfield discovery of protein synthesis. *Stud Hist Philos Biol Biomed Sci* **33**:1–28. http://dx.doi.org/10.1016/S1369-8486(01)00021-8.

9. **Medawar PB**. 1996. *The Strange Case of the Spotted Mice*. Oxford University Press, Oxford, United Kingdom.

10. **Bechtel W, Abrahamsen A**. 2005. Explanation: a mechanist alternative. *Stud Hist Philos Biol Biomed Sci* **36**:421–441. http://dx.doi.org/10.1016/j.shpsc.2005.03.010.

11. **Slonczewski JL, Kaneshiro ES**. 2009. Defining descriptive research. *Microbe* **4**:50.

12. **U.S. Supreme Court**. 1964. *Jacobellis v. Ohio*, 378 U.S. 184. http://cdn.loc.gov/service/ll/usrep/usrep378/usrep378184/usrep378184.pdf.

13. **Weinert F**. 2016. *The Demons of Science. What They Can and Cannot Tell Us about Our World*. Springer, Cham, Switzerland. http://dx.doi.org/10.1007/978-3-319-31708-3.

14. **Malhotra R, Holman M, Ito T**. 2001. Chaos and stability of the solar system. *Proc Natl Acad Sci U S A* **98**:12342–12343. http://dx.doi.org/10.1073/pnas.231384098.

15. **Suki B, Bates JH, Frey U**. 2011. Complexity and emergent phenomena. *Compr Physiol* **1**:995–1029. http://dx.doi.org/10.1002/cphy.c100022.

16. **Parrill AL, Reddy MR (ed)**. 1999. *Rational Drug Design: Novel Methodology and Practical Applications*. American Chemical Society, New York, NY. http://dx.doi.org/10.1021/bk-1999-0719.

17. **Kent R, Huber B**. 1999. Gertrude Belle Elion (1918-99). *Nature* **398**:380. http://dx.doi.org/10.1038/18790.

18. **Shader RI**. 2018. A tribute to Gertrude Belle Elion on the 100th anniversary of her birth. *Clin Ther* **40**:181–185. http://dx.doi.org/10.1016/j.clinthera.2018.01.008.

19. **Nobel Foundation**. 2006. Autobiography of Gertrude B. Elion, the Nobel Prize in Physiology or Medicine 1988. *Oncologist* **11**:966–968. http://dx.doi.org/10.1634/theoncologist.11-9-966.

20. **Keefer CS, Blake FG, Marshall EK, Jr, Lockwood JS, Wood WB**. 1943. Penicillin in the treatment of infections. *JAMA* **122**:1217–1224. http://dx.doi.org/10.1001/jama.1943.02840350001001.

Chapter 4 – Reductionistic and Holistic Science

1. **Dawkins R**. 1996. *The Blind Watchmaker*. W.W. Norton, New York, NY.

2. **Ideker T, Galitski T, Hood L**. 2001. A new approach to decoding life: systems biology. *Annu Rev Genomics Hum Genet* **2**:343–372. http://dx.doi.org/10.1146/annurev.genom.2.1.343.

3. **Kitano H**. 2002. Systems biology: a brief overview. *Science* **295**:1662–1664. http://dx.doi.org/10.1126/science.1069492.

4. **Adams D**. 1987. *Dirk Gently's Holistic Detective Agency*. Simon and Schuster, New York, NY.

5. **Raoult D**. 2010. Technology-driven research will dominate hypothesis-driven research: the future of microbiology. *Future Microbiol* **5**:135–137. http://dx.doi.org/10.2217/fmb.09.119.

6. **Mazzocchi F**. 2008. Complexity in biology. Exceeding the limits of reductionism and determinism using complexity theory. *EMBO Rep* **9**:10–14. http://dx.doi.org/10.1038/sj.embor.7401147.

7. **Brigandt I, Love A**. 2008. Reductionism in biology. *In* **Zalta EN** (ed), *The Stanford Encyclopedia of Philosophy*. http://plato.stanford.edu/archives/fall2008/entries/reduction-biology/.

8. **Crick F**. 1966. *Of Molecules and Man*. University of Washington Press, Seattle, WA.

9. **Sarukkai S**. 2005. Revisiting the "unreasonable effectiveness" of mathematics. *Curr Sci* **88**:415–423.

10. **Bacon F**. 1620. *Novum Organum*. John Bill, London, United Kingdom.

11. **Glass DJ, Hall N**. 2008. A brief history of the hypothesis. *Cell* **134**:378–381. http://dx.doi.org/10.1016/j.cell.2008.07.033.

12. **Descartes R**. 1637. *Discours de la Méthode Pour bien conduire sa raison, et chercher la vérité dans les sciences*. Jean Maire, Leiden, The Netherlands.

13. **DiRita VJ, Mekalanos JJ**. 1991. Periplasmic interaction between two membrane regulatory proteins, ToxR and ToxS, results in signal transduction and transcriptional activation. *Cell* **64**:29–37. http://dx.doi.org/10.1016/0092-8674(91)90206-E.

14. **Kanjilal S, Citorik R, LaRocque RC, Ramoni MF, Calderwood SB**. 2010. A systems biology approach to modeling *Vibrio cholerae* gene expression under virulence-inducing conditions. *J Bacteriol* **192**:4300–4310. http://dx.doi.org/10.1128/JB.00182-10.

15. **Lee SH, Hava DL, Waldor MK, Camilli A**. 1999. Regulation and temporal expression patterns of *Vibrio cholerae* virulence genes during infection. *Cell* **99**:625–634. http://dx.doi.org/10.1016/S0092-8674(00)81551-2.

16. **Fields PI, Swanson RV, Haidaris CG, Heffron F**. 1986. Mutants of *Salmonella typhimurium* that cannot survive within the macrophage are avirulent. *Proc Natl Acad Sci U S A* **83**:5189–5193. http://dx.doi.org/10.1073/pnas.83.14.5189.

17. **O'Brien AD, Rosenstreich DL, Scher I, Campbell GH, MacDermott RP, Formal SB**. 1980. Genetic control of susceptibility to *Salmonella typhimurium* in mice: role of the LPS gene. *J Immunol* **124**:20–24. http://dx.doi.org/10.4049/jimmunol.124.1.20.

18. **Vazquez-Torres A, Vallance BA, Bergman MA, Finlay BB, Cookson BT, Jones-Carson J, Fang FC**. 2004. Toll-like receptor 4 dependence of innate and adaptive immunity to *Salmonella*: importance of the Kupffer cell network. *J Immunol* **172**:6202–6208. http://dx.doi.org/10.4049/jimmunol.172.10.6202.

19. **Kushner DJ**. 1969. Self-assembly of biological structures. *Bacteriol Rev* **33**:302–345. http://dx.doi.org/10.1128/br.33.2.302-345.1969.

20. **Gibson DG, Glass JI, Lartigue C, Noskov VN, Chuang RY, Algire MA, Benders GA, Montague MG, Ma L, Moodie MM, Merryman C, Vashee S, Krishnakumar R, Assad-Garcia N, Andrews-Pfannkoch C, Denisova EA, Young L, Qi ZQ, Segall-Shapiro TH, Calvey CH, Parmar PP, Hutchison CA, III, Smith HO, Venter JC**. 2010. Creation of a bacterial cell controlled by a chemically synthesized genome. *Science* **329**:52–56. http://dx.doi.org/10.1126/science.1190719.

21. **Smuts JC**. 1926. *Holism and Evolution*. Macmillan, London, United Kingdom.

22. **Aggarwal K, Lee KH**. 2003. Functional genomics and proteomics as a foundation for systems biology. *Brief Funct Genomics Proteomics* **2**:175–184. http://dx.doi.org/10.1093/bfgp/2.3.175.

23. **Hood L, Heath JR, Phelps ME, Lin B**. 2004. Systems biology and new technologies enable predictive and preventative medicine. *Science* **306**:640–643. http://dx.doi.org/10.1126/science.1104635.

24. **Beresford MJ**. 2010. Medical reductionism: lessons from the great philosophers. *QJM* **103**:721–724. http://dx.doi.org/10.1093/qjmed/hcq057.

25. **Kauffman S**. 1993. *The Origins of Order: Self Organization and Selection in Evolution*. Oxford University Press, Oxford, United Kingdom.

26. **Jeffrey BM, Suchland RJ, Quinn KL, Davidson JR, Stamm WE, Rockey DD**. 2010. Genome sequencing of recent clinical *Chlamydia trachomatis* strains identifies loci associated with tissue tropism and regions of apparent recombination. *Infect Immun* **78**:2544–2553. http://dx.doi.org/10.1128/IAI.01324-09.

27. **McGill MA, Edmondson DG, Carroll JA, Cook RG, Orkiszewski RS, Norris SJ**. 2010. Characterization and serologic analysis of the *Treponema pallidum* proteome. *Infect Immun* **78**:2631–2643. http://dx.doi.org/10.1128/IAI.00173-10.

28. **Pearson MM, Rasko DA, Smith SN, Mobley HL**. 2010. Transcriptome of swarming *Proteus mirabilis*. *Infect Immun* **78**:2834–2845. http://dx.doi.org/10.1128/IAI.01222-09.

29. **Bruggeman FJ, Westerhoff HV**. 2007. The nature of systems biology. *Trends Microbiol* **15**:45–50. http://dx.doi.org/10.1016/j.tim.2006.11.003.

30. **Atkinson MR, Savageau MA, Myers JT, Ninfa AJ**. 2003. Development of genetic circuitry exhibiting toggle switch or oscillatory behavior in *Escherichia coli*. *Cell* **113**:597–607. http://dx.doi.org/10.1016/S0092-8674(03)00346-5.

31. **Covert MW, Knight EM, Reed JL, Herrgard MJ, Palsson BO**. 2004. Integrating high-throughput and computational data elucidates bacterial networks. *Nature* **429**:92–96. http://dx.doi.org/10.1038/nature02456.

32. **Ozbudak EM, Thattai M, Lim HN, Shraiman BI, Van Oudenaarden A**. 2004. Multistability in the lactose utilization network of *Escherichia coli*. *Nature* **427**:737–740. http://dx.doi.org/10.1038/nature02298.

33. **Park H, Pontius W, Guet CC, Marko JF, Emonet T, Cluzel P**. 2010. Interdependence of behavioural variability and response to small stimuli in bacteria. *Nature* **468**:819–823. http://dx.doi.org/10.1038/nature09551.

34. **Rosenfeld N, Young JW, Alon U, Swain PS, Elowitz MB**. 2005. Gene regulation at the single-cell level. *Science* **307**:1962–1965. http://dx.doi.org/10.1126/science.1106914.

35. **Früh K, Finlay B, McFadden G**. 2010. On the road to systems biology of host-pathogen interactions. *Future Microbiol* **5**:131–133. http://dx.doi.org/10.2217/fmb.09.130.

36. **Querec TD, Akondy RS, Lee EK, Cao W, Nakaya HI, Teuwen D, Pirani A, Gernert K, Deng J, Marzolf B, Kennedy K, Wu H, Bennouna S, Oluoch H, Miller J, Vencio RZ, Mulligan M, Aderem A, Ahmed R, Pulendran B**. 2009. Systems biology approach predicts immunogenicity of the yellow fever vaccine in humans. *Nat Immunol* **10**:116–125. http://dx.doi.org/10.1038/ni.1688.

37. **De Backer P, De Waele D, Van Speybroeck L**. 2010. Ins and outs of systems biology vis-à-vis molecular biology: continuation or clear cut? *Acta Biotheor* **58**:15–49. http://dx.doi.org/10.1007/s10441-009-9089-6.

38. **Casadevall A, Fang FC**. 2008. Descriptive science. *Infect Immun* **76**:3835–3836. http://dx.doi.org/10.1128/IAI.00743-08.

39. **Avery OT, Macleod CM, McCarty M**. 1944. Studies on the chemical nature of the substance inducing transformation of pneumococcal types: induction of transformation by a desoxyribonucleic acid fraction isolated from pneumococcus type III. *J Exp Med* **79**:137–158. http://dx.doi.org/10.1084/jem.79.2.137.

40. **Isberg RR, Falkow S**. 1985. A single genetic locus encoded by *Yersinia pseudotuberculosis* permits invasion of cultured animal cells by *Escherichia coli* K-12. *Nature* **317**:262–264. http://dx.doi.org/10.1038/317262a0.

41. **Lecuit M, Vandormael-Pournin S, Lefort J, Huerre M, Gounon P, Dupuy C, Babinet C, Cossart P**. 2001. A transgenic model for listeriosis: role of internalin in crossing the intestinal barrier. *Science* **292**:1722–1725. http://dx.doi.org/10.1126/science.1059852.

42. **Sorek R, Zhu Y, Creevey CJ, Francino MP, Bork P, Rubin EM**. 2007. Genome-wide experimental determination of barriers to horizontal gene transfer. *Science* **318**:1449–1452. http://dx.doi.org/10.1126/science.1147112.

43. **Sharma CM, Hoffmann S, Darfeuille F, Reignier J, Findeiss S, Sittka A, Chabas S, Reiche K, Hackermüller J, Reinhardt R, Stadler PF, Vogel J**. 2010. The primary transcriptome of the major human pathogen *Helicobacter pylori*. *Nature* **464**:250–255. http://dx.doi.org/10.1038/nature08756.

44. **Cash HL, Whitham CV, Behrendt CL, Hooper LV**. 2006. Symbiotic bacteria direct expression of an intestinal bactericidal lectin. *Science* **313**:1126–1130. http://dx.doi.org/10.1126/science.1127119.

45. **Lehotzky RE, Partch CL, Mukherjee S, Cash HL, Goldman WE, Gardner KH, Hooper LV**. 2010. Molecular basis for peptidoglycan recognition by a bactericidal lectin. *Proc Natl Acad Sci U S A* **107**:7722–7727. http://dx.doi.org/10.1073/pnas.0909449107.

46. **Karlas A, Machuy N, Shin Y, Pleissner KP, Artarini A, Heuer D, Becker D, Khalil H, Ogilvie LA, Hess S, Mäurer AP, Müller E, Wolff T, Rudel T, Meyer TF**. 2010. Genome-wide RNAi screen identifies human host factors crucial for influenza virus replication. *Nature* **463**:818–822. http://dx.doi.org/10.1038/nature08760.

47. **Mesarovic MD, Sreenath SN, Keene JD**. 2004. Search for organising principles: understanding in systems biology. *Syst Biol (Stevenage)* **1**:19–27. http://dx.doi.org/10.1049/sb:20045010.

48. **Dawkins R**. 2001. Eulogy for Douglas Adams, Church of Saint Martin in the Fields, London, UK. http://www.edge.org/documents/adams_index.html.

49. **Crick F**. 1970. Central dogma of molecular biology. *Nature* **227**:561–563. http://dx.doi.org/10.1038/227561a0.

50. **Thieffry D, Sarkar S**. 1998. Forty years under the central dogma. *Trends Biochem Sci* **23**:312–316. http://dx.doi.org/10.1016/S0968-0004(98)01244-4.

51. **Ideker T, Galitski T, Hood L**. 2001. A new approach to decoding life: systems biology. *Annu Rev Genomics Hum Genet* **2**:343–372. http://dx.doi.org/10.1146/annurev.genom.2.1.343.

Chapter 5 – Elegant Science

1. **Arp H**. 1991. Wrangling over the Bang. *Sci News* **140**:51. http://dx.doi.org/10.2307/3976114.

2. **House P**. 17 August 2015. What is elegance in science? *New Yorker*. https://www.newyorker.com/tech/elements/what-is-elegance-in-science.

3. **Ball P**. 11 August 2016. The tyranny of simple explanations. *Atlantic Monthly*. https://www.theatlantic.com/science/archive/2016/08/occams-razor/495332/.

4. **Johnson G**. 24 September 2002. Here they are, science's 10 most beautiful experiments. *New York Times*. http://www.nytimes.com/2002/09/24/science/here-they-are-science-s-10-most-beautiful-experiments.html.

5. **Online Etymology Dictionary**. 2022. https://www.etymonline.com/word/elegance.
6. **Merriam-Webster**. 2022. Merriam-Webster online dictionary. https://www.merriam-webster.com/dictionary/elegance.
7. **Glynn I**. 2010. *Elegance in Science*. Oxford University Press, Oxford, United Kingdom.
8. **Toumey C**. 2010. Elegance and empiricism. *Nat Nanotechnol* **5**:693–694. http://dx.doi.org/10.1038/nnano.2010.195.
9. **Nathan MJ, Brancaccio D**. 2013. The importance of being elegant: a discussion of elegance in nephrology and biomedical science. *Nephrol Dial Transplant* **28**:1385–1389. http://dx.doi.org/10.1093/ndt/gft005.
10. **Casadevall A, Fang FC**. 2016. Rigorous science: a how-to guide. *mBio* **7**:e01902-16. http://dx.doi.org/10.1128/mBio.01902-16.
11. **Watson JD, Crick FH**. 1953. Molecular structure of nucleic acids; a structure for deoxyribose nucleic acid. *Nature* **171**:737–738. http://dx.doi.org/10.1038/171737a0.
12. **Pauling L, Corey RB**. 1953. A proposed structure for the nucleic acids. *Proc Natl Acad Sci U S A* **39**:84–97. http://dx.doi.org/10.1073/pnas.39.2.84.
13. **Meselson M, Stahl FW**. 1958. The replication of DNA. *Cold Spring Harb Symp Quant Biol* **23**:9–12. http://dx.doi.org/10.1101/SQB.1958.023.01.004.
14. **Holmes FL**. 2001. *Meselson, Stahl and the Replication of DNA*. Yale University Press, New Haven, CT. http://dx.doi.org/10.12987/yale/9780300085402.001.0001.
15. **Hanawalt PC**. 2004. Density matters: the semiconservative replication of DNA. *Proc Natl Acad Sci U S A* **101**:17889–17894. http://dx.doi.org/10.1073/pnas.0407539101.
16. **Mullis KB**. 1990. The unusual origin of the polymerase chain reaction. *Sci Am* **262**:56–61, 64–65. http://dx.doi.org/10.1038/scientificamerican0490-56.
17. **Saiki RK, Gelfand DH, Stoffel S, Scharf SJ, Higuchi R, Horn GT, Mullis KB, Erlich HA**. 1988. Primer-directed enzymatic amplification of DNA with a thermostable DNA polymerase. *Science* **239**:487–491. http://dx.doi.org/10.1126/science.2448875.
18. **Casadevall A, Fang FC**. 2016. Revolutionary science. *mBio* **7**:e00158. http://dx.doi.org/10.1128/mBio.00158-16.
19. **Kleppe K, Ohtsuka E, Kleppe R, Molineux I, Khorana HG**. 1971. Studies on polynucleotides. XCVI. Repair replications of short synthetic DNA's as catalyzed by DNA polymerases. *J Mol Biol* **56**:341–361. http://dx.doi.org/10.1016/0022-2836(71)90469-4.
20. **Darwin C**. 1859. *On the Origin of Species by Means of Natural Selection*. John Murray, London, United Kingdom.
21. **Crick F**. 1988. *What Mad Pursuit: a Personal View of Scientific Discovery*. Basic Books, New York, NY.
22. **Tierney J**. 3 June 2008. The future is now? Pretty soon, at least. *New York Times*. https://www.nytimes.com/2008/06/03/science/03tier.html.
23. **Rabin M, Schrag JL**. 1999. First impressions matter: a model of confirmatory bias. *Q J Econ* **114**:37–82. http://dx.doi.org/10.1162/003355399555945.
24. **Einstein A**. 1934. On the method of theoretical physics. *Philos Sci* **1**:163–169. http://dx.doi.org/10.1086/286316.
25. **Pelletier C, Gates SJJ**. 2019. *Proving Einstein Right: the Daring Expeditions That Changed How We Look at the Universe*. PublicAffairs, New York, NY.
26. **Rigoni-Stern DA**. 1842. Fatti statistici relativi alle malattie cancerose. *G Serv Prog Pathol Terap Ser* **2**:507–517.
27. **Gasparini R, Panatto D**. 2009. Cervical cancer: from Hippocrates through Rigoni-Stern to zur Hausen. *Vaccine* **27**(Suppl 1):A4–A5. http://dx.doi.org/10.1016/j.vaccine.2008.11.069.
28. **Ramachandran VS**. 2011. *The Tell-Tale Brain: a Neuroscientist's Quest for What Makes Us Human*. W.W. Norton, New York, NY.
29. **Casadevall A, Freij JB, Hann-Soden C, Taylor J**. 2017. Continental drift and speciation of the *Cryptococcus neoformans* and *Cryptococcus gattii* species complexes. *mSphere* **2**(2):e00103-17. http://dx.doi.org/10.1128/mSphere.00103-17.

30. **zur Hausen H.** 1996. Papillomavirus infections—a major cause of human cancers. *Biochim Biophys Acta* **1288**:F55–F78.

31. **Thomas L.** 1971. The technology of medicine. *N Engl J Med* **285**:1366–1368. http://dx.doi.org/10.1056/NEJM197112092852411.

32. **Lei J, Ploner A, Elfström KM, Wang J, Roth A, Fang F, Sundström K, Dillner J, Sparén P.** 2020. HPV vaccination and the risk of invasive cervical cancer. *N Engl J Med* **383**:1340–1348. http://dx.doi.org/10.1056/NEJMoa1917338.

Chapter 6 – Rigorous Science

1. **Kline M.** 1980. *Mathematics: the Loss of Certainty*. Oxford University Press, Oxford, United Kingdom.

2. **Prinz F, Schlange T, Asadullah K.** 2011. Believe it or not: how much can we rely on published data on potential drug targets? *Nat Rev Drug Discov* **10**:712. http://dx.doi.org/10.1038/nrd3439-c1.

3. **Begley CG, Ellis LM.** 2012. Drug development: raise standards for preclinical cancer research. *Nature* **483**:531–533. http://dx.doi.org/10.1038/483531a.

4. **Fang FC, Steen RG, Casadevall A.** 2012. Misconduct accounts for the majority of retracted scientific publications. *Proc Natl Acad Sci U S A* **109**:17028–17033. http://dx.doi.org/10.1073/pnas.1212247109.

5. **Bik EM, Casadevall A, Fang FC.** 2016. The prevalence of inappropriate image duplication in biomedical research publications. *mBio* **7**:e00809-16. http://dx.doi.org/10.1128/mBio.00809-16.

6. **Bowen A, Casadevall A.** 2015. Increasing disparities between resource inputs and outcomes, as measured by certain health deliverables, in biomedical research. *Proc Natl Acad Sci U S A* **112**:11335–11340. http://dx.doi.org/10.1073/pnas.1504955112.

7. **Scannell JW, Bosley J.** 2016. When quality beats quantity: decision theory, drug discovery, and the reproducibility crisis. *PLoS One* **11**:e0147215. http://dx.doi.org/10.1371/journal.pone.0147215.

8. **ASCB Data Reproducibility Task Force.** 2014. How can scientists enhance rigor in conducting basic research and reporting research results? A white paper from the American Society for Cell Biology. American Society for Cell Biology, Bethesda, MD. http://www.ascb.org/wp-content/uploads/2015/11/How-can-scientist-enhance-rigor.pdf.

9. **National Institutes of Health.** 2015. Implementing rigor and transparency in NIH & AHRQ research grant applications. National Institutes of Health, Bethesda, MD. http://grants.nih.gov/grants/guide/notice-files/NOT-OD-16-011.html.

10. **Casadevall A, Ellis LM, Davies EW, McFall-Ngai M, Fang FC.** 2016. A framework for improving the quality of research in the biological sciences. *mBio* **7**:e01256-16. http://dx.doi.org/10.1128/mBio.01256-16.

11. **Online Etymology Dictionary.** 2022. https://www.etymonline.com/search?q=rigor.

12. **Merriam-Webster.** 2022. Merriam-Webster online dictionary. https://www.merriam-webster.com/dictionary/rigor.

13. **National Institutes of Health.** 2016. Frequently asked questions. Rigor and reproducibility. http://grants.nih.gov/reproducibility/faqs.htm.

14. **Lamb E.** 17 July 2012. 5 sigma what's that? *Scientific American Blogs*. http://blogs.scientificamerican.com/observations/five-sigmawhats-that.

15. **Johnson VE.** 2013. Revised standards for statistical evidence. *Proc Natl Acad Sci U S A* **110**:19313–19317. http://dx.doi.org/10.1073/pnas.1313476110.

16. **Casadevall A, Steen RG, Fang FC.** 2014. Sources of error in the retracted scientific literature. *FASEB J* **28**:3847–3855. http://dx.doi.org/10.1096/fj.14-256735.

17. **Popper KR, Hudson GE.** 1963. Conjectures and refutations. *Phys Today* **16**:33–39. http://dx.doi.org/10.1063/1.3050617.

18. **Wikipedia.** 2022. https://en.wikipedia.org/wiki/Intellectual_honesty.

19. **Casadevall A, Fang FC.** 2012. Reforming science: methodological and cultural reforms. *Infect Immun* **80**:891–896. http://dx.doi.org/10.1128/IAI.06183-11.

20. **Baker M**. 2016. 1,500 scientists lift the lid on reproducibility. *Nature* **533**:452–454. http://dx.doi.org/10.1038/533452a.

21. **Weissgerber TL, Garovic VD, Milin-Lazovic JS, Winham SJ, Obradovic Z, Trzeciakowski JP, Milic NM**. 2016. Reinventing biostatistics education for basic scientists. *PLoS Biol* **14**:e1002430. http://dx.doi.org/10.1371/journal.pbio.1002430.

22. **Leek JT, Peng RD**. 2015. Statistics: *P* values are just the tip of the iceberg. *Nature* **520**:612. http://dx.doi.org/10.1038/520612a.

23. **Baker M**. 2016. Statisticians issue warning over misuse of *P* values. *Nature* **531**:151. http://dx.doi.org/10.1038/nature.2016.19503.

24. **Wasserstein RL, Lazar NA**. 2016. The ASA's statement on *P*-values: context, process, and purpose. *Am Stat* **70**:129–133. http://dx.doi.org/10.1080/00031305.2016.1154108.

25. **Casadevall A, Fang FC**. 2015. Impacted science: impact is not importance. *mBio* **6**:e01593-15. http://dx.doi.org/10.1128/mBio.01593-15.

26. **Hofseth LJ**. 2018. Getting rigorous with scientific rigor. *Carcinogenesis* **39**:21–25. http://dx.doi.org/10.1093/carcin/bgx085.

27. **National Institutes of Health**. 2022. Grants & funding. https://grants.nih.gov/policy/reproducibility/index.htm.

28. **Webster CR, Mahaffy PR, Atreya SK, Flesch GJ, Mischna MA, Meslin PY, Farley KA, Conrad PG, Christensen LE, Pavlov AA, Martín-Torres J, Zorzano MP, McConnochie TH, Owen T, Eigenbrode JL, Glavin DP, Steele A, Malespin CA, Archer PD**, Jr, **Sutter B, Coll P, Freissinet C, McKay CP, Moores JE, Schwenzer SP, Bridges JC, Navarro-Gonzalez R, Gellert R, Lemmon MT, MSL Science Team**. 2015. Mars atmosphere. Mars methane detection and variability at Gale crater. *Science* **347**:415–417. http://dx.doi.org/10.1126/science.1261713.

29. **Korablev O, Vandaele AC, Montmessin F, Fedorova AA, Trokhimovskiy A, Forget F, Lefèvre F, Daerden F, Thomas IR, Trompet L, Erwin JT, Aoki S, Robert S, Neary L, Viscardy S, Grigoriev AV, Ignatiev NI, Shakun A, Patrakeev A, Belyaev DA, Bertaux JL, Olsen KS, Baggio L, Alday J, Ivanov YS, Ristic B, Mason J, Willame Y, Depiesse C, Hetey L, Berkenbosch S, Clairquin R, Queirolo C, Beeckman B, Neefs E, Patel MR, Bellucci G, López-Moreno JJ, Wilson CF, Etiope G, Zelenyi L, Svedhem H, Vago JL, ACS and NOMAD Science Teams**. 2019. No detection of methane on Mars from early ExoMars Trace Gas Orbiter observations. *Nature* **568**:517–520. http://dx.doi.org/10.1038/s41586-019-1096-4.

30. **Mumma MJ, Villanueva GL, Novak RE, Hewagama T, Bonev BP, Disanti MA, Mandell AM, Smith MD**. 2009. Strong release of methane on Mars in northern summer 2003. *Science* **323**:1041–1045. http://dx.doi.org/10.1126/science.1165243.

21. **Webster CR, Mahaffy PR, Atreya SK, Moores JE, Flesch GJ, Malespin C, McKay CP, Martinez G, Smith CL, Martin-Torres J, Gomez-Elvira J, Zorzano MP, Wong MH, Trainer MG, Steele A, Archer D**, Jr, **Sutter B, Coll PJ, Freissinet C, Meslin PY, Gough RV, House CH, Pavlov A, Eigenbrode JL, Glavin DP, Pearson JC, Keymeulen D, Christensen LE, Schwenzer SP, Navarro-Gonzalez R, Pla-García J, Rafkin SCR, Vicente-Retortillo Á, Kahanpää H, Viudez-Moreiras D, Smith MD, Harri AM, Genzer M, Hassler DM, Lemmon M, Crisp J, Sander SP, Zurek RW, Vasavada AR**. 2018. Background levels of methane in Mars' atmosphere show strong seasonal variations. *Science* **360**:1093–1096. http://dx.doi.org/10.1126/science.aaq0131.

32. **Leek JT, Peng RD**. 2015. Opinion: Reproducible research can still be wrong: adopting a prevention approach. *Proc Natl Acad Sci U S A* **112**:1645–1646. http://dx.doi.org/10.1073/pnas.1421412111.

Chapter 7 – Reproducible Science

1. **Popper KR**. 1959. *The Logic of Scientific Discovery*. Hutchinson, London, United Kingdom.

2. **Daeschler EB, Shubin NH, Jenkins FA, Jr**. 2006. A Devonian tetrapod-like fish and the evolution of the tetrapod body plan. *Nature* **440**:757–763. http://dx.doi.org/10.1038/nature04639.

3. **Shubin NH, Daeschler EB, Jenkins FA, Jr**. 2014. Pelvic girdle and fin of *Tiktaalik roseae*. *Proc Natl Acad Sci U S A* **111**:893–899. http://dx.doi.org/10.1073/pnas.1322559111.

4. **Gould SJ**. 1989. *Wonderful Life: the Burgess Shale and the Nature of History*. W.W. Norton, New York, NY.

5. **Blount ZD, Borland CZ, Lenski RE**. 2008. Historical contingency and the evolution of a key innovation in an experimental population of *Escherichia coli. Proc Natl Acad Sci U S A* **105**:7899–7906. http://dx.doi.org/10.1073/pnas.0803151105.

6. **Voss D**. 1 Mar 1999. Whatever happened to cold fusion? *Physics World* https://physicsworld.com/a/whatever-happened-to-cold-fusion/.

7. **American Physical Society**. 2000. This month in physics history: May 10, 1752: first experiment to draw electricity from lightning. *APS News* **9**(5). https://www.aps.org/publications/apsnews/200005/history.cfm.

8. **Drummond C**. 2009. Replicability is not reproducibility: nor is it good science. *Workshop 26th ICML*. PQ, Canada, Montreal. http://www.csi.uottawa.ca/~cdrummon/pubs/ICMLws09.pdf.

9. **Federal Register**. 1989. 1989 final rules: Animal welfare. 9 CFR parts 1 and 2. *Fed Regist* **54**:36112–36163.

10. **Nuffield Council on Bioethics**. 2005. The ethics of research involving animals: consensus statement by all members of the working party. Nuffield Council on Bioethics, London, United Kingdom.

11. **Prinz F, Schlange T, Asadullah K**. 2011. Believe it or not: how much can we rely on published data on potential drug targets? *Nat Rev Drug Discov* **10**:712. http://dx.doi.org/10.1038/nrd3439-c1.

12. **Begley CG, Ellis LM**. 2012. Drug development: raise standards for preclinical cancer research. *Nature* **483**:531–533. http://dx.doi.org/10.1038/483531a.

13. **Baker M**. 2016. 1,500 scientists lift the lid on reproducibility. *Nature* **533**:452–454. http://dx.doi.org/10.1038/533452a.

14. **Errington TM, Iorns E, Gunn W, Tan FE, Lomax J, Nosek BA**. 2014. An open investigation of the reproducibility of cancer biology research. *eLife* **3**:e04333. http://dx.doi.org/10.7554/eLife.04333.

15. **Errington TM, Mathur M, Soderberg CK, Denis A, Perfito N, Iorns E, Nosek BA**. 2021. Investigating the replicability of preclinical cancer biology. *eLife* **10**:e71601. http://dx.doi.org/10.7554/eLife.71601.

16. **Fanelli D**. 2018. Opinion: is science really facing a reproducibility crisis, and do we need it to? *Proc Natl Acad Sci U S A* **115**:2628–2631. http://dx.doi.org/10.1073/pnas.1708272114.

17. **Open Science Collaboration**. 2015. PSYCHOLOGY. Estimating the reproducibility of psychological science. *Science* **349**:aac4716. http://dx.doi.org/10.1126/science.aac4716.

18. **Hutson M**. 2018. Artificial intelligence faces reproducibility crisis. *Science* **359**:725–726. http://dx.doi.org/10.1126/science.359.6377.725.

19. **Tiwari K, Kananathan S, Roberts MG, Meyer JP, Sharif Shohan MU, Xavier A, Maire M, Zyoud A, Men J, Ng S, Nguyen TVN, Glont M, Hermjakob H, Malik-Sheriff RS**. 2021. Reproducibility in systems biology modelling. *Mol Syst Biol* **17**:e9982. http://dx.doi.org/10.15252/msb.20209982.

20. **Han R, Walton KS, Sholl DS**. 2019. Does chemical engineering research have a reproducibility problem? *Annu Rev Chem Biomol Eng* **10**:43–57. http://dx.doi.org/10.1146/annurev-chembioeng-060718-030323.

21. **Casadevall A, Steen RG, Fang FC**. 2014. Sources of error in the retracted scientific literature. *FASEB J* **28**:3847–3855. http://dx.doi.org/10.1096/fj.14-256735.

22. **França TF, Monserrat JM**. 2018. Reproducibility crisis in science or unrealistic expectations? *EMBO Rep* **19**:e46008. http://dx.doi.org/10.15252/embr.201846008.

23. **Friedl P**. 2019. Rethinking research into metastasis. *eLife* **8**:e53511. http://dx.doi.org/10.7554/eLife.53511.

24. **Ioannidis JP**. 2005. Why most published research findings are false. *PLoS Med* **2**:e124. http://dx.doi.org/10.1371/journal.pmed.0020124.

25. **Ioannidis JP, Ntzani EE, Trikalinos TA, Contopoulos-Ioannidis DG**. 2001. Replication validity of genetic association studies. *Nat Genet* **29**:306–309. http://dx.doi.org/10.1038/ng749.

26. **Lowenstein PR, Castro MG**. 2009. Uncertainty in the translation of preclinical experiments to clinical trials. Why do most phase III clinical trials fail? *Curr Gene Ther* **9**:368–374. http://dx.doi.org/10.2174/156652309789753392.

27. **Ioannidis JP, Allison DB, Ball CA, Coulibaly I, Cui X, Culhane AC, Falchi M, Furlanello C, Game L, Jurman G, Mangion J, Mehta T, Nitzberg M, Page GP, Petretto E, van Noort V**. 2009. Repeatability of published microarray gene expression analyses. *Nat Genet* **41**:149–155. http://dx.doi.org/10.1038/ng.295.

28. **Vieland VJ**. 2001. The replication requirement. *Nat Genet* **29**:244–245. http://dx.doi.org/10.1038/ng1101-244.

29. **Shahrezaei V, Swain PS**. 2008. The stochastic nature of biochemical networks. *Curr Opin Biotechnol* **19**:369–374. http://dx.doi.org/10.1016/j.copbio.2008.06.011.

30. **Barnes PD, Bergman MA, Mecsas J, Isberg RR**. 2006. *Yersinia pseudotuberculosis* disseminates directly from a replicating bacterial pool in the intestine. *J Exp Med* **203**:1591–1601. http://dx.doi.org/10.1084/jem.20060905.

31. **Kerr Bernal S**. 2006. A massive snowball of fraud and deceit. *J Androl* **27**:313–315. http://dx.doi.org/10.2164/jandrol.06007.

32. **Buck MJ, Lieb JD**. 2004. ChIP-chip: considerations for the design, analysis, and application of genome-wide chromatin immunoprecipitation experiments. *Genomics* **83**:349–360. http://dx.doi.org/10.1016/j.ygeno.2003.11.004.

33. **Glass DJ, Hall N**. 2008. A brief history of the hypothesis. *Cell* **134**:378–381. http://dx.doi.org/10.1016/j.cell.2008.07.033.

34. **Taleb N**. 2007. *The Black Swan: the Impact of the Highly Improbable*. Random House, New York, NY.

35. **Flier JS**. 2022. The problem of irreproducible bioscience research. *Perspect Biol Med* **65**:373–395. http://dx.doi.org/10.1353/pbm.2022.0032.

36. **Williams JL, Chu HC, Lown MK, Daniel J, Meckl RD, Patel D, Ibrahim R**. 2022. Weaknesses in experimental design and reporting decrease the likelihood of reproducibility and generalization of recent cardiovascular research. *Cureus* **14**:e21086. http://dx.doi.org/10.7759/cureus.21086.

37. **Souren NY, Fusenig NE, Heck S, Dirks WG, Capes-Davis A, Bianchini F, Plass C**. 2022. Cell line authentication: a necessity for reproducible biomedical research. *EMBO J* **41**:e111307. http://dx.doi.org/10.15252/embj.2022111307.

38. **Baker M**. 2015. Reproducibility crisis: blame it on the antibodies. *Nature* **521**:274–276. http://dx.doi.org/10.1038/521274a.

39. **Lloyd K, Franklin C, Lutz C, Magnuson T**. 2015. Reproducibility: use mouse biobanks or lose them. *Nature* **522**:151–153. http://dx.doi.org/10.1038/522151a.

40. **Halsey LG, Curran-Everett D, Vowler SL, Drummond GB**. 2015. The fickle *P* value generates irreproducible results. *Nat Methods* **12**:179–185. http://dx.doi.org/10.1038/nmeth.3288.

41. **An G**. 2018. The crisis of reproducibility, the denominator problem and the scientific role of multi-scale modeling. *Bull Math Biol* **80**:3071–3080. http://dx.doi.org/10.1007/s11538-018-0497-0.

42. **National Institutes of Health**. 2022. Grants & funding. https://grants.nih.gov/policy/reproducibility/index.htm.

Chapter 8 – Important Science

1. **Rescher N**. 2001. Importance in scientific discovery. http://philsci-archive.pitt.edu/archive/00000486/.

2. **Editors of the American Heritage Dictionary**. 2013. *American Heritage Dictionary of the English Language*. Houghton Mifflin Harcourt Publishers, Boston, MA.

3. **Olson CM, Rennie D, Cook D, Dickersin K, Flanagin A, Hogan JW, Zhu Q, Reiling J, Pace B**. 2002. Publication bias in editorial decision making. *JAMA* **287**:2825–2828. http://dx.doi.org/10.1001/jama.287.21.2825.

4. **Mendel JG**. 1866. Versuche über pflanzen-hybriden. *Verh Naturforsch Ver Brunn* **4**:3–47.

5. **Houben RM, Dodd PJ.** 2016. The global burden of latent tuberculosis infection: a re-estimation using mathematical modelling. *PLoS Med* **13**:e1002152. http://dx.doi.org/10.1371/journal. pmed.1002152.

6. **Rossman RE.** 1965. The history and significance of serendipity in medical discovery. *Trans Stud Coll Physicians Phila* **33**:104–120.

7. **Chu C, Schneerson R, Robbins JB, Rastogi SC.** 1983. Further studies on the immunogenicity of *Haemophilus influenzae* type b and pneumococcal type 6A polysaccharide-protein conjugates. *Infect Immun* **40**:245–256. http://dx.doi.org/10.1128/iai.40.1.245-256.1983.

8. **Eskola J, Peltola H, Takala AK, Käyhty H, Hakulinen M, Karanko V, Kela E, Rekola P, Rönnberg P-R, Samuelson JS, Gordon LK, Mäkelä PH.** 1987. Efficacy of *Haemophilus influenzae* type b polysaccharide-diphtheria toxoid conjugate vaccine in infancy. *N Engl J Med* **317**:717–722. http://dx.doi.org/10.1056/NEJM198709173171201.

9. **Robbins JB, Schneerson R, Anderson P, Smith DH.** 1996. The 1996 Albert Lasker Medical Research Awards. Prevention of systemic infections, especially meningitis, caused by *Haemophilus influenzae* type b. Impact on public health and implications for other polysaccharide-based vaccines. *JAMA* **276**:1181–1185. http://dx.doi.org/10.1001/jama.276.14.1181.

10. **Eskola J, Kilpi T, Palmu A, Jokinen J, Haapakoski J, Herva E, Takala A, Käyhty H, Karma P, Kohberger R, Siber G, Mäkelä PH, Finnish Otitis Media Study Group.** 2001. Efficacy of a pneumococcal conjugate vaccine against acute otitis media. *N Engl J Med* **344**:403–409. http://dx.doi.org/10.1056/NEJM200102083440602.

11. **Hook EB (ed).** 2002. *Prematurity in Scientific Discovery: On Resistance and Neglect.* University of California Press, Berkeley, CA. http://dx.doi.org/10.1525/9780520927735

12. **Glen W.** 2002. A triptych to Serendip: prematurity and resistance to discovery in the earth sciences, p 92–108. *In* **Hook EB** (ed), *Prematurity in Scientific Discovery: On Resistance and Neglect.* University of California Press, Berkeley, CA.

13. **Avery OT, Macleod CM, McCarty M.** 1944. Studies on the chemical nature of the substance inducing transformation of pneumococcal types: induction of transformation by a desoxyribonucleic acid fraction isolated from pneumococcus type III. *J Exp Med* **79**:137–158. http://dx.doi.org/10.1084/jem.79.2.137.

14. **Stent GS.** 2002. Prematurity in scientific discovery, p 22–23. *In* **Hook EB** (ed), *Prematurity in Scientific Discovery: On Resistance and Neglect.* University of California Press, Berkeley, CA.

15. **Chargaff E.** 1994. Chemical specificity of nucleic acids and mechanism of their enzymatic degradation. 1950. *Experientia* **50**:368–376.

16. **Hershey AD, Chase M.** 1952. Independent functions of viral protein and nucleic acid in growth of bacteriophage. *J Gen Physiol* **36**:39–56. http://dx.doi.org/10.1085/jgp.36.1.39.

17. **Culotta E, Koshland DE, Jr.** 1992. NO news is good news. *Science* **258**:1862–1865. http://dx.doi.org/10.1126/science.1361684.

18. **Stephan PE.** 1996. The economics of science. *J Econ Lit* **34**:1199–1235.

19. **Houbraken J, Frisvad JC, Samson RA.** 2011. Fleming's penicillin producing strain is not *Penicillium chrysogenum* but *P. rubens. IMA Fungus* **2**:87–95. http://dx.doi.org/10.5598/imafungus.2011.02.01.12.

20. **Nobel A.** 2022. Alfred Nobel's will. https://www.nobelprize.org/alfred-nobel/alfred-nobels-will/.

21. **Belter CW.** 2015. Bibliometric indicators: opportunities and limits. *J Med Libr Assoc* **103**:219–221. http://dx.doi.org/10.3163/1536-5050.103.4.014.

22. **Potenski C.** 8 March 2015. Celebrating 150 years of Mendelian genetics. https://blogs.biomedcentral.com/bmcseriesblog/2015/03/08/celebrating-150-years-mendelian-genetics/.

23. **Van Noorden R, Maher B, Nuzzo R.** 2014. The top 100 papers. *Nature* **514**:550–553. http://dx.doi.org/10.1038/514550a.

24. **Hutchins BI, Yuan X, Anderson JM, Santangelo GM.** 2016. Relative citation ratio (RCR): a new metric that uses citation rates to measure influence at the article level. *PLoS Biol* **14**:e1002541. http://dx.doi.org/10.1371/journal.pbio.1002541.

25. **Schlosshauer M, Kofler J, Zeilinger A.** 2013. A snapshot of foundational attitudes toward quantum mechanics. *Stud Hist Philos Sci Part Stud Hist Philos Mod Phys* **44**:222–230. http://dx.doi.org/10.1016/j.shpsb.2013.04.004.

26. **Bornmann L, Haunschild R.** 2018. Do altmetrics correlate with the quality of papers? A large-scale empirical study based on F1000Prime data. *PLoS One* **13**:e0197133. http://dx.doi.org/10.1371/journal.pone.0197133.

27. **Davis P.** 27 October 2016. The fallacy of "sound" science. https://scholarlykitchen.sspnet.org/2016/10/27/the-fallacy-of-sound-science/.

28. **Kierkegaard S.** 2015. *Kierkegaard's Journals and Notebooks.* Princeton University Press, Princeton, NJ. http://dx.doi.org/10.1515/9781400866342

29. **Harold FM.** 1986. *The Vital Force: a Study of Bioenergetics.* W.H. Freeman, New York, NY.

30. **ATLAS Collaboration.** 2012. A particle consistent with the Higgs boson observed with the ATLAS detector at the Large Hadron Collider. *Science* **338**:1576–1582. http://dx.doi.org/10.1126/science.1232005.

31. **CMS Collaboration.** 2012. A new boson with a mass of 125 GeV observed with the CMS experiment at the Large Hadron Collider. *Science* **338**:1569–1575. http://dx.doi.org/10.1126/science.1230816.

32. **Abbott BP, Abbott R, Abbott TD, et al; LIGO Scientific Collaboration and Virgo Collaboration.** 2016. Observation of gravitational waves from a binary black hole merger. *Phys Rev Lett* **116**:061102. http://dx.doi.org/10.1103/PhysRevLett.116.061102.

33. **Lampidis T, Barksdale L.** 1971. Park-Williams number 8 strain of *Corynebacterium diphtheriae.* *J Bacteriol* **105**:77–85. http://dx.doi.org/10.1128/jb.105.1.77-85.1971.

34. **Anonymous.** 24 March 1934. 94 retired by city; 208 more will go. *New York Times*, New York, NY.

35. **American Association of Immunology.** 2012. Anna Wessels Williams, M.D. Infectious disease pioneer and public health advocate. *Immunology* **2012**:50–52.

Chapter 9 – Historical Science

1. **Medawar PB.** 1996. *The Strange Case of the Spotted Mice.* Oxford University Press, Oxford, United Kingdom.

2. **Momigliano A.** 1958. The place of Herodotus in the history of historiography. *History (Lond)* **43**:1–13. http://dx.doi.org/10.1111/j.1468-229X.1958.tb02501.x.

3. **Merriam-Webster.** 2022. Merriam-Webster online dictionary. https://www.merriam-webster.com/dictionary/history.

4. **Hirshfeld AW.** 2001. *Parallax: the Race to Measure the Cosmos.* W.H. Freeman, New York, NY.

5. **Chargaff E.** 1968. A quick climb up Mount Olympus. *Science* **159**:1448–1449. http://dx.doi.org/10.1126/science.159.3822.1448.

6. **Medawar PB.** 1963. Is the scientific paper a fraud? *Listener* **70**:377–378.

7. **Carmody J.** 2001. Celebrating science. *Nature* **412**:383. http://dx.doi.org/10.1038/35086659.

8. **Metchnikoff E.** 1893. *Lectures on the Comparative Pathology of Inflammation.* Kegan, Paul, Trench, Trübner & Co., Ltd, London, United Kingdom.

9. **Metchnikoff E.** 1921. *Life of Elie Metchnikoff, 1845-1916.* Constable, London, United Kingdom. http://dx.doi.org/10.5962/bhl.title.28845.

10. **Howitt SM, Wilson AN.** 2014. Revisiting "Is the scientific paper a fraud?": the way textbooks and scientific research articles are being used to teach undergraduate students could convey a misleading image of scientific research. *EMBO Rep* **15**:481–484. http://dx.doi.org/10.1002/embr.201338302.

11. **Committee on Science, Engineering and Public Policy.** 1995. *On Being a Scientist: Responsible Conduct of Research.* National Academy Press, Washington, DC.

12. **Sawin VI, Robinson KA.** 2016. Biased and inadequate citation of prior research in reports of cardiovascular trials is a continuing source of waste in research. *J Clin Epidemiol* **69**:174–178. http://dx.doi.org/10.1016/j.jclinepi.2015.03.026.

13. **Cornell Law School**. 2022. Stare decisis. https://www.law.cornell.edu/wex/stare_decisis.

14. **Casadevall A, Dragotakes Q, Johnson PW, Senefeld JW, Klassen SA, Wright RS, Joyner MJ, Paneth N, Carter RE**. 2021. Convalescent plasma use in the USA was inversely correlated with COVID-19 mortality. *eLife* **10**:e69866. http://dx.doi.org/10.7554/eLife.69866.

15. **Libster R, Pérez Marc G, Wappner D, Coviello S, Bianchi A, Braem V, Esteban I, Caballero MT, Wood C, Berrueta M, Rondan A, Lescano G, Cruz P, Ritou Y, Fernández Viña V, Álvarez Paggi D, Esperante S, Ferreti A, Ofman G, Ciganda Á, Rodriguez R, Lantos J, Valentini R, Itcovici N, Hintze A, Oyarvide ML, Etchegaray C, Neira A, Name I, Alfonso J, López Castelo R, Caruso G, Rapelius S, Alvez F, Etchenique F, Dimase F, Alvarez D, Aranda SS, Sánchez Yanotti C, De Luca J, Jares Baglivo S, Laudanno S, Nowogrodzki F, Larrea R, Silveyra M, Leberzstein G, Debonis A, Molinos J, González M, Perez E, Kreplak N, Pastor Argüello S, Gibbons L, Althabe F, Bergel E, Polack FP, Fundación INFANT–COVID-19 Group**. 2021. Early high-titer plasma therapy to prevent severe Covid-19 in older adults. *N Engl J Med* **384**:610–618. http://dx.doi.org/10.1056/NEJMoa2033700.

16. **Sullivan DJ, Gebo KA, Shoham S, Bloch EM, Lau B, Shenoy AG, Mosnaim GS, Gniadek TJ, Fukuta Y, Patel B, Heath SL, Levine AC, Meisenberg BR, Spivak ES, Anjan S, Huaman MA, Blair JE, Currier JS, Paxton JH, Gerber JM, Petrini JR, Broderick PB, Rausch W, Cordisco ME, Hammel J, Greenblatt B, Cluzet VC, Cruser D, Oei K, Abinante M, Hammitt LL, Sutcliffe CG, Forthal DN, Zand MS, Cachay ER, Raval JS, Kassaye SG, Foster EC, Roth M, Marshall CE, Yarava A, Lane K, McBee NA, Gawad AL, Karlen N, Singh A, Ford DE, Jabs DA, Appel LJ, Shade DM, Ehrhardt S, Baksh SN, Laeyendecker O, Pekosz A, Klein SL, Casadevall A, Tobian AAR, Hanley DF**. 2022. Early outpatient treatment for Covid-19 with convalescent plasma. *N Engl J Med* **386**:1700–1711. http://dx.doi.org/10.1056/NEJMoa2119657.

17. **Galler S**. 2015. Forgotten research from 19th century: science should not follow fashion. *J Muscle Res Cell Motil* **36**:5–9. http://dx.doi.org/10.1007/s10974-014-9399-4.

18. **Yeatman JD, Weiner KS, Pestilli F, Rokem A, Mezer A, Wandell BA**. 2014. The vertical occipital fasciculus: a century of controversy resolved by in vivo measurements. *Proc Natl Acad Sci U S A* **111**:E5214–E5223. http://dx.doi.org/10.1073/pnas.1418503111.

19. **Brush SG**. 1974. Should the history of science be rated X?: the way scientists behave (according to historians) might not be a good model for students. *Science* **183**:1164–1172. http://dx.doi.org/10.1126/science.183.4130.1164.

20. **Burian RM**. 1977. More than a marriage of convenience: on the inextricability of history and philosophy of science. *Philos Sci* **44**:1–42. http://dx.doi.org/10.1086/288722.

21. **Toulmin S**. 1974. The Alexandrian trap. *Encounter* **42**:61–72.

22. **Casadevall A, Steen RG, Fang FC**. 2014. Sources of error in the retracted scientific literature. *FASEB J* **28**:3847–3855. http://dx.doi.org/10.1096/fj.14-256735.

23. **Atkinson JW**. 1979. The importance of the history of science to the American Society of Zoologists. *Am Zool* **19**:1243–1246. http://dx.doi.org/10.1093/icb/19.4.1243.

24. **Kuhn TS**. 1962. *The Structure of Scientific Revolutions*. University of Chicago Press, Chicago, IL.

25. **Marshall BJ, Warren JR**. 1984. Unidentified curved bacilli in the stomach of patients with gastritis and peptic ulceration. *Lancet* **1**:1311–1315. http://dx.doi.org/10.1016/S0140-6736(84)91816-6.

26. **Meyers MA**. 1995. Glen W. Hartman Lecture. Science, creativity, and serendipity. *AJR Am J Roentgenol* **165**:755–764. http://dx.doi.org/10.2214/ajr.165.4.7676963.

27. **Fang FC, Casadevall A**. 2010. Lost in translation—basic science in the era of translational research. *Infect Immun* **78**:563–566. http://dx.doi.org/10.1128/IAI.01318-09.

28. **Botstein D**. 2012. Why we need more basic biology research, not less. *Mol Biol Cell* **23**:4160–4161. http://dx.doi.org/10.1091/mbc.e12-05-0406.

29. **Schatz A, Robinson KA**. 1993. The true story of the discovery of streptomycin. *Actinomycetes* **4**:27–39.

30. **Lawrence PA**. 2002. Rank injustice. *Nature* **415**:835–836. http://dx.doi.org/10.1038/415835a.

31. **Pringle P**. 2012. *Experiment Eleven: Deceit and Betrayal in the Discovery of the Cure for Tuberculosis*. Bloomsbury UK, London, United Kingdom.

32. **Casadevall A, Fang FC**. 2013. Is the Nobel Prize good for science? *FASEB J* **27**:4682–4690. http://dx.doi.org/10.1096/fj.13-238758.

33. **Jones JH**. 1981. *Bad Blood: the Tuskegee Syphilis Experiment*. The Free Press, New York, NY.

34. **Geison GL**. 1995. *The Private Science of Louis Pasteur*. Princeton University Press, Princeton, NJ.

35. **Skloot R**. 2010. *The Immortal Life of Henrietta Lacks*. Broadway Books, New York, NY.

36. **Rodriguez MA, García R**. 2013. First, do no harm: the US sexually transmitted disease experiments in Guatemala. *Am J Public Health* **103**:2122–2126. http://dx.doi.org/10.2105/AJPH.2013.301520.

37. **Casadevall A, Fang FC**. 2009. Important science—it's all about the SPIN. *Infect Immun* **77**: 4177–4180. http://dx.doi.org/10.1128/IAI.00757-09.

38. **Casadevall A, Fang FC**. 2012. Winner takes all. *Sci Am* **307**:13. http://dx.doi.org/10.1038/scientificamerican0812-13.

39. **De Kruif P**. 1926. *Microbe Hunters*. Harcourt, Brace & Co, New York, NY. http://dx.doi.org/10.2307/3221690

40. **Watson JD**. 1968. *The Double Helix: a Personal Account of the Discovery of the Structure of DNA*. Atheneum, New York, NY.

41. **Judson HF**. 1979. *The Eighth Day of Creation: Makers of the Revolution in Biology*. Simon and Schuster, New York, NY.

42. **Chargaff E**. 1974. Building the tower of Babble. *Nature* **248**:776–779. http://dx.doi.org/10.1038/248776a0.

43. **Goldman DL, Fries BC, Franzot SP, Montella L, Casadevall A**. 1998. Phenotypic switching in the human pathogenic fungus *Cryptococcus neoformans* is associated with changes in virulence and pulmonary inflammatory response in rodents. *Proc Natl Acad Sci U S A* **95**:14967–14972. http://dx.doi.org/10.1073/pnas.95.25.14967.

44. **Darwin C**. 1859. *On the Origin of Species by Means of Natural Selection*. John Murray, London, United Kingdom.

45. **Luke TC, Kilbane EM, Jackson JL, Hoffman SL**. 2006. Meta-analysis: convalescent blood products for Spanish influenza pneumonia: a future H5N1 treatment? *Ann Intern Med* **145**:599–609. http://dx.doi.org/10.7326/0003-4819-145-8-200610170-00139.

46. **Luke TC, Casadevall A, Watowich SJ, Hoffman SL, Beigel JH, Burgess TH**. 2010. Hark back: passive immunotherapy for influenza and other serious infections. *Crit Care Med* **38**(Suppl):e66–e73. http://dx.doi.org/10.1097/CCM.0b013e3181d44c1e.

47. **Senefeld JW, Johnson PW, Kunze KL, Bloch EM, van Helmond N, Golafshar MA, Klassen SA, Klompas AM, Sexton MA, Diaz Soto JC, Grossman BJ, Tobian AAR, Goel R, Wiggins CC, Bruno KA, van Buskirk CM, Stubbs JR, Winters JL, Casadevall A, Paneth NS, Shaz BH, Petersen MM, Sachais BS, Buras MR, Wieczorek MA, Russoniello B, Dumont LJ, Baker SE, Vassallo RR, Shepherd JRA, Young PP, Verdun NC, Marks P, Haley NR, Rea RF, Katz L, Herasevich V, Waxman DA, Whelan ER, Bergman A, Clayburn AJ, Grabowski MK, Larson KF, Ripoll JG, Andersen KJ, Vogt MNP, Dennis JJ, Regimbal RJ, Bauer PR, Blair JE, Buchholtz ZA, Pletsch MC, Wright K, Greenshields JT, Joyner MJ, Wright RS, Carter RE, Fairweather D**. 2021. Access to and safety of COVID-19 convalescent plasma in the United States Expanded Access Program: a national registry study. *PLoS Med* **18**:e1003872. http://dx.doi.org/10.1371/journal.pmed.1003872.

48. **Joyner MJ, Carter RE, Senefeld JW, Klassen SA, Mills JR, Johnson PW, Theel ES, Wiggins CC, Bruno KA, Klompas AM, Lesser ER, Kunze KL, Sexton MA, Diaz Soto JC, Baker SE, Shepherd JRA, van Helmond N, Verdun NC, Marks P, van Buskirk CM, Winters JL, Stubbs JR, Rea RF, Hodge DO, Herasevich V, Whelan ER, Clayburn AJ, Larson KF, Ripoll JG, Andersen KJ, Buras MR, Vogt MNP, Dennis JJ, Regimbal RJ, Bauer PR, Blair JE, Paneth NS, Fairweather D, Wright RS, Casadevall A**. 2021. Convalescent plasma antibody levels and the risk of death from Covid-19. *N Engl J Med* **384**:1015–1027. http://dx.doi.org/10.1056/NEJMoa2031893.

49. **Cecil RL**. 1937. Effects of very early serum treatment in pneumococcus type I pneumonia. *JAMA* **108**:689–692. http://dx.doi.org/10.1001/jama.1937.02780090001001.

50. **Focosi D, Franchini M, Pirofski LA, Burnouf T, Paneth N, Joyner MJ, Casadevall A.** 2022. COVID-19 convalescent plasma and clinical trials: understanding conflicting outcomes. *Clin Microbiol Rev* **35**:e0020021. http://dx.doi.org/10.1128/cmr.00200-21.

Chapter 10 – Specialized Science

1. **Smith A.** 1776. *An Inquiry into the Nature and Causes of the Wealth of Nations*. Methuen & Co, London, United Kingdom.
2. **Wikipedia.** 2022. https://www.wikipedia.org/wiki/interservice_rivalry.
3. **Becker GS, Murphy KM.** 1992. The division of labor, coordination costs, and knowledge. *Q J Econ* **107**:1137–1160. http://dx.doi.org/10.2307/2118383.
4. **Biron CA, Casadevall A.** 2010. On immunologists and microbiologists: ground zero in the battle for interdisciplinary knowledge. *mBio* **1**:e00260-10. http://dx.doi.org/10.1128/mBio.00260-10.
5. **Wikipedia.** 2022. https://www.wikipedia.org/wiki/Outline_of_biology.
6. **Oxford English Dictionary.** *OED Online*. Oxford University Press, Oxford, United Kingdom.
7. **Kuhn TS.** 1962. *The Structure of Scientific Revolutions*. University of Chicago Press, Chicago, IL.
8. **Darden L.**1978. Discoveries and the emergence of new fields in science, p 149–160. *In* **Asquith PD, Hacking I** (ed), *PSA: Proceedings of the Biennial Meeting of the Philosophy of Science Association*. Philosophy of Science Association, East Lansing, MI.
9. **Darden L, Maull N.** 1977. Interfield theories. *Philos Sci* **44**:43–64. http://dx.doi.org/10.1086/288723.
10. **Findlay L.** 2014. Meet Alice C. *Evans Microbe* **9**:235–239.
11. **Bruce-Chwatt LJ.** 1981. Alphonse Laveran's discovery 100 years ago and today's global fight against malaria. *J R Soc Med* **74**:531–536. http://dx.doi.org/10.1177/014107688107400715.
12. **Glatman-Freedman A, Casadevall A.** 1998. Serum therapy for tuberculosis revisited: reappraisal of the role of antibody-mediated immunity against *Mycobacterium tuberculosis*. *Clin Microbiol Rev* **11**:514–532. http://dx.doi.org/10.1128/CMR.11.3.514.
13. **Achkar JM, Casadevall A.** 2013. Antibody-mediated immunity against tuberculosis: implications for vaccine development. *Cell Host Microbe* **13**:250–262. http://dx.doi.org/10.1016/j.chom.2013.02.009.
14. **Temin HM, Baltimore D.** 1972. RNA-directed DNA synthesis and RNA tumor viruses. *Adv Virus Res* **17**:129–186. http://dx.doi.org/10.1016/S0065-3527(08)60749-6.
15. **Prusiner SB.** 1982. Novel proteinaceous infectious particles cause scrapie. *Science* **216**:136–144. http://dx.doi.org/10.1126/science.6801762.
16. **Telesnitsky A, Goff SP.** 1997. Reverse transcriptase and the generation of retroviral DNA. *In* **Coffin JM, Hughes SH, Varmus HE** (ed), *Retroviruses*. Cold Spring Harbor Laboratory Press, Cold Spring Harbor, NY.
17. **Koonin EV.** 2012. Does the central dogma still stand? *Biol Direct* **7**:27. http://dx.doi.org/10.1186/1745-6150-7-27.
18. **Planck M.** 1948. *Wissenschaftliche Selbstbiographie. Mit einem Bildnis und der von Max von Laue gehaltenen Traueransprache*. Johann Ambrosius Barth Verlag, Leipzig, Germany.
19. **DiRita VJ.** 2013. Microbiology is an integrative field, so why are we a divided society? *Microbe* **8**:384–385. http://dx.doi.org/10.1128/microbe.8.384.1.
20. **Dunbar R.** 1998. The social brain hypothesis. *Evol Anthropol* **6**:178–190. http://dx.doi.org/10.1002/(SICI)1520-6505(1998)6:5<178::AID-EVAN5>3.0.CO;2-8.
21. **Dunbar RIM.** 1992. Neocortex size as a constraint on group size in primates. *J Hum Evol* **22**:369–493. http://dx.doi.org/10.1016/0047-2484(92)90081-J.
22. **Louveau A, Smirnov I, Keyes TJ, Eccles JD, Rouhani SJ, Peske JD, Derecki NC, Castle D, Mandell JW, Lee KS, Harris TH, Kipnis J.** 2015. Structural and functional features of central nervous system lymphatic vessels. *Nature* **523**:337–341. http://dx.doi.org/10.1038/nature14432.
23. **Eizenberg N.** 2015. Anatomy and its impact on medicine: will it continue? *Australas Med J* **8**:373–377. http://dx.doi.org/10.4066/AMJ.2015.2550.

24. **Ofir G, Sorek R**. 2018. Contemporary phage biology: from classic models to new insights. *Cell* **172**:1260–1270. http://dx.doi.org/10.1016/j.cell.2017.10.045.

25. **Kass EH, Hayes KM**. 1988. A history of the Infectious Diseases Society of America. *Rev Infect Dis* **10**(Suppl 2):1–159.

26. **Petersdorf RG**. 1978. The doctors' dilemma. *N Engl J Med* **299**:628–634. http://dx.doi.org/10.1056/NEJM197809212991204.

27. **Feynman RP, Leighton RB, Sands M, Treiman SB**. 1964. *The Feynman Lectures on Physics, 1.* Addison-Wesley, Boston, MA. http://dx.doi.org/10.1063/1.3051743

28. **Harman O, Dietrich MR**. 2013. *Outsider Scientists: Routes to Innovation in Biology.* University of Chicago Press, Chicago, IL. http://dx.doi.org/10.7208/chicago/9780226078540.001.0001

29. **Pulverer B**. 2013. Impact fact-or fiction? *EMBO J* **32**:1651–1652. http://dx.doi.org/10.1038/emboj.2013.126.

30. **Brembs B, Button K, Munafò M**. 2013. Deep impact: unintended consequences of journal rank. *Front Hum Neurosci* **7**:291. http://dx.doi.org/10.3389/fnhum.2013.00291.

31. **Garfield E**. 2006. The history and meaning of the journal impact factor. *JAMA* **295**:90–93. http://dx.doi.org/10.1001/jama.295.1.90.

32. **Casadevall A, Fang FC**. 2009. Important science—it's all about the SPIN. *Infect Immun* **77**:4177–4180. http://dx.doi.org/10.1128/IAI.00757-09.

33. **Fang FC, Casadevall A**. 2011. Retracted science and the retraction index. *Infect Immun* **79**:3855–3859. http://dx.doi.org/10.1128/IAI.05661-11.

34. **Fang FC, Casadevall A**. 2012. Reforming science: structural reforms. *Infect Immun* **80**:897–901. http://dx.doi.org/10.1128/IAI.06184-11.

35. **Cummings JN, Kiesler S, Bosagh Zadeh R, Balakrishnan AD**. 2013. Group heterogeneity increases the risks of large group size: a longitudinal study of productivity in research groups. *Psychol Sci* **24**:880–890. http://dx.doi.org/10.1177/0956797612463082.

36. **Melero E, Palomeras N**. 2012. The renaissance of the *Renaissance Man*?: specialists vs. generalists in teams of inventors. INDEM-Working Paper Business Economic Series, Instituto para el Desarrollo Empresarial (INDEM). https://e-archivo.uc3m.es/bitstream/handle/10016/14057/indemwp12_01.pdf?sequence=1.

37. **Levy SB**. 2002. *The Antibiotic Paradox: How the Misuse of Antibiotics Destroys Their Curative Powers.* Perseus, Cambridge, MA.

38. **Committee for a New Biology for the 21st Century**. 2009. *Ensuring the United States Leads the Coming Biology Revolution: A New Biology for the 21st Century.* National Academies Press, Washington, DC.

39. **Massachusetts Institute of Technology**. 2011. *The Third Revolution: the Convergence of the Life Sciences, Physical Sciences, and Engineering.* Massachusetts Institute of Technology, Cambridge, MA.

40. **Arise Committee II**. 2013. *Unleashing America's Research and Innovation Enterprise.* American Academy of Arts and Sciences, Cambridge, MA.

41. **Gray B**. 2008. Enhancing transdisciplinary research through collaborative leadership. *Am J Prev Med* **35**(Suppl):S124–S132. http://dx.doi.org/10.1016/j.amepre.2008.03.037.

42. **O'Brien T, Yamamoto K, Hawgood S**. 2013. Commentary: team science. *Acad Med* **88**:156–157. http://dx.doi.org/10.1097/ACM.0b013e31827c0e34.

43. **Bauer HH**. 1990. Barriers against interdisciplinarity: implications for studies of science, technology and society (STS). *Sci Technol Human Values* **15**:105–119. http://dx.doi.org/10.1177/016224399001500110.

44. **Wikipedia**. 2022. https://www.wikipedia.org/wiki/Dogma.

45. **Wikipedia**. 2022. https://www.wikipedia.org/wiki/Hilbert's_problems.

46. **Del Poeta M, Casadevall A**. 2012. Ten challenges on *Cryptococcus* and cryptococcosis. *Mycopathologia* **173**:303–310. http://dx.doi.org/10.1007/s11046-011-9473-z.

47. **Burroughs-Wellcome Fund**. 2022. http://www.bwfund.org.

48. **National Institutes of Health**. 2022. Research career development awards. https://researchtraining.nih.gov/programs/career-development.

49. **Casadevall A, Fang FC**. 2012. Reforming science: methodological and cultural reforms. *Infect Immun* **80**:891–896. http://dx.doi.org/10.1128/IAI.06183-11.

50. **Bosch G, Casadevall A**. 2017. Graduate biomedical science education needs a new philosophy. mBio **8**:e01539-17. http://dx.doi.org/10.1128/mBio.01539-17.

51. **Baekeland LH**. 1907. The danger of overspecialization. *Science* **25**:845–854. http://dx.doi.org/10.1126/science.25.648.845.

52. **Fahy D**. 30 April 2015. Carl Sagan, and the rise of the "celebrity scientist." Science Friday. https://www.sciencefriday.com/articles/carl-sagan-and-the-rise-of-the-celebrity-scientist/.

53. **Marsh O**. 2019. Life cycle of a star: Carl Sagan and the circulation of reputation. *Br J Hist Sci* **52**:467–486. http://dx.doi.org/10.1017/S0007087419000049.

Chapter 11 – Revolutionary Science

1. **Campbell J**. 2004. *Pathways to Bliss*. New World Library, Novato, CA.

2. **Kuhn TS**. 1962. *The Structure of Scientific Revolutions*. University of Chicago Press, Chicago, IL.

3. **Toulmin S**. 1970. Does the distinction between normal and revolutionary science hold water?, p 38–48. *In* **Lakatos I, Musgrave A** (ed), *Criticism and the Growth of Knowledge*. Cambridge University Press, Cambridge, United Kingdom. http://dx.doi.org/10.1017/CBO9781139171434.005

4. **Watkins J**. 1970. Against "normal science," p 25–38. *In* **Lakatos I, Musgrave A** (ed), *Criticism and the Growth of Knowledge*. Cambridge University Press, Cambridge, United Kingdom. http://dx.doi.org/10.1017/CBO9781139171434.004

5. **Wilkins AS**. 1996. Are there "Kuhnian" revolutions in biology? *BioEssays* **18**:695–696. http://dx.doi.org/10.1002/bies.950180902.

6. **Online Etymology Dictionary**. 2022. https://www.etymonline.com/search?q=revolutionary.

7. **Cohen IB**. 1987. *Revolution in Science*. Belknap Press of Harvard University Press, Cambridge, MA.

8. **Charlton BG**. 2007. Measuring revolutionary biomedical science 1992-2006 using Nobel prizes, Lasker (clinical medicine) awards and Gairdner awards (NLG metric). *Med Hypotheses* **69**:1–5. http://dx.doi.org/10.1016/j.mehy.2007.01.001.

9. **Charlton BG**. 2007. Scientometric identification of elite "revolutionary science" research institutions by analysis of trends in Nobel prizes 1947-2006. *Med Hypotheses* **68**:931–934. http://dx.doi.org/10.1016/j.mehy.2006.12.006.

10. **Charlton BG**. 2007. Which are the best nations and institutions for revolutionary science 1987-2006? Analysis using a combined metric of Nobel prizes, Fields medals, Lasker awards and Turing awards (NFLT metric). *Med Hypotheses* **68**:1191–1194. http://dx.doi.org/10.1016/j.mehy.2006.12.007.

11. **Casadevall A, Fang FC**. 2013. Is the Nobel Prize good for science? *FASEB J* **27**:4682–4690. http://dx.doi.org/10.1096/fj.13-238758.

12. **Stolley PD, Lasky T**. 1992. Johannes Fibiger and his Nobel Prize for the hypothesis that a worm causes stomach cancer. *Ann Intern Med* **116**:765–769. http://dx.doi.org/10.7326/0003-4819-116-9-765.

13. **Ceze L, Nivala J, Strauss K**. 2019. Molecular digital data storage using DNA. *Nat Rev Genet* **20**:456–466. http://dx.doi.org/10.1038/s41576-019-0125-3.

14. **Casadevall A, Fang FC**. 2009. Important science—it's all about the SPIN. *Infect Immun* **77**:4177–4180. http://dx.doi.org/10.1128/IAI.00757-09.

15. **Blaser MJ**. 2014. The microbiome revolution. *J Clin Invest* **124**:4162–4165. http://dx.doi.org/10.1172/JCI78366.

16. **Tiedje JM, Bruns MA, Casadevall A, Criddle CS, Eloe-Fadrosh E, Karl DM, Nguyen NK, Zhou J**. 2022. Microbes and climate change: a research prospectus for the future. *mBio* **13**:e0080022. http://dx.doi.org/10.1128/mbio.00800-22.

17. **Schrag DP, Hoffman PF**. 2001. Life, geology and snowball Earth. *Nature* **409**:306. http://dx.doi.org/10.1038/35053170.

18. **Casadevall A, Freij JB, Hann-Soden C, Taylor J**. 2017. Continental drift and speciation of the *Cryptococcus neoformans* and *Cryptococcus gattii* species complexes. *mSphere* **2**:e00103-17. http://dx.doi.org/10.1128/mSphere.00103-17.

19. **Gould SJ**. 1997. *Dinosaur in a Haystack: Reflections in Natural History*. Crown Publishing Company, New York, NY.

20. **Rothstein E**. 21 July 2001. Coming to blows over how valid science really is. *New York Times*. https://www.nytimes.com/2001/07/21/books/coming-to-blows-over-how-valid-science-really-is.html.

21. **Horgan J**. 23 May 2012. What Thomas Kuhn really thought about scientific "truth." *Scientific American Blogs*. https://blogs.scientificamerican.com/cross-check/what-thomas-kuhn-really-thought-about-scientific-truth/.

22. **Popper KR**. 1970. Normal science and its dangers, p 51–58. *In* **Lakatos I, Musgrave A** (ed), *Criticism and the Growth of Knowledge*. Cambridge University Press, Cambridge, United Kingdom. http://dx.doi.org/10.1017/CBO9781139171434.007

23. **Pirofski LA, Casadevall A**. 2012. Q and A: What is a pathogen? A question that begs the point. *BMC Biol* **10**:6. http://dx.doi.org/10.1186/1741-7007-10-6.

24. **Saiki RK, Gelfand DH, Stoffel S, Scharf SJ, Higuchi R, Horn GT, Mullis KB, Erlich HA**. 1988. Primer-directed enzymatic amplification of DNA with a thermostable DNA polymerase. *Science* **239**:487–491. http://dx.doi.org/10.1126/science.2448875.

25. **Kleppe K, Ohtsuka E, Kleppe R, Molineux I, Khorana HG**. 1971. Studies on polynucleotides. XCVI. Repair replications of short synthetic DNA's as catalyzed by DNA polymerases. *J Mol Biol* **56**:341–361. http://dx.doi.org/10.1016/0022-2836(71)90469-4.

26. **Chien A, Edgar DB, Trela JM**. 1976. Deoxyribonucleic acid polymerase from the extreme thermophile *Thermus aquaticus*. *J Bacteriol* **127**:1550–1557. http://dx.doi.org/10.1128/jb.127.3.1550-1557.1976.

27. **Brock TD**. 1967. Life at high temperatures. Evolutionary, ecological, and biochemical significance of organisms living in hot springs is discussed. *Science* **158**:1012–1019. http://dx.doi.org/10.1126/science.158.3804.1012.

28. **Mullis KB**. 1990. The unusual origin of the polymerase chain reaction. *Sci Am* **262**:56–61, 64–65. http://dx.doi.org/10.1038/scientificamerican0490-56.

29. **Ridley M**. 23 October 2015. The myth of basic science. *Wall Street Journal*. https://www.wsj.com/articles/the-myth-of-basic-science-1445613954.

Chapter 12 – Translational Science

1. **Untermeyer L**. 1964. *Robert Frost: a Backward Look*. Library of Congress, Washington, DC.

2. **Ahluwalia J, Tinker A, Clapp LH, Duchen MR, Abramov AY, Pope S, Nobles M, Segal AW**. 2004. The large-conductance Ca^{2+}-activated K^+ channel is essential for innate immunity. *Nature* **427**:853–858. http://dx.doi.org/10.1038/nature02356.

3. **Feldman AM**. 2015. Bench-to-bedside; clinical and translational research; personalized medicine; precision medicine—what's in a name? *Clin Transl Sci* **8**:171–173. http://dx.doi.org/10.1111/cts.12302.

4. **Woolf SH**. 2008. The meaning of translational research and why it matters. *JAMA* **299**:211–213. http://dx.doi.org/10.1001/jama.2007.26.

5. **Andrews N, Burris JE, Cech TR, Coller BS, Crowley WF, Jr, Gallin EK, Kelner KL, Kirch DG, Leshner AI, Morris CD, Nguyen FT, Oates J, Sung NS**. 2009. Translational careers. *Science* **324**:855. http://dx.doi.org/10.1126/science.1172137.

6. **Bowen A, Casadevall A**. 2015. Increasing disparities between resource inputs and outcomes, as measured by certain health deliverables, in biomedical research. *Proc Natl Acad Sci U S A* **112**:11335–11340. http://dx.doi.org/10.1073/pnas.1504955112.

7. **Casadevall A**. 2018. Is the pace of biomedical innovation slowing? *Perspect Biol Med* **61**:584–593. http://dx.doi.org/10.1353/pbm.2018.0067.

8. **Park M, Leahey E, Funk RJ.** 2023. Papers and patents are becoming less disruptive over time. *Nature* **613**:138–144. http://dx.doi.org/10.1038/s41586-022-05543-x.

9. **Hudson J, Khazragui HF.** 2013. Into the valley of death: research to innovation. *Drug Discov Today* **18**:610–613. http://dx.doi.org/10.1016/j.drudis.2013.01.012.

10. **Seok J, Warren HS, Cuenca AG, Mindrinos MN, Baker HV, Xu W, Richards DR, McDonald-Smith GP, Gao H, Hennessy L, Finnerty CC, López CM, Honari S, Moore EE, Minei JP, Cuschieri J, Bankey PE, Johnson JL, Sperry J, Nathens AB, Billiar TR, West MA, Jeschke MG, Klein MB, Gamelli RL, Gibran NS, Brownstein BH, Miller-Graziano C, Calvano SE, Mason PH, Cobb JP, Rahme LG, Lowry SF, Maier RV, Moldawer LL, Herndon DN, Davis RW, Xiao W, Tompkins RG, Inflammation and Host Response to Injury, Large Scale Collaborative Research Program.** 2013. Genomic responses in mouse models poorly mimic human inflammatory diseases. *Proc Natl Acad Sci U S A* **110**:3507–3512. http://dx.doi.org/10.1073/pnas.1222878110.

11. **Duyk G.** 2003. Attrition and translation. *Science* **302**:603–605. http://dx.doi.org/10.1126/science.1090521.

12. **Bush V.** 1945. *Science, the Endless Frontier—a Report to the President on a Program for Postwar Scientific Research.* United States Government Printing Office, Washington, DC.

13. **Nathan DG.** 2002. Careers in translational clinical research—historical perspectives, future challenges. *JAMA* **287**:2424–2427. http://dx.doi.org/10.1001/jama.287.18.2424.

14. **Koshland DE, Jr.** 1993. Basic research (I). *Science* **259**:291. http://dx.doi.org/10.1126/science.8419994.

15. **Casadevall A, Fang FC.** 2009. Important science—it's all about the SPIN. *Infect Immun* **77**:4177–4180. http://dx.doi.org/10.1128/IAI.00757-09.

16. **Johnson TE, Kirkland CL, Lu Y, Smithies RH, Brown M, Hartnady MIH.** 2022. Giant impacts and the origin and evolution of continents. *Nature* **608**:330–335. http://dx.doi.org/10.1038/s41586-022-04956-y.

17. **Shukla R, Lavore F, Maity S, Derks MGN, Jones CR, Vermeulen BJA, Melcrová A, Morris MA, Becker LM, Wang X, Kumar R, Medeiros-Silva J, van Beekveld RAM, Bonvin AMJJ, Lorent JH, Lelli M, Nowick JS, MacGillavry HD, Peoples AJ, Spoering AL, Ling LL, Hughes DE, Roos WH, Breukink E, Lewis K, Weingarth M.** 2022. Teixobactin kills bacteria by a two-pronged attack on the cell envelope. *Nature* **608**:390–396. http://dx.doi.org/10.1038/s41586-022-05019-y.

18. **Zakaria F.** 23 November 2009. Can America still innovate? *Newsweek.* https://www.newsweek.com/ambrid-can-america-still-innovate-77023.

19. **National Institutes of Health.** 2016. Definitions of criteria and considerations for research project grant critiques. https://grants.nih.gov/grants/peer/critiques/rpg.htm.

20. **Weissmann G.** 2005. Roadmaps, translational research, and childish curiosity. *FASEB J* **19**:1761–1762. http://dx.doi.org/10.1096/fj.05-1101ufm.

21. **Clabby C.** 2009. An interview with Harold Varmus. *American Scientist.* http://www.americanscientist.org/bookshelf/pub/an-interview-with-harold-varmus.

22. **Intersociety Working Group.** 2008. *AAAS report XXXIII: research and development FY 2009. American Association for the Advancement of Science*, Washington, DC. http://www.aaas.org/spp/rd/rd09main.htm.

23. **Hourihan M, Zimmerman A.** 2022. *U.S. R&D and innovation in a global context: 2022 data update.* American Association for the Advancement of Science, Washington, DC. https://www.aaas.org/news/Global-Context-2022.

24. **Mandel HG, Vesell ES.** 2008. Declines in NIH R01 research grant funding. *Science* **322**:189. http://dx.doi.org/10.1126/science.322.5899.189a.

25. **Manz MG, Di Santo JP.** 2009. Renaissance for mouse models of human hematopoiesis and immunobiology. *Nat Immunol* **10**:1039–1042. http://dx.doi.org/10.1038/ni1009-1039.

26. **Curie M.** 1921. The Discovery of Radium, Ellen S. Richards Monographs no. 2. Vassar College, Poughkeepsie, NY.

27. **Vasselon T, Detmers PA.** 2002. Toll receptors: a central element in innate immune responses. *Infect Immun* **70**:1033–1041. http://dx.doi.org/10.1128/IAI.70.3.1033-1041.2002.

28. **Nüsslein-Volhard C.** 10 December 1995. Nobel Prize Banquet speech. https://www.nobelprize.org/prizes/medicine/1995/nusslein-volhard/speech/.

29. **Rienzi G, Huang A.** 12 October 2009. Our newest Nobelist: Carol Greider. *JHU Gazette.* http://gazette.jhu.edu/2009/10/12/our-newest-nobelist-carol-greider.

30. **Van Epps HL.** 2005. Peyton Rous: father of the tumor virus. *J Exp Med* **201**:320. http://dx.doi.org/10.1084/jem.2013fta.

31. **Baltimore D.** 1995. Discovery of the reverse transcriptase. *FASEB J* **9**:1660–1663. http://dx.doi.org/10.1096/fasebj.9.15.8529847.

32. **Elemento O.** 2021. The road from Rous sarcoma virus to precision medicine. *J Exp Med* **218**:e20201754. http://dx.doi.org/10.1084/jem.20201754.

33. **Poiesz BJ, Ruscetti FW, Gazdar AF, Bunn PA, Minna JD, Gallo RC.** 1980. Detection and isolation of type C retrovirus particles from fresh and cultured lymphocytes of a patient with cutaneous T-cell lymphoma. *Proc Natl Acad Sci U S A* **77**:7415–7419. http://dx.doi.org/10.1073/pnas.77.12.7415.

34. **Centers for Disease Control (CDC).** 1981. Pneumocystis pneumonia—Los Angeles. *MMWR Morb Mortal Wkly Rep* **30**:250–252.

35. **Gallo RC, Montagnier L.** 2003. The discovery of HIV as the cause of AIDS. *N Engl J Med* **349**:2283–2285. http://dx.doi.org/10.1056/NEJMp038194.

36. **Tujios S, Fontana RJ.** 2011. Mechanisms of drug-induced liver injury: from bedside to bench. *Nat Rev Gastroenterol Hepatol* **8**:202–211. http://dx.doi.org/10.1038/nrgastro.2011.22.

37. **Rehman W, Arfons LM, Lazarus HM.** 2011. The rise, fall and subsequent triumph of thalidomide: lessons learned in drug development. *Ther Adv Hematol* **2**:291–308. http://dx.doi.org/10.1177/2040620711413165.

38. **Millrine D, Kishimoto T.** 2017. A brighter side to thalidomide: its potential use in immunological disorders. *Trends Mol Med* **23**:348–361. http://dx.doi.org/10.1016/j.molmed.2017.02.006.

39. **Congressional Budget Office.** 27 June 2019. Federal investment, 1962 to 2018. https://www.cbo.gov/publication/55375.

40. **Mandt R, Seetharam K, Cheng CHM.** 20 August 2020. Federal R&D funding: the bedrock of national innovation. MIT Science Policy Review. https://sciencepolicyreview.org/2020/08/federal-rd-funding-the-bedrock-of-national-innovation/.

Chapter 13 – Moonshot Science

1. **Eliason RT.** 2009. Shoot for the moon because even if you miss, you'll land among the stars. Success. http://www.success.com/article/shoot-for-the-moon.

2. **Mason JL, Johnston E, Berndt S, Segal K, Lei M, Wiest JS.** 2016. Labor and skills gap analysis of the biomedical research workforce. *FASEB J* **30**:2673–2683. http://dx.doi.org/10.1096/fj.201500067R.

3. **Breivik J.** 27 May 2016. We won't cure cancer. *New York Times.* https://www.nytimes.com/2016/05/27/opinion/obamas-pointless-cancer-moonshot.html.

4. **Cohen J.** 2021. A pandemic prevention moonshot. *Science* **373**:1183. http://dx.doi.org/10.1126/science.acx9044.

5. **Shivaram D.** 12 September 2022. Biden touts his "cancer moonshot" on the anniversary of JFK's "man on the moon" speech. National Public Radio. https://www.npr.org/2022/09/12/1122480037/biden-cancer-moonshot.

6. **Scudellari M.** 2017. Big science has a buzzword problem. *Nature* **541**:450–453. http://dx.doi.org/10.1038/541450a.

7. **Smyth HD.** 1945. *Atomic Energy for Military Purposes.* Princeton University Press, Princeton, NJ.

8. **Freeman M.** 1993. *How We Got to the Moon: the Story of the German Space Pioneers.* 21st Century Science Associates, Washington, DC.

9. **Hsiung GD.** 1987. Perspectives on retroviruses and the etiologic agent of AIDS. *Yale J Biol Med* **60**:505–514.

10. **Barré-Sinoussi F, Chermann JC, Rey F, Nugeyre MT, Chamaret S, Gruest J, Dauguet C, Axler-Blin C, Vézinet-Brun F, Rouzioux C, Rozenbaum W, Montagnier L**. 1983. Isolation of a T-lymphotropic retrovirus from a patient at risk for acquired immune deficiency syndrome (AIDS). *Science* **220**:868–871. http://dx.doi.org/10.1126/science.6189183.

11. **Kolata G**. 1987. FDA approves AZT. *Science* **235**:1570. http://dx.doi.org/10.1126/science.235. 4796.1570-b.

12. **Horwitz JP, Chua J, Noel M**. 1964. Nucleosides. V. The monomesylates of 1-(2-deoxy-β-D-lyxofuranosyl)thymine. *J Org Chem* **29**:2076–2078. http://dx.doi.org/10.1021/jo01030a546.

13. **Collins FS, Morgan M, Patrinos A**. 2003. The Human Genome Project: lessons from large-scale biology. *Science* **300**:286–290. http://dx.doi.org/10.1126/science.1084564.

14. **Horvath A, Rachlew E**. 2016. Nuclear power in the 21st century: challenges and possibilities. *Ambio* **45**(Suppl 1):S38–S49. http://dx.doi.org/10.1007/s13280-015-0732-y.

15. **Zylstra AB, et al**. 2022. Burning plasma achieved in inertial fusion. *Nature* **601**:542–548. http://dx.doi.org/10.1038/s41586-021-04281-w.

16. **Barbarino M**. 2020. A brief history of nuclear fusion. *Nat Phys* **16**:890–893. http://dx.doi.org/10.1038/s41567-020-0940-7.

17. **Hsu DC, O'Connell RJ**. 2017. Progress in HIV vaccine development. *Hum Vaccin Immunother* **13**:1018–1030. http://dx.doi.org/10.1080/21645515.2016.1276138.

18. **Haynes BF, Wiehe K, Borrrow P, Saunders KO, Korber B, Wagh K, McMichael AJ, Kelsoe G, Hahn BH, Alt F, Shaw GM**. 2023. Strategies for HIV-1 vaccines that induce broadly neutralizing antibodies. *Nat Rev Immunol* **23**:142–158. http://dx.doi.org/10.1038/s41577-022-00753-w.

19. **Peiris JS, Yuen KY, Osterhaus AD, Stöhr K**. 2003. The severe acute respiratory syndrome. *N Engl J Med* **349**:2431–2441. http://dx.doi.org/10.1056/NEJMra032498.

20. **Poon LL, Guan Y, Nicholls JM, Yuen KY, Peiris JS**. 2004. The aetiology, origins, and diagnosis of severe acute respiratory syndrome. *Lancet Infect Dis* **4**:663–671. http://dx.doi.org/10.1016/S1473-3099(04)01172-7.

21. **Sui J, Li W, Murakami A, Tamin A, Matthews LJ, Wong SK, Moore MJ, Tallarico AS, Olurinde M, Choe H, Anderson LJ, Bellini WJ, Farzan M, Marasco WA**. 2004. Potent neutralization of severe acute respiratory syndrome (SARS) coronavirus by a human mAb to S1 protein that blocks receptor association. *Proc Natl Acad Sci U S A* **101**:2536–2541. http://dx.doi.org/10.1073/pnas.0307140101.

22. **Chretien JP, Riley S, George DB**. 2015. Mathematical modeling of the West Africa Ebola epidemic. *eLife* **4**:e09186. http://dx.doi.org/10.7554/eLife.09186.

23. **Kennedy SB, Neaton JD, Lane HC, Kieh MW, Massaquoi MB, Touchette NA, Nason MC, Follmann DA, Boley FK, Johnson MP, Larson G, Kateh FN, Nyenswah TG**. 2016. Implementation of an Ebola virus disease vaccine clinical trial during the Ebola epidemic in Liberia: design, procedures, and challenges. *Clin Trials* **13**:49–56. http://dx.doi.org/10.1177/1740774515621037.

24. **Wikan N, Smith DR**. 2016. Zika virus: history of a newly emerging arbovirus. *Lancet Infect Dis* **16**:e119–e126. http://dx.doi.org/10.1016/S1473-3099(16)30010-X.

25. **Slaoui M, Hepburn M**. 2020. Developing safe and effective Covid vaccines—Operation Warp Speed's strategy and approach. *N Engl J Med* **383**:1701–1703. http://dx.doi.org/10.1056/NEJMp2027405.

26. **Casadevall A**. 2021. The mRNA vaccine revolution is the dividend from decades of basic science research. *J Clin Invest* **131**:e153721. http://dx.doi.org/10.1172/JCI153721.

27. **American Cancer Society**. 2016. Cancer facts and figures. http://www.cancer.org/acs/groups/content/@research/documents/document/acspc-047079.pdf.

28. **Siegel RL, Miller KD, Wagle NS, Jemal A**. 2023. Cancer statistics, 2023. *CA Cancer J Clin* **73**: 17–48. http://dx.doi.org/10.3322/caac.21763.

29. **Green RE, Malaspinas AS, Krause J, Briggs AW, Johnson PL, Uhler C, Meyer M, Good JM, Maricic T, Stenzel U, Prüfer K, Siebauer M, Burbano HA, Ronan M, Rothberg JM, Egholm M,**

Rudan P, Brajković D, Kućan Z, Gusić I, Wikström M, Laakkonen L, Kelso J, Slatkin M, Pääbo S. 2008. A complete Neandertal mitochondrial genome sequence determined by high-throughput sequencing. *Cell* **134**:416–426. http://dx.doi.org/10.1016/j.cell.2008.06.021.

30. Prüfer K, Racimo F, Patterson N, Jay F, Sankararaman S, Sawyer S, Heinze A, Renaud G, Sudmant PH, de Filippo C, Li H, Mallick S, Dannemann M, Fu Q, Kircher M, Kuhlwilm M, Lachmann M, Meyer M, Ongyerth M, Siebauer M, Theunert C, Tandon A, Moorjani P, Pickrell J, Mullikin JC, Vohr SH, Green RE, Hellmann I, Johnson PL, Blanche H, Cann H, Kitzman JO, Shendure J, Eichler EE, Lein ES, Bakken TE, Golovanova LV, Doronichev VB, Shunkov MV, Derevianko AP, Viola B, Slatkin M, Reich D, Kelso J, Pääbo S. 2014. The complete genome sequence of a Neanderthal from the Altai Mountains. *Nature* **505**:43–49. http://dx.doi.org/10.1038/nature12886.

31. Curry A. 2022. Ancient DNA pioneer Svante Pääbo wins Nobel. *Science* **378**:12. http://dx.doi.org/10.1126/science.adf1845.

32. Chan JF, Lau SK, To KK, Cheng VC, Woo PC, Yuen KY. 2015. Middle East respiratory syndrome coronavirus: another zoonotic betacoronavirus causing SARS-like disease. *Clin Microbiol Rev* **28**:465–522. http://dx.doi.org/10.1128/CMR.00102-14.

33. Sharpless NE, Singer DS. 2021. Progress and potential: the Cancer Moonshot. *Cancer Cell* **39**:889–894. http://dx.doi.org/10.1016/j.ccell.2021.04.015.

34. Logsdon JM. 1989. Evaluating Apollo. *Space Policy* **5**:188–192. http://dx.doi.org/10.1016/0265-9646(89)90085-4.

35. Potter S. 5 June 2019. Exploring the Moon promises innovation and benefit at home. National Aeronautics and Space Administration. https://www.nasa.gov/feature/exploring-the-moon-promises-innovation-and-benefit-at-home.

36. DiCicco M. 15 July 2019. Going to the Moon was hard—but the benefits were huge, for all of us. National Aeronautics and Space Administration. https://www.nasa.gov/directorates/spacetech/feature/Going_to_the_Moon_Was_Hard_But_the_Benefits_Were_Huge.

37. Wiechert U, Halliday AN, Lee DC, Snyder GA, Taylor LA, Rumble D. 2001. Oxygen isotopes and the moon-forming giant impact. *Science* **294**:345–348. http://dx.doi.org/10.1126/science.1063037.

38. Wood JA, Dickey JS, Jr, Marvin UB, Powell BN. 1970. Lunar anorthosites. *Science* **167**:602–604. http://dx.doi.org/10.1126/science.167.3918.602.

39. Taylor GJ. 16 July 2019. Scientific discoveries from the Apollo 11 mission. Planetary Science Research Discoveries. http://www.psrd.hawaii.edu/July19/Apollo11-discoveries.html.

40. Verne J. 1877. *From the Earth to the Moon Direct and Round It*. George Routledge and Sons, London, United Kingdom.

Chapter 14 – Serendipitous Science

1. Hauptman J. 2011. *Particle Physics Experiments at High Energy Colliders*. Wiley-VCH, Weinheim, Germany.

2. Shulman JL, Barber E, Merton RI. 2011. *The Travels and Adventures of Serendipity: a Study in Sociological Semantics and the Sociology of Science*. Princeton University Press, Princeton, NJ.

3. Dictionary.com. 2022. https://www.dictionary.com/browse/serendipity.

4. Gal J. 2008. The discovery of biological enantioselectivity: Louis Pasteur and the fermentation of tartaric acid, 1857—a review and analysis 150 yr later. *Chirality* **20**:5–19. http://dx.doi.org/10.1002/chir.20494.

5. Vantomme G, Crassous J. 2021. Pasteur and chirality: a story of how serendipity favors the prepared minds. *Chirality* **33**:597–601. http://dx.doi.org/10.1002/chir.23349.

6. Becquerel J, Crowther JA. 1948. Discovery of radioactivity. *Nature* **161**:609. http://dx.doi.org/10.1038/161609b0.

7. American Physical Society. March 2008. This month in physics history: March 1, 1986: Henri Becquerel discovers radioactivity. *APS News*. https://www.aps.org/publications/apsnews/200803/physicshistory.cfm.

8. **Myers WG**. 1976. Becquerel's discovery of radioactivity in 1896. *J Nucl Med* **17**:579–582.

9. **Cantey JB, Doern CD**. 2015. "A defective and imperfect" method: H. Christian Gram and the history of the Gram stain. *Pediatr Infect Dis J* 34:848. http://dx.doi.org/10.1097/INF.0000000000000749.

10. **Houbraken J, Frisvad JC, Samson RA**. 2011. Fleming's penicillin producing strain is not *Penicillium chrysogenum* but *P. rubens*. *IMA Fungus* **2**:87–95. http://dx.doi.org/10.5598/imafungus.2011.02.01.12.

11. **Henderson JW**. 1997. The yellow brick road to penicillin: a story of serendipity. *Mayo Clin Proc* **72**:683–687. http://dx.doi.org/10.1016/S0025-6196(11)63577-5.

12. **Hare R**. 1970. New light on the discovery of penicillin. *Med Leg J* **38**:31–41. http://dx.doi.org/10.1177/002581727003800202.

13. **Bone WA**. 1927. The centenary of the friction match. *Nature* **119**:495–496. http://dx.doi.org/10.1038/119495a0.

14. **Wotiz JH**. 1978. The discovery of saccharin. *J Chem Educ* **55**:161. http://dx.doi.org/10.1021/ed055p161.

15. **Roberts RM**. 1989. *Serendipity: Accidental Discoveries in Science*. Wiley, New York, NY.

16. **Meyers MA**. 1995. Glen W. Hartman Lecture. Science, creativity, and serendipity. *AJR Am J Roentgenol* **165**:755–764. http://dx.doi.org/10.2214/ajr.165.4.7676963.

17. **Schwager E**. 1998. From petroleum jelly to riches. *Drug News Perspect* **11**:127–128.

18. **American Physical Society**. November 2001. This month in physics history: November 8, 1895: Roentgen's discovery of X-rays. *APS News*. https://www.aps.org/publications/apsnews/200111/history.cfm.

19. **Alderden RA, Hall MD, Hambley TW**. 2006. The discovery and development of cisplatin. *J Chem Educ* **83**:728. http://dx.doi.org/10.1021/ed083p728.

20. **Warner DJ**. 2008. Ira Remsen, saccharin, and the linear model. *Ambix* **55**:50–61. http://dx.doi.org/10.1179/174582308X255415.

21. **Ancuceanu R**. 2011. Saccharin—urban myths and scientific data. *Pract Farm* **4**:69–78.

22. **Yao JS, Eskandari MK**. 2012. Accidental discovery: the polytetrafluoroethylene graft. *Surgery* **151**:126–128. http://dx.doi.org/10.1016/j.surg.2011.09.036.

23. **Markel H**. 2017. *The Kelloggs: the Battling Brothers of Battle Creek*. Penguin Random House, New York, NY.

24. **Goldstein I, Burnett AL, Rosen RC, Park PW, Stecher VJ**. 2019. The serendipitous story of sildenafil: an unexpected oral therapy for erectile dysfunction. *Sex Med Rev* **7**:115–128. http://dx.doi.org/10.1016/j.sxmr.2018.06.005.

25. **McKenzie J**. 2020. Serendipity in action. *Phys World* **33**:23.

26. **American Physical Society**. 2021. This month in physics history: April 6, 1938: discovery of Teflon. *APS News*. https://www.aps.org/publications/apsnews/202104/history.cfm.

27. **Wilson RW**. 1979. The cosmic microwave background radiation. *Science* **205**:866–874. http://dx.doi.org/10.1126/science.205.4409.866.

28. **Kipnis N**. 2005. Chance in science: the discovery of electromagnetism by H.C. Oersted. *Sci Educ* **14**:1–28. http://dx.doi.org/10.1007/s11191-004-3286-0.

29. **Goldman DL, Fries BC, Franzot SP, Montella L, Casadevall A**. 1998. Phenotypic switching in the human pathogenic fungus *Cryptococcus neoformans* is associated with changes in virulence and pulmonary inflammatory response in rodents. *Proc Natl Acad Sci U S A* **95**:14967–14972. http://dx.doi.org/10.1073/pnas.95.25.14967.

30. **Alvarez M, Casadevall A**. 2006. Phagosome extrusion and host-cell survival after *Cryptococcus neoformans* phagocytosis by macrophages. *Curr Biol* **16**:2161–2165. http://dx.doi.org/10.1016/j.cub.2006.09.061.

31. **Dorman CJ, Ni Bhriain N, Higgins CF**. 1990. DNA supercoiling and environmental regulation of virulence gene expression in *Shigella flexneri*. *Nature* **344**:789–792. http://dx.doi.org/10.1038/344789a0.

32. **Fang FC, Krause M, Roudier C, Fierer J, Guiney DG**. 1991. Growth regulation of a *Salmonella* plasmid gene essential for virulence. *J Bacteriol* **173**:6783–6789. http://dx.doi.org/10.1128/ jb.173.21.6783-6789.1991.

33. **Copeland S.** 2019. On serendipity in science: discovery at the intersection of chance and wisdom. *Synthese* **196**:2385–2406. http://dx.doi.org/10.1007/s11229-017-1544-3.

34. **Yaqub O.** 2018. Serendipity: towards a taxonomy and a theory. *Res Policy* **47**:169–179. http:// dx.doi.org/10.1016/j.respol.2017.10.007.

35. **Fleming A.** 1932. Lysozyme: president's address. *Proc R Soc Med* **26**:71–84. http://dx.doi.org/ 10.1177/003591573202600201.

36. **Goldsworthy PD, McFarlane AC**. 2002. Howard Florey, Alexander Fleming and the fairy tale of penicillin. *Med J Aust* **176**:176–178. http://dx.doi.org/10.5694/j.1326-5377.2002.tb04349.x.

37. **Pasteur L.** 7 December 1854. Presentation at the Université de Lille, France.

38. **Ariel G, Rabani A, Benisty S, Partridge JD, Harshey RM, Be'er A**. 2015. Swarming bacteria migrate by Lévy Walk. *Nat Commun* **6**:8396. http://dx.doi.org/10.1038/ncomms9396.

39. **Tolkien JRR.** 1954. *The Fellowship of the Ring*. George Allen & Unwin, London, United Kingdom.

40. **Colman DR.** 2006. The three princes of Serendip: notes on a mysterious phenomenon. *McGill J Med* **9**:161–163.

41. **Casadevall A, Fang FC**. 2009. Important science—it's all about the SPIN. *Infect Immun* **77**: 4177–4180. http://dx.doi.org/10.1128/IAI.00757-09.

42. **Wikiquote.** 2022. https://en.wikiquote.org/wiki/Alexander_Fleming.

43. **MacCallum P, Tolhurst JC, Buckle G, Sissons HA**. 1948. A new mycobacterial infection in man. *J Pathol Bacteriol* **60**:93–122. http://dx.doi.org/10.1002/path.1700600111.

44. **Parson W.** 1999. *Mycobacterium ulcerans*. *Lancet* **354**:2171. http://dx.doi.org/10.1016/S0140-6736(05)77082-3.

45. **Anonymous.** 1876. Three Cinghalese chiefs waiting for the Prince of Wales at Kandy, Ceylon. *Illustrated London News*. **68**:49.

Chapter 15 – Unequal Science

1. **Burns B.** 21 January 2021. Acting with authenticity and grit: a conversation with Freeman Hrabowski. University Innovation Alliance. https://theuia.org/blog/post/ weekly-wisdom-episode-9-freeman-hrabowski-university-of-maryland-baltimore-county.

2. **Benson RA.** 19 March 2006. Executive suite—advice from the top. *USA Today*. https://usatoday30. usatoday.com/money/companies/management/2006-03-19-sally-ride_x.htm.

3. **Graves JL Jr, Kearney M, Barabino G, Malcom S**. 2022. Inequality in science and the case for a new agenda. *Proc Natl Acad Sci U S A* **119**:e2117831119.

4. **Anonymous.** 2016. Is science only for the rich? *Nature* **537**:466–470.

5. **Cech EA, Waidzunas TJ**. 2021. Systemic inequalities for LGBTQ professionals in STEM. *Sci Adv* **7**:eabe0933. http://dx.doi.org/10.1126/sciadv.abe0933.

6. **Funk C, Parker K**. 2018. *Women and Men in STEM Often at Odds over Workplace Equity*. Pew Research Center, Washington, DC. https://www.pewresearch.org/social-trends/wp-content/ uploads/sites/3/2018/01/PS_2018.01.09_STEM_FINAL.pdf.

7. **Colwell R.** 30 August 2020. Women scientists have the evidence about sexism. *The Atlantic*. https:// www.theatlantic.com/ideas/archive/2020/08/women-scientists-have-evidence-about-sexism-science/615823/.

8. **Huang J, Gates AJ, Sinatra R, Barabási AL**. 2020. Historical comparison of gender inequality in scientific careers across countries and disciplines. *Proc Natl Acad Sci U S A* **117**:4609–4616.

9. **Oliveira DFM, Ma Y, Woodruff TK, Uzzi B**. 2019. Comparison of National Institutes of Health grant amounts to first-time male and female principal investigators. *JAMA* **321**:898–900. http:// dx.doi.org/10.1001/jama.2018.21944.

10. **Head MG, Fitchett JR, Cooke MK, Wurie FB, Atun R**. 2013. Differences in research funding for women scientists: a systematic comparison of UK investments in global infectious disease research during 1997-2010. *BMJ Open* **3**:e003362. http://dx.doi.org/10.1136/bmjopen-2013-003362.

11. **Pohlhaus JR, Jiang H, Wagner RM, Schaffer WT, Pinn VW**. 2011. Sex differences in application, success, and funding rates for NIH extramural programs. *Acad Med* **86**:759–767. http://dx.doi.org/10.1097/ACM.0b013e31821836ff.

12. **Safdar B, Naveed S, Chaudhary AMD, Saboor S, Zeshan M, Khosa F**. 2021. Gender disparity in grants and awards at the National Institute of Health. *Cureus* **13**:e14644. http://dx.doi.org/10.7759/cureus.14644.

13. **Rørstad K, Aksnes DW**. 2015. Publication rate expressed by age, gender and academic position—a large-scale analysis of Norwegian academic staff. *J Informetrics* **9**:317–333. http://dx.doi.org/10.1016/j.joi.2015.02.003.

14. **van der Wal JEM, Thorogood R, Horrocks NPC**. 2021. Collaboration enhances career progression in academic science, especially for female researchers. *Proc Biol Sci* **288**:20210219. http://dx.doi.org/10.1098/rspb.2021.0219.

15. **Chatterjee P, Werner RM**. 2021. Gender disparity in citations in high-impact journal articles. *JAMA Netw Open* **4**:e2114509. http://dx.doi.org/10.1001/jamanetworkopen.2021.14509.

16. **Massen JJM, Bauer L, Spurny B, Bugnyar T, Kret ME**. 2017. Sharing of science is most likely among male scientists. *Sci Rep* **7**:12927. http://dx.doi.org/10.1038/s41598-017-13491-0.

17. **Ross MB, Glennon BM, Murciano-Goroff R, Berkes EG, Weinberg BA, Lane JI**. 2022. Women are credited less in science than men. *Nature* **608**:135–145. http://dx.doi.org/10.1038/s41586-022-04966-w.

18. **Volerman A, Arora VM, Cursio JF, Wei H, Press VG**. 2021. Representation of women on National Institutes of Health study sections. *JAMA Netw Open* **4**:e2037346. http://dx.doi.org/10.1001/jamanetworkopen.2020.37346.

19. **Fang FC, Bennett JW, Casadevall A**. 2013. Males are overrepresented among life science researchers committing scientific misconduct. *mBio* **4**:e00640-12. http://dx.doi.org/10.1128/mBio.00640-12.

20. **Rosenzweig EQ, Hecht CA, Priniski SJ, Canning EA, Asher MW, Tibbetts Y, Hyde JS, Harackiewicz JM**. 2021. Inside the STEM pipeline: changes in students' biomedical career plans across the college years. *Sci Adv* **7**:7. http://dx.doi.org/10.1126/sciadv.abe0985.

21. **Lauer M.** 17 November 2020. NIH challenges academia to share strategies to strengthen gender diversity. NIH Extramural Nexus. https://nexus.od.nih.gov/all/2020/11/16/nih-challenges-academia-to-share-strategies-to-strengthen-gender-diversity/.

22. **Alegria SN, Branch EH**. 2015. Causes and consequences of inequality in STEM: diversity and its discontents. *Int J Gend Sci Technol* **7**:321–342.

23. **Mason MA, Goulden M, Frasch K**. 2011. Keeping women in the science pipeline. *Ann Am Acad Pol Soc Sci* **638**:141–162. http://dx.doi.org/10.1177/0002716211416925.

24. **Ledin A, Bornmann L, Gannon F, Wallon G**. 2007. A persistent problem. Traditional gender roles hold back female scientists. *EMBO Rep* **8**:982–987. http://dx.doi.org/10.1038/sj.embor.7401109.

25. **National Academies of Sciences, Engineering, and Medicine**. 2021. *The Impact of COVID-19 on the Careers of Women in Academic Sciences, Engineering, and Medicine*. National Academies Press, Washington, DC.

26. **Yudell M, Roberts D, DeSalle R, Tishkoff S**. 2016. SCIENCE AND SOCIETY. Taking race out of human genetics. *Science* **351**:564–565.

27. **Hoover EL**. 2007. There is no scientific rationale for race-based research. *J Natl Med Assoc* **99**:690–692.

28. **Meyers LC, Brown AM, Moneta-Koehler L, Chalkley R**. 2018. Survey of checkpoints along the pathway to diverse biomedical research faculty. *PloS One* **13**:e0190606. http://dx.doi.org/10.1371/journal.pone.0190606.

29. **Gibbs KD, Jr, Basson J, Xierali IM, Broniatowski DA**. 2016. Decoupling of the minority PhD talent pool and assistant professor hiring in medical school basic science departments in the US. *eLife* **5**:e21393. http://dx.doi.org/10.7554/eLife.21393.

30. **Ginther DK, Schaffer WT, Schnell J, Masimore B, Liu F, Haak LL, Kington R**. 2011. Race, ethnicity, and NIH research awards. *Science* **333**:1015–1019. http://dx.doi.org/10.1126/science.1196783.

31. **Oh SS, Galanter J, Thakur N, Pino-Yanes M, Barcelo NE, White MJ, de Bruin DM, Greenblatt RM, Bibbins-Domingo K, Wu AH, Borrell LN, Gunter C, Powe NR, Burchard EG**. 2015. Diversity in clinical and biomedical research: a promise yet to be fulfilled. *PloS Med* **12**:e1001918. http://dx.doi.org/10.1371/journal.pmed.1001918.

32. **Bernard MA, Johnson AC, Hopkins-Laboy T, Tabak LA**. 2021. The US National Institutes of Health approach to inclusive excellence. *Nat Med* **27**:1861–1864. http://dx.doi.org/10.1038/s41591-021-01532-1.

33. **Lauer MS, Roychowdhury D**. 2021. Inequalities in the distribution of National Institutes of Health research project grant funding. *eLife* **10**:e71712. http://dx.doi.org/10.7554/eLife.71712.

34. **Chen CY, Kahanamoku SS, Tripati A, Alegado RA, Morris VR, Andrade K, Hosbey J**. 2022. Systemic racial disparities in funding rates at the National Science Foundation. *eLife* **11**:e83071. http://dx.doi.org/10.7554/eLife.83071.

35. **Ginther DK, Basner J, Jensen U, Schnell J, Kington R, Schaffer WT**. 2018. Publications as predictors of racial and ethnic differences in NIH research awards. *PloS One* **13**:e0205929. http://dx.doi.org/10.1371/journal.pone.0205929.

36. **Rainey K, Dancy M, Mickelson R, Stearns E, Moller S**. 2018. Race and gender differences in how sense of belonging influences decisions to major in STEM. *Int J STEM Educ* **5**:10. http://dx.doi.org/10.1186/s40594-018-0115-6.

37. **Chrousos GP, Mentis AA**. 2020. Imposter syndrome threatens diversity. *Science* **367**:749–750. http://dx.doi.org/10.1126/science.aba8039.

38. **Ciocca DR, Delgado G**. 2017. The reality of scientific research in Latin America; an insider's perspective. *Cell Stress Chaperones* **22**:847–852. http://dx.doi.org/10.1007/s12192-017-0815-8.

39. **Rodrigues ML, Morel CM**. 2016. The Brazilian dilemma: increased scientific production and high publication costs during a global health crisis and major economic downturn. *mBio* **7**:e00907-16. http://dx.doi.org/10.1128/mBio.00907-16.

40. **Gibbs WW**. 1995. Lost science in the third world. *Sci Am* **273**:92–99. http://dx.doi.org/10.1038/scientificamerican0895-92.

41. **Cañizares-Esguerra J**. 2004. Iberian science in the Renaissance: ignored how much longer? *Perspect Sci* **12**:86–124. http://dx.doi.org/10.1162/106361404773843355.

42. **Poulin R, McDougall C, Presswell B**. 2022. What's in a name? Taxonomic and gender biases in the etymology of new species names. *Proc Biol Sci* **289**:20212708. http://dx.doi.org/10.1098/rspb.2021.2708.

43. **Moss-Racusin CA, Dovidio JF, Brescoll VL, Graham MJ, Handelsman J**. 2012. Science faculty's subtle gender biases favor male students. *Proc Natl Acad Sci U S A* **109**:16474–16479. http://dx.doi.org/10.1073/pnas.1211286109.

44. **Holman L, Stuart-Fox D, Hauser CE**. 2018. The gender gap in science: how long until women are equally represented? *PloS Biol* **16**:e2004956. http://dx.doi.org/10.1371/journal.pbio.2004956.

45. **Bendels MHK, Müller R, Brueggmann D, Groneberg DA**. 2018. Gender disparities in high-quality research revealed by Nature Index journals. *PloS One* **13**:e0189136. http://dx.doi.org/10.1371/journal.pone.0189136.

46. **West JD, Jacquet J, King MM, Correll SJ, Bergstrom CT**. 2013. The role of gender in scholarly authorship. *PloS One* **8**:e66212. http://dx.doi.org/10.1371/journal.pone.0066212.

47. **Aakhus E, Mitra N, Lautenbach E, Joffe S**. 2018. Gender and byline placement of co-first authors in clinical and basic science journals with high impact factors. *JAMA* **319**:610–611. http://dx.doi.org/10.1001/jama.2017.18672.

48. **Broderick NA, Casadevall A**. 2019. Gender inequalities among authors who contributed equally. *eLife* **8**:e36399. http://dx.doi.org/10.7554/eLife.36399.

49. **Ni C, Smith E, Yuan H, Larivière V, Sugimoto CR**. 2021. The gendered nature of authorship. *Sci Adv* **7**:eabe4639. http://dx.doi.org/10.1126/sciadv.abe4639.

50. **Casadevall A, Schloss P**. 2019. Explaining order among those who share positions in the author byline. *mBio* **10**:e01989-19. http://dx.doi.org/10.1128/mBio.01981-19.

51. **Casadevall A, Semenza GL, Jackson S, Tomaselli G, Ahima RS**. 2019. Reducing bias: accounting for the order of co-first authors. *J Clin Invest* **129**:2167–2168. http://dx.doi.org/10.1172/JCI128764.

52. **Stadtfeld C, Takács K, Vörös A**. 2020. The emergence and stability of groups in social networks. *Social Networks* **60**:129–145.

53. **Jan YN**. 2022. Underrepresentation of Asian awardees of United States biomedical research prizes. *Cell* **185**:407–410. http://dx.doi.org/10.1016/j.cell.2022.01.004.

54. **Lurie E**. 1954. Louis Agassiz and the races of Man. *Isis* **45**:227–242. http://dx.doi.org/10.1086/348335.

55. **Solly M**. 15 January 2019. DNA pioneer James Watson loses honorary titles over racist comments. *Smithsonian Magazine*. https://www.smithsonianmag.com/smart-news/dna-pioneer-james-watson-loses-honorary-titles-over-racist-comments-180971266/.

56. **Brush SG**. 1991. Women in science and engineering. *Am Sci* **79**:404–419.

57. **Gornick V**. 1990. *Women in Science*. Touchstone, New York, NY.

58. **McBride J**. 20 October 2018. Nobel laureate Donna Strickland: "I see myself as a scientist, not a woman in science." *The Guardian*. https://www.theguardian.com/science/2018/oct/20/nobel-laureate-donna-strickland-i-see-myself-as-a-scientist-not-a-woman-in-science.

59. **Mackenzie D**. January 2009. What Larry Summers said—and didn't say. *Swarthmore College Bulletin*. https://www.swarthmore.edu/bulletin/archive/wp/january-2009_what-larry-summers-said-and-didnt-say.html.

60. **Hyde JS, Lindberg SM, Linn MC, Ellis AB, Williams CC**. 2008. Diversity. Gender similarities characterize math performance. *Science* **321**:494–495. http://dx.doi.org/10.1126/science.1160364.

61. **Alec M**. 1986. *Hypatia's Heritage: A History of Women in Science from Antiquity to the Late Nineteenth Century*. Beacon Press, Boston, MA.

62. **Mercer C**. 2018. The philosophical roots of Western misogyny. *Philos Topics* **46**:183–208.

63. **Seifert V**. 22 February 2022. Sexist science. *Chemistry World*. https://www.chemistryworld.com/opinion/the-different-shades-of-sexist-science/4015190.article.

64. **Dupree CH, Boykin CM**. 2021. Racial inequality in academia: systemic origins, modern challenges, and policy recommendations. *Behav Brain Sci* **8**:11–18.

65. **Kessler JH, Kidd JS, Morin KA, Kidd RA**. 1996. *Distinguished African American Scientists of the 20th Century*. Oryx Press, Phoenix, AZ.

66. **Warren W**. 1999. *Black Women Scientists in the United States*. Indiana University Press, Bloomington, IN.

67. **Ricks D**. 10 April 2023. Overlooked no more: Alice Ball, chemist who created a treatment for leprosy, p B8. *New York Times*, New York, NY.

68. **Fry R, Kennedy B, Funk C**. 1 April 2021. STEM jobs see uneven progress in increasing gender, racial and ethnic diversity. Pew Research Scented. https://www.pewresearch.org/science/2021/04/01/stem-jobs-see-uneven-progress-in-increasing-gender-racial-and-ethnic-diversity/.

69. **Hoppe TA, Litovitz A, Willis KA, Meseroll RA, Perkins MJ, Hutchins BI, Davis AF, Lauer MS, Valantine HA, Anderson JM, Santangelo GM**. 2019. Topic choice contributes to the lower rate of NIH awards to African-American/black scientists. *Sci Adv* **5**:eaaw7238. http://dx.doi.org/10.1126/sciadv.aaw7238.

70. **Kozlowski D, Larivière V, Sugimoto CR, Monroe-White T**. 2022. Intersectional inequalities in science. *Proc Natl Acad Sci U S A* **119**:e2113067119. http://dx.doi.org/10.1073/pnas.2113067119.

71. **Mervis J**. 2019. Topic choice works against black NIH applicants. *Science* **366**:164–165. http://dx.doi.org/10.1126/science.366.6462.164.

72. **Ceci SJ, Ginther DK, Kahn S, Williams WM**. 2014. Women in academic science: a changing landscape. *Psychol Sci Public Interest* **15**:75–141. http://dx.doi.org/10.1177/1529100614541236.

73. **Educational Opportunity Monitoring Project**. 2022. Racial and ethnic achievement gaps. Stanford Center for Education Policy Analysis. https://cepa.stanford.edu/educational-opportunity-monitoring-project/achievement-gaps/race/.

74. **Cowtan K.** 2020. Structural barriers to scientific progress. *Acta Crystallogr D Struct Biol* **76**:908–911.

75. **Stillman B.** 2018. Why has America been such a magnet for immigrant scientists? *Scientific American Blogs*. https://blogs.scientificamerican.com/voices/why-has-america-been-such-a-magnet-for-immigrant-scientists/.

76. **Hrabowski FA,** III. 2011. Boosting minorities in science. *Science* **331**:125. http://dx.doi.org/10.1126/science.1202388.

77. **Del Giudice M.** 2014. Why it's crucial to get more women into science. *National Geographic*. http://news.nationalgeographic.com/news/2014/11/141107-gender-studies-women-scientific-research-feminist/.

78. **Koning R, Samila S, Ferguson JP**. 2021. Who do we invent for? Patents by women focus more on women's health, but few women get to invent. *Science* **372**:1345–1348. http://dx.doi.org/10.1126/science.aba6990.

79. **Bear JB, Woolley AW**. 2011. The role of gender in team collaboration and performance. *Interdiscip Sci Rev* **36**:146–153. http://dx.doi.org/10.1179/030801811X13013181961473.

80. **Freeman RB, Huang W**. 2014. Collaboration: strength in diversity. *Nature* **513**:305. http://dx.doi.org/10.1038/513305a.

81. **Campbell LG, Mehtani S, Dozier ME, Rinehart J**. 2013. Gender-heterogeneous working groups produce higher quality science. *PLoS One* **8**:e79147. http://dx.doi.org/10.1371/journal.pone.0079147.

82. **Holman L, Morandin C**. 2019. Researchers collaborate with same-gendered colleagues more often than expected across the life sciences. *PLoS One* **14**:e0216128. http://dx.doi.org/10.1371/journal.pone.0216128.

83. **Yang Y, Tian TY, Woodruff TK, Jones BF, Uzzi B**. 2022. Gender-diverse teams produce more novel and higher-impact scientific ideas. *Proc Natl Acad Sci U S A* **119**:e2200841119. http://dx.doi.org/10.1073/pnas.2200841119.

84. **National Academies of Sciences, Engineering, and Medicine Policy, and Global Affairs Committee on Women in Science, Engineering, and Medicine Committee, on the Impacts of Sexual Harassment in Academia**. 2018. *Sexual Harassment of Women: Climate, Culture, and Consequences in Academic Sciences, Engineering, and Medicine*. National Academies Press, Washington, DC.

85. **Swartz TH, Palermo AS, Masur SK, Aberg JA**. 2019. The science and value of diversity: closing the gaps in our understanding of inclusion and diversity. *J Infect Dis* **220**(Suppl 2):S33–S41. http://dx.doi.org/10.1093/infdis/jiz174.

86. **Kozłowski D, Larivière V, Sugimoto CR, Monroe-White T**. 2022. Intersectional inequalities in science. *Proc Natl Acad Sci USA* **119**:e2113067119.

87. **Padilla AM**. 1994. Ethnic minority scholars, research, and mentoring: current and future issues. *Educ Res* **23**:24–27.

88. **Joseph TD, Hirshfield LE**. 2011. "Why don't you get somebody new to do it?" Race and cultural taxation in the academy. *Ethn Racial Stud* **34**:121–141. http://dx.doi.org/10.1080/01419870.2010.496489.

89. **Hirshfield LE, Joseph TD**. 2012. "We need a woman, we need a black woman": gender, race, and identity taxation in the academy. *Gend Educ* **24**:213–227. http://dx.doi.org/10.1080/09540253.2011.606208.

90. **Guillaume RO, Apodaca EC**. 2022. Early career faculty of color and promotion and tenure: the intersection of advancement in the academy and cultural taxation. *Race Ethn Educ* **25**:546–563. http://dx.doi.org/10.1080/13613324.2020.1718084.

91. **Adamo SA**. 2013. Attrition of women in the biological sciences: workload, motherhood, and other explanations revisited. *Bioscience* **63**:43–48. http://dx.doi.org/10.1525/bio.2013.63.1.9.

92. **Einaudi P, Gordon J, Kang K**. 9 August 2022. Baccalaureate origins of underrepresented minority research doctorate recipients. National Science Foundation. https://ncses.nsf.gov/pubs/nsf22335.

93. **Widener A.** 2020. Who has the most success preparing Black students for careers in science? Historically Black colleges and universities. *Chemical Engineering News*. https://cen.acs.org/articles/98/i34/success-preparing-Black-students-careers.html.

94. **Gewin V, Payne D**. 2021. The heart of the Black STEM pipeline. *Nature* **597**:435–438. http://dx.doi.org/10.1038/d41586-021-02487-6.

95. **Kuehn BM**. 2018. Looking for the best of both worlds. *eLife* **7**:e36366. http://dx.doi.org/10.7554/eLife.36366.

96. **Dobbin F, Kalev A**. 2016. Why diversity programs fail. *Harv Bus Rev* **94**:52–60.

97. **Carnes M, Fine E, Sheridan J**. 2019. Promises and pitfalls of diversity statements: proceed with caution. *Acad Med* **94**:20–24 http://dx.doi.org/10.1097/ACM.0000000000002388.

98. **Guterl F**. 2014. The inclusion equation. *Sci Am* **311**:38–41. http://dx.doi.org/10.1038/scientificamerican1014-38.

99. **Dennehy TC, Dasgupta N**. 2017. Female peer mentors early in college increase women's positive academic experiences and retention in engineering. *Proc Natl Acad Sci U S A* **114**:5964–5969 http://dx.doi.org/10.1073/pnas.1613117114.

100. **Windsor LC, Crawford KF**. 2021. Women and minorities encouraged to apply (not stay). *Trends Genet* **37**:491–493 http://dx.doi.org/10.1016/j.tig.2021.03.003.

101. **Martinez LR, Boucaud DW, Casadevall A, August A**. 2018. Factors contributing to the success of NIH-designated underrepresented minorities in academic and nonacademic research positions. *CBE Life Sci Educ* **17**:ar32. http://dx.doi.org/10.1187/cbe.16-09-0287.

102. **Casadevall A, Handelsman J**. 2014. The presence of female conveners correlates with a higher proportion of female speakers at scientific symposia. *mBio* **5**:e00846-13. http://dx.doi.org/10.1128/mBio.00846-13.

103. **Casadevall A**. 2015. Achieving speaker gender equity at the American Society for Microbiology General Meeting. *mBio* **6**:e01146. http://dx.doi.org/10.1128/mBio.01146-15.

Chapter 16 – Pseudoscience

1. **Astin JA, Harkness E, Ernst E**. 2000. The efficacy of "distant healing": a systematic review of randomized trials. *Ann Intern Med* **132**:903–910. http://dx.doi.org/10.7326/0003-4819-132-11-200006060-00009.

2. **Hammerschlag R, Marx BL, Aickin M**. 2014. Nontouch biofield therapy: a systematic review of human randomized controlled trials reporting use of only nonphysical contact treatment. *J Altern Complement Med* **20**:881–892. http://dx.doi.org/10.1089/acm.2014.0017.

3. **Hansson SO**. 2021. Science and pseudo-science. Stanford Encyclopedia of Philosophy. https://plato.stanford.edu/archives/fall2021/entries/pseudo-science/.

4. **Hansson SO**. 2013. Defining pseudoscience and science. *In* **Pigliucci M, Boudry M** (ed), *Philosophy of Pseudoscience: Reconsidering the Demarcation Problem*. University of Chicago Press, Chicago, IL. http://dx.doi.org/10.7208/chicago/9780226051826.003.0005

5. **Lakatos I**. 30 June 1973. Science and pseudoscience. Radio lecture, broadcast by Open University. https://www.youtube.com/watch?v=_YBrhzqKJWo.

6. **Turro NJ**. 1998. Toward a general theory of pathological science. 21st C. Vol. 3, issue 4. http://www.columbia.edu/cu/21stC/issue-3.4/turro.html.

7. **Elton DC, Spencer PD**. 2021. Pathological water science—four examples and what they have in common. *In* **Gadomski A** (ed), *Water in Biomechanical and Related Systems*. Springer Nature, Cham, Switzerland. http://dx.doi.org/10.1007/978-3-030-67227-0_8

8. **Klotz IM**. 1980. The N-ray affair. *Sci Am* **242**:168–175. http://dx.doi.org/10.1038/scientificamerican0580-168.

9. **Maddox J, Randi J, Stewart WW**. 1988. "High-dilution" experiments a delusion. *Nature* **334**:287–291. http://dx.doi.org/10.1038/334287a0.

10. **Diamond AMJ, Jr**. 2009. The career consequences of a mistaken research project: the case of polywater. *Am J Econ Sociol* **68**:387–412. http://dx.doi.org/10.1111/j.1536-7150.2009.00633.x.

11. **Lilienfeld SO**. 1 September 2005. The 10 commandments of helping students distinguish science from pseudoscience in psychology. *APS Observer*. https://www.psychologicalscience.org/observer/the-10-commandments-of-helping-students-distinguish-science-from-pseudoscience-in-psychology.

12. **Beyerstein BL**. 2001. Alternative medicine and common errors of reasoning. *Acad Med* **76**:230–237. http://dx.doi.org/10.1097/00001888-200103000-00009.

13. **Winnick TA**. 2005. From quackery to "complementary" medicine: the American medical profession confronts alternative therapies. *Soc Probl* **52**:38–61. http://dx.doi.org/10.1525/sp.2005.52.1.38.

14. **Diamond J.** 2000. Quacks on the rack. *The Observer*. https://www.theguardian.com/observer/comment/story/0,,406126,00.html.
15. **Lewontin RC.** 1972. The apportionment of human diversity. *Evol Biol* **6**:381–398.
16. **Witherspoon DJ, Wooding S, Rogers AR, Marchani EE, Watkins WS, Batzer MA, Jorde LB.** 2007. Genetic similarities within and between human populations. *Genetics* **176**:351–359. http://dx.doi.org/10.1534/genetics.106.067355.
17. **Maglo KN, Mersha TB, Martin LJ.** 2016. Population genomics and the statistical values of race: an interdisciplinary perspective on the biological classification of human populations and implications for clinical genetic epidemiological research. *Front Genet* **7**:22. http://dx.doi.org/10.3389/fgene.2016.00022.
18. **Templeton AR.** 2013. Biological races in humans. *Stud Hist Philos Biol Biomed Sci* **44**:262–271. http://dx.doi.org/10.1016/j.shpsc.2013.04.010.
19. **Raff J.** 2014. Nicholas Wade and race: building a scientific façade. *Hum Biol* **86**:227–232. http://dx.doi.org/10.13110/humanbiology.86.3.0227.
20. **Hansson SO.** 2017. Science denial as a form of pseudoscience. *Stud Hist Philos Sci* **63**:39–47. http://dx.doi.org/10.1016/j.shpsa.2017.05.002.
21. **Garrett BM, Cutting RL.** 2017. Magical beliefs and discriminating science from pseudoscience in undergraduate professional students. *Heliyon* **3**:e00433. http://dx.doi.org/10.1016/j.heliyon.2017.e00433.
22. **Blancke S, Boudry M, Pigliucci M.** 2007. Why do irrational beliefs mimic science? The cultural evolution of pseudoscience. *Theoria* **83**:78–97. http://dx.doi.org/10.1111/theo.12109.
23. **McIntyre L.** 2019. *The Scientific Attitude: Defending Science from Denial, Fraud, and Pseudoscience.* MIT Press, Cambridge, MA. http://dx.doi.org/10.7551/mitpress/12203.001.0001

Chapter 17 – Duplicated Science

1. **Turan J.** 2020. The science and art of detecting data manipulation and fraud: an interview with Elisabeth Bik. *Physiol News Mag* **118**:10–12. http://dx.doi.org/10.36866/pn.118.10.
2. **Fang FC, Steen RG, Casadevall A.** 2012. Misconduct accounts for the majority of retracted scientific publications. *Proc Natl Acad Sci U S A* **109**:17028–17033. http://dx.doi.org/10.1073/pnas.1212247109.
3. **Bik EM, Casadevall A, Fang FC.** 2016. The prevalence of inappropriate image duplication in biomedical research publications. *mBio* **7**:e00809-16. http://dx.doi.org/10.1128/mBio.00809-16.
4. **Bik EM, Fang FC, Kullas AL, Davis RJ, Casadevall A.** 2018. Analysis and correction of inappropriate image duplication: the *Molecular and Cellular Biology* experience. *Mol Cell Biol* **38**:e00309–e00318. http://dx.doi.org/10.1128/MCB.00309-18.
5. **Fanelli D, Costas R, Fang FC, Casadevall A, Bik EM.** 2019. Testing hypotheses on risk factors for scientific misconduct via matched-control analysis of papers containing problematic image duplications. *Sci Eng Ethics* **25**:771–789. http://dx.doi.org/10.1007/s11948-018-0023-7.
6. **Fanelli D, Schleicher M, Fang FC, Casadevall A, Bik EM.** 2022. Do individual and institutional predictors of misconduct vary by country? Results of a matched-control analysis of problematic image duplications. *PLoS One* **17**:e0255334. http://dx.doi.org/10.1371/journal.pone.0255334.
7. **Casadevall A, Steen RG, Fang FC.** 2014. Sources of error in the retracted scientific literature. *FASEB J* **28**:3847–3855. http://dx.doi.org/10.1096/fj.14-256735.
8. **Steneck NH.** 2006. Fostering integrity in research: definitions, current knowledge, and future directions. *Sci Eng Ethics* **12**:53–74. http://dx.doi.org/10.1007/s11948-006-0006-y.
9. **Fanelli D.** 2009. How many scientists fabricate and falsify research? A systematic review and meta-analysis of survey data. *PLoS One* **4**:e5738. http://dx.doi.org/10.1371/journal.pone.0005738.
10. **Lüscher TF.** 2013. The codex of science: honesty, precision, and truth—and its violations. *Eur Heart J* **34**:1018–1023. http://dx.doi.org/10.1093/eurheartj/eht063.
11. **Geison GL.** 1996. *The Private Science of Louis Pasteur*, revised. Princeton University Press, Princeton, NJ.
12. **Alberts B, Cicerone RJ, Fienberg SE, Kamb A, McNutt M, Nerem RM, Schekman R, Shiffrin R, Stodden V, Suresh S, Zuber MT, Pope BK, Jamieson KH.** 2015. SCIENTIFIC INTEGRITY. Self-correction in science at work. *Science* **348**:1420–1422. http://dx.doi.org/10.1126/science.aab3847.

13. **Steen RG, Casadevall A, Fang FC**. 2013. Why has the number of scientific retractions increased? *PLoS One* **8**:e68397. http://dx.doi.org/10.1371/journal.pone.0068397.

14. **Begley CG, Ellis LM**. 2012. Drug development: raise standards for preclinical cancer research. *Nature* **483**:531–533. http://dx.doi.org/10.1038/483531a.

15. **Prinz F, Schlange T, Asadullah K**. 2011. Believe it or not: how much can we rely on published data on potential drug targets? *Nat Rev Drug Discov* **10**:712. http://dx.doi.org/10.1038/nrd3439-c1.

16. **Open Science Collaboration**. 2015. PSYCHOLOGY. Estimating the reproducibility of psychological science. *Science* **349**:aac4716. http://dx.doi.org/10.1126/science.aac4716.

17. **Flaherty DK**. 2011. The vaccine-autism connection: a public health crisis caused by unethical medical practices and fraudulent science. *Ann Pharmacother* **45**:1302–1304. http://dx.doi.org/10.1345/aph.1Q318.

18. **Bowen A, Casadevall A**. 2015. Increasing disparities between resource inputs and outcomes, as measured by certain health deliverables, in biomedical research. *Proc Natl Acad Sci U S A* **112**:11335–11340. http://dx.doi.org/10.1073/pnas.1504955112.

19. **Rossner M**. 2002. Figure manipulation: assessing what is acceptable. *J Cell Biol* **158**:1151. http://dx.doi.org/10.1083/jcb.200209084.

20. **Rossner M, Yamada KM**. 2004. What's in a picture? The temptation of image manipulation. *J Cell Biol* **166**:11–15. http://dx.doi.org/10.1083/jcb.200406019.

21. **Neill U, Turka LA**. 2007. Navigating through the gray (and CMYK) areas of figure manipulation: rules at the *JCI*. *J Clin Invest* **117**:2736. http://dx.doi.org/10.1172/JCI33829.

22. **Fanelli D, Costas R, Larivière V**. 2015. Misconduct policies, academic culture and career stage, not gender or pressures to publish, affect scientific integrity. *PLoS One* **10**:e0127556. http://dx.doi.org/10.1371/journal.pone.0127556.

23. **Pulverer B**. 2014. STAP dance. *EMBO J* **33**:1285–1286. http://dx.doi.org/10.15252/embj.201489076.

24. **Williams CL, Casadevall A, Jackson S**. 2019. Figure errors, sloppy science, and fraud: keeping eyes on your data. *J Clin Invest* **129**:1805–1807. http://dx.doi.org/10.1172/JCI128380.

25. **Kroon C, Breuer L, Jones L, An J, Akan A, Mohamed Ali EA, Busch F, Fislage M, Ghosh B, Hellrigel-Holderbaum M, Kazezian V, Koppold A, Moreira Restrepo CA, Riedel N, Scherschinski L, Urrutia Gonzalez FR, Weissgerber TL**. 2022. Blind spots on western blots: assessment of common problems in western blot figures and methods reporting with recommendations to improve them. *PLoS Biol* **20**:e3001783. http://dx.doi.org/10.1371/journal.pbio.3001783.

26. **Oksvold MP**. 2016. Incidence of data duplications in a randomly selected pool of life science publications. *Sci Eng Ethics* **22**:487–496. http://dx.doi.org/10.1007/s11948-015-9668-7.

27. **Kullas AL, Davis RJ**. 2017. Setting the (scientific) record straight: *Molecular and Cellular Biology* responds to postpublication review. *Mol Cell Biol* **37**:e00199-17. http://dx.doi.org/10.1128/MCB.00199-17.

28. **Yamada KM, Hall A**. 2015. Reproducibility and cell biology. *J Cell Biol* **209**:191–193. http://dx.doi.org/10.1083/jcb.201503036.

29. **Bonetta L**. 2006. The aftermath of scientific fraud. *Cell* **124**:873–875. http://dx.doi.org/10.1016/j.cell.2006.02.032.

30. **Glonti K, Cauchi D, Cobo E, Boutron I, Moher D, Hren D**. 2017. A scoping review protocol on the roles and tasks of peer reviewers in the manuscript review process in biomedical journals. *BMJ Open* **7**:e017468. http://dx.doi.org/10.1136/bmjopen-2017-017468.

31. **Kronick DA**. 1990. Peer review in 18th-century scientific journalism. *JAMA* **263**:1321–1322. http://dx.doi.org/10.1001/jama.1990.03440100021002.

32. **Babalola O, Grant-Kels JM, Parish LC**. 2012. Ethical dilemmas in journal publication. *Clin Dermatol* **30**:231–236. http://dx.doi.org/10.1016/j.clindermatol.2011.06.013.

33. **Collaborative Working Group from the conference "Keeping the Pool Clean: Prevention and Management of Misconduct Related Retractions."** 2018. RePAIR consensus guidelines: responsibilities of Publishers, Agencies, Institutions, and Researchers in protecting the integrity of the research record. *Res Integr Peer Rev* **3**:15. http://dx.doi.org/10.1186/s41073-018-0055-1.

34. **Taylor DB**. 2017. Plagiarism in manuscripts submitted to the *AJR*: development of an optimal screening algorithm and management pathways. *AJR Am J Roentgenol* **209**:W56. http://dx.doi.org/10.2214/AJR.17.18078.

35. **Butler D.** 2018. Researchers have finally created a tool to spot duplicated images across thousands of papers. *Nature* **555**:18. http://dx.doi.org/10.1038/d41586-018-02421-3.

36. **Anderson MS, Ronning EA, De Vries R, Martinson BC**. 2007. The perverse effects of competition on scientists' work and relationships. *Sci Eng Ethics* **13**:437–461. http://dx.doi.org/10.1007/s11948-007-9042-5.

37. **Cordero RJ, de León-Rodriguez CM, Alvarado-Torres JK, Rodriguez AR, Casadevall A**. 2016. Life science's average publishable unit (APU) has increased over the past two decades. *PLoS One* **11**:e0156983. http://dx.doi.org/10.1371/journal.pone.0156983.

38. **Collins S, Gemayel R, Chenette EJ**. 2017. Avoiding common pitfalls of manuscript and figure preparation. *FEBS J* **284**:1262–1266. http://dx.doi.org/10.1111/febs.14020.

39. **ASM Journals**. 2022. Publishing ethics. https://journals.asm.org/publishing-ethics.

40. **Proofig.** 2022. https://www.proofig.com/.

41. **Jackson S, Williams CL, Collins KL, McNally EM**. 2022. Data we can trust. *J Clin Invest* **132**:e162884. http://dx.doi.org/10.1172/JCI162884.

42. **Gautret P, Lagier JC, Parola P, Hoang VT, Meddeb L, Mailhe M, Doudier B, Courjon J, Giordanengo V, Vieira VE, Tissot Dupont H, Honoré S, Colson P, Chabrière E, La Scola B, Rolain JM, Brouqui P, Raoult D**. 2020. Hydroxychloroquine and azithromycin as a treatment of COVID-19: results of an open-label non-randomized clinical trial. *Int J Antimicrob Agents* **56**:105949. http://dx.doi.org/10.1016/j.ijantimicag.2020.105949.

43. **Bik E.** 24 March 2020. Thoughts on the Gautret et al. paper about hydroxychloroquine and azithromycin treatment of COVID-19 infections. *Science Integrity Digest.* https://scienceintegritydigest.com/2020/03/24/thoughts-on-the-gautret-et-al-paper-about-hydroxychloroquine-and-azithromycin-treatment-of-covid-19-infections/.

44. **O'Grady C.** 2021. Scientists rally around misconduct consultant facing lawsuit after challenging COVID-19 drug researcher. *Science.* https://www.science.org/content/article/scientists-rally-around-misconduct-consultant-facing-legal-threat-after-challenging.

45. **Code of Federal Regulations**. 2009. Title 42. Public health. Chapter I. Public Health Service, Department of Health and Human Services. Subchapter H. Health assessments and health effects studies of hazardous substances releases and facilities. Part 93. Public Health Service policies on research misconduct. 42 CFR 93.105. https://www.gpo.gov/fdsys/granule/CFR-2009-title42-vol1/CFR-2009-title42-vol1-sec93-105.

46. **Ortega JL**. 2022. Classification and analysis of PubPeer comments: how a web journal club is used. *J Assoc Inf Sci Technol* **73**:655–670. http://dx.doi.org/10.1002/asi.24568.

47. **Blatt MR**. 2015. Vigilante science. *Plant Physiol* **169**:907–909. http://dx.doi.org/10.1104/pp.15.01443.

48. **Wikimedia Commons**. Headshot of Elisabeth Bik. https://commons.wikimedia.org/wiki/File:Square_Headshot_Elisabeth_Bik.png.

49. **Zou X, Sorenson BS, Ross KF, Herzberg MC**. 2013. Augmentation of epithelial resistance to invading bacteria by using mRNA transfections. *Infect Immun* **81**:3975–3983. http://dx.doi.org/10.1128/IAI.00539-13.

50. **Zou X, Sorenson BS, Ross KF, Herzberg MC**. 2015. Correction for Zou et al., Augmentation of epithelial resistance to invading bacteria by using mRNA transfections. *Infect Immun* **83**:1226–1227. http://dx.doi.org/10.1128/IAI.02991-14.

51. **Liu J, Xu P, Collins C, Liu H, Zhang J, Keblesh JP, Xiong H**. 2013. HIV-1 Tat protein increases microglial outward K$^+$ current and resultant neurotoxic activity. *PLoS One* **8**:e64904. http://dx.doi.org/10.1371/journal.pone.0064904.

52. **PLoS One Staff**. 2014. Correction: HIV-1 Tat protein increases microglial outward K current and resultant neurotoxic activity. *PLoS One* **9**:e109218. http://dx.doi.org/10.1371/journal.pone.0109218.

53. **Liu Y, Liu Y, Sun C, Gan L, Zhang L, Mao A, Du Y, Zhou R, Zhang H**. 2014. Carbon ion radiation inhibits glioma and endothelial cell migration induced by secreted VEGF. *PLoS One* **9**:e98448. http://dx.doi.org/10.1371/journal.pone.0098448.

54. **Liu Y, Liu Y, Sun C, Gan L, Zhang L, Mao A, Du Y, Zhou R, Zhang H**. 2015. Correction: Carbon ion radiation inhibits glioma and endothelial cell migration induced by secreted VEGF. *PLoS One* **10**:e0135508. http://dx.doi.org/10.1371/journal.pone.0135508.

55. **Pulloor NK, Nair S, McCaffrey K, Kostic AD, Bist P, Weaver JD, Riley AM, Tyagi R, Uchil PD, York JD, Snyder SH, García-Sastre A, Potter BV, Lin R, Shears SB, Xavier RJ, Krishnan MN**. 2014. Human genome-wide RNAi screen identifies an essential role for inositol pyrophosphates in Type-I interferon response. *PLoS Pathog* **10**:e1003981. http://dx.doi.org/10.1371/journal.ppat.1003981.

56. *PLoS Pathogens* Staff. 2014. Correction: Human genome-wide RNAi screen identifies an essential role for inositol pyrophosphates in type-I interferon response. *PLoS Pathog* **10**:e1004519.

57. **Xu X, Liu T, Zhang A, Huo X, Luo Q, Chen Z, Yu L, Li Q, Liu L, Lun ZR, Shen J**. 2012. Reactive oxygen species-triggered trophoblast apoptosis is initiated by endoplasmic reticulum stress via activation of caspase-12, CHOP, and the JNK pathway in *Toxoplasma gondii* infection in mice. *Infect Immun* **80**:2121–2132. http://dx.doi.org/10.1128/IAI.06295-11.

58. **Xu X, Liu T, Zhang A, Huo X, Luo Q, Chen Z, Yu L, Li Q, Liu L, Lun ZR, Shen J**. 2015. Retraction for Xu et al., Reactive oxygen species-triggered trophoblast apoptosis is initiated by endoplasmic reticulum stress via activation of caspase-12, CHOP, and the JNK pathway in *Toxoplasma gondii* infection in mice. *Infect Immun* **83**:1735. http://dx.doi.org/10.1128/IAI.00118-15.

59. **Friedman-Levi Y, Mizrahi M, Frid K, Binyamin O, Gabizon R**. 2013. PrP(ST), a soluble, protease resistant and truncated PrP form features in the pathogenesis of a genetic prion disease. *PLoS One* **8**:e69583. http://dx.doi.org/10.1371/journal.pone.0069583.

60. **Friedman-Levi Y, Mizrahi M, Frid K, Binyamin O, Gabizon R**. 2015. Correction: PrPST, a soluble, protease resistant and truncated PrP form features in the pathogenesis of a genetic prion disease. *PLoS One* **10**:e0133911. http://dx.doi.org/10.1371/journal.pone.0133911.

Chapter 18 – Fraudulent Science

1. **Fenning TM**. 2004. Fraud offers big rewards for relatively little risk. *Nature* **427**:393. http://dx.doi.org/10.1038/427393a.

2. **Koshland DE, Jr**. 1987. Fraud in science. *Science* **235**:141. http://dx.doi.org/10.1126/science.3798097.

3. **Callahan JC**. 1994. Professions, institutions, and moral risk, p 243–270. *In* **Wueste DE** (ed), *Professional Ethics and Social Responsibility*. Rowman & Littlefield, Lanham, MD.

4. **Medawar PB**. 1996. *The Strange Case of the Spotted Mice*. Oxford University Press, Oxford, United Kingdom.

5. **Racker E**. 1989. A view of misconduct in science. *Nature* **339**:91–93. http://dx.doi.org/10.1038/339091a0.

6. **Relman AS**. 1983. Lessons from the Darsee affair. *N Engl J Med* **308**:1415–1417. http://dx.doi.org/10.1056/NEJM198306093082311.

7. **Friedman PJ**. 1990. Correcting the literature following fraudulent publication. *JAMA* **263**:1416–1419. http://dx.doi.org/10.1001/jama.1990.03440100136019.

8. **Service RF**. 2003. Scientific misconduct. More of Bell Labs physicist's papers retracted. *Science* **299**:31. http://dx.doi.org/10.1126/science.299.5603.31b.

9. **Check E, Cyranoski D**. 2005. Korean scandal will have global fallout. *Nature* **438**:1056–1057. http://dx.doi.org/10.1038/4381056a.

10. **Interlandi J**. 22 October 2006. An unwelcome discovery. Walter DeNino was a young lab technician who analyzed data for his mentor, Eric Poehlman. What he found was that Poehlman was not the scientist he appeared to be, p 98–103. *New York Times Magazine*. https://www.ncbi.nlm.nih.gov/pubmed/17115505.

11. **Mori N, Oishi K, Sar B, Mukaida N, Nagatake T, Matsushima K, Yamamoto N**. 1999. Essential role of transcription factor nuclear factor-kappaB in regulation of interleukin-8 gene expression by nitrite reductase from *Pseudomonas aeruginosa* in respiratory epithelial cells. *Infect Immun* **67**:3872–3878. http://dx.doi.org/10.1128/IAI.67.8.3872-3878.1999.

12. **Mori N, Wada A, Hirayama T, Parks TP, Stratowa C, Yamamoto N**. 2000. Activation of intercellular adhesion molecule 1 expression by *Helicobacter pylori* is regulated by NF-kappaB in gastric epithelial cancer cells. *Infect Immun* **68**:1806–1814. http://dx.doi.org/10.1128/IAI.68.4.1806-1814.2000.

13. **Mori N, Oishi K, Sar B, Mukaida N, Nagatake T, Matsushima K, Yamamoto N**. 2011. Retraction. Essential role of transcription factor nuclear factor-κB in regulation of interleukin-8 gene expression by nitrite reductase from *Pseudomonas aeruginosa* in respiratory epithelial cells. *Infect Immun* **79**:3473. http://dx.doi.org/10.1128/IAI.05483-11.

14. **Mori N, Wada A, Hirayama T, Parks TP, Stratowa C, Yamamoto N**. 2011. Retraction. Activation of intercellular adhesion molecule 1 expression by *Helicobacter pylori* is regulated by NF-κB in gastric epithelial cancer cells. *Infect Immun* **79**:542. http://dx.doi.org/10.1128/IAI.01065-10.

15. **Martinson BC, Anderson MS, de Vries R**. 2005. Scientists behaving badly. *Nature* **435**:737–738. http://dx.doi.org/10.1038/435737a.

16. **Gopalakrishna G, Ter Riet G, Vink G, Stoop I, Wicherts JM, Bouter LM**. 2022. Prevalence of questionable research practices, research misconduct and their potential explanatory factors: a survey among academic researchers in The Netherlands. *PLoS One* **17**:e0263023. http://dx.doi.org/10.1371/journal.pone.0263023.

17. **Fang FC, Bennett JW, Casadevall A**. 2013. Males are overrepresented among life science researchers committing scientific misconduct. *mBio* **4**:e00640-12. http://dx.doi.org/10.1128/mBio.00640-12.

18. **Pfeifer MP, Snodgrass GL**. 1990. The continued use of retracted, invalid scientific literature. *JAMA* **263**:1420–1423. http://dx.doi.org/10.1001/jama.1990.03440100140020.

19. **Campanario JM**. 2000. Fraud: retracted articles are still being cited. *Nature* **408**:288. http://dx.doi.org/10.1038/35042753.

20. **Resnik DB, Dinse GE**. 2013. Scientific retractions and corrections related to misconduct findings. *J Med Ethics* **39**:46–50. http://dx.doi.org/10.1136/medethics-2012-100766.

21. **Casadevall A, Steen RG, Fang FC**. 2014. Sources of error in the retracted scientific literature. *FASEB J* **28**:3847–3855. http://dx.doi.org/10.1096/fj.14-256735.

22. **Allison DB, Brown AW, George BJ, Kaiser KA**. 2016. Reproducibility: a tragedy of errors. *Nature* **530**:27–29. http://dx.doi.org/10.1038/530027a.

23. **Hagberg JM**. 2020. The unfortunately long life of some retracted biomedical research publications. *J Appl Physiol 1985* **128**:1381–1391. http://dx.doi.org/10.1152/japplphysiol.00003.2020.

24. **Collaborative Working Group from the conference "Keeping the Pool Clean: Prevention and Management of Misconduct Related Retractions."** 2018. RePAIR consensus guidelines: responsibilities of Publishers, Agencies, Institutions, and Researchers in protecting the integrity of the research record. *Res Integr Peer Rev* **3**:15. http://dx.doi.org/10.1186/s41073-018-0055-1.

25. **Stern AM, Casadevall A, Steen RG, Fang FC**. 2014. Financial costs and personal consequences of research misconduct resulting in retracted publications. *eLife* **3**:e02956. http://dx.doi.org/10.7554/eLife.02956.

26. **Fang FC, Casadevall A.** 1 May 2013. Why we cheat. *Scientific American Mind.* https://www.scientificamerican.com/article/why-we-cheat/.

27. **Byrne RW, Corp N**. 2004. Neocortex size predicts deception rate in primates. *Proc Biol Sci* **271**:1693–1699. http://dx.doi.org/10.1098/rspb.2004.2780.

28. **Dunbar RI, Shultz S**. 2007. Understanding primate brain evolution. *Philos Trans R Soc Lond B Biol Sci* **362**:649–658. http://dx.doi.org/10.1098/rstb.2006.2001.

29. **Ariely D**. 2012. *The (Honest) Truth About Dishonesty: How We Lie to Everyone—Especially Ourselves.* HarperCollins, New York, NY.

30. **Fanelli D, Costas R, Fang FC, Casadevall A, Bik EM**. 2019. Testing hypotheses on risk factors for scientific misconduct via matched-control analysis of papers containing problematic image duplications. *Sci Eng Ethics* **25**:771–789. http://dx.doi.org/10.1007/s11948-018-0023-7.

31. **Fanelli D, Schleicher M, Fang FC, Casadevall A, Bik EM**. 2022. Do individual and institutional predictors of misconduct vary by country? Results of a matched-control analysis of problematic image duplications. *PLoS One* **17**:e0255334. http://dx.doi.org/10.1371/journal.pone.0255334.

32. **Rick S, Loewenstein G**. 2008. Hypermotivation. *J Mark Res* **45**:645–653.

33. **Anonymous**. 2007. Breeding cheats. *Nature* **445**:242–243. http://dx.doi.org/10.1038/445242a.

34. **Weaver D, Reis MH, Albanese C, Costantini F, Baltimore D, Imanishi-Kari T**. 1986. Altered repertoire of endogenous immunoglobulin gene expression in transgenic mice containing a rearranged mu heavy chain gene. *Cell* **45**:247–259. http://dx.doi.org/10.1016/0092-8674(86)90389-2.

35. **Associated Press**. 2 August 1996. Cleared of fraud, scientist is rehired. *New York Times*. https://www.nytimes.com/1996/08/02/us/cleared-of-fraud-scientist-is-rehired.html.
36. **Anonymous**. 24 July 1996. Imanishi-Kari case ends, but debate on scientific conduct continues. MIT News. https://news.mit.edu/1996/imanishi-0724.
37. **Department of Health and Human Services**. 1996. Thereza Imanishi-Kari, Ph.D., DAB No. 1582 (1996), p. Decision No. 1582. https://www.hhs.gov/sites/default/files/static/dab/decisions/board-decisions/1996/dab1582.html.
38. **Lang S**. 1993. Questions of scientific responsibility: the Baltimore case. *Ethics Behav* **3**:3–72. http://dx.doi.org/10.1207/s15327019eb0301_1.
39. **Kevles D**. 1998. *The Baltimore Case: a Trial of Politics, Science, and Character*. W.W. Norton, New York, NY.
40. **Gunsalus CK**. 1999. Book Review: *The Baltimore case: A trial of politics, science, and character*. *N Engl J Med* **340**:242. http://dx.doi.org/10.1056/NEJM199901213400320.
41. **Shashok K**. 1999. The Baltimore affair: a different view. *Int Microbiol* **2**:275–278.
42. **Sibbison JB**. 1989. The Baltimore dispute. *Lancet* **1**:1148–1149.
43. **Anderson C**. 1991. Imanishi-Kari affair. Baltimore resigns. *Nature* **354**:341.

Chapter 19 – Dismal Science

1. **Stephan PE**. 1996. The economics of science. *J Econ Lit* **34**:1199–1235.
2. **Carlyle T**. 1849. Occasional discourse on the Negro question. XL:670–679. *Fraser's Magazine for Town and Country*.
3. **Epstein J**. 3 September 2019. Economics is dismal, but is it science? *Wall Street Journal*. https://www.wsj.com/articles/economics-is-dismal-but-is-it-science-11567551425.
4. **Levy DM, Peart SJ**. 22 January 2001. The secret history of the dismal science. Part I. Economics, religion and race in the 19th century. EconLib. https://www.econlib.org/library/Columns/Levy Peartdismal.html.
5. **Wikipedia**. 2022. https://en.wikipedia.org/wiki/Economic_system.
6. **Diamond AM, Jr**. 2008. Economics of science. Economics Faculty Publications 36. https://digitalcommons.unomaha.edu/econrealestatefacpub/36.
7. **van Elk R, ter Weel B, van der Wiel K, Wouterse B**. 2019. Estimating the returns to public R&D investments: evidence from production function models. *Economist* **167**:45–87. http://dx.doi.org/10.1007/s10645-019-09331-3.
8. **Penzias AA, Wilson RW**. 1965. A measurement of excess antenna temperature at 4080 Mc/s. *Astrophys J* **142**:419–421. http://dx.doi.org/10.1086/148307.
9. **Bush V**. 1945. *Science, the Endless Frontier—a Report to the President on a Program for Postwar Scientific Research*. United States Government Printing Office, Washington, DC.
10. **Arrow K**. 1962. Economic welfare and the allocation of resources for invention, p 609–626. In **Nelson R** (ed), *The Rate and Direction of Inventive Activity: Economic and Social Factors*. National Bureau of Economic Research, Inc, Washington, DC. http://dx.doi.org/10.1515/9781400879762-024
11. **Flier JS**. 2019. Academia and industry: allocating credit for discovery and development of new therapies. *J Clin Invest* **129**:2172–2174. http://dx.doi.org/10.1172/JCI129122.
12. **Stigler GJ**. 8 December 1982. The process and progress of economics. Nobel Memorial Lecture. https://www.jstor.org/stable/1831067.
13. **Partha D, David PA**. 1994. Towards a new economics of science. *Res Policy* **23**:487–521. http://dx.doi.org/10.1016/0048-7333(94)01002-1.
14. **Merton RK**. 1957. Priorities in scientific discovery: a chapter in the sociology of science. *Am Sociol Rev* **22**:653–659. http://dx.doi.org/10.2307/2089193.
15. **Strevens M**. 2003. The role of the priority rule in science. *J Philos* **100**:55–79. http://dx.doi.org/10.5840/jphil2003100224.
16. **Stephan P**. 2012. *How Economics Shapes Science*. Harvard University Press, Cambridge, MA. http://dx.doi.org/10.4159/harvard.9780674062757
17. **Flier JS**. 2019. Credit and priority in scientific discovery: a scientist's perspective. *Perspect Biol Med* **62**:189–215. http://dx.doi.org/10.1353/pbm.2019.0010.

18. **Casadevall A, Fang FC**. 2015. Field science—the nature and utility of scientific fields. *mBio* **6**:e01259-15. http://dx.doi.org/10.1128/mBio.01259-15.

19. **Moustafa K.** 2014. Don't fall in common science pitfall! *Front Plant Sci* **5**:536. http://dx.doi.org/10.3389/fpls.2014.00536.

20. **Casadevall A, Fang FC**. 2009. Important science—it's all about the SPIN. *Infect Immun* **77**:4177–4180. http://dx.doi.org/10.1128/IAI.00757-09.

21. **Cole S, Cole JR**. 1967. Scientific output and recognition: a study in the operation of the reward system in science. *Am Sociol Rev* **32**:377–390. http://dx.doi.org/10.2307/2091085.

22. **Casadevall A, Fang FC**. 2013. Is the Nobel Prize good for science? *FASEB J* **27**:4682–4690. http://dx.doi.org/10.1096/fj.13-238758.

23. **Azoulay P, Fons-Rosen C, Zivin JSG**. 2019. Does science advance one funeral at a time? *Am Econ Rev* **109**:2889–2920. http://dx.doi.org/10.1257/aer.20161574.

24. **Planck M.** 1948. *Wissenschaftliche Selbstbiographie. Mit einem Bildnis und der von Max von Laue gehaltenen Traueransprache.* Johann Ambrosius Barth Verlag, Leipzig, Germany.

25. **Hirsch JE**. 2005. An index to quantify an individual's scientific research output. *Proc Natl Acad Sci U S A* **102**:16569–16572. http://dx.doi.org/10.1073/pnas.0507655102.

26. **Ioannidis JPA, Baas J, Klavans R, Boyack KW**. 2019. A standardized citation metrics author database annotated for scientific field. *PLoS Biol* **17**:e3000384. http://dx.doi.org/10.1371/journal.pbio.3000384.

27. **Fire M, Guestrin C**. 2019. Over-optimization of academic publishing metrics: observing Goodhart's Law in action. *Gigascience* **8**:giz053. http://dx.doi.org/10.1093/gigascience/giz053.

28. **Brainard J.** May 2001. Elitism, excellence, or both at the National Academy of Sciences? *Chronicle of Higher Education* **47**:A24–A26.

29. **Breithaupt H.** 2005. The prize of discovery. *EMBO Rep* **6**:810–813. http://dx.doi.org/10.1038/sj.embor.7400523.

30. **Tenner E.** 2012. The Nobel Prize as some competition. *Atlantic Monthly.* https://www.theatlantic.com/international/archive/2012/08/the-nobel-prize-has-some-competition/260558/.

31. **Mason JL, Johnston E, Berndt S, Segal K, Lei M, Wiest JS**. 2016. Labor and skills gap analysis of the biomedical research workforce. *FASEB J* **30**:2673–2683. http://dx.doi.org/10.1096/fj.201500067R.

32. **Wosen J.** 2022. "The tipping point is coming": unprecedented exodus of young life scientists is shaking up academia. *STAT.* https://www.statnews.com/2022/11/10/tipping-point-is-coming-unprecedented-exodus-of-young-life-scientists-shaking-up-academia/.

33. **Daniels RJ**. 2015. A generation at risk: young investigators and the future of the biomedical workforce. *Proc Natl Acad Sci U S A* **112**:313–318. http://dx.doi.org/10.1073/pnas.1418761112.

34. **Carroll L.** 1872. *Through the Looking-Glass, and What Alice Found There.* Macmillan and Co, London, United Kingdom.

35. **Casadevall A, Fang FC**. 2014. Causes for the persistence of impact factor mania. *mBio* **5**:e00064-14. http://dx.doi.org/10.1128/mBio.00064-14.

36. **Reich ES**. 2013. Science publishing: the golden club. *Nature* **502**:291–293. http://dx.doi.org/10.1038/502291a.

37. **Oishi S, Kesebir S, Diener E**. 2011. Income inequality and happiness. *Psychol Sci* **22**:1095–1100. http://dx.doi.org/10.1177/0956797611417262.

38. **Merton RK**. 1968. The Matthew effect in science. The reward and communication systems of science are considered. *Science* **159**:56–63. http://dx.doi.org/10.1126/science.159.3810.56.

39. **Fang FC, Steen RG, Casadevall A**. 2012. Misconduct accounts for the majority of retracted scientific publications. *Proc Natl Acad Sci U S A* **109**:17028–17033. http://dx.doi.org/10.1073/pnas.1212247109.

40. **Begley CG, Ellis LM**. 2012. Drug development: raise standards for preclinical cancer research. *Nature* **483**:531–533. http://dx.doi.org/10.1038/483531a.

41. **Prinz F, Schlange T, Asadullah K**. 2011. Believe it or not: how much can we rely on published data on potential drug targets? *Nat Rev Drug Discov* **10**:712. http://dx.doi.org/10.1038/nrd3439-c1.

42. **Bik EM, Casadevall A, Fang FC**. 2016. The prevalence of inappropriate image duplication in biomedical research publications. *mBio* **7**:e00809-16. http://dx.doi.org/10.1128/mBio.00809-16.

43. **Casadevall A, Fang FC**. 2015. Impacted science: impact is not importance. *mBio* **6**:e01593-15. http://dx.doi.org/10.1128/mBio.01593-15.

44. **Alberts B, Kirschner MW, Tilghman S, Varmus H**. 2014. Rescuing US biomedical research from its systemic flaws. *Proc Natl Acad Sci U S A* **111**:5773–5777. http://dx.doi.org/10.1073/pnas.1404402111.

45. **Fang FC, Casadevall A**. 2015. Competitive science: is competition ruining science? *Infect Immun* **83**:1229–1233. http://dx.doi.org/10.1128/IAI.02939-14.

46. **Conway C.** Spring 2015. Science interrupted: how young investigators must sideline science to compete for scarce funding. *UCSF Magazine*. https://magazine.ucsf.edu/science-interrupted.

47. **Woolston C.** 2019. PhDs: the tortuous truth. *Nature* **575**:403–406. http://dx.doi.org/10.1038/d41586-019-03459-7.

48. **Lawrence PA**. 2007. The mismeasurement of science. *Curr Biol* **17**:R583–R585. http://dx.doi.org/10.1016/j.cub.2007.06.014.

49. **Casadevall A, Fang FC**. 2016. Rigorous science: a how-to guide. *mBio* **7**:e1902-16. http://dx.doi.org/10.1128/mBio.01902-16.

50. **Edwards MA, Roy S**. 2017. Academic research in the 21st century: maintaining scientific integrity in a climate of perverse incentives and hypercompetition. *Environ Eng Sci* **34**:51–61. http://dx.doi.org/10.1089/ees.2016.0223.

51. **Wadman M.** 2010. Study says middle sized labs do best. *Nature* **468**:356–357. http://dx.doi.org/10.1038/468356a.

52. **Solly M.** 15 January 2019. DNA pioneer James Watson loses honorary titles over racist comments. *Smithsonian Magazine*. https://www.smithsonianmag.com/smart-news/dna-pioneer-james-watson-loses-honorary-titles-over-racist-comments-180971266/.

53. **Ortega RP**. 2021. National Academy of Sciences ejects biologist Francisco Ayala in the wake of sexual harassment findings. *Science*. https://www.science.org/content/article/national-academy-sciences-ejects-biologist-francisco-ayala-wake-sexual-harassment.

54. **Kaiser J.** 2021. Astronomer Geoff Marcy booted from National Academy of Sciences in wake of sexual harassment. *Science*. https://www.science.org/content/article/astronomer-geoff-marcy-booted-national-academy-sciences-wake-sexual-harassment.

55. **Frontier Economics**. July 2014. *Rates of Return to Investment in Science and Innovation. A Report Prepared for the Department for Business, Innovation and Skills (BIS)*. Frontier Economics, Ltd., London, United Kingdom. https://assets.publishing.service.gov.uk/government/uploads/system/uploads/attachment_data/file/333006/bis-14-990-rates-of-return-to-investment-in-science-and-innovation-revised-final-report.pdf.

56. **Romer PM**. 2001. Should the government subsidize supply or demand in the market for scientists and engineers? *In* **Jaffe AB, Lerner J, Stern S** (ed), *Innovation Policy and the Economy*, vol **1**. MIT Press, Cambridge, MA.

57. **Stevens AJ, Jensen JJ, Wyller K, Kilgore PC, Chatterjee S, Rohrbaugh ML**. 2011. The role of public-sector research in the discovery of drugs and vaccines. *N Engl J Med* **364**:535–541. http://dx.doi.org/10.1056/NEJMsa1008268.

58. **Singer PL**. February 2014. Federally supported innovations: 22 examples of major technology advances that stem from federal research support. Information Technology & Innovation Foundation. http://www2.itif.org/2014-federally-supported-innovations.pdf.

59. **Patridge EV, Gareiss PC, Kinch MS, Hoyer DW**. 2015. An analysis of original research contributions toward FDA-approved drugs. *Drug Discov Today* **20**:1182–1187. http://dx.doi.org/10.1016/j.drudis.2015.06.006.

60. **Kocher MG, Luptacik M, Sutter M**. 2001. Measuring productivity of research in economics. A cross-country study using DEA. Department of Economics Working Paper Series 77. WU Vienna University of Economics and Business, Vienna, Austria.

61. **Wootton R.** 2013. A simple, generalizable method for measuring individual research productivity and its use in the long-term analysis of departmental performance, including between-country comparisons. *Health Res Policy Syst* **11**:2. http://dx.doi.org/10.1186/1478-4505-11-2.

62. **Casadevall A, Fang FC**. 2010. Reproducible science. *Infect Immun* **78**:4972–4975. http://dx.doi. org/10.1128/IAI.00908-10.

63. **Casadevall A, Ellis LM, Davies EW, McFall-Ngai M, Fang FC**. 2016. A framework for improving the quality of research in the biological sciences. *mBio* **7**:e01256-16. http://dx.doi.org/10.1128/ mBio.01256-16.

64. **Casadevall A, Fang FC**. 2016. Revolutionary science. *mBio* **7**:e00158. http://dx.doi.org/10.1128/ mBio.00158-16.

65. **Abbott BP, et al; LIGO Scientific Collaboration and Virgo Collaboration**. 2016. Observation of gravitational waves from a binary black hole merger. *Phys Rev Lett* **116**:061102. http://dx.doi. org/10.1103/PhysRevLett.116.061102.

66. **Lawrence PA**. 2002. Rank injustice. *Nature* **415**:835–836. http://dx.doi.org/10.1038/415835a.

67. **Enserink M, Travis J**. 2011. Nobel prize for immunologists provokes yet another debate. *Science*. https://www.science.org/content/article/nobel-prize-immunologists-provokes-yet-another-debate.

68. **Lederman L, Teresi D**. 1993. *The God Particle: If the Universe Is the Answer, What Is the Question?* Dell Publishing, New York, NY. http://dx.doi.org/10.1063/1.2808974

69. **Dickerson K**. 20 May 2015. Here's what scientists really wanted to call the world's most famous particle. *Business Insider*. https://www.businessinsider.com/why-the-higgs-is-called-the-god-particle-2015-5.

70. **Eveleth R**. 23 April 2013. Should the Higgs boson be renamed to credit more scientists? *Smithsonian Magazine*. https://www.smithsonianmag.com/smart-news/should-the-higgs-boson-be-renamed-to-credit-more-scientists-39147793/.

71. **Higgs P**. 1964. Broken symmetries and the masses of gauge bosons. *Phys Rev Lett* **13**:508–509. http://dx.doi.org/10.1103/PhysRevLett.13.508.

72. **Ellis J**. 2012. Credit due to Peter Higgs. *Science* **338**:740–741. http://dx.doi.org/10.1126/science. 338.6108.740-b.

73. **Arvan M**. 31 May 2019. Eight papers rejected before winning the Nobel Prize. The Philosopher's Cocoon. https://philosopherscocoon.typepad.com/blog/2019/05/8-papers-rejected-before-winning-the-nobel-prize.html.

74. **Aitkenhead D**. 6 December 2013. Peter Higgs interview: "I have this kind of underlying incompetence," *The Guardian*. https://www.theguardian.com/science/2013/dec/06/peter-higgs-interview-underlying-incompetence.

Chapter 20 – Competitive Science

1. **Gleick J**. 1987. *Chaos: Making a New Science*. Vintage, New York, NY.

2. **Merton RK**. 1957. Priorities in scientific discovery: a chapter in the sociology of science. *Am Sociol Rev* **22**:653–659. http://dx.doi.org/10.2307/2089193.

3. **Stephan P**. 2012. *How Economics Shapes Science*. Harvard University Press, Cambridge, MA. http://dx.doi.org/10.4159/harvard.9780674062757

4. **Ridley M**. 27 July 2012. Three cheers for scientific backbiting. *Wall Street Journal*. http://online.wsj. com/articles/SB10001424052702304039104577534830901741156.

5. **Hagstrom WO**. 1974. Competition in science. *Am Sociol Rev* **39**:1–18. http://dx.doi.org/10.2307/ 2094272.

6. **Sullivan D**. 1975. Competition in bio-medical science: extent, structure, and consequences. *Sociol Educ* **48**:223–241. http://dx.doi.org/10.2307/2112477.

7. **McCain KW**. 1991. Communication, competition and secrecy: the production and dissemination of research-related information in genetics. *Sci Technol Human Values* **16**:491–516. http://dx.doi. org/10.1177/016224399101600404.

8. **Anderson MS, Ronning EA, De Vries R, Martinson BC**. 2007. The perverse effects of competition on scientists' work and relationships. *Sci Eng Ethics* **13**:437–461. http://dx.doi.org/10.1007/ s11948-007-9042-5.

9. **Alberts B, Kirschner MW, Tilghman S, Varmus H**. 2014. Rescuing US biomedical research from its systemic flaws. *Proc Natl Acad Sci U S A* **111**:5773–5777. http://dx.doi.org/10.1073/pnas.1404402111.

10. **Casadevall A, Fang FC.** 2012. Winner takes all. *Sci Am* **307**:13. http://dx.doi.org/10.1038/scientificamerican0812-13.

11. **Adamo SA.** 2013. Attrition of women in the biological sciences: workload, motherhood, and other explanations revisited. *Bioscience* **63**:43–48. http://dx.doi.org/10.1525/bio.2013.63.1.9.

12. **Fang FC, Casadevall A.** 2012. Reforming science: structural reforms. *Infect Immun* **80**:897–901. http://dx.doi.org/10.1128/IAI.06184-11.

13. **Lober Newsome J.** 2008. *The Chemistry PhD: the Impact on Women's Retention.* Royal Society of Chemistry, London, United Kingdom.

14. **Heffernan M.** 2014. *A Bigger Prize: How We Can Do Better Than the Competition.* Public Affairs, New York, NY.

15. **Batts S.** 21 November 2006. An in-depth look at MIT's TonegawaGate emails. Science Blogs. http://scienceblogs.com/retrospectacle/2006/11/21/an-indepth-look-at-mits-tonega/.

16. **Anonymous.** 2011. Combating scientific misconduct. *Nat Cell Biol* **13**:1. http://dx.doi.org/10.1038/ncb0111-1.

17. **Interlandi J.** 22 October 2006. An unwelcome discovery. Walter DeNino was a young lab technician who analyzed data for his mentor, Eric Poehlman. What he found was that Poehlman was not the scientist he appeared to be, p 98–103. *New York Times Magazine.* https://www.ncbi.nlm.nih.gov/pubmed/17115505.

18. **Fanelli D.** 2010. Do pressures to publish increase scientists' bias? An empirical support from US States Data. *PLoS One* **5**:e10271. http://dx.doi.org/10.1371/journal.pone.0010271.

19. **Ioannidis JP.** 2011. An epidemic of false claims. Competition and conflicts of interest distort too many medical findings. *Sci Am* **304**:16. http://dx.doi.org/10.1038/scientificamerican0611-16.

20. **Martinson BC, Crain AL, Anderson MS, De Vries R.** 2009. Institutions' expectations for researchers' self-funding, federal grant holding, and private industry involvement: manifold drivers of self-interest and researcher behavior. *Acad Med* **84**:1491–1499. http://dx.doi.org/10.1097/ACM.0b013e3181bb2ca6.

21. **Tijdink JK, Verbeke R, Smulders YM.** 2014. Publication pressure and scientific misconduct in medical scientists. *J Empir Res Hum Res Ethics* **9**:64–71. http://dx.doi.org/10.1177/1556264614552421.

22. **von Behring E, Kitasato S.** 1991. The mechanism of diphtheria immunity and tetanus immunity in animals. 1890. *Mol Immunol* **28**:1317, 1319–1320. (In German.)

23. **Griffith F.** 1928. The significance of pneumococcal types. *J Hyg (Lond)* **27**:113–159. http://dx.doi.org/10.1017/S0022172400031879.

24. **Mullis K, Faloona F, Scharf S, Saiki R, Horn G, Erlich H.** 1986. Specific enzymatic amplification of DNA *in vitro*: the polymerase chain reaction. *Cold Spring Harb Symp Quant Biol* **51**:263–273. http://dx.doi.org/10.1101/SQB.1986.051.01.032.

25. **Roe B.** 2014. We need undirected research. *APS News* **23**:8.

26. **White M.** 2001. *Acid Tongues and Tranquil Dreamers.* HarperCollins, New York, NY.

27. **Wallace AR.** 1889. *Darwinism: an Exposition on the Theory of Natural Selection with Some of Its Applications.* Macmillan & Co, London, United Kingdom. http://dx.doi.org/10.5962/bhl.title.2472.

28. **Livio M.** 2013. *Brilliant Blunders: from Darwin to Einstein.* Simon and Schuster, New York, NY.

29. **Watson JD.** 1968. *The Double Helix: a Personal Account of the Discovery of the Structure of DNA.* Atheneum, New York, NY.

30. **Weiner J.** 2012. Laboratory confidential. *Columbia J Rev* **51**:48.

31. **Byrne JP.** 2012. *Encyclopedia of the Black Death.* ABC-CLIO, Santa Barbara, CA.

32. **Kitasato S.** 1894. The bacillus of bubonic plague. *Lancet* **ii**:428–430. http://dx.doi.org/10.1016/S0140-6736(01)58670-5.

33. **Bibel DJ, Chen TH.** 1976. Diagnosis of plague: an analysis of the Yersin-Kitasato controversy. *Bacteriol Rev* **40**:633–651. http://dx.doi.org/10.1128/br.40.3.633-651.1976.

34. **Solomon T.** 1997. Hong Kong, 1894: the role of James A Lowson in the controversial discovery of the plague bacillus. *Lancet* **350**:59–62. http://dx.doi.org/10.1016/S0140-6736(97)01438-4.

35. **Yersin A.** 1894. Le peste bubonique a Hong Kong. *Ann Inst Pasteur (Paris)* **8**:662–667.
36. **Cakulev I, Efimov IR, Waldo AL.** 2009. Cardioversion: past, present, and future. *Circulation* **120**:1623–1632. http://dx.doi.org/10.1161/CIRCULATIONAHA.109.865535.
37. **Casadevall A, Fang FC.** 2013. Is the Nobel Prize good for science? *FASEB J* **27**:4682–4690. http://dx.doi.org/10.1096/fj.13-238758.
38. **Shendure J, Mitra RD, Varma C, Church GM.** 2004. Advanced sequencing technologies: methods and goals. *Nat Rev Genet* **5**:335–344. http://dx.doi.org/10.1038/nrg1325.
39. **Templeton G.** 23 July 2013. Challenges unmet: unclaimed prizes in science and technology. Geek. http://www.geek.com/science/challenges-unmet-unclaimed-prizes-in-science-and-technology-1562333/.
40. **Disis ML, Slattery JT.** 2010. The road we must take: multidisciplinary team science. *Sci Transl Med* **2**:22cm9. http://dx.doi.org/10.1126/scitranslmed.3000421.
41. **Balch C, Arias-Pulido H, Banerjee S, Lancaster AK, Clark KB, Perilstein M, Hawkins B, Rhodes J, Sliz P, Wilkins J, Chittenden TW.** 2015. Science and technology consortia in U.S. biomedical research: a paradigm shift in response to unsustainable academic growth. *BioEssays* **37**:119–122. http://dx.doi.org/10.1002/bies.201400167.
42. **Milojević S.** 2014. Principles of scientific research team formation and evolution. *Proc Natl Acad Sci U S A* **111**:3984–3989. http://dx.doi.org/10.1073/pnas.1309723111.
43. **Shen HW, Barabási AL.** 2014. Collective credit allocation in science. *Proc Natl Acad Sci U S A* **111**:12325–12330. http://dx.doi.org/10.1073/pnas.1401992111.
44. **Sawyer RK.** 2012. *Explaining Creativity: the Science of Human Innovation.* Oxford University Press, Oxford, United Kingdom.
45. **Friedman J.** 1 May 1999. Creativity in science. American Council of Learned Societies Occasional Paper No. 47. http://archives.acls.org/op/op47-2.htm.
46. **Andreasen NC, Ramchandran K.** 2012. Creativity in art and science: are there two cultures? *Dialogues Clin Neurosci* **14**:49–54. http://dx.doi.org/10.31887/DCNS.2012.14.1/nandreasen.
47. **Jabr F.** 2013. Why your brain needs more downtime. *Scientific American.* http://www.scientificamerican.com/article/mental-downtime/.
48. **Hennessey BA, Amabile TM.** 2010. Creativity. *Annu Rev Psychol* **61**:569–598. http://dx.doi.org/10.1146/annurev.psych.093008.100416.
49. **Amabile TM.** 1996. *Creativity in Context.* Westview Press, Boulder, CO.
50. **Amabile TM, Hennessey BA, Grossman BS.** 1986. Social influences on creativity: the effects of contracted-for reward. *J Pers Soc Psychol* **50**:14–23. http://dx.doi.org/10.1037/0022-3514.50.1.14.
51. **Amabile TM.** 1998. How to kill creativity. *Harv Bus Rev* **76**:76–87, 186.
52. **Amabile TM, Hadley CN, Kramer SJ.** 2002. Creativity under the gun. *Harv Bus Rev* **80**:52–61, 147.
53. **Rick S, Loewenstein G.** 2008. Hypermotivation. *J Mark Res* **45**:645–653.
54. **Byron K, Khazanchi S, Nazarian D.** 2010. The relationship between stressors and creativity: a meta-analysis examining competing theoretical models. *J Appl Psychol* **95**:201–212. http://dx.doi.org/10.1037/a0017868.
55. **Baer M, Vadera AK, Leenders RT, Oldham GR.** 2014. Intergroup competition as a double-edged sword: how sex composition regulates the effects of competition on group creativity. *Organizat Sci* **25**:892–908. http://dx.doi.org/10.1287/orsc.2013.0878.
56. **McDonald MM, Navarrete CD, Van Vugt M.** 2012. Evolution and the psychology of intergroup conflict: the male warrior hypothesis. *Philos Trans R Soc Lond B Biol Sci* **367**:670–679. http://dx.doi.org/10.1098/rstb.2011.0301.
57. **Del Giudice M.** 8 November 2014. Why it's crucial to get more women into science. *National Geographic.* https://www.nationalgeographic.com/culture/article/141107-gender-studies-women-scientific-research-feminist.
58. **Johnson S.** 2010. *Where Good Ideas Come From: the Natural History of Innovation.* Penguin, New York, NY.
59. **Alon U.** 2010. How to build a motivated research group. *Mol Cell* **37**:151–152. http://dx.doi.org/10.1016/j.molcel.2010.01.011.

60. **Casadevall A, Fang FC**. 2014. Causes for the persistence of impact factor mania. *mBio* **5**:e00064-14. http://dx.doi.org/10.1128/mBio.00064-14.

61. **Navarre WW, Porwollik S, Wang Y, McClelland M, Rosen H, Libby SJ, Fang FC**. 2006. Selective silencing of foreign DNA with low GC content by the H-NS protein in *Salmonella. Science* **313**:236–238. http://dx.doi.org/10.1126/science.1128794.

62. **Lucchini S, Rowley G, Goldberg MD, Hurd D, Harrison M, Hinton JC**. 2006. H-NS mediates the silencing of laterally acquired genes in bacteria. *PLoS Pathog* **2**:e81. http://dx.doi.org/10.1371/journal.ppat.0020081.

63. **Alvarez M, Casadevall A**. 2006. Phagosome extrusion and host-cell survival after *Cryptococcus neoformans* phagocytosis by macrophages. *Curr Biol* **16**:2161–2165. http://dx.doi.org/10.1016/j.cub.2006.09.061.

64. **Ma H, Croudace JE, Lammas DA, May RC**. 2006. Expulsion of live pathogenic yeast by macrophages. *Curr Biol* **16**:2156–2160. http://dx.doi.org/10.1016/j.cub.2006.09.032.

65. **Alvarez M, Casadevall A**. 2007. Cell-to-cell spread and massive vacuole formation after *Cryptococcus neoformans* infection of murine macrophages. *BMC Immunol* **8**:16. http://dx.doi.org/10.1186/1471-2172-8-16.

66. **Ma H, Croudace JE, Lammas DA, May RC**. 2007. Direct cell-to-cell spread of a pathogenic yeast. *BMC Immunol* **8**:15. http://dx.doi.org/10.1186/1471-2172-8-15.

67. **Park SY, Fung P, Nishimura N, Jensen DR, Fujii H, Zhao Y, Lumba S, Santiago J, Rodrigues A, Chow TF, Alfred SE, Bonetta D, Finkelstein R, Provart NJ, Desveaux D, Rodriguez PL, McCourt P, Zhu JK, Schroeder JI, Volkman BF, Cutler SR**. 2009. Abscisic acid inhibits type 2C protein phosphatases via the PYR/PYL family of START proteins. *Science* **324**:1068–1071. http://dx.doi.org/10.1126/science.1173041.

68. **Zaragoza O, García-Rodas R, Nosanchuk JD, Cuenca-Estrella M, Rodríguez-Tudela JL, Casadevall A**. 2010. Fungal cell gigantism during mammalian infection. *PLoS Pathog* **6**:e1000945. http://dx.doi.org/10.1371/journal.ppat.1000945.

69. **Okagaki LH, Strain AK, Nielsen JN, Charlier C, Baltes NJ, Chrétien F, Heitman J, Dromer F, Nielsen K**. 2010. Cryptococcal cell morphology affects host cell interactions and pathogenicity. *PLoS Pathog* **6**:e1000953. http://dx.doi.org/10.1371/journal.ppat.1000953.

70. **Barbosa-Morais NL, Irimia M, Pan Q, Xiong HY, Gueroussov S, Lee LJ, Slobodeniuc V, Kutter C, Watt S, Colak R, Kim T, Misquitta-Ali CM, Wilson MD, Kim PM, Odom DT, Frey BJ, Blencowe BJ**. 2012. The evolutionary landscape of alternative splicing in vertebrate species. *Science* **338**:1587–1593. http://dx.doi.org/10.1126/science.1230612.

71. **Merkin J, Russell C, Chen P, Burge CB**. 2012. Evolutionary dynamics of gene and isoform regulation in Mammalian tissues. *Science* **338**:1593–1599. http://dx.doi.org/10.1126/science.1228186.

72. **Vernot B, Akey JM**. 2014. Resurrecting surviving Neandertal lineages from modern human genomes. *Science* **343**:1017–1021. http://dx.doi.org/10.1126/science.1245938.

73. **Sankararaman S, Mallick S, Dannemann M, Prüfer K, Kelso J, Pääbo S, Patterson N, Reich D**. 2014. The genomic landscape of Neanderthal ancestry in present-day humans. *Nature* **507**:354–357. http://dx.doi.org/10.1038/nature12961.

74. **Azoulay P, Graff Zivin JS, Manso G**. 2009. Incentives and creativity: evidence from the academic life sciences. National Bureau of Economic Research Working Paper Series No. 15466.

75. **Danthi N, Wu CO, Shi P, Lauer M**. 2014. Percentile ranking and citation impact of a large cohort of National Heart, Lung, and Blood Institute-funded cardiovascular R01 grants. *Circ Res* **114**:600–606. http://dx.doi.org/10.1161/CIRCRESAHA.114.302656.

76. **Kaltman JR, Evans FJ, Danthi NS, Wu CO, DiMichele DM, Lauer MS**. 2014. Prior publication productivity, grant percentile ranking, and topic-normalized citation impact of NHLBI cardiovascular R01 grants. *Circ Res* **115**:617–624. http://dx.doi.org/10.1161/CIRCRESAHA.115.304766.

77. **Offit P**. 2006. *The Cutter Incident: How America's First Polio Vaccine Led to a Growing Vaccine Crisis.* Yale University Press, New Haven, CT.

78. **Ciapponi A, Bardach A, Rey Ares L, Glujovsky D, Cafferata ML, Cesaroni S, Bhatti A.** 2019. Sequential inactivated (IPV) and live oral (OPV) poliovirus vaccines for preventing poliomyelitis. *Cochrane Database Syst Rev* **12**:CD011260. http://dx.doi.org/10.1002/14651858.CD011260.pub2.
79. **Planck M.** 1948. *Wissenschaftliche Selbstbiographie. Mit einem Bildnis und der von Max von Laue gehaltenen Traueransprache.* Johann Ambrosius Barth Verlag, Leipzig, Germany.

Chapter 21 – Prized Science

1. **Friedman RM.** 2001. *The Politics of Excellence: Behind the Nobel Prize in Science.* Times Books, New York, NY.
2. **Casadevall A, Fang FC.** 2012. Winner takes all. *Sci Am* **307**:13. http://dx.doi.org/10.1038/scientificamerican0812-13.
3. **Sobel D.** 1995. *Longitude: the True Story of a Lone Genius Who Solved the Greatest Scientific Problem of His Time.* Walker & Co, New York, NY.
4. **Ma Y, Uzzi B.** 2018. Scientific prize network predicts who pushes the boundaries of science. *Proc Natl Acad Sci U S A* **115**:12608–12615. http://dx.doi.org/10.1073/pnas.1800485115.
5. **Popkin G.** 16 June 2022. Prizes are not always a win for science. *Physics.* https://physics.aps.org/articles/v15/86?fbclid=IwAR3RrhfO3PAZj4tdaY3CmRyKxfYIhGowh_w4RbW92f3ij1RrBHqS03QnX8w.
6. **Andrews E.** Update 23 July 2020. Did a premature obituary inspire the Nobel Prize? History.com. https://www.history.com/news/did-a-premature-obituary-inspire-the-nobel-prize.
7. **Nobel A.** 2007. Alfred Nobel's will. *Ann Noninvasive Electrocardiol* **12**:81–82. http://dx.doi.org/10.1111/j.1542-474X.2007.00142.x.
8. **Moss AJ.** 2007. Introductory note to Alfred Nobel's will. *Ann Noninvasive Electrocardiol* **12**:79–80. http://dx.doi.org/10.1111/j.1542-474X.2007.00141.x.
9. **Casadevall A, Fang FC.** 2009. Important science—it's all about the SPIN. *Infect Immun* **77**: 4177–4180. http://dx.doi.org/10.1128/IAI.00757-09.
10. **Heilbron JL.** 1987. H. Moseley and the Nobel Prize. *Nature* **330**:694. http://dx.doi.org/10.1038/330694a0.
11. **Kahn CR, Roth J.** 2004. Berson, Yalow, and the JCI: the agony and the ecstasy. *J Clin Invest* **114**: 1051–1054. http://dx.doi.org/10.1172/JCI23316.
12. **Allison JP, Benoist C, Chervonsky AV.** 2011. Nobels: Toll pioneers deserve recognition. *Nature* **479**:178. http://dx.doi.org/10.1038/479178a.
13. **Purkis C.** 2012. The commemoration of the Strindberg centenary of 1949 in Britain, vol 2. Swedish Book Review.
14. **Feynman R.** 24 January 1989. The last journey of a genius. *NOVA.* Season 16, episode 13.
15. **Mazloumian A, Eom YH, Helbing D, Lozano S, Fortunato S.** 2011. How citation boosts promote scientific paradigm shifts and Nobel Prizes. *PLoS One* **6**:e18975. http://dx.doi.org/10.1371/journal.pone.0018975.
16. **Huber J, Inoua S, Kerschbamer R, König-Kersting C, Palan S, Smith VL.** 2022. Nobel and novice: author prominence affects peer review. *Proc Natl Acad Sci U S A* **119**:e2205779119. http://dx.doi.org/10.1073/pnas.2205779119.
17. **Rablen MD, Oswald AJ.** 2008. Mortality and immortality: the Nobel Prize as an experiment into the effect of status upon longevity. *J Health Econ* **27**:1462–1471. http://dx.doi.org/10.1016/j.jhealeco.2008.06.001.
18. **Ploegh HL.** 2012. A dark edge to the glory. *Nature* **486**:318–319. http://dx.doi.org/10.1038/486318a.
19. **Coffey P.** 2008. *Cathedrals of Science.* Oxford University Press, Oxford, United Kingdom.
20. **Chan HF, Gleeson L, Torgler B.** 2014. Awards before and after the Nobel Prize: a Matthew effect and/or a ticket to one's own funeral? *Res Eval* **23**:210–220. http://dx.doi.org/10.1093/reseval/rvu011.

21. **Hamming R.** 7 March 1986. You and your research. Transcription of the Bell Communications Research Colloquium Seminar, 7 March 1986. https://www.cs.virginia.edu/~robins/YouAndYourResearch.html.

22. **Diamandis EP.** 2013. Nobelitis: a common disease among Nobel laureates? *Clin Chem Lab Med* **51**:1573–1574. http://dx.doi.org/10.1515/cclm-2013-0273.

23. **Basterfield C, Lilienfeld SO, Bowes SM, Costello TH.** 2020. The Nobel disease: when intelligence fails to protect against irrationality. *Skeptical Inquirer* **44**:32–37.

24. **Saxon W.** 14 August 1989. Obituary: William B. Shockley, 79, creator of transistor and theory on race. *New York Times.* http://www.nytimes.com/learning/general/onthisday/bday/0213.html.

25. **Johnson G.** 28 October 2007. Bright scientists, dim notions. *New York Times.* http://www.nytimes.com/2007/10/28/weekin review/28johnson.html?_r=0.

26. **Solly M.** 15 January 2019. DNA pioneer James Watson loses honorary titles over racist comments. *Smithsonian Magazine.* https://www.smithsonianmag.com/smart-news/dna-pioneer-james-watson-loses-honorary-titles-over-racist-comments-180971266/.

27. **Strauss M.** 6 October 2015. Nobel Laureates who were not always noble. *National Geographic.* https://www.nationalgeographic.com/science/article/151005-nobel-laureates-forget-racist-sexist-science.

28. **Engineering and Public Policy Committee on Science.** 2004. *Facilitating Interdisciplinary Research.* National Academies Press, Washington, DC.

29. **Casadevall A, Pirofski LA.** 2006. A reappraisal of humoral immunity based on mechanisms of antibody-mediated protection against intracellular pathogens. *Adv Immunol* **91**:1–44. http://dx.doi.org/10.1016/S0065-2776(06)91001-3.

30. **Rhodes R.** 1998. *Deadly Feasts.* Simon and Schuster, New York, NY.

31. **Coutinho M, Freire O, Jr, Dias JC.** 1999. The noble enigma: Chagas' nominations for the Nobel prize. *Mem Inst Oswaldo Cruz* **94**(Suppl 1):123–129. http://dx.doi.org/10.1590/S0074-02761999000700012.

32. **Bestetti RB, Cardinalli-Neto A.** 2013. Dissecting slander and crying for justice: Carlos Chagas and the Nobel Prize of 1921. *Int J Cardiol* **168**:2328–2334. http://dx.doi.org/10.1016/j.ijcard.2013.01.048.

33. **Anonymous.** 1975. The reward system needs overhauling. *Nature* **254**:277. http://dx.doi.org/10.1038/254277a0.

34. **Meyers MA.** 2012. *Prize Fight: the Race and the Rivalry to be the First in Science,* Palgrave Macmillan, New York, NY.

35. **Kantha SS.** 1991. A centennial review; the 1890 tetanus antitoxin paper of von Behring and Kitasato and the related developments. *Keio J Med* **40**:35–39. http://dx.doi.org/10.2302/kjm.40.35.

36. **Eisenberg RL.** 1992. Cathode rays and controversy. *AJR Am J Roentgenol* **159**:996. http://dx.doi.org/10.2214/ajr.159.5.1414813.

37. **Lawrence PA.** 2002. Rank injustice. *Nature* **415**:835–836. http://dx.doi.org/10.1038/415835a.

38. **Wainwright M.** 2005. A response to William Kingston, "Streptomycin, Schatz versus Waksman, and the balance of credit for discovery." *J Hist Med Allied Sci* **60**:218–220, discussion 221. http://dx.doi.org/10.1093/jhmas/jri024.

39. **Pringle P.** 2012. *Experiment Eleven: Deceit and Betrayal in the Discovery of the Cure for Tuberculosis.* Bloomsbury UK, London, United Kingdom.

40. **Wainwright M.** 1989. Nobel infallibility. *Nature* **342**:336. http://dx.doi.org/10.1038/342336d0.

41. **Lawrence PA.** 2012. Rank, invention and the Nobel Prize. *Curr Biol* **22**:R214–R216. http://dx.doi.org/10.1016/j.cub.2012.02.053.

42. **Broad WJ.** 1982. Toying with the truth to win a Nobel. *Science* **217**:1120–1122. http://dx.doi.org/10.1126/science.7051287.

43. **Dreizen P.** 2004. The Nobel prize for MRI: a wonderful discovery and a sad controversy. *Lancet* **363**:78. http://dx.doi.org/10.1016/S0140-6736(03)15182-3.

44. **Macchia RJ, Termine JE, Buchen CD.** 2007. Raymond V. Damadian, M.D.: magnetic resonance imaging and the controversy of the 2003 Nobel Prize in Physiology or Medicine. *J Urol* **178**:783–785. http://dx.doi.org/10.1016/j.juro.2007.05.019.

45. **Pearson H.** 2003. Physician launches public protest over medical Nobel. *Nature* **425**:648. http://dx.doi.org/10.1038/425648b.

46. **Zuckerman H.** 1977. *Scientific Elite: Nobel Laureates in the United States.* Free Press, New York, NY.

47. **Faerstein E, Winkelstein W, Jr.** 2010. Carlos Juan Finlay: rejected, respected, and right. *Epidemiology* **21**:158. http://dx.doi.org/10.1097/EDE.0b013e3181c308e0.

48. **Jorgensen TJ.** 7 February 2019. Lise Meitner—the forgotten woman of nuclear physics who deserved a Nobel Prize. The Conversation. https://theconversation.com/lise-meitner-the-forgotten-woman-of-nuclear-physics-who-deserved-a-nobel-prize-106220.

49. **Reichard P.** 2002. Osvald T. Avery and the Nobel Prize in medicine. *J Biol Chem* **277**:13355–13362. http://dx.doi.org/10.1074/jbc.R200002200.

50. **Portugal F.** 2010. Oswald T. Avery: Nobel Laureate or noble luminary? *Perspect Biol Med* **53**:558–570. http://dx.doi.org/10.1353/pbm.2010.0014.

51. **Christianson GE.** 1996. *Edwin Hubble: Mariner of the Nebulae.* University of Chicago Press, Chicago, IL.

52. **Holden C.** 1989. Chauvinism in Nobel nominations. *Science* **243**:471. http://dx.doi.org/10.1126/science.243.4890.471.b.

53. **Kuhn TS.** 1962. *The Structure of Scientific Revolutions.* University of Chicago Press, Chicago, IL.

54. **Feynman R.**1999. *The Pleasure of Finding Things Out.* Perseus, New York, NY.

55. **Rees M.** 27 October 2022. The problem with the Nobel Prizes. *Time.* https://time.com/6225572/nobel-prizes-problem/.

56. **Yong E.** 3 October 2017. The absurdity of the Nobel Prizes in science. *Atlantic Monthly.* https://www.theatlantic.com/science/archive/2017/10/the-absurdity-of-the-nobel-prizes-in-science/541863/.

57. **Raven P, King D, De Waal F, Brilliant L, Brooks R, Diamandis P, Hunt T, Margulis L, Pinker S, Wilson EO.** 30 September 2009. Open letter to the Nobel prize committee. *New Scientist.* https://www.newscientist.com/article/dn17863-open-letter-to-the-nobel-prize-committee/.

58. **Popkin G.** 3 October 2016. Update the Nobel Prizes. *New York Times.* https://www.nytimes.com/2016/10/03/opinion/update-the-nobel-prizes.html.

59. **Anonymous.** 2012. Solve the Nobel Prize dilemma. Now that teams, not individuals, drive high-impact science, the Nobel Foundation should change how it awards its prize. *Sci Am* **307**:12.

60. **Abad-Santos A.** 2012. How could the Higgs boson lose the Nobel Prize? Atlantic Monthly. http://m.theatlanticwire.com/ global/2012/10/how-could-higgs-boson-lose-nobel-prize/322734/.

61. **Al-Khalili J.** 8 October 2012. Why the Nobel Prizes need a shakeup. *The Guardian.* https://www.theguardian.com/commentisfree/2012/oct/08/nobel-prizes-need-shakeup.

62. **Blow NS.** 2012. Prizes in an age of collaborative research. *Biotechniques* **52**:11. http://dx.doi.org/10.2144/000113792.

63. **Clery D.** 2013. Higgs theorists win physics Nobel in overtime. *Science.* https://www.science.org/content/article/higgs-theorists-win-physics-nobel-overtime.

64. **Devlin H, Sample I.** 3 October 2017. Nobel prize in physics awarded for discovery of gravitational waves. *The Guardian.* https://www.theguardian.com/science/2017/oct/03/nobel-prize-physics-discovery-gravitational-waves-ligo.

65. **Ball P.** 2021. The CRISPR wars. *Lancet* **397**:1340–1341. http://dx.doi.org/10.1016/S0140-6736(21)00774-1.

66. **Su XZ, Miller LH.** 2015. The discovery of artemisinin and the Nobel Prize in Physiology or Medicine. *Sci China Life Sci* **58**:1175–1179. http://dx.doi.org/10.1007/s11427-015-4948-7.

67. **Mueller B.** 3 October 2022. Nobel Prize awarded to scientist who sequenced Neanderthal genome. *New York Times.* https://www.nytimes.com/2022/10/03/health/nobel-prize-medicine-physiology-winner.html.

68. **Gozum M.** 2009. An award for science is an obsolete notion. *Science* **323**:207–208. http://dx.doi.org/10.1126/science.323.5911.207.

69. **Anonymous.** 1971. Three research teams receive Rumford award. *Phys Today* **24**:69. http://dx.doi.org/10.1063/1.3022823.

70. **Fundación Princesa de Asturias**. 2023. Regulations. http://www.fpa.es/en/prince-of-asturias-awards/regulations/.
71. **Doherty PC.** 2 October 2011. Nobel Prize means more for media than for science—a personal account. The Conversation. https://theconversation.com/nobel-prize-means-more-for-media-than-for-science-a-personal-account-3599.
72. **Chargaff E.** 1994. Chemical specificity of nucleic acids and mechanism of their enzymatic degradation. 1950. *Experientia* **50**:368–376.
73. **Avery OT, Macleod CM, McCarty M**. 1944. Studies on the chemical nature of the substance inducing transformation of pneumococcal types: induction of transformation by a desoxyribonucleic acid fraction isolated from pneumococcus type III. *J Exp Med* **79**:137–158. http://dx.doi.org/10.1084/jem.79.2.137.
74. **Watson JD, Crick FH**. 1953. Molecular structure of nucleic acids; a structure for deoxyribose nucleic acid. *Nature* **171**:737–738. http://dx.doi.org/10.1038/171737a0.
75. **Franklin RE, Gosling RG**. 1953. Molecular configuration in sodium thymonucleate. *Nature* **171**:740–741. http://dx.doi.org/10.1038/171740a0.
76. **Wilkins MH, Stokes AR, Wilson HR**. 1953. Molecular structure of deoxypentose nucleic acids. *Nature* **171**:738–740. http://dx.doi.org/10.1038/171738a0.
77. **Astbury WT, Bell FO**. 1938. X-ray study of thymonucleic acid. *Nature* **141**:747–748. http://dx.doi.org/10.1038/141747b0.
78. **Pray LA**. 2008. Discovery of DNA structure and function: Watson and Crick. *Nat Educ* **1**:100. https://www.nature.com/scitable/topicpage/discovery-of-dna-structure-and-function-watson-397/.
79. **Lee JJ.** 19 May 2013. Six women scientists who were snubbed due to sexism. *National Geographic.* https://www.nationalgeographic.com/culture/article/130519-women-scientists-overlooked-dna-history-science.
80. **Anonymous**. 22 March 1975. Hoyle disputes Nobel physics award. *New York Times.* https://www.nytimes.com/1975/03/22/archives/hoyle-disputes-nobel-physics-award.html.
81. **McKie R.** 2 October 2010. Fred Hoyle: the scientist whose rudeness cost him a Nobel Prize. *The Guardian.* https://www.theguardian.com/science/2010/oct/03/fred-hoyle-nobel-prize.
82. **Sample I.** 6 September 2018. British astrophysicist overlooked by Nobels wins $3M award for pulsar work. *The Guardian.* https://www.theguardian.com/science/2ambrid/06/jocelyn-bell-burnell-british-astrophysicist-overlooked-by-nobels-3m-award-pulsars.
83. **Baloh RW**. 2002. Robert Bárány and the controversy surrounding his discovery of the caloric reaction. *Neurology* **58**:1094–1099. http://dx.doi.org/10.1212/WNL.58.7.1094.
84. **Habashi F.** 2004. The social responsibilities of scientists. *CIM Bull* **97**:77–82.
85. **Rosenfeld L**. 2002. Insulin: discovery and controversy. *Clin Chem* **48**:2270–2288. http://dx.doi.org/10.1093/clinchem/48.12.2270.
86. **Broad WJ**. 1982. Sharing credit for the Nobel. *Science* **218**:549. http://dx.doi.org/10.1126/science.218.4572.549.
87. **Fletcher H.** 1982. My work with Millikan on the oil-drop experiment. *Phys Today* **35**:43–47. http://dx.doi.org/10.1063/1.2915126.
88. **Goodstein D.** 2001. In defense of Robert Andrews Millikan. *Am Sci* **89**:54–60. http://dx.doi.org/10.1511/2001.14.724.
89. **Stolt CM, Klein G, Jansson AT**. 2004. An analysis of a wrong Nobel Prize—Johannes Fibiger, 1926: a study in the Nobel archives. *Adv Cancer Res* **92**:1–12. http://dx.doi.org/10.1016/S0065-230X(04)92001-5.
90. **Singh R, Riess F**. 2001. The 1930 Nobel Prize for physics: a close decision? *Notes Rec R Soc Lond* **55**:267–283. http://dx.doi.org/10.1098/rsnr.2001.0143.
91. **Crawford E, Sime RL, Walker M**. 1996. A Nobel tale of wartime injustice. *Nature* **382**:393–395. http://dx.doi.org/10.1038/382393a0.
92. **Palmer B.** 8 October 2012. The worst Nobel Prize: who is the least deserving Nobel laureate of all time? *Slate.* http://www.slate.com/articles/health_and_science/explainer/2012/10/who_is_the_least_deserving_winner_of_a_nobel_prize_.html.

93. **Wainwright M.** 1991. Streptomycin: discovery and resultant controversy. *Hist Philos Life Sci* **13**:97–124.

94. **Hargittai M.** 2012. Credit where credit's due? *Phys World* **38**:38–43. http://dx.doi.org/10.1088/2058-7058/25/09/37.

95. **Benson AA.** 2010. Last days in the old radiation laboratory (ORL), Berkeley, California, 1954. *Photosynth Res* **105**:209–212. http://dx.doi.org/10.1007/s11120-010-9592-2.

96. **Wright P.** 2 July 2002. Obituary: Erwin Chargaff. *The Guardian*. https://www.theguardian.com/news/2002/jul/02/guardianobituaries.obituaries.

97. **Townes CH.** 2007. Obituary: Theodore H. Maiman (1927-2007). *Nature* **447**:654. http://dx.doi.org/10.1038/447654a.

98. **Judson HF.** 2004. *The Great Betrayal: Fraud in Science*. Houghton Mifflin Harcourt, New York, NY.

99. **Eisenberg L.** 2005. Which image for Lorenz? *Am J Psychiatry* **162**:1760. http://dx.doi.org/10.1176/appi.ajp.162.9.1760.

100. **Feldman B.** 2000. *The Nobel Prize: a History of Genius, Controversy, and Prestige*. Arcade, New York, NY.

101. **Satir P.** 1997. Keith Roberts Porter: 1912-1997. *J Cell Biol* **138**:223–224. http://dx.doi.org/10.1083/jcb.138.2.223.

102. **Wilford JN.** 18 August 2007. Ralph Alpher, 86, expert in work on the Big Bang, dies. *New York Times*. http://www.nytimes.com/2007/08/18/us/18alpher.html.

103. **Broad WJ.** 1980. Riddle of the Nobel debate. *Science* **207**:37–38. http://dx.doi.org/10.1126/science.6985744.

104. **Anonymous.** 1980. How to change the Nobel Prize rules. *Nature* **287**:667–668. http://dx.doi.org/10.1038/287667a0.

105. **Anonymous.** 1983. Where was Hoyle? *Nature* **305**:750. http://dx.doi.org/10.1038/305750b0.

106. **Wade N.** 1975. Discovery of pulsars: a graduate student's story. *Science* **189**:358–364. http://dx.doi.org/10.1126/science.189.4200.358.

107. **Abbott A.** 2009. Neuroscience: one hundred years of Rita. *Nature* **458**:564–567. http://dx.doi.org/10.1038/458564a.

108. **Newmark P.** 1989. Nobel dispute continues. *Nature* **342**:329. http://dx.doi.org/10.1038/342329b0.

109. **McElheny VK.** 2004. *Watson and DNA: Making a Scientific Revolution*. Basic Books, New York, NY.

110. **Coles H.** 1997. Nobel panel rewards prion theory after years of heated debate. *Nature* **389**:529. http://dx.doi.org/10.1038/39120.

111. **Guillén JB, Fiallos EA, Castillo AB, Vallejo GA, Lancaster JR, Jr, Malmström BG.** 1998. Protest at Nobel omission of Moncada. *Nature* **396**:614. http://dx.doi.org/10.1038/25215.

112. **Howlett R.** 1998. Nobel award stirs up debate on nitric oxide breakthrough. *Nature* **395**:625–626. http://dx.doi.org/10.1038/27019.

113. **Lloyd E.** May 2004. Walking before dawn. *PittMed*. https://www.pittmed.health.pitt.edu/May_2004/walking.pdf.

114. **Anonymous.** 1999. A Nobel Prize for cell biology. *Nat Cell Biol* **1**:E169. http://dx.doi.org/10.1038/15602.

115. **Helmuth L.** 2001. Nobel prize. Researcher overlooked for 2000 Nobel. *Science* **291**:567–569. http://dx.doi.org/10.1126/science.291.5504.567.

116. **Goodman S.** 2001. French Nobel protest makes chemist a cause célèbre. *Nature* **414**:239. http://dx.doi.org/10.1038/35104765.

117. **Service RF.** 2001. Nobel prize in chemistry. Chemists hear one hand clapping. *Science* **294**:503–505. http://dx.doi.org/10.1126/science.294.5542.503b.

118. **Holley J.** 20 August 2005. Solar physics expert John Bahcall dies. *Washington Post*. https://www.washingtonpost.com/archive/local/2005/08/20/solar-physics-expert-john-bahcall-dies/1ae65712-810b-4f88-97bb-3efb8616c199/.

119. **Baumeister W, Bachmair A, Chau V, Cohen R, Coffino P, Demartino G, Deshaies R, Dohmen J, Emr S, Finley D, Hampton R, Hill C, Hochstrasser M, Huber R, Jackson P, Jentsch S, Johnson E,**

Kwon YT, Pagano M, Pickart C, Rechsteiner M, Scheffner M, Sommer T, Tansey W, Tyers M, Vierstra R, Weissman A, Wilkinson KD, Wolf D. 2004. Varshavsky's contributions. *Science* **306**:1290–1292. http://dx.doi.org/10.1126/science.306.5700.1290.

120. **Bots M, Maughan S, Nieuwland J.** 2006. RNAi Nobel ignores vital groundwork on plants. *Nature* **443**:906. http://dx.doi.org/10.1038/443906a.

121. **Nair P.** 2011. Profile of Gary Ruvkun. *Proc Natl Acad Sci U S A* **108**:15043–15045. http://dx.doi.org/10.1073/pnas.1111960108.

122. **Abbadessa G, et al.** 2009. Unsung hero Robert C. Gallo. *Science* **323**:206–207. http://dx.doi.org/10.1126/science.323.5911.206.

123. **Piqueras M.** 2008. Year's comments for 2008. *Int Microbiol* **11**:227–229.

124. **Service RF.** 2008. Nobel Prize in chemistry. Three scientists bask in prize's fluorescent glow. *Science* **322**:361. http://dx.doi.org/10.1126/science.322.5900.361.

125. **Jamieson V.** 2008. Physics Nobel snubs key researcher. *New Scientist*. http://www.newscientist.com/article/ dn14885-physics-nobel-snubs-key-researcher.html#.UiFf37x4OgI.

126. **Martin D.** 9 May 2011. Willard S. Boyle, father of digital eye, dies at 86. *New York Times*. http://www.nytimes.com/2011/05/ 10/science/space/10boyle.html.

127. **Reich ES.** 2010. Nobel document triggers debate. *Nature* **468**:486. http://dx.doi.org/10.1038/468486a.

128. **Wagner H.** 2012. Innate immunity's path to the Nobel Prize 2011 and beyond. *Eur J Immunol* **42**:1089–1092. http://dx.doi.org/10.1002/eji.201242404.

129. **Merali Z.** 2010. Physicists get political over Higgs. *Nature* **4**:10.1038/news.2010.390.

130. **Patel NV.** 2014. Nobel shocker: RCA had the first blue LED in 1972. *IEEE Spectrum*. https://spectrum.ieee.org/rcas-forgotten-work-on-the-blue-led#toggle-gdpr.

131. **Gallagher P.** 8 October 2014. Nobel Prize 2014: inventor of the red LED hits out at committee for "overlooking" his seminal 1960s work. *The Independent*. https://www.independent.co.uk/news/science/nobel-prize-2014-inventor-of-the-red-led-hits-out-at-committee-for-overlooking-his-seminal-1960s-work-9782948.html.

132. **Guarino B.** 3 October 2017. Three Americans win Nobel Prize in physics for gravitational wave discovery. *Washington Post*. https://www.washingtonpost.com/news/speaking-of-science/wp/2017/10/03/nobel-prize-in-physics-won-by-rainer-weiss-barry-barish-and-kip-thorne/.

133. **Zimmer M.** 7 October 2020. Nobel prize for CRISPR honors two great scientists—and leaves out many others. The Conversation. https://theconversation.com/nobel-prize-for-crispr-honors-two-great-scientists-and-leaves-out-many-others-147730.

134. **Lander ES.** 2016. The heroes of CRISPR. *Cell* **164**:18–28. http://dx.doi.org/10.1016/j.cell.2015.12.041.

Chapter 22 – Rejected Science

1. **Mole.** 2007. Rebuffs and rebuttals. I: How rejected is rejected? *J Cell Sci* **120**:1143–1144.

2. **Mole.** 2007. Rebuffs and rebuttals. II: Take me back! *J Cell Sci* **120**:1311–1313.

3. **ASM Journals.** 2021. *2021 Annual Report*. American Society for Microbiology, Washington, DC. https://journals.asm.org/pb-assets/pdf-text-excel-files/ASM-Journals-Annual-Report-1655922103097.pdf.

4. **Couzin J, Miller G.** 2007. NIH budget. Boom and bust. *Science* **316**:356–361. http://dx.doi.org/10.1126/science.316.5823.356.

5. **Petsko GA.** 2006. The system is broken. *Genome Biol* **7**:105. http://dx.doi.org/10.1186/gb-2006-7-3-105.

6. **Colwell AS, Wong FK, Chung KC.** 2022. Why do manuscripts get rejected? *Plast Reconstr Surg* **150**:1169–1173. http://dx.doi.org/10.1097/PRS.0000000000009627.

7. **Putman M, Berquist JB, Ruderman EM, Sparks JA.** 2022. Any given Monday: association between desk rejections and weekend manuscript submissions to rheumatology journals. *J Rheumatol* **49**:652–653. http://dx.doi.org/10.3899/jrheum.220099.

8. **Zoccali C, Amodeo D, Argiles A, Arici M, D'arrigo G, Evenepoel P, Fliser D, Fox J, Gesualdo L, Jadoul M, Ketteler M, Malyszko J, Massy Z, Mayer G, Ortiz A, Sever M, Vanholder R, Vinck C, Wanner C, Więcek A**. 2015. The fate of triaged and rejected manuscripts. *Nephrol Dial Transplant* **30**:1947–1950. http://dx.doi.org/10.1093/ndt/gfv387.

9. **Schroter S, Weber WEJ, Loder E, Wilkinson J, Kirkham JJ**. 2022. Evaluation of editors' abilities to predict the citation potential of research manuscripts submitted to *The BMJ*: a cohort study. *BMJ* **379**:e073880. http://dx.doi.org/10.1136/bmj-2022-073880.

10. **Maddox J, Randi J, Stewart WW**. 1988. "High-dilution" experiments a delusion. *Nature* **334**: 287–291. http://dx.doi.org/10.1038/334287a0.

11. **Wolpert L**. 1994. *The Unnatural Nature of Science*. Harvard University Press, Cambridge, MA.

12. **Cromer AH**. 1993. *Uncommon Sense: the Heretical Nature of Science*. Oxford University Press, Oxford, United Kingdom.

13. **Polak JF**. 1995. The role of the manuscript reviewer in the peer review process. *AJR Am J Roentgenol* **165**:685–688. http://dx.doi.org/10.2214/ajr.165.3.7645496.

14. **Shahar E**. 2007. On editorial practice and peer review. *J Eval Clin Pract* **13**:699–701. http://dx.doi.org/10.1111/j.1365-2753.2007.00800.x.

15. **Copernicus N**. 1543. *De Revolutionibus Orbium Coelestium*. Warnock Library, Nuremberg, Germany.

16. **Merriam-Webster**. 2022. Merriam-Webster online dictionary. https://www.merriam-webster.com/dictionary/censor.

17. **Benos DJ, Bashari E, Chaves JM, Gaggar A, Kapoor N, LaFrance M, Mans R, Mayhew D, McGowan S, Polter A, Qadri Y, Sarfare S, Schultz K, Splittgerber R, Stephenson J, Tower C, Walton RG, Zotov A**. 2007. The ups and downs of peer review. *Adv Physiol Educ* **31**:145–152. http://dx.doi.org/10.1152/advan.00104.2006.

18. **Brown T**. 2004. *Peer Review and the Acceptance of New Scientific Ideas*. Sense about Science, London, United Kingdom.

19. **Burnham JC**. 1990. The evolution of editorial peer review. *JAMA* **263**:1323–1329. http://dx.doi.org/10.1001/jama.1990.03440100023003.

20. **Spier R**. 2002. The history of the peer-review process. *Trends Biotechnol* **20**:357–358. http://dx.doi.org/10.1016/S0167-7799(02)01985-6.

21. **Baldwin M**. 2015. Credibility, peer review, and *Nature*, 1945-1990. *Notes Rec R Soc Lond* **69**:337–352. http://dx.doi.org/10.1098/rsnr.2015.0029.

22. **Garcia-Costa D, Forte A, Lòpez-Iñesta E, Squazzoni F, Grimaldo F**. 2022. Does peer review improve the statistical content of manuscripts? A study on 27 467 submissions to four journals. *R Soc Open Sci* **9**:210681. http://dx.doi.org/10.1098/rsos.210681.

23. **Cobo E, Selva-O'Callagham A, Ribera JM, Cardellach F, Dominguez R, Vilardell M**. 2007. Statistical reviewers improve reporting in biomedical articles: a randomized trial. *PLoS One* **2**:e332. http://dx.doi.org/10.1371/journal.pone.0000332.

24. **Hopewell S, Witt CM, Linde K, Icke K, Adedire O, Kirtley S, Altman DG**. 2018. Influence of peer review on the reporting of primary outcome(s) and statistical analyses of randomised trials. *Trials* **19**:30. http://dx.doi.org/10.1186/s13063-017-2395-4.

25. **Blanco D, Schroter S, Aldcroft A, Moher D, Boutron I, Kirkham JJ, Cobo E**. 2020. Effect of an editorial intervention to improve the completeness of reporting of randomised trials: a randomised controlled trial. *BMJ Open* **10**:e036799. http://dx.doi.org/10.1136/bmjopen-2020-036799.

26. **Horrobin DF**. 1990. The philosophical basis of peer review and the suppression of innovation. *JAMA* **263**:1438–1441. http://dx.doi.org/10.1001/jama.1990.03440100162024.

27. **Rothwell PM, Martyn CN**. 2000. Reproducibility of peer review in clinical neuroscience. Is agreement between reviewers any greater than would be expected by chance alone? *Brain* **123**:1964–1969. http://dx.doi.org/10.1093/brain/123.9.1964.

28. **Gannon F**. 2001. The essential role of peer review. *EMBO Rep* **2**:743. http://dx.doi.org/10.1093/embo-reports/kve188.

29. **Atkinson M.** 2001. "Peer review" culture. *Sci Eng Ethics* **7**:193–204. http://dx.doi.org/10.1007/s11948-001-0040-8.

30. **Jefferson T, Wager E, Davidoff F.** 2002. Measuring the quality of editorial peer review. *JAMA* **287**:2786–2790. http://dx.doi.org/10.1001/jama.287.21.2786.

31. **Smith R.** 2006. Peer review: a flawed process at the heart of science and journals. *J R Soc Med* **99**: 178–182. http://dx.doi.org/10.1177/014107680609900414.

32. **Hojat M, Gonnella JS, Caelleigh AS.** 2003. Impartial judgment by the "gatekeepers" of science: fallibility and accountability in the peer review process. *Adv Health Sci Educ Theory Pract* **8**:75–96. http://dx.doi.org/10.1023/A:1022670432373.

33. **Triggle CR, Triggle DJ.** 2007. What is the future of peer review? Why is there fraud in science? Is plagiarism out of control? Why do scientists do bad things? Is it all a case of: "all that is necessary for the triumph of evil is that good men do nothing"? *Vasc Health Risk Manag* **3**:39–53.

34. **Huber J, Inoua S, Kerschbamer R, König-Kersting C, Palan S, Smith VL.** 2022. Nobel and novice: author prominence affects peer review. *Proc Natl Acad Sci U S A* **119**:e2205779119. http://dx.doi.org/10.1073/pnas.2205779119.

35. **Cole S, Cole JR, Simon GA.** 1981. Chance and consensus in peer review. *Science* **214**:881–886. http://dx.doi.org/10.1126/science.7302566.

36. **Bauer HH.** 2004. Science in the 21st century: knowledge monopolies and research cartels. *J Sci Explor* **18**:643–660.

37. **Gannon F.** 2005. Is the system dumbing down research? *EMBO Rep* **6**:387. http://dx.doi.org/10.1038/sj.embor.7400407.

38. **Godlee F, Gale CR, Martyn CN.** 1998. Effect on the quality of peer review of blinding reviewers and asking them to sign their reports: a randomized controlled trial. *JAMA* **280**:237–240. http://dx.doi.org/10.1001/jama.280.3.237.

39. **Isenberg SJ, Sanchez E, Zafran KC.** 2009. The effect of masking manuscripts for the peer-review process of an ophthalmic journal. *Br J Ophthalmol* **93**:881–884. http://dx.doi.org/10.1136/bjo.2008.151886.

40. **Giles J.** 2007. Open-access journal will publish first, judge later. *Nature* **445**:9. http://dx.doi.org/10.1038/445009a.

41. **Pulverer B, Lemberger T.** 2019. Peer review beyond journals. *EMBO J* **38**:e103998. http://dx.doi.org/10.15252/embj.2019103998.

42. **Ross-Hellauer T, Görögh E.** 2019. Guidelines for open peer review implementation. *Res Integr Peer Rev* **4**:4. http://dx.doi.org/10.1186/s41073-019-0063-9.

43. **Anonymous.** 1999. Pros and cons of open peer review. *Nat Neurosci* **2**:197–198. http://dx.doi.org/10.1038/6295.

44. **Eisen MB, Akhmanova A, Behrens TE, Diedrichsen J, Harper DM, Iordanova MD, Weigel D, Zaidi M.** 2022. *Peer review without gatekeeping. eLife* **11**:83889. http://dx.doi.org/10.7554/eLife.83889.

45. **Else H.** 2022. *eLife* won't reject papers once they are under review—what researchers think. *Nature* doi:10.1038/d41586-022-03534-6. http://dx.doi.org/10.1038/d41586-022-03534-6.

46. **Cohen J.** 1994. The Duesberg phenomenon. *Science* **266**:1642–1644. http://dx.doi.org/10.1126/science.7992043.

47. **Forrest B, Gross PR.** 2004. *Creationism's Trojan Horse: the Wedge of Intelligent Design.* Oxford University Press, Oxford, United Kingdom. http://dx.doi.org/10.1093/acprof:oso/9780195157420.001.0001

48. **Smith TC, Novella SP.** 2007. HIV denial in the Internet era. *PLoS Med* **4**:e256. http://dx.doi.org/10.1371/journal.pmed.0040256.

49. **Revkin AC.** 9 July 2008. Cheney's office said to edit draft testimony on warming. *New York Times.* https://www.nytimes.com/2008/07/09/washington/09enviro.html.

50. **Harris G.** 11 July 2007. Surgeon general sees 4-year term as compromised. *New York Times.* https://www.nytimes.com/2007/07/11/washington/11surgeon.html .

51. **Macilwain C, Brumfiel G**. 2006. US scientists fight political meddling. *Nature* **439**:896–897. http://dx.doi.org/10.1038/439896a.

52. **Salyers A**. 2002. Science, censorship, and public health. *Science* **296**:617. http://dx.doi.org/10.1126/science.296.5568.617.

53. **Crijns TJ, Ottenhoff JSE, Ring D**. 2021. The effect of peer review on the improvement of rejected manuscripts. *Account Res* **28**:517–527. http://dx.doi.org/10.1080/08989621.2020.1869547.

54. **Galilei G**. 1957. *Letter to Don Virginio Cesarini*. Doubleday, New York, NY.

55. **Anonymous**. 2003. Coping with peer rejection. *Nature* **425**:645. http://dx.doi.org/10.1038/425645a.

56. **Campanario JM**. 2009. Rejecting and resisting Nobel class discoveries: accounts by Nobel laureates. *Scientometrics* **81**:549–565. http://dx.doi.org/10.1007/s11192-008-2141-5.

57. **Dowdy SF**. 2006. The anonymous American Idol manuscript reviewer. *Cell* **127**:662–663, author reply 664–665. http://dx.doi.org/10.1016/j.cell.2006.11.007.

58. **Anonymous**. 2002. When the hunter becomes the hunted. By Caveman. *J Cell Sci* **115**:237–238.

59. **Sever R, Eisen M, Inglis J**. 2019. Plan U: universal access to scientific and medical research via funder preprint mandates. *PLoS Biol* **17**:e3000273. http://dx.doi.org/10.1371/journal.pbio.3000273.

60. **Krumholz HM, Bloom T, Sever R, Rawlinson C, Inglis JR, Ross JS**. 2020. Submissions and downloads of preprints in the first year of medRxiv. *JAMA* **324**:1903–1905. http://dx.doi.org/10.1001/jama.2020.17529.

61. **Fraser N, Brierley L, Dey G, Polka JK, Pálfy M, Nanni F, Coates JA**. 2021. The evolving role of preprints in the dissemination of COVID-19 research and their impact on the science communication landscape. *PLoS Biol* **19**:e3000959. http://dx.doi.org/10.1371/journal.pbio.3000959.

62. **Vale RD**. 2015. Accelerating scientific publication in biology. *Proc Natl Acad Sci U S A* **112**:13439–13446. http://dx.doi.org/10.1073/pnas.1511912112.

63. **Marder E**. 2017. Beyond scoops to best practices. *eLife* **6**:e30076. http://dx.doi.org/10.7554/eLife.30076.

64. **Ray J, Berkwits M, Davidoff F**. 2000. The fate of manuscripts rejected by a general medical journal. *Am J Med* **109**:131–135. http://dx.doi.org/10.1016/S0002-9343(00)00450-2.

65. **McDonald RJ, Cloft HJ, Kallmes DF**. 2007. Fate of submitted manuscripts rejected from the *American Journal of Neuroradiology*: outcomes and commentary. *AJNR Am J Neuroradiol* **28**:1430–1434. http://dx.doi.org/10.3174/ajnr.A0766.

66. **McDonald RJ, Cloft HJ, Kallmes DF**. 2009. Fate of manuscripts previously rejected by the *American Journal of Neuroradiology*: a follow-up analysis. *AJNR Am J Neuroradiol* **30**:253–256. http://dx.doi.org/10.3174/ajnr.A1366.

67. **Earnshaw CH, Edwin C, Bhat J, Krishnan M, Mamais C, Somashekar S, Sunil A, Williams SP, Leong SC**. 2017. An analysis of the fate of 917 manuscripts rejected from *Clinical Otolaryngology*. *Clin Otolaryngol* **42**:709–714. http://dx.doi.org/10.1111/coa.12820.

68. **Docherty AB, Klein AA**. 2017. The fate of manuscripts rejected from *Anaesthesia*. *Anaesthesia* **72**:427–430. http://dx.doi.org/10.1111/anae.13829.

69. **Citerio G, Deutsch E, Sala E, Lavillonnière M, Perner A, Jaber S, Timsit JF, Azoulay E**. 2018. Fate of manuscripts rejected by *Intensive Care Medicine* from 2013 to 2016: a follow-up analysis. *Intensive Care Med* **44**:2300–2301. http://dx.doi.org/10.1007/s00134-018-5407-2.

70. **Barwich AS**. 2019. The value of failure in science: the story of grandmother cells in neuroscience. *Front Neurosci* **13**:1121. http://dx.doi.org/10.3389/fnins.2019.01121.

71. **Taneja R**. 7 November 2019. CV Raman birthday: know about India's greatest physicist who discovered "Raman effect." NDTV. https://www.ndtv.com/india-news/cv-raman-birthday-know-about-raman-effect-greatest-physicist-quotes-tributes-on-twitter-2128614 .

72. **Barry JM**. 2004. *The Great Influenza*. Penguin, New York, NY.

73. **Parkes E**. 10 January 2019. Scientific progress is built on failure. *Nature*. https://www.nature.com/articles/d41586-019-00107-y .

74. **Dreyfuss E**. 25 April 2019. Scientists need to talk more about failure. *Wired*. https://www.wired.com/story/scientists-need-more-failure-talk/.

Chapter 23 – Unfunded Science

1. **Guthrie S.** 2019. Innovating in the research funding process: peer review alternatives and adaptations. Academy Health. https://academyhealth.org/publications/2019-11/innovating-research-funding-process-peer-review-alternatives-and-adaptations .

2. **Siliciano R.** 19 March 2007. *Testimony before the Senate Committee on Appropriations: Subcommittee on Labor, Health and Human Services, Education, and Related Agencies.* US Senate Committee on Appropriations, Washington, DC.

3. **Intersociety Working Group.** 2008. AAAS report XXXIII: research and development FY 2009, American Association for the Advancement of Science, Washington, DC. http://www.aaas.org/spp/rd/rd09main.htm .

4. **AAAS.** 2022. Historical trends in federal R&D. American Association for the Advancement of Science, Washington, DC. https://www.aaas.org/programs/r-d-budget-and-policy/historical-trends-federal-rd .

5. **Mandel HG, Vesell ES.** 2008. Declines in NIH R01 research grant funding. *Science* **322**:189. http://dx.doi.org/10.1126/science.322.5899.189a.

6. **Anonymous.** 2008. A metareview at the NIH. *Nat Med* **14**:351. http://dx.doi.org/10.1038/nm0408-351.

7. **Anonymous.** 2008. Research and recovery. *Nat Med* **14**:1129. http://dx.doi.org/10.1038/nm1108-1129.

8. **Benderly BL.** 2008. Taken for granted: lost in space. *Sci Careers* **7**: 10.1126/science.caredit.a0800064.

9. **Van Wart M.** 2005. *Dynamics of Leadership in Public Service: Theory and Practice.* M.E. Sharpe, Armonk, NY.

10. **Bonetta L.** 2008. Enhancing NIH grant peer review: a broader perspective. *Cell* **135**:201–204. http://dx.doi.org/10.1016/j.cell.2008.09.051.

11. **Finn R.** 1995. NIH study section members acknowledge major flaws in the reviewing system. *The Scientist* **9**:7.

12. **Wessely S.** 1998. Peer review of grant applications: what do we know? *Lancet* **352**:301–305. http://dx.doi.org/10.1016/S0140-6736(97)11129-1.

13. **Fang FC, Casadevall A.** 2009. NIH peer review reform—change we need, or lipstick on a pig? *Infect Immun* **77**:929–932. http://dx.doi.org/10.1128/IAI.01567-08.

14. **Pagano M.** 2006. *American Idol* and NIH grant review. *Cell* **126**:637–638. http://dx.doi.org/10.1016/j.cell.2006.08.004.

15. **Petsko GA.** 2006. The system is broken. *Genome Biol* **7**:105. http://dx.doi.org/10.1186/gb-2006-7-3-105.

16. **Johnson VE.** 2008. Statistical analysis of the National Institutes of Health peer review system. *Proc Natl Acad Sci U S A.* **105**:11076–11080 http://dx.doi.org/10.1073/pnas.0804538105.

17. **Kaplan D, Lacetera N, Kaplan C.** 2008. Sample size and precision in NIH peer review. *PLoS One* **3**:e2761. http://dx.doi.org/10.1371/journal.pone.0002761.

18. **Cole S, Cole JR, Simon GA.** 1981. Chance and consensus in peer review. *Science* **214**:881–886. http://dx.doi.org/10.1126/science.7302566.

19. **Pier EL, Brauer M, Filut A, Kaatz A, Raclaw J, Nathan MJ, Ford CE, Carnes M.** 2018. Low agreement among reviewers evaluating the same NIH grant applications. *Proc Natl Acad Sci U S A* **115**:2952–2957. http://dx.doi.org/10.1073/pnas.1714379115.

20. **Hannun YA.** 2008. NIH: grants revamp needs grounding in evidence. *Nature* **452**:811. http://dx.doi.org/10.1038/452811c.

21. **Li D, Agha L.** 2015. Research funding. Big names or big ideas: do peer-review panels select the best science proposals? *Science* **348**:434–438. http://dx.doi.org/10.1126/science.aaa0185.

22. **Fang FC, Bowen A, Casadevall A.** 2016. NIH peer review percentile scores are poorly predictive of grant productivity. *eLife* **5**:e13323. http://dx.doi.org/10.7554/eLife.13323.

23. **Danthi N, Wu CO, Shi P, Lauer M.** 2014. Percentile ranking and citation impact of a large cohort of National Heart, Lung, and Blood Institute-funded cardiovascular R01 grants. *Circ Res* **114**:600–606. http://dx.doi.org/10.1161/CIRCRESAHA.114.302656.

24. **Berg J.** 2013. On deck chairs and lifeboats. *ASBMB Today* **12**:3–6.

25. **Mayo NE, Brophy J, Goldberg MS, Klein MB, Miller S, Platt RW, Ritchie J.** 2006. Peering at peer review revealed high degree of chance associated with funding of grant applications. *J Clin Epidemiol* **59**:842–848. http://dx.doi.org/10.1016/j.jclinepi.2005.12.007.

26. **Graves N, Barnett AG, Clarke P.** 2011. Funding grant proposals for scientific research: retrospective analysis of scores by members of grant review panel. *BMJ* **343**:d4797. http://dx.doi.org/10.1136/bmj.d4797.

27. **Marsh HW, Jayasinghe UW, Bond NW.** 2008. Improving the peer-review process for grant applications: reliability, validity, bias, and generalizability. *Am Psychol* **63**:160–168. http://dx.doi.org/10.1037/0003-066X.63.3.160.

28. **Fogelholm M, Leppinen S, Auvinen A, Raitanen J, Nuutinen A, Väänänen K.** 2012. Panel discussion does not improve reliability of peer review for medical research grant proposals. *J Clin Epidemiol* **65**:47–52. http://dx.doi.org/10.1016/j.jclinepi.2011.05.001.

29. **Abdoul H, Perrey C, Amiel P, Tubach F, Gottot S, Durand-Zaleski I, Alberti C.** 2012. Peer review of grant applications: criteria used and qualitative study of reviewer practices. *PLoS One* **7**:e46054. http://dx.doi.org/10.1371/journal.pone.0046054.

30. **Tetlock P.** 2005. *Expert Political Judgment: How Good Is It? How Can We Know?* Princeton University Press, Princeton, NJ.

31. **Nicholson JM, Ioannidis JP.** 2012. Research grants: conform and be funded. *Nature* **492**:34–36. http://dx.doi.org/10.1038/492034a.

32. **Kuhn TS.** 1962. *The Structure of Scientific Revolutions*. University of Chicago Press, Chicago, IL.

33. **Lee C.** 28 May 2007. Slump in NIH funding is taking a toll on research, p A06. *Washington Post*.

34. **Casadevall A, Fang FC.** 2016. Revolutionary science. *mBio* **7**:e00158. http://dx.doi.org/10.1128/mBio.00158-16.

35. **Bornmann L, Mutz R, Daniel D.** 2007. Gender differences in grant peer review: a meta-analysis. *J Informetrics* **1**:226–238. http://dx.doi.org/10.1016/j.joi.2007.03.001.

36. **Pohlhaus JR, Jiang H, Wagner RM, Schaffer WT, Pinn VW.** 2011. Sex differences in application, success, and funding rates for NIH extramural programs. *Acad Med* **86**:759–767. http://dx.doi.org/10.1097/ACM.0b013e31821836ff.

37. **Ginther DK, Schaffer WT, Schnell J, Masimore B, Liu F, Haak LL, Kington R.** 2011. Race, ethnicity, and NIH research awards. *Science* **333**:1015–1019. http://dx.doi.org/10.1126/science.1196783.

38. **Erosheva EA, Grant S, Chen MC, Lindner MD, Nakamura RK, Lee CJ.** 2020. NIH peer review: criterion scores completely account for racial disparities in overall impact scores. *Sci Adv* **6**:eaaz4868. http://dx.doi.org/10.1126/sciadv.aaz4868.

39. **Check Hayden E.** 2015. Racial bias continues to haunt NIH grants. *Nature* **527**:286–287. http://dx.doi.org/10.1038/527286a.

40. **Taffe MA, Gilpin NW.** 2021. Racial inequity in grant funding from the US National Institutes of Health. *eLife* **10**:e74744. http://dx.doi.org/10.7554/eLife.74744.

41. **Kozlov M.** 9 December 2022. NIH plans grant-review overhaul to reduce bias. *Nature*. https://www.ncbi.nlm.nih.gov/pubmed/36494447 .

42. **Day TE.** 2015. The big consequences of small biases: a simulation of peer review. *Res Policy* **44**:1266–1270. http://dx.doi.org/10.1016/j.respol.2015.01.006.

43. **Hargrave-Thomas E, Yu B, Reynisson J.** 2012. Serendipity in anticancer drug discovery. *World J Clin Oncol* **3**:1–6. http://dx.doi.org/10.5306/wjco.v3.i1.1.

44. **Conix S, De Block A, Vaesen K.** 2021. Grant writing and grant peer review as questionable research practices. *F1000 Res* **10**:1126. https:doi.org/10.12688/f1000research.73893.2.

45. **Price M.** 4 June 2013. A shortcut to better grantsmanship. *Science*. http://sciencecareers.sciencemag.org/career_magazine/previous_issues/articles/2013_06_04/caredit.a1300119.

46. **Kaiser J.** 2008. National Institutes of Health. Two strikes and you're out, grant applicants learn. *Science* **322**:358. http://dx.doi.org/10.1126/science.322.5900.358b .

47. **Benezra R.** 2013. Grant applications: undo NIH policy to ease effect of cuts. *Nature* **493**:480. http://dx.doi.org/10.1038/493480e.

48. **Kaiser J.** 2014. Biomedical funding. At NIH, two strikes policy is out. *Science* **344**:350. http://dx.doi.org/10.1126/science.344.6182.350.

49. **Atkinson RD, Ezell SJ, Giddings LV, Stewart LA, Andes SM**. 2012. *Leadership in Decline: Assessing U.S. International Competitiveness in Biomedical Research.* Information Technology and Innovation Foundation and United for Medical Research, Washington, DC. http://www2.itif.org/2012-leadership-in-decline.pdf.

50. **von Hippel T, von Hippel C**. 2015. To apply or not to apply: a survey analysis of grant writing costs and benefits. *PLoS One* **10**:e0118494. http://dx.doi.org/10.1371/journal.pone.0118494.

51. **Clotfelter CT, Cook PJ**. 1989. *Selling Hope: State Lotteries in America.* Harvard University Press, Cambridge, MA.

52. **Basken P, Voosen P**. 24 February 2014. Strapped scientists abandon research and students. *Chronicle of Higher Education.* http://chronicle.com/article/Strapped-Scientists-Abandon/144921/.

53. **Staw BM**. 1976. Knee-deep in the big muddy: a study of escalating commitment to a chosen course of action. *Organ Behav Hum Perform* **16**:27–44. http://dx.doi.org/10.1016/0030-5073(76)90005-2.

54. **Casadevall A, Fang FC**. 2014. Specialized science. *Infect Immun* **82**:1355–1360. http://dx.doi.org/10.1128/IAI.01530-13.

55. **Xue G, He Q, Lei X, Chen C, Liu Y, Chen C, Lu ZL, Dong Q, Bechara A**. 2012. The gambler's fallacy is associated with weak affective decision making but strong cognitive ability. *PLoS One* **7**:e47019. http://dx.doi.org/10.1371/journal.pone.0047019.

56. **Foster R**. 26 September 2013. Optimism is a sine qua non for scientists. *Times Higher Education.* https://www.timeshighereducation.com/features/optimism-is-a-sine-qua-non-for-scientists/2007563.article.

57. **Sharot T, Korn CW, Dolan RJ**. 2011. How unrealistic optimism is maintained in the face of reality. *Nat Neurosci* **14**:1475–1479. http://dx.doi.org/10.1038/nn.2949.

58. **Brumfiel G**. 2008. Older scientists publish more papers. *Nature.* **455**:1161.

59. **Larsson U**. 2005. *Cultures of Creativity: Birth of a 21st Century Museum.* Science History Publications, Sagamore Beach, MA.

60. **Meyers MA**. 1995. Glen W. Hartman Lecture. Science, creativity, and serendipity. *AJR Am J Roentgenol* **165**:755–764. http://dx.doi.org/10.2214/ajr.165.4.7676963.

61. **Lewison G, Anderson J, Jack J**. 1995. Assessing track records. *Nature* **377**:671. http://dx.doi.org/10.1038/377671a0.

62. **Hand E**. 2008. 222 NIH grants: 22 researchers. *Nature* **452**:258–259. http://dx.doi.org/10.1038/452258a.

63. **Wadman M**. 2010. Study says middle sized labs do best. *Nature* **468**:356–357. http://dx.doi.org/10.1038/468356a.

64. **Leshner AI**. 2008. Reduce administrative burden. *Science* **322**:1609. http://dx.doi.org/10.1126/science.1168345.

65. **Barnett AG, Graves N, Clarke P, Herbert D**. 2015. The impact of a streamlined funding application process on application time: two cross-sectional surveys of Australian researchers. *BMJ Open* **5**:e006912. http://dx.doi.org/10.1136/bmjopen-2014-006912.

66. **Mervis J**. 2007. Scientific publishing. U.S. output flattens, and NSF wonders why. *Science* **317**:582. http://dx.doi.org/10.1126/science.317.5838.582.

67. **Moore JP**. 2007. Speaking out about U.S. science output. *Science* **318**:913. http://dx.doi.org/10.1126/science.318.5852.913b.

68. **Fang FC, Casadevall A**. 2016. Research funding: the case for a modified lottery. *mBio* **7**:e00422–e16. http://dx.doi.org/10.1128/mBio.00422-16.

69. **Mervis J**. 2014. Peering into peer review. *Science* **343**:596–598. http://dx.doi.org/10.1126/science.343.6171.596.

70. **Bollen J, Crandall D, Junk D, Ding Y, Börner K**. 2014. From funding agencies to scientific agency: collective allocation of science funding as an alternative to peer review. *EMBO Rep* **15**:131–133. http://dx.doi.org/10.1002/embr.201338068.

71. **Kaltman JR, Evans FJ, Danthi NS, Wu CO, DiMichele DM, Lauer MS**. 2014. Prior publication productivity, grant percentile ranking, and topic-normalized citation impact of NHLBI cardiovascular R01 grants. *Circ Res* **115**:617–624. http://dx.doi.org/10.1161/CIRCRESAHA.115.304766.

72. **Ioannidis JP**. 2011. More time for research: fund people not projects. *Nature* **477**:529–531. http://dx.doi.org/10.1038/477529a.

73. **Fang FC, Casadevall A**. 14 April 2014. Taking the Powerball approach to funding medical research. *Wall Street Journal*. http://www.wsj.com/articles/SB10001424052702303532704579477530153771424.

74. **Fang FC, Casadevall A**. 2016. Grant funding: playing the odds. *Science* **352**:158. http://dx.doi.org/10.1126/science.352.6282.158-a.

75. **Malkiel BG**. 1973. *A Random Walk Down Wall Street*. W.W. Norton, New York, NY.

76. **Sommer J**. 14 March 2015. How many mutual funds routinely rout the market? Zero. *New York Times*. https://www.nytimes.com/2015/03/15/your-money/how-many-mutual-funds-routinely-rout-the-market-zero.html.

77. **Buffett WE**. 2014. *Berkshire Hathaway 2013 Annual Report*. Berkshire Hathaway, Omaha, NE.

78. **Taleb N**. 2007. *The Black Swan: the Impact of the Highly Improbable*. Random House, New York, NY.

79. **Biondo AE, Pluchino A, Rapisarda A, Helbing D**. 2013. Are random trading strategies more successful than technical ones? *PLoS One* **8**:e68344. http://dx.doi.org/10.1371/journal.pone.0068344.

80. **Knapp A**. 2013. Computer simulation suggests that the best investment strategy is a random one. *Forbes*. 22 March 2013. https://www.forbes.com/sites/alexknapp/2013/03/22/computer-simulation-suggests-that-the-best-investment-strategy-is-a-random-one/?sh=a6b183f51362.

81. **Scannell JW, Blanckley A, Boldon H, Warrington B**. 2012. Diagnosing the decline in pharmaceutical R&D efficiency. *Nat Rev Drug Discov* **11**:191–200. http://dx.doi.org/10.1038/nrd3681.

82. **Bowen A, Casadevall A**. 2015. Increasing disparities between resource inputs and outcomes, as measured by certain health deliverables, in biomedical research. *Proc Natl Acad Sci U S A* **112**:11335–11340. http://dx.doi.org/10.1073/pnas.1504955112.

83. **National Institutes of Health**. 2016. Early stage investigator policies. https://grants.nih.gov/policy/early-stage/index.htm.

84. **Druss BG, Marcus SC**. 2005. Tracking publication outcomes of National Institutes of Health grants. *Am J Med* **118**:658–663. http://dx.doi.org/10.1016/j.amjmed.2005.02.015.

85. **Merton RK**. 1968. The Matthew effect in science. *The reward and communication systems of science are considered. Science* **159**:56–63. http://dx.doi.org/10.1126/science.159.3810.56.

86. **Chabrier J, Cohodes S, Oreopoulos P**. 2016. What can we learn from charter school lotteries? *J Econ Perspect* **30**:57–84. http://dx.doi.org/10.1257/jep.30.3.57.

87. **Persad G, Wertheimer A, Emanuel EJ**. 2009. Principles for allocation of scarce medical interventions. *Lancet* **373**:423–431. http://dx.doi.org/10.1016/S0140-6736(09)60137-9.

88. **Annas GJ**. 1985. The prostitute, the playboy, and the poet: rationing schemes for organ transplantation. *Am J Public Health* **75**:187–189. http://dx.doi.org/10.2105/AJPH.75.2.187.

89. **McLachlan HV**. 2012. A proposed non-consequentialist policy for the ethical distribution of scarce vaccination in the face of an influenza pandemic. *J Med Ethics* **38**:317–318. http://dx.doi.org/10.1136/medethics-2011-100031.

90. **Greenberg D**. 11 February 2008. Peer review at NIH: a lottery would be better. *Chronicle of Higher Education*. https://www.chronicle.com/blogs/brainstorm/peer-review-at-nih-a-lottery-would-be-better.

91. **National Science Board**. 2013. *Report to the National Science Board on the National Science Foundation's Merit Review Process, Fiscal Year 2012*. National Science Board, Arlington, VA.

92. **Adam D**. 2019. Science funders gamble on grant lotteries. *Nature* **575**:574–575. http://dx.doi.org/10.1038/d41586-019-03572-7.

93. **Piper K.** 18 January 2019. Science funding is a mess. Could grant lotteries make it better? Vox. https://www.vox.com/future-perfect/2019/1/18/18183939/science-funding-grant-lotteries-research.

94. **Page L, Barnett A.** 21 October 2022. Research funding is broken. Using a lottery approach could fix it. *STAT*. https://www.statnews.com/2022/10/21/research-funding-broken-lottery-approach-could-fix-it/.

95. **Anonymous**. 2022. The case for lotteries as a tiebreaker of quality in research funding. *Nature* **609**:653. http://dx.doi.org/10.1038/d41586-022-02959-3.

96. **Gross K, Bergstrom CT**. 2019. Contest models highlight inherent inefficiencies of scientific funding competitions. *PLoS Biol* **17**:e3000065. http://dx.doi.org/10.1371/journal.pbio.3000065.

97. **Avin S**. 2019. Mavericks and lotteries. *Stud Hist Philos Sci* **76**:13–23. http://dx.doi.org/10.1016/j.shpsa.2018.11.006.

98. **Roumbanis L**. 2019. Peer review or lottery? A critical analysis of two different forms of decision-making mechanisms for allocation of research grants. *Sci Technol Human Values* **44**:994–1019. http://dx.doi.org/10.1177/0162243918822744.

99. **De Peuter S, Conix S**. 2022. The modified lottery: formalizing the intrinsic randomness of research funding. *Account Res* **29**:324–345. http://dx.doi.org/10.1080/08989621.2021.1927727.

100. **Liu M, Choy V, Clarke P, Barnett A, Blakely T, Pomeroy L**. 2020. The acceptability of using a lottery to allocate research funding: a survey of applicants. *Res Integr Peer Rev* **5**:3. http://dx.doi.org/10.1186/s41073-019-0089-z.

101. **Swiss National Science Foundation**. 30 March 2021. Drawing lots as a tie-breaker. https://www.snf.ch/en/JyifP2I9SUo8CPxI/news/news-210331-drawing-lots-as-a-tie-breaker.

102. **Matthews D**. 26 November 2020. German funder sees early success in grant-by-lottery trial. Times Higher Education. https://www.timeshighereducation.com/news/german-funder-sees-early-success-grant-lottery-trial.

103. **VolkswagenStiftung**. 19 November 2020. Partially randomized procedure—lottery and peer review. https://www.volkswagenstiftung.de/en/partially-randomized-procedure-lottery-and-peer-review.

104. **The British Academy**. 2021. BA/Leverhulme Small Research Grants. https://www.thebritishacademy.ac.uk/funding/ba-leverhulme-small-research-grants/.

105. **Mesa N.** 25 February 2022. Q&A: A randomized approach to awarding grants. *The Scientist*. https://www.the-scientist.com/news-opinion/q-a-a-randomized-approach-to-awarding-grants-69741.

106. **Wilson EO**. 2014. *The Meaning of Human Existence*. Liveright, New York, NY.

107. **Freedman DH**. November 2010. Lies, damned lies, and medical science. *The Atlantic*. http://www.theatlantic.com/magazine/archive/2010/11/lies-damned-lies-and-medical-science/308269/.

108. **Fang FC, Casadevall A**. 2012. Reforming science: structural reforms. *Infect Immun* **80**:897–901. http://dx.doi.org/10.1128/IAI.06184-11.

109. **Casadevall A, Fang FC**. 2012. Reforming science: methodological and cultural reforms. *Infect Immun* **80**:891–896. http://dx.doi.org/10.1128/IAI.06183-11.

110. **Severin A, Martins J, Heyard R, Delavy F, Jorstad A, Egger M**. 2020. Gender and other potential biases in peer review: cross-sectional analysis of 38 250 external peer review reports. *BMJ Open* **10**:e035058. http://dx.doi.org/10.1136/bmjopen-2019-035058.

111. **van der Lee R, Ellemers N**. 2015. Gender contributes to personal research funding success in The Netherlands. *Proc Natl Acad Sci U S A* **112**:12349–12353. http://dx.doi.org/10.1073/pnas.1510159112.

112. **Murray DL, Morris D, Lavoie C, Leavitt PR, MacIsaac H, Masson ME, Villard MA**. 2016. Bias in research grant evaluation has dire consequences for small universities. *PLoS One* **11**:e0155876. http://dx.doi.org/10.1371/journal.pone.0155876.

113. **Wahls WP**. 2016. Biases in grant proposal success rates, funding rates and award sizes affect the geographical distribution of funding for biomedical research. *PeerJ* **4**:e1917. http://dx.doi.org/10.7717/peerj.1917.

114. **Bisson LF**. 2014. Bias in grant review: the fault in our metrics? UC Davis Advance. https://ucd-advance.ucdavis.edu/sites/main/files/file-attachments/bisson_advance_roundtable_presentation.pdf.

115. **Banal-Estañol A, Macho-Stadler I, Pérez-Castrillo D**. 2019. Evaluation in research funding agencies: are structurally diverse teams biased against? *Res Policy* **48**:1823–1840. http://dx.doi.org/10.1016/j.respol.2019.04.008.

116. **Bromham L, Dinnage R, Hua X.** 2016. Interdisciplinary research has consistently lower funding success. *Nature* **534**:684–687. http://dx.doi.org/10.1038/nature18315.

117. **Lanei JN, Teplitskiy M, Gray G, Ranu H, Menietti M, Guinan E, Lakhani KR.** 2022. Conservatism gets funded? A field experiment on the role of negative information in novel project evaluation. *Manage Sci* **68**:4478–4495. http://dx.doi.org/10.1287/mnsc.2021.4107.

118. **Brainard J.** 29 October 2021. Funding agency's reviewers were biased against scientists with novel ideas. *Science*. https://www.science.org/content/article/funding-agency-s-reviewers-were-biased-against-scientists-novel-ideas.

119. **Singh Chawla D.** 17 April 2019. "Friendly" reviewers rate grant applications more highly. *Nature*. https://www.nature.com/articles/d41586-019-01198-3.

120. **Nair P.** 2021. QnAs with Katalin Karikó. *Proc Natl Acad Sci U S A* **118**:e2119757118. http://dx.doi.org/10.1073/pnas.2119757118.

121. **Buck S.** 1 February 2022. The Karikó problem: lessons for funding basic research. *STAT*. https://www.statnews.com/ambri2/01/kariko-problem-lessons-funding-basic-research/.

Chapter 24 – Retracted Science

1. **Oransky I, Fremes SE, Kurlansky P, Gaudino M**. 2021. Retractions in medicine: the tip of the iceberg. *Eur Heart J* **42**:4205–4206. http://dx.doi.org/10.1093/eurheartj/ehab398.

2. **Greenhalgh T.** 2010. Why did the *Lancet* take so long? *BMJ* **340**:c644. http://dx.doi.org/10.1136/bmj.c644.

3. **Mintz AP.** 2002. *Web of Deception: Misinformation on the Internet.* CyberAge Books, Medford, NJ.

4. **Del Vicario M, Bessi A, Zollo F, Petroni F, Scala A, Caldarelli G, Stanley HE, Quattrociocchi W.** 2016. The spreading of misinformation online. *Proc Natl Acad Sci U S A* **113**:554–559. http://dx.doi.org/10.1073/pnas.1517441113.

5. **Bak-Coleman JB, Alfano M, Barfuss W, Bergstrom CT, Centeno MA, Couzin ID, Donges JF, Galesic M, Gersick AS, Jacquet J, Kao AB, Moran RE, Romanczuk P, Rubenstein DI, Tombak KJ, Van Bavel JJ, Weber EU.** 2021. Stewardship of global collective behavior. *Proc Natl Acad Sci U S A* **118**:e2025764118. http://dx.doi.org/10.1073/pnas.2025764118.

6. **Suelzer EM, Deal J, Hanus KL, Ruggeri B, Sieracki R, Witkowski E.** 2019. Assessment of citations of the retracted article by Wakefield et al with fraudulent claims of an association between vaccination and autism. *JAMA Netw Open* **2**:e1915552. http://dx.doi.org/10.1001/jamanetworkopen.2019.15552.

7. **Woolf PK.** 1986. Pressure to publish and fraud in science. *Ann Intern Med* **104**:254–256. http://dx.doi.org/10.7326/0003-4819-104-2-254.

8. **Steen RG.** 2011. Retractions in the scientific literature: is the incidence of research fraud increasing? *J Med Ethics* **37**:249–253. http://dx.doi.org/10.1136/jme.2010.040923.

9. **Oransky I, Marcus A.** 3 August 2010. Launch date. Retraction Watch. http://retractionwatch.wordpress.com.

10. **Steen RG.** 2011. Misinformation in the medical literature: what role do error and fraud play? *J Med Ethics* **37**:498–503. http://dx.doi.org/10.1136/jme.2010.041830.

11. **Mori N, Oishi K, Sar B, Mukaida N, Nagatake T, Matsushima K, Yamamoto N.** 1999. Essential role of transcription factor nuclear factor-kappaB in regulation of interleukin-8 gene expression by nitrite reductase from *Pseudomonas aeruginosa* in respiratory epithelial cells. *Infect Immun* **67**: 3872–3878. http://dx.doi.org/10.1128/IAI.67.8.3872-3878.1999.

12. **Mori N, Wada A, Hirayama T, Parks TP, Stratowa C, Yamamoto N.** 2000. Activation of intercellular adhesion molecule 1 expression by *Helicobacter pylori* is regulated by NF-kappaB in gastric epithelial cancer cells. *Infect Immun* **68**:1806–1814. http://dx.doi.org/10.1128/IAI.68.4.1806-1814.2000.

13. **Mori N, Krensky AM, Geleziunas R, Wada A, Hirayama T, Sasakawa C, Yamamoto N.** 2003. *Helicobacter pylori* induces RANTES through activation of NF-kappa B. *Infect Immun* **71**:3748–3756. http://dx.doi.org/10.1128/IAI.71.7.3748-3756.2003.

14. **Mori N, Ueda A, Geleziunas R, Wada A, Hirayama T, Yoshimura T, Yamamoto N.** 2001. Induction of monocyte chemoattractant protein 1 by *Helicobacter pylori* involves NF-kappaB. *Infect Immun* **69**:1280–1286. http://dx.doi.org/10.1128/IAI.69.3.1280-1286.2001.

15. **Takeshima E, Tomimori K, Teruya H, Ishikawa C, Senba M, D'Ambrosio D, Kinjo F, Mimuro H, Sasakawa C, Hirayama T, Fujita J, Mori N.** 2009. *Helicobacter pylori*-induced interleukin-12 p40 expression. *Infect Immun* **77**:1337–1348. http://dx.doi.org/10.1128/IAI.01456-08.

16. **Tomimori K, Uema E, Teruya H, Ishikawa C, Okudaira T, Senba M, Yamamoto K, Matsuyama T, Kinjo F, Fujita J, Mori N.** 2007. *Helicobacter pylori* induces CCL20 expression. *Infect Immun* **75**:5223–5232. http://dx.doi.org/10.1128/IAI.00731-07.

17. **McNally A, Roe AJ, Simpson S, Thomson-Carter FM, Hoey DE, Currie C, Chakraborty T, Smith DG, Gally DL.** 2001. Differences in levels of secreted locus of enterocyte effacement proteins between human disease-associated and bovine *Escherichia coli* O157. *Infect Immun* **69**:5107–5114. http://dx.doi.org/10.1128/IAI.69.8.5107-5114.2001.

18. **Shin JJ, Bryksin AV, Godfrey HP, Cabello FC.** 2004. Localization of BmpA on the exposed outer membrane of *Borrelia burgdorferi* by monospecific anti-recombinant BmpA rabbit antibodies. *Infect Immun* **72**:2280–2287. http://dx.doi.org/10.1128/IAI.72.4.2280-2287.2004.

19. **Kalia A, Enright MC, Spratt BG, Bessen DE.** 2001. Directional gene movement from human-pathogenic to commensal-like streptococci. *Infect Immun* **69**:4858–4869. http://dx.doi.org/10.1128/IAI.69.8.4858-4869.2001.

20. **Reynaud A, Federighi M, Licois D, Guillot JF, Joly B.** 1991. R plasmid in *Escherichia coli* O103 coding for colonization of the rabbit intestinal tract. *Infect Immun* **59**:1888–1892. http://dx.doi.org/10.1128/iai.59.6.1888-1892.1991.

21. **Ismail SO, Skeiky YA, Bhatia A, Omara-Opyene LA, Gedamu L.** 1994. Molecular cloning, characterization, and expression in *Escherichia coli* of iron superoxide dismutase cDNA from *Leishmania donovani chagasi*. *Infect Immun* **62**:657–664. http://dx.doi.org/10.1128/iai.62.2.657-664.1994.

22. **Cue DR, Cleary PP.** 1997. High-frequency invasion of epithelial cells by *Streptococcus pyogenes* can be activated by fibrinogen and peptides containing the sequence RGD. *Infect Immun* **65**:2759–2764. http://dx.doi.org/10.1128/iai.65.7.2759-2764.1997.

23. **Marcato P, Mulvey G, Armstrong GD.** 2002. Cloned Shiga toxin 2 B subunit induces apoptosis in Ramos Burkitt's lymphoma B cells. *Infect Immun* **70**:1279–1286. http://dx.doi.org/10.1128/IAI.70.3.1279-1286.2002.

24. **Orme IM, Furney SK, Skinner PS, Roberts AD, Brennan PJ, Russell DG, Shiratsuchi H, Ellner JJ, Weiser WY.** 1993. Inhibition of growth of *Mycobacterium avium* in murine and human mononuclear phagocytes by migration inhibitory factor. *Infect Immun* **61**:338–342. http://dx.doi.org/10.1128/iai.61.1.338-342.1993.

25. **Khatua B, Ghoshal A, Bhattacharya K, Mandal C, Crocker PR, Mandal C.** 2008. Sialic acids, important constituents and selective recognition factors of *Pseudomonas aeruginosa*, an opportunistic pathogen. *Infect Immun* doi:10.1128/IAI.01083-08.

26. **Wager E, Barbour V, Yentis S, Kleinert S, COPE Council.** 2010. Retractions: guidance from the Committee on Publication Ethics (COPE). *Obes Rev* **11**:64–66. http://dx.doi.org/10.1111/j.1467-789X.2009.00702.x.

27. **Wager E, Williams P.** 2011. Why and how do journals retract articles? An analysis of Medline retractions 1988-2008. *J Med Ethics* **37**:567–570. http://dx.doi.org/10.1136/jme.2010.040964.

28. **Smith R.** 2005. Investigating the previous studies of a fraudulent author. *BMJ* **331**:288–291. http://dx.doi.org/10.1136/bmj.331.7511.288.

29. **Sox HC, Rennie D.** 2006. Research misconduct, retraction, and cleansing the medical literature: lessons from the Poehlman case. *Ann Intern Med* **144**:609–613. http://dx.doi.org/10.7326/0003-4819-144-8-200604180-00123.

30. **National Science Foundation.** 2002. *Research misconduct, 45 C.F.R. part 689.1.* National Science Foundation, Arlington, VA.

31. **Bird SJ**. 2002. Self-plagiarism and dual and redundant publications: what is the problem? Commentary on "Seven ways to plagiarize: handling real allegations of research misconduct." *Sci Eng Ethics* **8**:543–544. http://dx.doi.org/10.1007/s11948-002-0007-4.

32. **Kühberger A, Streit D, Scherndl T**. 2022. Self-correction in science: the effect of retraction on the frequency of citations. *PLoS One* **17**:e0277814. http://dx.doi.org/10.1371/journal.pone.0277814.

33. **Frasco PE, Smith BB, Murray AW, Khurmi N, Mueller JT, Poterack KA**. 2022. Context analysis of continued citation of retracted manuscripts published in anesthesiology journals. *Anesth Analg* **135**:1011–1020. http://dx.doi.org/10.1213/ANE.0000000000006195.

34. **Schneider J, Woods ND, Proescholdt R, Burns H, Howell K, Campbell MT, Hsiao T-K, Yip YYV, Fu Y, Arianlou Y, RISRS Team**. 2022. Reducing the inadvertent spread of retracted science: recommendations from the RISRS report. *Res Integr Peer Rev* **7**:6. http://dx.doi.org/10.1186/s41073-022-00125-x.

35. **Atlas MC**. 2004. Retraction policies of high-impact biomedical journals. *J Med Libr Assoc* **92**:242–250.

36. **American Society for Microbiology**. 2022. Publishing ethics. https://journals.asm.org/publishing-ethics.

37. **Marcus E**. 2005. Retraction controversy. *Cell* **123**:173–175. http://dx.doi.org/10.1016/j.cell.2005.10.007.

38. **Resnik DB, Wager E, Kissling GE**. 2015. Retraction policies of top scientific journals ranked by impact factor. *J Med Libr Assoc* **103**:136–139. http://dx.doi.org/10.3163/1536-5050.103.3.006.

39. **Souder L**. 2010. A rhetorical analysis of apologies for scientific misconduct: do they really mean it? *Sci Eng Ethics* **16**:175–184. http://dx.doi.org/10.1007/s11948-009-9149-y.

40. **Oransky I**. 14 March 2011. As last of 12 promised Bulfone-Paus retractions appears, a (disappointing) report card on journal transparency. Retraction Watch. http://retractionwatch.wordpress.com/2011/03/14/as-last-of-12-promised-bulfone-paus-retractions-appears-a-disappointing-report-card-on-journal-transparency/.

41. **Brooks D, Collins G**. 27 January 2010. What Obama should say tonight. *New York Times*. https://archive.nytimes.com/opinionator.blogs.nytimes.com/2010/01/27/what-obama-should-say-tonight-2/.

42. **Jacobson H**. 1955. Information, reproduction and the origin of life. *Am Sci* **43**:125.

43. **Dean C**. 25 October 2007. '55 "Origin of life" paper is retracted. *New York Times*. https://www.nytimes.com/2007/10/25/science/25jacobson.html.

44. **Anonymous**. 12 January 1920. Believes rocket can reach moon; Smithsonian Institution tells of Prof. Goddard's invention to explore upper air. Multiple-charge system instruments could go up 200 miles, and bigger rocket might land on satellite. *New York Times*. https://www.nytimes.com/1920/01/12/archives/believes-rocket-can-reach-moon-smithsonian-institution-tells-of.html.

45. *The Lancet*. 2015. Correcting the scientific literature: retraction and republication. *Lancet* **385**:394. http://dx.doi.org/10.1016/S0140-6736(15)60137-4.

46. **Bauchner H, Golub RM**. 2019. Ensuring an accurate scientific record—retraction and republication. *JAMA* **322**:1380. http://dx.doi.org/10.1001/jama.2019.14503.

47. **Cittadini E, Goadsby PJ**. 2005. Psychiatric side effects during methysergide treatment. *J Neurol Neurosurg Psychiatry* **76**:1037–1038. http://dx.doi.org/10.1136/jnnp.2004.048363.

48. **Fontanarosa PB, DeAngelis CD**. 2005. Correcting the literature—retraction and republication. *JAMA* **293**:2536. http://dx.doi.org/10.1001/jama.293.20.2536.

49. **Khatua B, Ghoshal A, Bhattacharya K, Mandal C, Saha B, Crocker PR, Mandal C**. 2010. Sialic acids acquired by *Pseudomonas aeruginosa* are involved in reduced complement deposition and siglec mediated host-cell recognition. *FEBS Lett* **584**:555–561. http://dx.doi.org/10.1016/j.febslet.2009.11.087.

50. **Levin LI, Munger KL, Rubertone MV, Peck CA, Lennette ET, Spiegelman D, Ascherio A**. 2005. Temporal relationship between elevation of Epstein-Barr virus antibody titers and initial onset of

neurological symptoms in multiple sclerosis. *JAMA* **293**:2496–2500. http://dx.doi.org/10.1001/jama.293.20.2496.

51. **Levin LI, Munger KL, Rubertone MV, Peck CA, Lennette ET, Spiegelman D, Ascherio A**. 2003. Multiple sclerosis and Epstein-Barr virus. *JAMA* **289**:1533–1536. http://dx.doi.org/10.1001/jama.289.12.1533.

52. **Liu SV**. 2006. Top journals' top retraction rates. *Sci Ethics* **1**:91–93.

53. **Cokol M, Iossifov I, Rodriguez-Esteban R, Rzhetsky A**. 2007. How many scientific papers should be retracted? *EMBO Rep* **8**:422–423. http://dx.doi.org/10.1038/sj.embor.7400970.

54. **Bloch S, Walter G**. 2001. The Impact Factor: time for change. *Aust N Z J Psychiatry* **35**:563–568. http://dx.doi.org/10.1080/0004867010060502.

55. **Fersht A**. 2009. The most influential journals: Impact Factor and Eigenfactor. *Proc Natl Acad Sci U S A* **106**:6883–6884. http://dx.doi.org/10.1073/pnas.0903307106.

56. **Hansson S**. 1995. Impact factor as a misleading tool in evaluation of medical journals. *Lancet* **346**:906. http://dx.doi.org/10.1016/S0140-6736(95)92749-2.

57. **Seglen PO**. 1997. Why the impact factor of journals should not be used for evaluating research. *BMJ* **314**:498–502. http://dx.doi.org/10.1136/bmj.314.7079.497.

58. **Smith R**. 2008. Beware the tyranny of impact factors. *J Bone Joint Surg Br* **90**:125–126. http://dx.doi.org/10.1302/0301-620X.90B2.20258.

59. **Szklo M**. 2008. Impact factor: good reasons for concern. *Epidemiology* **19**:369. http://dx.doi.org/10.1097/EDE.0b013e31816b6a7a.

60. **Fanelli D**. 2009. How many scientists fabricate and falsify research? A systematic review and meta-analysis of survey data. *PLoS One* **4**:e5738. http://dx.doi.org/10.1371/journal.pone.0005738.

61. **Zimmer C**. 16 April 2012. A sharp rise in retractions prompts calls for reform. *New York Times*. https://carlzimmer.com/a-sharp-rise-in-retractions-prompts-calls-for-reform-264/.

62. **Steen RG, Casadevall A, Fang FC**. 2013. Why has the number of scientific retractions increased? *PLoS One* **8**:e68397. http://dx.doi.org/10.1371/journal.pone.0068397.

63. **Ahluwalia J, Tinker A, Clapp LH, Duchen MR, Abramov AY, Pope S, Nobles M, Segal AW**. 2004. The large-conductance Ca^{2+}-activated K^+ channel is essential for innate immunity. *Nature* **427**:853–858. http://dx.doi.org/10.1038/nature02356.

64. **Hwang WS, Ryu YJ, Park JH, Park ES, Lee EG, Koo JM, Jeon HY, Lee BC, Kang SK, Kim SJ, Ahn C, Hwang JH, Park KY, Cibelli JB, Moon SY**. 2004. Evidence of a pluripotent human embryonic stem cell line derived from a cloned blastocyst. *Science* **303**:1669–1674. http://dx.doi.org/10.1126/science.1094515.

65. **Hwang WS, Roh SI, Lee BC, Kang SK, Kwon DK, Kim S, Kim SJ, Park SW, Kwon HS, Lee CK, Lee JB, Kim JM, Ahn C, Paek SH, Chang SS, Koo JJ, Yoon HS, Hwang JH, Hwang YY, Park YS, Oh SK, Kim HS, Park JH, Moon SY, Schatten G**. 2005. Patient-specific embryonic stem cells derived from human SCNT blastocysts. *Science* **308**:1777–1783. http://dx.doi.org/10.1126/science.1112286.

66. **Potti A, Mukherjee S, Petersen R, Dressman HK, Bild A, Koontz J, Kratzke R, Watson MA, Kelley M, Ginsburg GS, West M, Harpole DH, Jr, Nevins JR**. 2006. A genomic strategy to refine prognosis in early-stage non-small-cell lung cancer. *N Engl J Med* **355**:570–580. http://dx.doi.org/10.1056/NEJMoa060467.

67. **Schon JH, Berg S, Kloc C, Batlogg B**. 2000. Ambipolar pentacene field-effect transistors and inverters. *Science* **287**:1022–1023. http://dx.doi.org/10.1126/science.287.5455.1022.

68. **Schön JH, Dodabalapur A, Kloc C, Batlogg B**. 2000. A light-emitting field-effect transistor. *Science* **290**:963–966. http://dx.doi.org/10.1126/science.290.5493.963.

69. **Schon JH, Kloc C, Batlogg B**. 2000. Fractional quantum hall effect in organic molecular semiconductors. *Science* **288**:2339–2340. http://dx.doi.org/10.1126/science.288.5475.2338.

70. **Schon JH, Kloc C, Dodabalapur A, Batlogg B**. 2000. An organic solid state injection laser. *Science* **289**:599–601. http://dx.doi.org/10.1126/science.289.5479.599.

71. **Schon JH, Kloc C, Haddon RC, Batlogg B**. 2000. A superconducting field-effect switch. *Science* **288**:656–658. http://dx.doi.org/10.1126/science.288.5466.656.

72. **Schön JH, Kloc C, Batlogg B**. 2001. High-temperature superconductivity in lattice-expanded C60. *Science* **293**:2432–2434. http://dx.doi.org/10.1126/science.1064773.

73. **Schön JH, Kloc C, Hwang HY, Batlogg B**. 2001. Josephson junctions with tunable weak links. *Science* **292**:252–254. http://dx.doi.org/10.1126/science.1058812.

74. **Schön JH, Meng H, Bao Z**. 2001. Field-effect modulation of the conductance of single molecules. *Science* **294**:2138–2140. http://dx.doi.org/10.1126/science.1066171.

75. **Wakefield AJ, Murch SH, Anthony A, Linnell J, Casson DM, Malik M, Berelowitz M, Dhillon AP, Thomson MA, Harvey P, Valentine A, Davies SE, Walker-Smith JA**. 1998. Ileal-lymphoid-nodular hyperplasia, non-specific colitis, and pervasive developmental disorder in children. *Lancet* **351**:637–641. http://dx.doi.org/10.1016/S0140-6736(97)11096-0.

76. **Abbott A, Schwarz J**. 2002. Dubious data remain in print two years after misconduct inquiry. *Nature* **418**:113. http://dx.doi.org/10.1038/418113a.

77. **Couzin J, Unger K**. 2006. Scientific misconduct. Cleaning up the paper trail. *Science* **312**:38–43. http://dx.doi.org/10.1126/science.312.5770.38.

78. **DeCoursey TE**. 2006. It's difficult to publish contradictory findings. *Nature* **439**:784. http://dx.doi.org/10.1038/439784b.

79. **Trikalinos NA, Evangelou E, Ioannidis JP**. 2008. Falsified papers in high-impact journals were slow to retract and indistinguishable from nonfraudulent papers. *J Clin Epidemiol* **61**:464–470. http://dx.doi.org/10.1016/j.jclinepi.2007.11.019.

80. **Zimmer C**. 25 June 2011. It's science, but not necessarily right. *New York Times*. https://www.nytimes.com/2011/06/26/opinion/sunday/26ideas.html.

81. **Parrish DM**. 1999. Scientific misconduct and correcting the scientific literature. *Acad Med* **74**: 221–230. http://dx.doi.org/10.1097/00001888-199903000-00009.

82. **Budd JM, Sievert M, Schultz TR**. 1998. Phenomena of retraction: reasons for retraction and citations to the publications. *JAMA* **280**:296–297. http://dx.doi.org/10.1001/jama.280.3.296.

83. **Drury NE, Karamanou DM**. 2009. Citation of retracted articles: a call for vigilance. *Ann Thorac Surg* **87**:670. http://dx.doi.org/10.1016/j.athoracsur.2008.07.108.

84. **Korpela KM**. 2010. How long does it take for the scientific literature to purge itself of fraudulent material?: the Breuning case revisited. *Curr Med Res Opin* **26**:843–847. http://dx.doi.org/10.1185/03007991003603804.

85. **Peterson GM**. 2010. The effectiveness of the practice of correction and republication in the biomedical literature. *J Med Libr Assoc* **98**:135–139. http://dx.doi.org/10.3163/1536-5050.98.2.005.

86. **Redman BK, Yarandi HN, Merz JF**. 2008. Empirical developments in retraction. *J Med Ethics* **34**:807–809. http://dx.doi.org/10.1136/jme.2007.023069.

87. **Tatsioni A, Bonitsis NG, Ioannidis JP**. 2007. Persistence of contradicted claims in the literature. *JAMA* **298**:2517–2526. http://dx.doi.org/10.1001/jama.298.21.2517.

88. **Whitely WP, Rennie D, Hafner AW**. 1994. The scientific community's response to evidence of fraudulent publication. The Robert Slutsky case. *JAMA* **272**:170–173. http://dx.doi.org/10.1001/jama.1994.03520020096029.

89. **Iyengar R, Wang Y, Chow J, Charney DS**. 2009. An integrated approach to evaluate faculty members' research performance. *Acad Med* **84**:1610–1616. http://dx.doi.org/10.1097/ACM.0b013e3181bb2364.

90. **Funk CL, Barrett KA, Macrina FL**. 2007. Authorship and publication practices: evaluation of the effect of responsible conduct of research instruction to postdoctoral trainees. *Account Res* **14**: 269–305. http://dx.doi.org/10.1080/08989620701670187.

91. **Buranen L, Roy AM**. 1999. *Perspectives on Plagiarism and Intellectual Property in a Postmodern World*. SUNY Press, Albany, NY.

92. **Errami M, Garner H**. 2008. A tale of two citations. *Nature* **451**:397–399. http://dx.doi.org/10.1038/451397a.

93. **Giles J**. 2005. Special report: taking on the cheats. *Nature* **435**:258–259. http://dx.doi.org/10.1038/435258a.

94. **Long TC, Errami M, George AC, Sun Z, Garner HR**. 2009. Scientific integrity. Responding to possible plagiarism. *Science* **323**:1293–1294. http://dx.doi.org/10.1126/science.1167408.

95. **Vihinen M.** 2009. Problems with anti-plagiarism database. *Nature* **457**:26. http://dx.doi. org/10.1038/457026b.

96. **Brentlinger PE, Behrens CB, Micek MA, Steketee RW, Andrews KT, Skinner-Adams TS, Gardiner DL, McCarthy JS, Parikh S, ter Kuile F, Ayisi J**. 2009. Plagiarism. *Trans R Soc Trop Med Hyg* **103**:855, author reply 855–856. http://dx.doi.org/10.1016/j.trstmh.2009.05.006.

97. **Derby B.** 2008. Duplication and plagiarism increasing among students. *Nature* **452**:29. http:// dx.doi.org/10.1038/452029c.

98. **Jackson AC, Ronald AR, Steiner I**. 2010. Plagiarism. *Trans R Soc Trop Med Hyg* **104**:173, discussion 173–174. http://dx.doi.org/10.1016/j.trstmh.2009.10.002.

99. **Vessal K, Habibzadeh F**. 2007. Rules of the game of scientific writing: fair play and plagiarism. *Lancet* **369**:641. http://dx.doi.org/10.1016/S0140-6736(07)60307-9.

100. **Wicker P.** 2007. Plagiarism: understanding and management. *J Perioper Pract* **17**:372–382, 377–382. http://dx.doi.org/10.1177/175045890701700802.

101. **Yilmaz I.** 2007. Plagiarism? No, we're just borrowing better English. *Nature* **449**:658. http:// dx.doi.org/10.1038/449658a.

102. **Zhang Y.** 2010. Chinese journal finds 31% of submissions plagiarized. *Nature* **467**:153. http:// dx.doi.org/10.1038/467153d.

103. **Stern AM, Casadevall A, Steen RG, Fang FC**. 2014. Financial costs and personal consequences of research misconduct resulting in retracted publications. *eLife* **3**:e02956. http://dx.doi. org/10.7554/eLife.02956.

104. **van der Vet PE, Nijveen H**. 2016. Propagation of errors in citation networks: a study involving the entire citation network of a widely cited paper published in, and later retracted from, the journal Nature. *Res Integr Peer Rev* **1**:3. http://dx.doi.org/10.1186/s41073-016-0008-5.

105. **Oransky I.** 24 September 2012. Iranian mathematicians latest to have papers retracted for fake email addresses to get better reviews. Retraction Watch. https://retractionwatch. com/2012/09/24/iranian-mathematicians-latest-to-have-papers-retracted-for-fake-email-addresses-to-get-better-reviews/.

106. **Wilford JN.** 13 July 1985. Ex-Columbia worker takes blame for errors. *New York Times.* https:// www.nytimes.com/1985/07/13/us/ex-columbia-worker-takes-blame-for-errors.html.

107. **King GL, Hu KQ**. 1991. Endothelin stimulates a sustained 1,2-diacylglycerol increase and protein kinase C activation in bovine aortic smooth muscle cells. *Biochem Biophys Res Commun* **180**:1164. http://dx.doi.org/10.1016/S0006-291X(05)81189-1.

108. Office of Research Integrity. 1993. NIH guide. https://grants.nih.gov/grants/guide/notice-les/ not93-177.html.

109. **Bois PR, Izeradjene K, Houghton PJ, Cleveland JL, Houghton JA, Grosveldz GC**. 2007. FOXO1a acts as a selective tumor suppressor in alveolar rhabdomyosarcoma. *J Cell Biol* **177**:563. http://dx.doi.org/10.1083/jcb.20050104020070420r.

110. **Office of Research Integrity**. 2011. Annual Reports (2001-2010). U.S. Department of Health and Human Services, Washington, DC.

111. **Gaudino M, Robinson NB, Audisio K, Rahouma M, Benedetto U, Kurlansky P, Fremes SE**. 2021. Trends and characteristics of retracted articles in the biomedical literature, 1971 to 2020. *JAMA Intern Med* **181**:1118–1121. https://doi.org/10.1001/jamainternmed.2021.1807.

112. **Oransky I.** 2022. Nearing 5,000 retractions: a review of 2022. *Retraction Watch.* https://retractionwatch. com/2022/12/27/nearing-5000-retractions-a-review-of-2022/.

113. **Horton R.** 2005. Expression of concern: Indo-Mediterranean Diet Heart Study. *Lancet* **366**:354–356. http://dx.doi.org/10.1016/S0140-6736(05)67006-7.

114. **Editors of** *BMJ*. 2005. Expression of concern. *BMJ* **331**:266. http://dx.doi.org/10.1136/ bmj.331.7511.266.

115. **Singh RB, Rastogi SS, Verma R, Laxmi B, Singh R, Ghosh S, Niaz MA**. 1992. Randomised controlled trial of cardioprotective diet in patients with recent acute myocardial infarction: results of one year follow up. *BMJ* **304**:1015–1019. http://dx.doi.org/10.1136/ bmj.304.6833.1015.

116. **Singh RB, Dubnov G, Niaz MA, Ghosh S, Singh R, Rastogi SS, Manor O, Pella D, Berry EM**. 2002. Effect of an Indo-Mediterranean diet on progression of coronary artery disease in high risk patients (Indo-Mediterranean Diet Heart Study): a randomised single-blind trial. *Lancet* **360**: 1455–1461. http://dx.doi.org/10.1016/S0140-6736(02)11472-3.

117. **Spector M, O'Neal S, Racker E**. 1980. Phosphorylation of the beta subunit of Na⁺K⁺-ATPase in Ehrlich ascites tumor by a membrane-bound protein kinase. *J Biol Chem* **255**:8370–8373. http://dx.doi.org/10.1016/S0021-9258(18)43499-0.

118. **Spector M, Pepinsky RB, Vogt VM, Racker E**. 1981. A mouse homolog to the avian sarcoma virus src protein is a member of a protein kinase cascade. *Cell* **25**:9–21. http://dx.doi.org/10.1016/0092-8674(81)90227-0.

119. **Racker E**. 1989. A view of misconduct in science. *Nature* **339**:91–93. http://dx.doi.org/10.1038/339091a0.

120. **Meguid MM**. 2005. Retraction. *Nutrition* **21**:286. http://dx.doi.org/10.1016/j.nut.2004.12.002.

121. **Marcus A, Oransky I**. 2014. What studies of retractions tell us *J Microbiol Biol Educ* **15**:151–154. https://doi.org/10.1128/jmbe.v15i2.855.

122. **Nath SB, Marcus SC, Druss BG**. 2006. Retractions in the research literature: misconduct or mistakes? *Med J Aust* **185**:152–154. http://dx.doi.org/10.5694/j.1326-5377.2006.tb00504.x.

123. **Kornfeld DS**. 2012. Perspective: research misconduct: the search for a remedy. *Acad Med* **87**:877–882. http://dx.doi.org/10.1097/ACM.0b013e318257ee6a.

124. **Campos-Varela I, Ruano-Raviña A**. 2019. Misconduct as the main cause for retraction. A descriptive study of retracted publications and their authors. *Gac Sanit* **33**:356–360. http://dx.doi.org/10.1016/j.gaceta.2018.01.009.

125. **Kocyigit BF, Akyol A**. 2022. Analysis of retracted publications in the biomedical literature from Turkey. *J Korean Med Sci* **37**:e142. http://dx.doi.org/10.3346/jkms.2022.37.e142.

126. **Stavale R, Ferreira GI, Galvão JAM, Zicker F, Novaes MRCG, Oliveira CM, Guilhem D**. 2019. Research misconduct in health and life sciences research: a systematic review of retracted literature from Brazilian institutions. *PLoS One* **14**:e0214272. http://dx.doi.org/10.1371/journal.pone.0214272.

127. **Kamali N, Talebi Bezmin Abadi A, Rahimi F**. 2020. Plagiarism, fake peer-review, and duplication: predominant reasons underlying retractions of Iran-affiliated scientific papers. *Sci Eng Ethics* **26**:3455–3463. http://dx.doi.org/10.1007/s11948-020-00274-6.

128. **Steen RG**. 2011. Retractions in the scientific literature: do authors deliberately commit research fraud? *J Med Ethics* **37**:113–117. http://dx.doi.org/10.1136/jme.2010.038125.

129. **Poulton A**. 2007. Mistakes and misconduct in the research literature: retractions just the tip of the iceberg. *Med J Aust* **186**:323–324. http://dx.doi.org/10.5694/j.1326-5377.2007.tb00917.x.

130. **Brainard J, You J**. 25 October 2018. What a massive database of retracted papers reveals about science publishing's "death penalty." *Science*. https://www.science.org/content/article/what-massive-database-retracted-papers-reveals-about-science-publishing-s-death-penalty.

131. **Collaborative Working Group from the conference "Keeping the Pool Clean: Prevention and Management of Misconduct Related Retractions."** 2018. RePAIR consensus guidelines: responsibilities of Publishers, Agencies, Institutions, and Researchers in protecting the integrity of the research record. *Res Integr Peer Rev* **3**:15. http://dx.doi.org/10.1186/s41073-018-0055-1.

132. **Casadevall A, Fang FC**. 2012. Reforming science: methodological and cultural reforms. *Infect Immun* **80**:891–896. http://dx.doi.org/10.1128/IAI.06183-11.

133. **Ioannidis JP**. 2005. Why most published research findings are false. *PLoS Med* **2**:e124. http://dx.doi.org/10.1371/journal.pmed.0020124.

134. **Begley CG, Ellis LM**. 2012. Drug development: raise standards for preclinical cancer research. *Nature* **483**:531–533. http://dx.doi.org/10.1038/483531a.

135. **Martinson BC, Anderson MS, de Vries R**. 2005. Scientists behaving badly. *Nature* **435**:737–738. http://dx.doi.org/10.1038/435737a.

136. **Stephan P**. 2012. Research efficiency: perverse incentives. *Nature* **484**:29–31. http://dx.doi.org/10.1038/484029a.

137. **Fang FC, Casadevall A**. 2012. Reforming science: structural reforms. *Infect Immun* **80**:897–901. http://dx.doi.org/10.1128/IAI.06184-11.

138. **Casadevall A, Fang FC**. 2012. Winner takes all. *Sci Am* **307**:13. http://dx.doi.org/10.1038/scientificamerican0812-13.

139. **Avery S**. 9 January 2011. Flawed research appalls cancer patient. *Raleigh News and Observer.*

140. **Steen RG**. 2011. Retractions in the medical literature: how many patients are put at risk by flawed research? *J Med Ethics* **37**:688–692. http://dx.doi.org/10.1136/jme.2011.043133.

141. **Arst HN, Jr**. 2000. Apathy rewards misconduct—and everybody suffers. *Nature* **403**:478. http://dx.doi.org/10.1038/35000742.

142. **Institute for Scientific Information**. 2011. *ISI Journal Citation Reports 2010*. Thomson Reuters, New York, NY.

143. **Bell K, Kingori P, Mills D**. 17 July 2022. Scholarly publishing, boundary processes, and the problem of fake peer reviews. *Science, Technology, & Human Values*. http://dx.doi.org/10.1177/01622439221112463.

Chapter 25 – Erroneous Science

1. **Jefferson T**. 16 May 1792. Letter to George Washington. Philadelphia, PA.

2. **Brown AW, Kaiser KA, Allison DB**. 2018. Issues with data and analyses: errors, underlying themes, and potential solutions. *Proc Natl Acad Sci U S A* **115**:2563–2570. http://dx.doi.org/10.1073/pnas.1708279115.

3. **Solomon SM, Hackett EJ**. 1996. Setting boundaries between science and law: lessons from Daubert v. Merrell Dow Pharmaceuticals, Inc. *Sci Technol Hum Values* **21**:131–156.

4. **Merriam-Webster**. 2022. Merriam-Webster online dictionary. https://www.merriam-webster.com/dictionary/error.

5. **Fang FC, Steen RG, Casadevall A**. 2012. Misconduct accounts for the majority of retracted scientific publications. *Proc Natl Acad Sci U S A* **109**:17028–17033. http://dx.doi.org/10.1073/pnas.1212247109.

6. **Struys MM, Fechner J, Schüttler J, Schwilden H**. 2010. Erroneously published fospropofol pharmacokinetic-pharmacodynamic data and retraction of the affected publications. *Anesthesiology* **112**:1056–1057. http://dx.doi.org/10.1097/ALN.0b013e3181d536df.

7. **Quik M, Cook RG, Revah F, Changeux JP, Patrick J**. 1993. Presence of alpha-cobratoxin and phospholipase A2 activity in thymopoietin preparations. *Mol Pharmacol* **44**:678–679.

8. **Nardone RM**. 2007. Eradication of cross-contaminated cell lines: a call for action. *Cell Biol Toxicol* **23**:367–372. http://dx.doi.org/10.1007/s10565-007-9019-9.

9. **Rojas A, Gonzalez I, Figueroa H**. 2008. Cell line cross-contamination in biomedical research: a call to prevent unawareness. *Acta Pharmacol Sin* **29**:877–880. http://dx.doi.org/10.1111/j.1745-7254.2008.00809.x.

10. **Knight LA, Cree IA**. 2011. Quality assurance and good laboratory practice. *Methods Mol Biol* **731**:115–124. http://dx.doi.org/10.1007/978-1-61779-080-5_10.

11. **Barallon R, Bauer SR, Butler J, Capes-Davis A, Dirks WG, Elmore E, Furtado M, Kline MC, Kohara A, Los GV, MacLeod RA, Masters JR, Nardone M, Nardone RM, Nims RW, Price PJ, Reid YA, Shewale J, Sykes G, Steuer AF, Storts DR, Thomson J, Taraporewala Z, Alston-Roberts C, Kerrigan L**. 2010. Recommendation of short tandem repeat profiling for authenticating human cell lines, stem cells, and tissues. *In Vitro Cell Dev Biol Anim* **46**:727–732. http://dx.doi.org/10.1007/s11626-010-9333-z.

12. **Nelson-Rees WA, Daniels DW, Flandermeyer RR**. 1981. Cross-contamination of cells in culture. *Science* **212**:446–452. http://dx.doi.org/10.1126/science.6451928.

13. **Drexler HG, Dirks WG, Matsuo Y, MacLeod RA**. 2003. False leukemia-lymphoma cell lines: an update on over 500 cell lines. *Leukemia* **17**:416–426. http://dx.doi.org/10.1038/sj.leu.2402799.

14. **MacLeod RA, Dirks WG, Matsuo Y, Kaufmann M, Milch H, Drexler HG**. 1999. Widespread intraspecies cross-contamination of human tumor cell lines arising at source. *Int J Cancer* **83**: 555–563.http://dx.doi.org/10.1002/(SICI)1097-0215(19991112)83:4<555::AID-IJC19>3.0.CO;2-2.

15. **Capes-Davis A, Theodosopoulos G, Atkin I, Drexler HG, Kohara A, MacLeod RA, Masters JR, Nakamura Y, Reid YA, Reddel RR, Freshney RI**. 2010. Check your cultures! A list of cross-contaminated or misidentified cell lines. *Int J Cancer* **127**:1–8. http://dx.doi.org/10.1002/ijc.25242.

16. **Phuchareon J, Ohta Y, Woo JM, Eisele DW, Tetsu O**. 2009. Genetic profiling reveals cross-contamination and misidentification of 6 adenoid cystic carcinoma cell lines: ACC2, ACC3, ACCM, ACCNS, ACCS and CAC2. *PLoS One* **4**:e6040. http://dx.doi.org/10.1371/journal.pone.0006040.

17. **Neimark J**. 2015. Line of attack. *Science* **347**:938–940. http://dx.doi.org/10.1126/science.347.6225.938.

18. **Begley CG, Ellis LM**. 2012. Drug development: raise standards for preclinical cancer research. *Nature* **483**:531–533. http://dx.doi.org/10.1038/483531a.

19. **Miller G**. 2006. Scientific publishing. A scientist's nightmare: software problem leads to five retractions. *Science* **314**:1856–1857. http://dx.doi.org/10.1126/science.314.5807.1856.

20. **Rekers H, Affandi B**. 2004. Implanon studies conducted in Indonesia. *Contraception* **70**:433.

21. **Joppa LN, McInerny G, Harper R, Salido L, Takeda K, O'Hara K, Gavaghan D, Emmott S**. 2013. Computational science. Troubling trends in scientific software use. *Science* **340**:814–815. http://dx.doi.org/10.1126/science.1231535.

22. **Anonymous**. 18 October 2013. Unreliable research: trouble at the lab. *Economist*. https://www.economist.com/briefing/2013/10/18/trouble-at-the-lab.

23. **Anonymous**. 21 October 2013. Problems with scientific research: how science goes wrong. *Economist*. https://www.economist.com/leaders/2013/10/21/how-science-goes-wrong.

24. **Johnson G**. 20 January 2014. New truths that only one can see. *New York Times*. http://www.nytimes.com/2014/01/21/science/new-truths-that-only-one-can-see.html?_r=0.

25. **Conroy G**. 3 March 2020. Scientists reveal what they learnt from their biggest mistakes. *Nature*. https://www.nature.com/nature-index/news-blog/scientists-reveal-what-they-learnt-from-their-biggest-mistakes.

26. **Vasilevsky NA, Brush MH, Paddock H, Ponting L, Tripathy SJ, Larocca GM, Haendel MA**. 2013. On the reproducibility of science: unique identification of research resources in the biomedical literature. *PeerJ* **1**:e148. http://dx.doi.org/10.7717/peerj.148.

27. **Ram K**. 2013. Git can facilitate greater reproducibility and increased transparency in science. *Source Code Biol Med* **8**:7. http://dx.doi.org/10.1186/1751-0473-8-7.

28. ter **Riet G, Korevaar DA, Leenaars M, Sterk PJ, Van Noorden CJ, Bouter LM, Lutter R, Elferink RP, Hooft L**. 2012. Publication bias in laboratory animal research: a survey on magnitude, drivers, consequences and potential solutions. *PLoS One* **7**:e43404. http://dx.doi.org/10.1371/journal.pone.0043404.

29. **Loscalzo J**. 2012. Irreproducible experimental results: causes, (mis)interpretations, and consequences. *Circulation* **125**:1211–1214. http://dx.doi.org/10.1161/CIRCULATIONAHA.112.098244.

30. **Johnson VE**. 2013. Revised standards for statistical evidence. *Proc Natl Acad Sci U S A* **110**: 19313–19317. http://dx.doi.org/10.1073/pnas.1313476110.

31. **McShane BB, Gal D, Gelman A, Robert C, Tackett JL**. 2019. Abandon statistical significance. *Am Stat* **73**:235–245. http://dx.doi.org/10.1080/00031305.2018.1527253.

32. **Couzin-Frankel J**. 2012. Research quality. Service offers to reproduce results for a fee. *Science* **337**:1031. http://dx.doi.org/10.1126/science.337.6098.1031.

33. **Prinz F, Schlange T, Asadullah K**. 2011. Believe it or not: how much can we rely on published data on potential drug targets? *Nat Rev Drug Discov* **10**:712. http://dx.doi.org/10.1038/nrd3439-c1.

34. **Brown EN, Ramaswamy S**. 2007. Quality of protein crystal structures. *Acta Crystallogr D Biol Crystallogr* **63**:941–950. http://dx.doi.org/10.1107/S0907444907033847.

35. **Macarthur D.** 2012. Methods: face up to false positives. *Nature* **487**:427–428. http://dx.doi. org/10.1038/487427a.

36. **Wager E, Williams P.** 2011. Why and how do journals retract articles? An analysis of Medline retractions 1988-2008. *J Med Ethics* **37**:567–570. http://dx.doi.org/10.1136/jme.2010.040964.

37. **Grcar J.** 2013. Comments and corrigenda in scientific literature. *Am Sci* **101**:16. http://dx.doi. org/10.1511/2013.100.16.

38. **Chow YK, Hirsch MS, Merrill DP, Bechtel LJ, Eron JJ, Kaplan JC, D'Aquila RT.** 1993. Use of evolutionary limitations of HIV-1 multidrug resistance to optimize therapy. *Nature* **361**:650–654. http://dx.doi.org/10.1038/361650a0.

39. **Larder BA, Kellam P, Kemp SD.** 1993. Convergent combination therapy can select viable multidrug-resistant HIV-1 *in vitro*. *Nature* **365**:451–453. http://dx.doi.org/10.1038/365451a0.

40. **Chow YK, Hirsch MS, Kaplan JC, D'Aquila RT.** 1993. HIV-1 error revealed. *Nature* **364**:679. http://dx.doi.org/10.1038/364679a0.

41. **Wolfe-Simon F, Switzer Blum J, Kulp TR, Gordon GW, Hoeft SE, Pett-Ridge J, Stolz JF, Webb SM, Weber PK, Davies PC, Anbar AD, Oremland RS.** 2011. A bacterium that can grow by using arsenic instead of phosphorus. *Science* **332**:1163–1166. http://dx.doi.org/10.1126/science.1197258.

42. **Erb TJ, Kiefer P, Hattendorf B, Günther D, Vorholt JA.** 2012. GFAJ-1 is an arsenate-resistant, phosphate-dependent organism. *Science* **337**:467–470. http://dx.doi.org/10.1126/science.1218455.

43. **Reaves ML, Sinha S, Rabinowitz JD, Kruglyak L, Redfield RJ.** 2012. Absence of detectable arsenate in DNA from arsenate-grown GFAJ-1 cells. *Science* **337**:470–473. http://dx.doi.org/10.1126/ science.1219861.

44. **Alberts B.** 2011. Editor's note. *Science* **332**:1149. http://dx.doi.org/10.1126/science.1208877.

45. **Tomkins JL, Penrose MA, Greeff J, Lebas NR.** 2011. Retraction. *Science* **333**:1220. http://dx.doi. org/10.1126/science.333.6047.1220-a.

46. **Casadevall A, Fang FC.** 2012. Reforming science: methodological and cultural reforms. *Infect Immun* **80**:891–896. http://dx.doi.org/10.1128/IAI.06183-11.

47. **Ellison SLR, Hardcastle WA.** 2012. Causes of error in analytical chemistry: results of a web-based survey of proficiency testing participants. *Accredit Qual Assur* **17**:453–464. http://dx.doi. org/10.1007/s00769-012-0894-2.

48. **Young P.** 1990. Pulsar mystery ends: the TV camera did it. *Sci News* **137**:119. http://dx.doi. org/10.2307/3974642.

49. **Powell D.** 23 February 2012. Loose cable blamed for speedy neutrinos. *Science News*. https://www. sciencenews.org/article/loose-cable-blamed-speedy-neutrinos.

50. **Grant A.** 30 January 2015. Dust erases evidence for gravity wave detection. *Science News*. https:// www.sciencenews.org/article/dust-erases-evidence-gravity-wave-detection.

51. **Kerr RA.** 30 September 1999. English-metric miscue doomed Mars mission. *Science*. https://www. science.org/content/article/english-metric-miscue-doomed-mars-mission.

52. **Klassen M.** 28 August 2020. A software bug caused by the phase of the moon. Medium. https:// mikhailklassen.medium.com/a-software-bug-caused-by-the-phase-of-the-moon-998136f31eff.

53. **Lamb E.** 17 July 2012. 5 sigma what's that? Scientific American Blogs. http://blogs. scientificamerican.com/observations/five-sigmawhats-that.

54. **Conroy G.** 9 September 2019. Q&A Daniele Fanelli: Retracting your own paper can lead to a spike in citations. *Nature Index*. https://www.nature.com/nature-index/news-blog/ daniele-fanelli-do-retractions-make-a-difference.

55. **Curnoe D.** 19 December 2016. The biggest mistake in the history of science. The Conversation. https://theconversation.com/the-biggest-mistake-in-the-history-of-science-70575.

56. **Wagner JK, Yu JH, Ifekwunigwe JO, Harrell TM, Bamshad MJ, Royal CD.** 2017. Anthropologists' views on race, ancestry, and genetics. *Am J Phys Anthropol* **162**:318–327. http:// dx.doi.org/10.1002/ajpa.23120.

57. **Traver JR.** 1951. Unusual scalp dermatitis in humans caused by the mite, Dermatophagoides. (Acarine, Epidermoptidae). *Proc Entomol Soc Wash* **53**:1–25.

58. **Shelomi M.** 2013. Mad scientist: the unique case of a published delusion. *Sci Eng Ethics* **19**:381–388. http://dx.doi.org/10.1007/s11948-011-9339-2.

59. **Steinschneider A.** 1972. Prolonged apnea and the sudden infant death syndrome: clinical and laboratory observations. *Pediatrics* **50**:646–654. http://dx.doi.org/10.1542/peds.50.4.646.

60. **Pinholster G.** 1994. SIDS paper triggers a murder charge. *Science* **264**:197–198. http://dx.doi.org/10.1126/science.8146647.

61. **Davenas E, Beauvais F, Amara J, Oberbaum M, Robinzon B, Miadonnai A, Tedeschi A, Pomeranz B, Fortner P, Belon P, Sainte-Laudy J, Poitevin B, Benveniste J**. 1988. Human basophil degranulation triggered by very dilute antiserum against IgE. *Nature* **333**:816–818. http://dx.doi.org/10.1038/333816a0.

62. **Maddox J, Randi J, Stewart WW**. 1988. "High-dilution" experiments a delusion. *Nature* **334**:287–291. http://dx.doi.org/10.1038/334287a0.

63. **Hirst SJ, Hayes NA, Burridge J, Pearce FL, Foreman JC**. 1993. Human basophil degranulation is not triggered by very dilute antiserum against human IgE. *Nature* **366**:525–527. http://dx.doi.org/10.1038/366525a0.

64. **Benveniste J, Ducot B, Spira A**. 1994. Memory of water revisited. *Nature* **370**:322. http://dx.doi.org/10.1038/370322a0.

65. **Fleischmann M, Pons S**. 1989. Electrochemically induced nuclear fusion of deuterium. *J Electroanal Chem* **261**:301–308. http://dx.doi.org/10.1016/0022-0728(89)80006-3.

66. **Fleischmann M, Pons S, Hawkins M**. 1989. Errata. *J Electroanal Chem* **263**:187–188.

67. **Miskelly GM, Heben MJ, Kumar A, Penner RM, Sailor MJ, Lewis NS**. 1989. Analysis of the published calorimetric evidence for electrochemical fusion of deuterium in palladium. *Science* **246**:793–796. http://dx.doi.org/10.1126/science.246.4931.793.

68. **Salamon MH, Wrenn ME, Bergeson HE, Crawford HC, Delaney WH, Henderson CL, Li YQ, Rusho JA, Sandquist GM, Seltzer SM**. 1990. Limits on the emission of neutrons, γ-rays, electrons and protons from Pons/Fleischmann electrolytic cells. *Nature* **344**:401–405. http://dx.doi.org/10.1038/344401a0.

69. **Bagenal FS, Easton DF, Harris E, Chilvers CE, McElwain TJ**. 1990. Survival of patients with breast cancer attending Bristol Cancer Help Centre. *Lancet* **336**:606–610. http://dx.doi.org/10.1016/0140-6736(90)93402-B.

70. **Goodare H.** 1994. The scandal of poor medical research. Wrong results should be withdrawn. *BMJ* **308**:593.

71. **Emini EA, Graham DJ, Gotlib L, Condra JH, Byrnes VW, Schleif WA**. 1993. HIV and multidrug resistance. *Nature* **364**:679. http://dx.doi.org/10.1038/364679b0.

72. **Bellgrau D, Gold D, Selawry H, Moore J, Franzusoff A, Duke RC**. 1995. A role for CD95 ligand in preventing graft rejection. *Nature* **377**:630–632. http://dx.doi.org/10.1038/377630a0.

73. **Yagita H, Seino K, Kayagaki N, Okumura K**. 1996. CD95 ligand in graft rejection. *Nature* **379**:682. http://dx.doi.org/10.1038/379682a0.

74. **Allison J, Georgiou HM, Strasser A, Vaux DL**. 1997. Transgenic expression of CD95 ligand on islet beta cells induces a granulocytic infiltration but does not confer immune privilege upon islet allografts. *Proc Natl Acad Sci U S A* **94**:3943–3947. http://dx.doi.org/10.1073/pnas.94.8.3943.

75. **Kang SM, Schneider DB, Lin Z, Hanahan D, Dichek DA, Stock PG, Baekkeskov S**. 1997. Fas ligand expression in islets of Langerhans does not confer immune privilege and instead targets them for rapid destruction. *Nat Med* **3**:738–743. http://dx.doi.org/10.1038/nm0797-738.

76. **Vaux DL.** 1998. Immunology. Ways around rejection. *Nature* **394**:133. http://dx.doi.org/10.1038/28067.

77. **Gugliotti LA, Feldheim DL, Eaton BE**. 2004. RNA-mediated metal-metal bond formation in the synthesis of hexagonal palladium nanoparticles. *Science* **304**:850–852. http://dx.doi.org/10.1126/science.1095678.

78. **Franzen S, Cerruti M, Leonard DN, Duscher G**. 2007. The role of selection pressure in RNA-mediated evolutionary materials synthesis. *J Am Chem Soc* **129**:15340–15346. http://dx.doi.org/10.1021/ja076054r.

79. **Neff J.** 19 January 2014. State professor uncovers problems in lab journal. *News & Observer*, Raleigh, NC. http://www.newsobserver.com/2014/01/19/3544566/in-a-notebook-at-nc-state-a-smoking.html.

80. **Labandeira-Rey M, Couzon F, Boisset S, Brown EL, Bes M, Benito Y, Barbu EM, Vazquez V, Höök M, Etienne J, Vandenesch F, Bowden MG.** 2007. *Staphylococcus aureus* Panton-Valentine leukocidin causes necrotizing pneumonia. *Science* **315**:1130–1133. http://dx.doi.org/10.1126/science.1137165.

81. **Villaruz AE, Bubeck Wardenburg J, Khan BA, Whitney AR, Sturdevant DE, Gardner DJ, DeLeo FR, Otto M.** 2009. A point mutation in the *agr* locus rather than expression of the Panton-Valentine leukocidin caused previously reported phenotypes in *Staphylococcus aureus* pneumonia and gene regulation. *J Infect Dis* **200**:724–734. http://dx.doi.org/10.1086/604728.

82. **Wyatt MA, Wang W, Roux CM, Beasley FC, Heinrichs DE, Dunman PM, Magarvey NA.** 2010. *Staphylococcus aureus* nonribosomal peptide secondary metabolites regulate virulence. *Science* **329**:294–296. http://dx.doi.org/10.1126/science.1188888.

83. **Wyatt MA, Wang W, Roux CM, Beasley FC, Heinrichs DE, Dunman PM, Magarvey NA.** 2011. Clarification of "*Staphylococcus aureus* nonribosomal peptide secondary metabolites regulate virulence." *Science* **333**:1381.

84. **Sun F, Cho H, Jeong DW, Li C, He C, Bae T.** 2010. Aureusimines in *Staphylococcus aureus* are not involved in virulence. *PLoS One* **5**:e15703. http://dx.doi.org/10.1371/journal.pone.0015703.

85. **Regan MM, Leyland-Jones B, Bouzyk M, Pagani O, Tang W, Kammler R, Dell'orto P, Biasi MO, Thürlimann B, Lyng MB, Ditzel HJ, Neven P, Debled M, Maibach R, Price KN, Gelber RD, Coates AS, Goldhirsch A, Rae JM, Viale G, Breast International Group (BIG) 1-98 Collaborative Group.** 2012. CYP2D6 genotype and tamoxifen response in postmenopausal women with endocrine-responsive breast cancer: the Breast International Group 1-98 trial. *J Natl Cancer Inst* **104**:441–451. http://dx.doi.org/10.1093/jnci/djs125.

86. **Atkinson I.** 2012. Accuracy of data transfer: double data entry and estimating levels of error. *J Clin Nurs* **21**:2730–2735. http://dx.doi.org/10.1111/j.1365-2702.2012.04353.x.

Chapter 26 – Impacted Science

1. **Garfield E.** 2006. The history and meaning of the journal impact factor. *JAMA* **295**:90–93. http://dx.doi.org/10.1001/jama.295.1.90.

2. **Seglen PO.** 1997. Why the impact factor of journals should not be used for evaluating research. *BMJ* **314**:498–502. http://dx.doi.org/10.1136/bmj.314.7079.497.

3. **Colquhoun D.** 2003. Challenging the tyranny of impact factors. *Nature* **423**:479, discussion 480. http://dx.doi.org/10.1038/423479a.

4. **Lawrence PA.** 2007. The mismeasurement of science. *Curr Biol* **17**:R583–R585. http://dx.doi.org/10.1016/j.cub.2007.06.014.

5. **Brembs B, Button K, Munafò M.** 2013. Deep impact: unintended consequences of journal rank. *Front Hum Neurosci* **7**:291. http://dx.doi.org/10.3389/fnhum.2013.00291.

6. **Paulus FM, Cruz N, Krach S.** 2018. The impact factor fallacy. *Front Psychol* **9**:1487. http://dx.doi.org/10.3389/fpsyg.2018.01487.

7. **Bainbridge C.** 20 July 2022. What is a social construct? Why every part of society is a social construct. VeryWell Mind. https://www.verywellmind.com/definition-of-social-construct-1448922.

8. **Casadevall A, Fang FC.** 2014. Causes for the persistence of impact factor mania. *mBio* **5**:e00064-14. http://dx.doi.org/10.1128/mBio.00064-14.

9. **Editors of the American Heritage Dictionary.** 2013. *American Heritage Dictionary of the English Language.* Houghton Mifflin Harcourt Publishers, Boston, MA. https://www.ahdictionary.com/word/search.html?q=mania.

10. **Sekercioğlu CH.** 2013. Citation opportunity cost of the high impact factor obsession. *Curr Biol* **23**:R701–R702. http://dx.doi.org/10.1016/j.cub.2013.07.065.

11. **van Diest PJ, Holzel H, Burnett D, Crocker J**. 2001. Impactitis: new cures for an old disease. *J Clin Pathol* **54**:817–819. http://dx.doi.org/10.1136/jcp.54.11.817.

12. **Casadevall A, Fang FC**. 2014. Specialized science. *Infect Immun* **82**:1355–1360. http://dx.doi.org/10.1128/IAI.01530-13.

13. **Casadevall A, Fang FC**. 2009. Important science—it's all about the SPIN. *Infect Immun* **77**:4177–4180. http://dx.doi.org/10.1128/IAI.00757-09.

14. **Anonymous**. 2005. Not-so-deep impact. *Nature* **435**:1003–1004. http://dx.doi.org/10.1038/4351003b.

15. **Moher D, Naudet F, Cristea IA, Miedema F, Ioannidis JPA, Goodman SN**. 2018. Assessing scientists for hiring, promotion, and tenure. *PLoS Biol* **16**:e2004089. http://dx.doi.org/10.1371/journal.pbio.2004089.

16. **Steck N, Stalder L, Egger M**. 2020. Journal- or article-based citation measure? A study of academic promotion at a Swiss university. *F1000 Res* **9**:1188. http://dx.doi.org/10.12688/f1000research.26579.1.

17. **Cyranoski D, Gilbert N, Ledford H, Nayar A, Yahia M**. 2011. Education: the PhD factory. *Nature* **472**:276–279. http://dx.doi.org/10.1038/472276a.

18. **Stephan P**. 2012. Research efficiency: perverse incentives. *Nature* **484**:29–31. http://dx.doi.org/10.1038/484029a.

19. **Alberts B**. 2013. Am I wrong? *Science* **339**:1252. http://dx.doi.org/10.1126/science.1237434.

20. **Iyengar R, Wang Y, Chow J, Charney DS**. 2009. An integrated approach to evaluate faculty members' research performance. *Acad Med* **84**:1610–1616. http://dx.doi.org/10.1097/ACM.0b013e3181bb2364.

21. **Young NS, Ioannidis JP, Al-Ubaydli O**. 2008. Why current publication practices may distort science. *PLoS Med* **5**:e201. http://dx.doi.org/10.1371/journal.pmed.0050201.

22. **Petsko GA**. 2011. The one new journal we might actually need. *Genome Biol* **12**:129. http://dx.doi.org/10.1186/gb-2011-12-9-129.

23. **Callaham M, Wears RL, Weber E**. 2002. Journal prestige, publication bias, and other characteristics associated with citation of published studies in peer-reviewed journals. *JAMA* **287**:2847–2850. http://dx.doi.org/10.1001/jama.287.21.2847.

24. **Ioannidis JP**. 2006. Concentration of the most-cited papers in the scientific literature: analysis of journal ecosystems. *PLoS One* **1**:e5. http://dx.doi.org/10.1371/journal.pone.0000005.

25. **Maillard A, Delory T**. 2022. Blockbuster effect of COVID-19 on the impact factor of infectious disease journals. *Clin Microbiol Infect* **28**:1536–1538. http://dx.doi.org/10.1016/j.cmi.2022.08.011.

26. **Reich ES**. 2013. Science publishing: the golden club. *Nature* **502**:291–293. http://dx.doi.org/10.1038/502291a.

27. **Serenko A, Cox R, Bontis BL, Booker LD**. 2011. The superstar phenomenon in the knowledge management and intellectual capital academic discipline. *J Informetrics* **5**:333–345. http://dx.doi.org/10.1016/j.joi.2011.01.005.

28. **Merton RK**. 1968. The Matthew effect in science. The reward and communication systems of science are considered. *Science* **159**:56–63. http://dx.doi.org/10.1126/science.159.3810.56.

29. **Ferreira RC, Antoneli F, Briones MR**. 2013. The hidden factors in impact factors: a perspective from Brazilian science. *Front Genet* **4**:130. http://dx.doi.org/10.3389/fgene.2013.00130.

30. **Huggett S**. 2012. Impact factors: cash puts publishing ethics at risk in China. *Nature* **490**:342. http://dx.doi.org/10.1038/490342c.

31. **Franzoni C, Scellato G, Stephan P**. 2011. Science policy. Changing incentives to publish. *Science* **333**:702–703. http://dx.doi.org/10.1126/science.1197286.

32. **Jiménez-Contreras E, Delgado López-Cózar E, Ruiz-Pérez R, Fernández VM**. 2002. Impact-factor rewards affect Spanish research. *Nature* **417**:898. http://dx.doi.org/10.1038/417898b.

33. **Anonymous**. 2015. Cambridge Dictionaries Online, Cambridge University Press, Cambridge, UK. http://dictionary.cambridge.org/.

34. **Merriam-Webster**. 2022. Merriam-Webster online dictionary. https://www.merriam-webster.com/dictionary/impacted.

35. **Hicks D, Wouters P, Waltman L, de Rijcke S, Rafols I**. 2015. Bibliometrics: the Leiden Manifesto for research metrics. *Nature* **520**:429–431. http://dx.doi.org/10.1038/520429a.

36. **Van Noorden R, Maher B, Nuzzo R**. 2014. The top 100 papers. *Nature* **514**:550–553. http://dx.doi.org/10.1038/514550a.

37. **Ioannidis JP, Boyack KW, Small H, Sorensen AA, Klavans R**. 2014. Bibliometrics: is your most cited work your best? *Nature* **514**:561–562. http://dx.doi.org/10.1038/514561a.

38. **Watson JD, Crick FH**. 1953. Molecular structure of nucleic acids; a structure for deoxyribose nucleic acid. *Nature* **171**:737–738. http://dx.doi.org/10.1038/171737a0.

39. **Mendel JG**. 1866. Versuche über pflanzen-hybriden. *Verh Naturforsch Ver Brunn* **4**:3–47.

40. **Edwardson JR**. 1962. Another reference to Mendel before 1900. *J Hered* **53**:152. http://dx.doi.org/10.1093/oxfordjournals.jhered.a107154.

41. **Wolfe-Simon F, Switzer Blum J, Kulp TR, Gordon GW, Hoeft SE, Pett-Ridge J, Stolz JF, Webb SM, Weber PK, Davies PC, Anbar AD, Oremland RS**. 2011. A bacterium that can grow by using arsenic instead of phosphorus. *Science* **332**:1163–1166. http://dx.doi.org/10.1126/science.1197258.

42. **Reaves ML, Sinha S, Rabinowitz JD, Kruglyak L, Redfield RJ**. 2012. Absence of detectable arsenate in DNA from arsenate-grown GFAJ-1 cells. *Science* **337**:470–473. http://dx.doi.org/10.1126/science.1219861.

43. **Basturea GN, Harris TK, Deutscher MP**. 2012. Growth of a bacterium that apparently uses arsenic instead of phosphorus is a consequence of massive ribosome breakdown. *J Biol Chem* **287**:28816–28819. http://dx.doi.org/10.1074/jbc.C112.394403.

44. **Stern RE**. 1990. Uncitedness in the biomedical literature. *J Am Soc Inf Sci* **41**:193–196. http://dx.doi.org/10.1002/(SICI)1097-4571(199004)41:3<193::AID-ASI5>3.0.CO;2-B.

45. **Larivière V, Archambault E, Gingras Y, Wallace ML**. 2008. The fall of uncitedness, p 279–282. Abstr 10th Intl Conf Sci Technol Indicators, Vienna, Austria, 17 to 20 September 2008.

46. **Bornmann L, Haunschild R**. 2018. Do altmetrics correlate with the quality of papers? A large-scale empirical study based on F1000Prime data. *PLoS One* **13**:e0197133. http://dx.doi.org/10.1371/journal.pone.0197133.

47. **Ioannidis JP**. 2015. Is it possible to recognize a major scientific discovery? *JAMA* **314**:1135–1137. http://dx.doi.org/10.1001/jama.2015.9629.

48. **Brenner S**. 2014. Retrospective. Frederick Sanger (1918-2013). *Science* **343**:262. http://dx.doi.org/10.1126/science.1249912.

49. **Pinto ÂP, Mejdalani G, Mounce R, Silveira LF, Marinoni L, Rafael JA**. 2021. Are publications on zoological taxonomy under attack? *R Soc Open Sci* **8**:201617. http://dx.doi.org/10.1098/rsos.201617.

50. **Anonymous**. 2003. Coping with peer rejection. *Nature* **425**:645. http://dx.doi.org/10.1038/425645a.

51. **Berg J**. 2013. On deck chairs and lifeboats. *ASBMB Today* **12**:3–6.

52. **Kravitz DJ, Baker CI**. 2011. Toward a new model of scientific publishing: discussion and a proposal. *Front Comput Neurosci* **5**:55. http://dx.doi.org/10.3389/fncom.2011.00055.253463

53. **Kiesslich T, Beyreis M, Zimmermann G, Traweger A**. 2021. Citation inequality and the journal impact factor: median, mean, (does it) matter? *Scientometrics* **126**:1249–1269. http://dx.doi.org/10.1007/s11192-020-03812-y.

54. **Brito R, Rodríguez-Navarro A**. 2019. Evaluating research and researchers by the journal impact factor: is it better then coin flipping? *J Informetrics* **13**:314–324. http://dx.doi.org/10.1016/j.joi.2019.01.009.

55. **Falagas ME, Charitidou E, Alexiou VG**. 2008. Article and journal impact factor in various scientific fields. *Am J Med Sci* **335**:188–191. http://dx.doi.org/10.1097/MAJ.0b013e318145abb9.

56. **Ioannidis JPA, Thombs BD**. 2019. A user's guide to inflated and manipulated impact factors. *Eur J Clin Invest* **49**:e13151. http://dx.doi.org/10.1111/eci.13151.

57. **Hickman CF, Fong EA, Wilhite AW, Lee Y.** 2019. Academic misconduct and criminal liability: manipulating academic journal impact factors. *Sci Public Policy* **46**:661–667. http://dx.doi.org/10.1093/scipol/scz019.

58. **Mandelbrot B.** 1982. *The Fractal Geometry of Nature.* W.H. Freeman, New York, NY.

59. **Humphries NE, Queiroz N, Dyer JR, Pade NG, Musyl MK, Schaefer KM, Fuller DW, Brunnschweiler JM, Doyle TK, Houghton JD, Hays GC, Jones CS, Noble LR, Wearmouth VJ, Southall EJ, Sims DW.** 2010. Environmental context explains Lévy and Brownian movement patterns of marine predators. *Nature* **465**:1066–1069. http://dx.doi.org/10.1038/nature09116.

60. **Bush V.** 1945. *Science, the Endless Frontier—a Report to the President on a Program for Postwar Scientific Research.* United States Government Printing Office, Washington, DC.

61. **Rostami-Hodjegan A, Tucker GT.** 2001. Journal impact factors: a "bioequivalence" issue? *Br J Clin Pharmacol* **51**:111–117. http://dx.doi.org/10.1111/j.1365-2125.2001.01349.x.

62. **Lozano G, Larivière V, Gingras GS.** 2012. The weakening relationship between the impact factor and papers' citations in the digital age. *J Am Soc Inf Sci Technol* **63**:2140–2145. http://dx.doi.org/10.1002/asi.22731.

63. **Fang FC, Casadevall A.** 2011. Retracted science and the retraction index. *Infect Immun* **79**:3855–3859. http://dx.doi.org/10.1128/IAI.05661-11.

64. **Fang FC, Steen RG, Casadevall A.** 2012. Misconduct accounts for the majority of retracted scientific publications. *Proc Natl Acad Sci U S A* **109**:17028–17033. http://dx.doi.org/10.1073/pnas.1212247109.

65. **Brown EN, Ramaswamy S.** 2007. Quality of protein crystal structures. *Acta Crystallogr D Biol Crystallogr* **63**:941–950. http://dx.doi.org/10.1107/S0907444907033847.

66. **Hernán MA.** 2008. Epidemiologists (of all people) should question journal impact factors. *Epidemiology* **19**:366–368. http://dx.doi.org/10.1097/EDE.0b013e31816a9e28.

67. **Sutherland WJ, Goulson D, Potts SG, Dicks LV.** 2011. Quantifying the impact and relevance of scientific research. *PLoS One* **6**:e27537. http://dx.doi.org/10.1371/journal.pone.0027537.

68. **Eyre-Walker A, Stoletzki N.** 2013. The assessment of science: the relative merits of post-publication review, the impact factor, and the number of citations. *PLoS Biol* **11**:e1001675. http://dx.doi.org/10.1371/journal.pbio.1001675.

69. **Hecht F, Hecht BK, Sandberg AA.** 1998. The journal "impact factor": a misnamed, misleading, misused measure. *Cancer Genet Cytogenet* **104**:77–81. http://dx.doi.org/10.1016/S0165-4608(97)00459-7.

70. **Bernstein J, Gray CF.** 2012. Content Factor: a measure of a journal's contribution to knowledge. *PLoS One* **7**:e41554. http://dx.doi.org/10.1371/journal.pone.0041554.

71. **Van Noorden R.** 2013. Brazilian citation scheme outed. *Nature* **500**:510–511. http://dx.doi.org/10.1038/500510a.

72. **Lowe D.** 11 September 2019. Cite my papers. Or else. *Science.* https://www.science.org/content/blog-post/cite-my-papers-else.

73. **Baas J, Fennell C.** 6 September 2019. When peer reviewers go rogue—estimated prevalence of citation manipulation by reviewers based on the citation patterns of 69,000 reviewers. SSRN preprint. https://papers.ssrn.com/sol3/papers.cfm?abstract_id=3339568.

74. **Wren JD, Valencia A, Kelso J.** 2019. Reviewer-coerced citation: case report, update on journal policy and suggestions for future prevention. *Bioinformatics* **35**:3217–3218. http://dx.doi.org/10.1093/bioinformatics/btz071.

75. **Casadevall A, Fang FC.** 2012. Winner takes all. *Sci Am* **307**:13. http://dx.doi.org/10.1038/scientificamerican0812-13.

76. **Hardin G.** 1968. The tragedy of the commons. The population problem has no technical solution; it requires a fundamental extension in morality. *Science* **162**:1243–1248. http://dx.doi.org/10.1126/science.162.3859.1243.

77. **Arlinghaus R.** 2014. Are current research evaluation metrics causing a tragedy of the scientific commons and the extinction of university-based fisheries programs? *Fisheries* **39**:212–215. http://dx.doi.org/10.1080/03632415.2014.903837.

78. **Alberts B.** 2013. Impact factor distortions. *Science* **340**:787. http://dx.doi.org/10.1126/science.1240319.

79. **Walter G, Bloch S, Hunt G, Fisher K.** 2003. Counting on citations: a flawed way to measure quality. *Med J Aust* **178**:280–281. http://dx.doi.org/10.5694/j.1326-5377.2003.tb05196.x.

80. The *PLoS Medicine* **Editors.** 2006. The impact factor game. It is time to find a better way to assess the scientific literature. *PLoS Med* **3**:e291.

81. **Rossner M, Van Epps H, Hill E.** 2007. Show me the data. *J Exp Med* **204**:3052–3053. http://dx.doi.org/10.1084/jem.20072544.

82. **Simons K.** 2008. The misused impact factor. *Science* **322**:165. http://dx.doi.org/10.1126/science.1165316.

83. **Szklo M.** 2008. Impact factor: good reasons for concern. *Epidemiology* **19**:369. http://dx.doi.org/10.1097/EDE.0b013e31816b6a7a.

84. **Smith R.** 2008. Beware the tyranny of impact factors. *J Bone Joint Surg Br* **90**:125–126. http://dx.doi.org/10.1302/0301-620X.90B2.20258.

85. 2013. Beware the impact factor. *Nat Mater* **12**:89. http://dx.doi.org/10.1038/nmat3566.

86. **Schekman R, Patterson M.** 2013. Reforming research assessment. *eLife* **2**:e00855. http://dx.doi.org/10.7554/eLife.00855.

87. **Kaiser J.** 2013. Varmus's second act. *Science* **342**:416–419. http://dx.doi.org/10.1126/science.342.6157.416.

88. **Schmid SL.** 2017. Five years post-DORA: promoting best practices for research assessment. *Mol Biol Cell* **28**:2941–2944. http://dx.doi.org/10.1091/mbc.e17-08-0534.

89. **Patterson M.** 14 June 2012. eLife—an author's new best friend. *eLife*. https://elifesciences.org/inside-elife/d2970264/elife-an-author-s-new-best-friend.

90. **Berenbaum MR.** 2019. Impact factor impacts on early-career scientist careers. *Proc Natl Acad Sci U S A* **116**:16659–16662. http://dx.doi.org/10.1073/pnas.1911911116.

91. **Cagan R.** 2013. The San Francisco Declaration on Research Assessment. *Dis Model Mech* **6**:869–870. http://dx.doi.org/10.1242/dmm.012955.

92. **Johnston M.** 2013. We have met the enemy, and it is us. *Genetics* **194**:791–792. http://dx.doi.org/10.1534/genetics.113.153486.

93. **Russell R, Singh D.** 2009. Impact factor and its role in academic promotion. *Int J Chron Obstruct Pulmon Dis* **4**:265–266. http://dx.doi.org/10.2147/COPD.S6533.

94. **Misteli T.** 2013. Eliminating the impact of the Impact Factor. *J Cell Biol* **201**:651–652. http://dx.doi.org/10.1083/jcb.201304162.

95. **Falagas ME, Alexiou VG.** 2008. The top-ten in journal impact factor manipulation. *Arch Immunol Ther Exp (Warsz)* **56**:223–226. http://dx.doi.org/10.1007/s00005-008-0024-5.

96. **Larivière V, Gingras Y.** 2010. The impact factor's Matthew Effect: a natural experiment in bibliometrics. *J Assoc Inf Sci Technol* **61**:424–427.

97. **Fang FC, Casadevall A.** 1 May 2013. Why we cheat. *Scientific American Mind*. https://www.scientificamerican.com/article/why-we-cheat/.

98. **Casadevall A, Steen RG, Fang FC.** 2014. Sources of error in the retracted scientific literature. *FASEB J* **28**:3847–3855. http://dx.doi.org/10.1096/fj.14-256735.

99. **Begley CG, Ellis LM.** 2012. Drug development: raise standards for preclinical cancer research. *Nature* **483**:531–533. http://dx.doi.org/10.1038/483531a.

100. **Bowen A, Casadevall A.** 2015. Increasing disparities between resource inputs and outcomes, as measured by certain health deliverables, in biomedical research. *Proc Natl Acad Sci U S A* **112**:11335–11340. http://dx.doi.org/10.1073/pnas.1504955112.

101. **Woolston C.** 2021. Impact factor abandoned by Dutch university in hiring and promotion decisions. *Nature* **595**:462. http://dx.doi.org/10.1038/d41586-021-01759-5.

102. **Chawla DS.** 9 August 2021. Scientists at odds on Utrecht University reforms to hiring and promotion criteria. *Nature Index*. https://www.nature.com/nature-index/news-blog/scientists-argue-over-use-of-impact-factors-for-evaluating-research.

103. **Woolston C.** 2022. Grants and hiring: will impact factors and h-indices be scrapped? *Nature.* https://www.nature.com/articles/d41586-022-02984-2.

104. **Larivière V, Lozano GA, Gingras Y.** 2014. Are elite journals declining? *J Assoc Inf Sci* **65**:649–655. http://dx.doi.org/10.1002/asi.23005.

105. **Mull A.** 18 December 2019. Human experience, ranked. *The Atlantic.* https://www.theatlantic.com/health/archive/2019/12/enough-with-the-rankings/603829/.

106. **Bergstrom CT, West JD, Wiseman MA.** 2008. The Eigenfactor metrics. *J Neurosci* **28**:11433–11434. http://dx.doi.org/10.1523/JNEUROSCI.0003-08.2008.

107. **Hirsch JE.** 2007. Does the H index have predictive power? *Proc Natl Acad Sci U S A* **104**:19193–19198. http://dx.doi.org/10.1073/pnas.0707962104.

108. **Bi HH.** 2022. Four problems of the *h*-index for assessing the research productivity and impact of individual authors. *Scientometrics.* doi:10.1007/s11192-022-04323-8.

109. **Ioannidis JPA, Baas J, Klavans R, Boyack KW.** 2019. A standardized citation metrics author database annotated for scientific field. *PLoS Biol* **17**:e3000384. http://dx.doi.org/10.1371/journal.pbio.3000384.

110. **Hutchins BI, Yuan X, Anderson JM, Santangelo GM.** 2016. Relative citation ratio (RCR): a new metric that uses citation rates to measure influence at the article level. *PLoS Biol* **14**:e1002541. http://dx.doi.org/10.1371/journal.pbio.1002541.

Chapter 27 – Risky Science

1. **Green C.** 1976. *The Decline and Fall of Science.* Hamish Hamilton, London, United Kingdom.

2. **Asimov I.** 1973. *Today and Tomorrow.* Doubleday and Co, Garden City, NY.

3. **Weeks ME.** 1932. The discovery of the elements. XVII. The halogen family. *J Chem Educ* **9**:1915. http://dx.doi.org/10.1021/ed009p1915.

4. **Pike RM.** 1979. Laboratory-associated infections: incidence, fatalities, causes, and prevention. *Annu Rev Microbiol* **33**:41–66. http://dx.doi.org/10.1146/annurev.mi.33.100179.000353.

5. **Sewell DL.** 1995. Laboratory-associated infections and biosafety. *Clin Microbiol Rev* **8**:389–405. http://dx.doi.org/10.1128/CMR.8.3.389.

6. **Rozo M, Gronvall GK.** 2015. The reemergent 1977 H1N1 strain and the gain-of-function debate. *mBio* **6**:e01013-15. http://dx.doi.org/10.1128/mBio.01013-15.

7. **Bloom JD, Chan YA, Baric RS, Bjorkman PJ, Cobey S, Deverman BE, Fisman DN, Gupta R, Iwasaki A, Lipsitch M, Medzhitov R, Neher RA, Nielsen R, Patterson N, Stearns T, van Nimwegen E, Worobey M, Relman DA.** 2021. Investigate the origins of COVID-19. *Science* **372**:694. http://dx.doi.org/10.1126/science.abj0016.

8. **Casadevall A, Weiss SR, Imperiale MJ.** 2021. Can science help resolve the controversy on the origins of the SARS-CoV-2 pandemic? *mBio* **12**:e0194821. http://dx.doi.org/10.1128/mBio.01948-21.

9. **Blaizot JP, Ross GG, Iliopoulos J, Madsen J, Sonderegger P, Specht HJ.** 2003. *Study of Potentially Dangerous Events during Heavy-Ion Collisions at the LHC: Report of the LHC Safety Study Group.* CERN, Geneva, Switzerland.

10. **Buchanan M.** 2009. Is the LHC safe? *New Sci* **201**:32–33. http://dx.doi.org/10.1016/S0262-4079(09)60230-X.

11. **Alphey LS, Crisanti A, Randazzo FF, Akbari OS.** 2020. Opinion: standardizing the definition of gene drive. *Proc Natl Acad Sci U S A* **117**:30864–30867. http://dx.doi.org/10.1073/pnas.2020417117.

12. **Kahl LJ, Endy D.** 2013. A survey of enabling technologies in synthetic biology. *J Biol Eng* **7**:13. http://dx.doi.org/10.1186/1754-1611-7-13.

13. **Dyer GA, Serratos-Hernández JA, Perales HR, Gepts P, Piñeyro-Nelson A, Chávez A, Salinas-Arreortua N, Yúnez-Naude A, Taylor JE, Alvarez-Buylla ER.** 2009. Dispersal of transgenes through maize seed systems in Mexico. *PLoS One* **4**:e5734. http://dx.doi.org/10.1371/journal.pone.0005734.

14. **Wegier A, Piñeyro-Nelson A, Alarcón J, Gálvez-Mariscal A, Alvarez-Buylla ER, Piñero D**. 2011. Recent long-distance transgene flow into wild populations conforms to historical patterns of gene flow in cotton (*Gossypium hirsutum*) at its centre of origin. *Mol Ecol* **20**:4182–4194. http://dx.doi.org/10.1111/j.1365-294X.2011.05258.x.

15. **Wu F, Wesseler J, Zilberman D, Russell RM, Chen C, Dubock AC**. 2021. Opinion: Allow Golden Rice to save lives. *Proc Natl Acad Sci U S A* **118**:e2120901118. http://dx.doi.org/10.1073/pnas.2120901118.

16. **Osoba OA, Welser W**. 2017. *The Risks of Artificial Intelligence to Security and the Future of Work*. RAND Corporation, Santa Monica, CA. http://dx.doi.org/10.7249/PE237.

17. **Sridharan R, Monisha B, Kumar PS, Gayathri KV**. 2022. Carbon nanomaterials and its applications in pharmaceuticals: a brief review. *Chemosphere* **294**:133731. http://dx.doi.org/10.1016/j.chemosphere.2022.133731.

18. **Bergamaschi E, Garzaro G, Wilson Jones G, Buglisi M, Caniglia M, Godono A, Bosio D, Fenoglio I, Guseva Canu I**. 2021. Occupational exposure to carbon nanotubes and carbon nanofibres: more than a cobweb. *Nanomaterials (Basel)* **11**:745. http://dx.doi.org/10.3390/nano11030745.

19. **Imai M, Watanabe T, Hatta M, Das SC, Ozawa M, Shinya K, Zhong G, Hanson A, Katsura H, Watanabe S, Li C, Kawakami E, Yamada S, Kiso M, Suzuki Y, Maher EA, Neumann G, Kawaoka Y**. 2012. Experimental adaptation of an influenza H5 HA confers respiratory droplet transmission to a reassortant H5 HA/H1N1 virus in ferrets. *Nature* **486**:420–428. http://dx.doi.org/10.1038/nature10831.

20. **Herfst S, Schrauwen EJ, Linster M, Chutinimitkul S, de Wit E, Munster VJ, Sorrell EM, Bestebroer TM, Burke DF, Smith DJ, Rimmelzwaan GF, Osterhaus AD, Fouchier RA**. 2012. Airborne transmission of influenza A/H5N1 virus between ferrets. *Science* **336**:1534–1541. http://dx.doi.org/10.1126/science.1213362.

21. **Jackson RJ, Ramsay AJ, Christensen CD, Beaton S, Hall DF, Ramshaw IA**. 2001. Expression of mouse interleukin-4 by a recombinant ectromelia virus suppresses cytolytic lymphocyte responses and overcomes genetic resistance to mousepox. *J Virol* **75**:1205–1210. http://dx.doi.org/10.1128/JVI.75.3.1205-1210.2001.

22. **Müllbacher A, Lobigs M**. 2001. Creation of killer poxvirus could have been predicted. *J Virol* **75**:8353–8355. http://dx.doi.org/10.1128/JVI.75.18.8353-8355.2001.

23. **Finkel E**. 2001. Australia. Engineered mouse virus spurs bioweapon fears. *Science* **291**:585. http://dx.doi.org/10.1126/science.291.5504.585.

24. **Dunlop LR, Oehlberg KA, Reid JJ, Avci D, Rosengard AM**. 2003. Variola virus immune evasion proteins. *Microbes Infect* **5**:1049–1056. http://dx.doi.org/10.1016/S1286-4579(03)00194-1.

25. **Wein LM, Liu Y**. 2005. Analyzing a bioterror attack on the food supply: the case of botulinum toxin in milk. *Proc Natl Acad Sci U S A* **102**:9984–9989. http://dx.doi.org/10.1073/pnas.0408526102.

26. **National Science Advisory Board for Biosecurity**. 2007. Proposed framework for the oversight of dual use life sciences research: strategies for minimizing the potential misuse of research information. https://osp.od.nih.gov/wp-content/uploads/Proposed-Oversight-Framework-for-Dual-Use-Research.pdf.

27. **Keim PS**. 2012. The NSABB recommendations: rationale, impact, and implications. *mBio* **3**:300021-12. http://dx.doi.org/10.1128/mBio.00021-12.

28. **Berns KI, Casadevall A, Cohen ML, Ehrlich SA, Enquist LW, Fitch JP, Franz DR, Fraser-Liggett CM, Grant CM, Imperiale MJ, Kanabrocki J, Keim PS, Lemon SM, Levy SB, Lumpkin JR, Miller JF, Murch R, Nance ME, Osterholm MT, Relman DA, Roth JA, Vidaver AK**. 2012. Public health and biosecurity. Adaptations of avian flu virus are a cause for concern. *Science* **335**:660–661. http://dx.doi.org/10.1126/science.1217994.

29. **Dover N, Barash JR, Hill KK, Xie G, Arnon SS**. 2014. Molecular characterization of a novel botulinum neurotoxin type H gene. *J Infect Dis* **209**:192–202. http://dx.doi.org/10.1093/infdis/jit450.

30. **Hooper DC, Hirsch MS**. 2014. Novel *Clostridium botulinum* toxin and dual use research of concern issues. *J Infect Dis* **209**:167. http://dx.doi.org/10.1093/infdis/jit528.

31. **Casadevall A, Enquist L, Imperiale MJ, Keim P, Osterholm MT, Relman DA**. 2013. Redaction of sensitive data in the publication of dual use research of concern. *mBio* 5:e00991-13.

32. **Bush LM, Perez MT**. 2012. The anthrax attacks 10 years later. *Ann Intern Med* **156**:41–44. http://dx.doi.org/10.7326/0003-4819-155-12-201112200-00373.

33. **Willman D**. 2011. *The Mirage Man: Bruce Ivins, the Anthrax Attacks, and America's Rush to War*. Bantam, New York, NY.

34. **Reardon S**. 2014. "Forgotten" NIH smallpox virus languishes on death row. *Nature* **514**:544. http://dx.doi.org/10.1038/514544a.

35. **Kaiser J**. 16 July 2014. Congress asks why CDC safety problems persist. *Science*. https://www.science.org/content/article/congress-asks-why-cdc-safety-problems-persist.

36. **Lipsitch M, Bloom BR**. 2012. Rethinking biosafety in research on potential pandemic pathogens. *mBio* 3:e00360-12. http://dx.doi.org/10.1128/mBio.00360-12.

37. **Lipsitch M, Galvani AP**. 2014. Ethical alternatives to experiments with novel potential pandemic pathogens. *PLoS Med* **11**:e1001646. http://dx.doi.org/10.1371/journal.pmed.1001646.

38. **Casadevall A, Imperiale MJ**. 2014. Risks and benefits of gain-of-function experiments with pathogens of pandemic potential, such as influenza virus: a call for a science-based discussion. *mBio* **5**:e01730-14. http://dx.doi.org/10.1128/mBio.01730-14.

39. **Casadevall A, Howard D, Imperiale MJ**. 2014. An epistemological perspective on the value of gain-of-function experiments involving pathogens with pandemic potential. *mBio* **5**:e01875-14. http://dx.doi.org/10.1128/mBio.01875-14.

40. **Klotz LC, Sylvester EJ**. 2014. The consequences of a lab escape of a potential pandemic pathogen. *Front Public Health* **2**:116. http://dx.doi.org/10.3389/fpubh.2014.00116.

41. **Fouchier RA**. 2015. Studies on influenza virus transmission between ferrets: the public health risks revisited. *mBio* **6**:e02560-14. http://dx.doi.org/10.1128/mBio.02560-14.

42. **Kaiser J**. 2014. Biosecurity. U.S. halts two dozen risky virus studies. *Science* **346**:404. http://dx.doi.org/10.1126/science.346.6208.404.

43. **Kaiser J**. 2019. Controversial flu studies can resume, U.S. panel says. *Science* **363**:676–677. http://dx.doi.org/10.1126/science.363.6428.676.

44. **Imperiale MJ, Casadevall A**. 2020. Rethinking gain-of-function experiments in the context of the COVID-19 pandemic. *mBio* **11**:e01868-20. http://dx.doi.org/10.1128/mBio.01868-20.

45. **Rogin J**. 14 April 2020. State department cables warned of safety issues at Wuhan laboratory studying bat coronaviruses. *Washington Post*. https://www.washingtonpost.com/opinions/2020/04/14/state-department-cables-warned-safety-issues-wuhan-lab-studying-bat-coronaviruses/.

46. **Kaiser J**. 21 October 2021. NIH says grantee failed to report experiment in Wuhan that created a bat virus that made mice sicker. *Science*. https://www.science.org/content/article/nih-says-grantee-failed-report-experiment-wuhan-created-bat-virus-made-mice-sicker.

47. **Imperiale MJ, Casadevall A**. 2015. A new synthesis for dual use research of concern. *PLoS Med* **12**:e1001813. http://dx.doi.org/10.1371/journal.pmed.1001813.

48. **Casadevall A, Dermody TS, Imperiale MJ, Sandri-Goldin RM, Shenk T**. 2014. On the need for a national board to assess dual use research of concern. *J Virol* **88**:6535–6537. http://dx.doi.org/10.1128/JVI.00875-14.

49. **Johnson IS**. 1984. Role of the Recombinant Advisory Committee. *Science* **224**:243. http://dx.doi.org/10.1126/science.6710142.

50. **Tollefson J**. 2017. Iron-dumping ocean experiment sparks controversy. *Nature* **545**:393–394. http://dx.doi.org/10.1038/545393a.

51. **Li JR, Walker S, Nie JB, Zhang XQ**. 2019. Experiments that led to the first gene-edited babies: the ethical failings and the urgent need for better governance. *J Zhejiang Univ Sci B* **20**:32–38. http://dx.doi.org/10.1631/jzus.B1800624.

52. **Berg P, Baltimore D, Brenner S, Roblin RO, Singer MF**. 1975. Summary statement of the Asilomar conference on recombinant DNA molecules. *Proc Natl Acad Sci U S A* **72**:1981–1984. http://dx.doi.org/10.1073/pnas.72.6.1981.

53. **Mallapaty S, Maxmen A, Callaway E**. 2021. "Major stones unturned": COVID origin search must continue after WHO report, say scientists. *Nature* **590**:371–372. http://dx.doi.org/10.1038/d41586-021-00375-7.

54. **Maxmen A**. 2021. Divisive COVID "lab leak" debate prompts dire warnings from researchers. *Nature* **594**:15–16. http://dx.doi.org/10.1038/d41586-021-01383-3.

55. **Rozell DJ**. 2020. *Dangerous Science: Science Policy and Risk Analysis for Scientists and Engineers*. Ubiquity Press, London, United Kingdom.

56. **Palmer MJ, Fukuyama F, Relman DA**. 2015. SCIENCE GOVERNANCE. A more systematic approach to biological risk. *Science* **350**:1471–1473. http://dx.doi.org/10.1126/science.aad8849.

57. **Herington J, Tanona S**. 2020. The social risks of science. *Hastings Cent Rep* **50**:27–38. http://dx.doi.org/10.1002/hast.1196.

58. **Goodrum F, Lowen AC, Lakdawala S, Alwine J, Casadevall A, Imperiale MJ, et al**. 2023. Virology under the microscope: A call for rational discourse. *mBio* **14**:e0018823. doi: 10.1128/mbio.00188-23.

59. **Merriam-Webster**. 2022. Merriam-Webster online dictionary. https://www.merriam-webster.com/dictionary/biosafety.https://www.merriam-webster.com/dictionary/biosecurity.

60. **Steensma DP, Montori VM, Shampo MA, Kyle RA**. 2014. Stamp vignette on medical science. Daniel Alcides Carrión—Peruvian hero and medical martyr. *Mayo Clin Proc* **89**:e55–e56. http://dx.doi.org/10.1016/j.mayocp.2013.08.025.

61. **Annas GJ**. 2010. Self experimentation and the Nuremberg Code. *BMJ.* **341**:c7103. doi: 10.1136/bmj.c7103.

62. **Weisse AB**. 2012. Self-experimentation and its role in medical research. *Tex Heart Inst J.* **39**:51-54.

63. **Hanley BP, Bains W, Church G**. 2019. Review of scientific self-experimentation: Ethics history, regulation, scenarios, and views among ethics committees and prominent scientists. *Rejuvenation Res.* **22**:31-42. doi: 10.1089/rej.2018.2059.

64. **Kryzhanovsky LN**. 1990. The lightning rod in 18th-century St. Petersburg: a note on the occasion of the bicentennial of the death of Benjamin Franklin. *Technol Cult* **31**:813–817. http://dx.doi.org/10.2307/3105908.

65. **West JB**. 2014. Carl Wilhelm Scheele, the discoverer of oxygen, and a very productive chemist. *Am J Physiol Lung Cell Mol Physiol* **307**:L811–L816. http://dx.doi.org/10.1152/ajplung.00223.2014.

66. **Marsh N, Marsh A**. 2000. A short history of nitroglycerine and nitric oxide in pharmacology and physiology. *Clin Exp Pharmacol Physiol* **27**:313–319. http://dx.doi.org/10.1046/j.1440-1681.2000.03240.x.

67. **Elder ES, Lazzerini SD**. 1979. The deadly outcome of chance—Vera Estaf'evna Bogdanovskaia. *J Chem Educ* **56**:251–252. http://dx.doi.org/10.1021/ed056p251.

68. **Chaves-Carballo E**. 2013. Clara Maass, yellow fever and human experimentation. *Mil Med* **178**:557–562. http://dx.doi.org/10.7205/MILMED-D-12-00430.

69. **Sansare K, Khanna V, Karjodkar F**. 2011. Early victims of X-rays: a tribute and current perception. *Dentomaxillofac Radiol* **40**:123–125. http://dx.doi.org/10.1259/dmfr/73488299.

70. **Brown P**. 1995. American martyrs to radiology. Elizabeth Fleischman Ascheim (1859-1905). 1936. ss *AJR Am J Roentgenol* **164**:497–499. http://dx.doi.org/10.2214/ajr.164.2.7839997.

71. **Huestis DW**. 2007. Alexander Bogdanov: the forgotten pioneer of blood transfusion. *Transfus Med Rev* **21**:337–340. http://dx.doi.org/10.1016/j.tmrv.2007.05.008.

72. **Anwar Y, Lowenstein EJ**. 2015. Radium: Curie's perpetual sunshine. *JAMA Dermatol* **151**:801. http://dx.doi.org/10.1001/jamadermatol.2015.52.

73. **Oettingen M**. 2020. A criticality study on the LA-1 accident using Monte Carlo methods. *Nucl Eng Des* **359**:110467. http://dx.doi.org/10.1016/j.nucengdes.2019.110467.

74. **Anderson HL, Novick A, Morrison P**. 1946. Louis A. Slotin 1912-1946. *Science* **104**:182–183. http://dx.doi.org/10.1126/science.104.2695.182.

75. **Chai PR, Hayes BD, Erickson TB, Boyer EW**. 2018. Novichok agents: a historical, current, and toxicological perspective. *Toxicol Commun* **2**:45–48. http://dx.doi.org/10.1080/24734306.2018.1475151.

76. **Holden C**. 1997. Death from lab poisoning. *Science* **276**:1797. http://dx.doi.org/10.1126/science.276.5320.1797a.

Chapter 28 – Authoritarian Science

1. **Planck MK**. 1950. *Scientific Autobiography and Other Papers*. Philosophical Library, New York, NY. http://dx.doi.org/10.1119/1.1932511

2. **Anonymous**. 2022. Dictionary.com. https://www.dictionary.com/browse/authoritarian.

3. **Graham LR**. 1977. Political ideology and genetic theory: Russia and Germany in the 1920's. *Hastings Cent Rep* **7**:30–39. http://dx.doi.org/10.2307/3560721.

4. **Soyfer VN**. 2001. The consequences of political dictatorship for Russian science. *Nat Rev Genet* **2**:723–729. http://dx.doi.org/10.1038/35088598.

5. **Einstein A**. 2 August 1939. Letter to President Roosevelt—1939. https://www.atomicarchive.com/resources/documents/beginnings/einstein.html.

6. **Schultz M**. 2008. Rudolf Virchow. *Emerg Infect Dis* **14**:1480–1481. http://dx.doi.org/10.3201/eid1409.086672.

7. **Morabia A**. 2007. Epidemiologic interactions, complexity, and the lonesome death of Max von Pettenkofer. *Am J Epidemiol* **166**:1233–1238. http://dx.doi.org/10.1093/aje/kwm279.

8. **Lange KW**. 2021. Rudolf Virchow, poverty and global health: from "politics as medicine on a grand scale" to "health in all policies." *Glob Health J* **5**:149–154. http://dx.doi.org/10.1016/j.glohj.2021.07.003.

9. **Locher WG**. 2007. Max von Pettenkofer (1818-1901) as a pioneer of modern hygiene and preventive medicine. *Environ Health Prev Med* **12**:238–245. http://dx.doi.org/10.1007/BF02898030.

10. **Evans AS**. 1985. Two errors in enteric epidemiology: the stories of Austin Flint and Max von Pettenkofer. *Rev Infect Dis* **7**:434–440. http://dx.doi.org/10.1093/clinids/7.3.434.

11. **Basterfield C, Lilienfeld SO, Bowes SM, Costello TH**. 2020. The Nobel disease: when intelligence fails to protect against irrationality. *Skeptical Inquirer* **44**:32–37.

12. **Diamandis EP**. 2013. Nobelitis: a common disease among Nobel laureates? *Clin Chem Lab Med* **51**:1573–1574. http://dx.doi.org/10.1515/cclm-2013-0273.

13. **Ball P**. 13 February 2015. How 2 pro-Nazi Nobelists attacked Einstein's "Jewish science." *Scientific American*. https://www.scientificamerican.com/article/how-2-pro-nazi-nobelists-attacked-einstein-s-jewish-science-excerpt1/.

14. **Harmon A**. 1 January 2019. James Watson had a chance to salvage his reputation on race. He made things worse. *New York Times*. https://www.nytimes.com/2019/01/01/science/watson-dna-genetics-race.html.

15. **Tsay CJ**. 2013. Julius Wagner-Jauregg and the legacy of malarial therapy for the treatment of general paresis of the insane. *Yale J Biol Med* **86**:245–254.

16. **McGee EO**. 2022. Dismantle racism in science. *Science* **375**:937. http://dx.doi.org/10.1126/science.abo7849.

17. **Casadevall A, Fang FC**. 2015. Field science—the nature and utility of scientific fields. *mBio* **6**:e01259-15. http://dx.doi.org/10.1128/mBio.01259-15.

18. **Schlosshauer M, Kofler J, Zeilinger A**. 2013. A snapshot of foundational attitudes toward quantum mechanics. *Stud Hist Philos Sci Part Stud Hist Philos Mod Phys* **44**:222–230. http://dx.doi.org/10.1016/j.shpsb.2013.04.004.

19. **Azoulay P, Fons-Rosen C, Zivin JSG**. 2019. Does science advance one funeral at a time? *Am Econ Rev* **109**:2889–2920. http://dx.doi.org/10.1257/aer.20161574.

20. **Kuhn TS**. 1962. *The Structure of Scientific Revolutions*. University of Chicago Press, Chicago, IL.

21. **Plato**. 1892. *The Apology of Socrates*. (**Jowett B**, translator.) Oxford University Press, London, United Kingdom.

22. **Confucius**. As quoted by Henry David Thoreau in Walden, 1854. Ticknor & Fields, Boston, MA.

23. **da Vinci L, Wells T, Richter I**. 2008. *Leonardo da Vinci Notebooks*. Oxford University Press, Oxford, United Kingdom.

24. **Galilei G, Drake S**. 1957. *Discoveries and Opinions of Galileo*. Anchor Books, New York, NY.

25. **Habashi F**. 2012. The case of Nobelists: Philipp Lenard and Johannes Stark. *Chem Educ* **17**:78–79.

26. **Vernon G.** 2019. Alexis Carrel: "father of transplant surgery" and supporter of eugenics. *Br J Gen Pract* **69**:352. http://dx.doi.org/10.3399/bjgp19X704441.

27. **Donati M.** 2004. Beyond synchronicity: the worldview of Carl Gustav Jung and Wolfgang Pauli. *J Anal Psychol* **49**:707–728. http://dx.doi.org/10.1111/j.0021-8774.2004.00496.x.

28. **Zabriskie B.** 2014. Psychic energy and synchronicity. *J Anal Psychol* **59**:157–164. http://dx.doi.org/10.1111/1468-5922.12065.

29. **Hemilä H.** 1997. Vitamin C supplementation and the common cold—was Linus Pauling right or wrong? *Int J Vitam Nutr Res* **67**:329–335.

30. **George A.** 2006. Take nobody's word for it. *New Sci* **192**:56–57.

31. **Weigmann K.** 2018. The genesis of a conspiracy theory: why do people believe in scientific conspiracy theories and how do they spread? *EMBO Rep* **19**:e45935. http://dx.doi.org/10.15252/embr.201845935.

32. **Wanjek C**. 7 October 2003. Nitric oxide now—ask me how. *Washington Post*. https://www.washingtonpost.com/archive/lifestyle/wellness/2003/10/07/nitric-oxide-now-ask-me-how/531c9d82-31c3-4bff-874a-6437934d3d9f/.

33. **Montagnier L.** 2010. Newsmaker interview: Luc Montagnier. French Nobelist escapes "intellectual terror" to pursue radical ideas in China. Interview by Martin Enserink. *Science* **330**:1732. http://dx.doi.org/10.1126/science.330.6012.1732.

34. **Ledford H.** 2022. Luc Montagnier (1932-2022). *Nature* **603**:223. http://dx.doi.org/10.1038/d41586-022-00653-y.

35. **Steinsmith W.** 2000. Planck family paid a high price for opposing Hitler. *Nature* **405**:116. http://dx.doi.org/10.1038/35012165.

Chapter 29 – Deplorable Science

1. **von Hayek F.** 11 December 1974. The pretence of knowledge. Nobel Prize lecture. https://www.nobelprize.org/prizes/economic-sciences/1974/hayek/lecture/.

2. **Daempfle P.** 2012. *Science & Society: Scientific Thought and Education for the 21st Century*. Jones & Bartlett Learning, Burlington, MA.

3. **Chotiner I.** 28 October 2013. Richard Dawkins keeps making new enemies. *New Republic*. https://newrepublic.com/article/115339/richard-dawkins-interview-archbishop-atheism.

4. **Merriam-Webster**. 2022. Merriam-Webster online dictionary. https://www.merriam-webster.com/dictionary/deplorable. **Dictionary.com**. https://www.dictionary.com/browse/deplorable .

5. **Cohen MS, Cannon JG, Jerse AE, Charniga LM, Isbey SF, Whicker LG**. 1994. Human experimentation with *Neisseria gonorrhoeae*: rationale, methods, and implications for the biology of infection and vaccine development. *J Infect Dis* **169**:532–537. http://dx.doi.org/10.1093/infdis/169.3.532.

6. **Cohen MS, Cannon JG**. 1999. Human experimentation with *Neisseria gonorrhoeae*: progress and goals. *J Infect Dis* **179**(Suppl 2):S375–S379. http://dx.doi.org/10.1086/513847.

7. **Callaway E.** 2022. Scientists deliberately gave people COVID—here's what they learnt. *Nature* **602**:191–192. http://dx.doi.org/10.1038/d41586-022-00319-9.

8. **Vasconcelos SMR, Esher A, Penido C, Lima C, Rocha KA, Antunes MJM, Ribeiro MD, Pedrotti M**. 2022. An ongoing science-society-ethics experiment: the human challenge trial debate in COVID-19 pandemic. *EMBO Rep* **23**:e54184. http://dx.doi.org/10.15252/embr.202154184.

9. **Dowling HF.** 1966. Human experimentation in infectious diseases. *JAMA* **198**:997–999. http://dx.doi.org/10.1001/jama.1966.03110220081028.

10. **Beecher HK**. 1966. Ethics and clinical research. *N Engl J Med* **274**:1354–1360. http://dx.doi.org/10.1056/NEJM196606162742405.

11. **Nix E.** 16 May 2017. Tuskegee experiment: the infamous syphilis study. History.com. https://www.history.com/news/the-infamous-40-year-tuskegee-study.

12. **Zenilman JM.** 2013. Ethics gone awry: the US Public Health Service studies in Guatemala; 1946-1948. *Sex Transm Infect* **89**:295–300. http://dx.doi.org/10.1136/sextrans-2012-050741.

13. **Baader G, Lederer SE, Low M, Schmaltz F, Schwerin AV**. 2005. Pathways to human experimentation, 1933-1945: Germany, Japan, and the United States. *Osiris* **20**:205–231. http://dx.doi.org/10.1086/649419.

14. **Batt S.** 2018. Revisiting New Zealand's "Unfortunate Experiment": is medical ethics ever a thing done? *Indian J Med Ethics* **3**:142–146.

15. **Lowry S.** 1988. New Zealand smear trial risked lives. *BMJ* **297**:507–508.

16. **Ramanathan M, Jesani A.** 2012. The legacy of scandals and non-scandals in research and its lessons for bioethics in India. *Indian J Med Ethics* **9**:4–6. http://dx.doi.org/10.20529/IJME.2012.002.

17. **Paul C.** 1988. The New Zealand cervical cancer study: could it happen again? *BMJ* **297**:533–539. http://dx.doi.org/10.1136/bmj.297.6647.533.

18. **Lantos J.** 2016. Henry K. Beecher and the oversight of research in children. *Perspect Biol Med* **59**:95–106. http://dx.doi.org/10.1353/pbm.2016.0017.

19. **Krugman S.** 1986. The Willowbrook hepatitis studies revisited: ethical aspects. *Rev Infect Dis* **8**:157–162. http://dx.doi.org/10.1093/clinids/8.1.157.

20. **Rothman DJ.** 1982. Were Tuskegee & Willowbrook "studies in nature"? *Hastings Cent Rep* **12**:5–7. http://dx.doi.org/10.2307/3561798.

21. **Casadevall A, Fang FC.** 2015. (A)Historical science. *Infect Immun* **83**:4460–4464. http://dx.doi.org/10.1128/IAI.00921-15.

22. **Associated Press.** 7 October 2022. Philadelphia apologizes for experiments on Black inmates. National Public Radio. https://www.npr.org/2022/10/07/1127406363/philadelphia-apologizes-experiments-black-inmates.

23. **Adamson AS, Lipoff JB.** 2021. Reconsidering named honorifics in medicine—the troubling legacy of dermatologist Albert Kligman. *JAMA Dermatol* **157**:153–155. http://dx.doi.org/10.1001/jamadermatol.2020.4570.

24. **Carlson RV, Boyd KM, Webb DJ.** 2004. The revision of the Declaration of Helsinki: past, present and future. *Br J Clin Pharmacol* **57**:695–713. http://dx.doi.org/10.1111/j.1365-2125.2004.02103.x.

25. **Adashi EY, Walters LB, Menikoff JA.** 2018. The Belmont Report at 40: reckoning with time. *Am J Public Health* **108**:1345–1348. http://dx.doi.org/10.2105/AJPH.2018.304580.

26. **Casadevall A, Fang FC.** 2018. Making the scientific literature fail-safe. *J Clin Invest* **128**:4243–4244. http://dx.doi.org/10.1172/JCI123884.

27. **Neufeld MJ.** 20 May 2019. Wernher von Braun and the Nazis. American Experience, Public Broadcasting System. https://www.pbs.org/wgbh/americanexperience/features/chasing-moon-wernher-von-braun-and-nazis/.

28. **Wikimedia Commons.** Saturn V space vehicle lift off for the Apollo 11 mission. https://commons.wikimedia.org/wiki/File:Apollo_11_Launch_-_GPN-2000-000630.jpg

29. **Caplan AL, Danovitch G, Shapiro M, Lavee J, Epstein M.** 2011. Time for a boycott of Chinese science and medicine pertaining to organ transplantation. *Lancet* **378**:1218. http://dx.doi.org/10.1016/S0140-6736(11)61536-5.

30. **Furmanski M.** 2002. Citation of unethical research. *JAMA* **287**:452–453. http://dx.doi.org/10.1001/jama.287.4.452.

31. **Dennis DT, Inglesby TV, O'Toole T, Henderson DA.** 2002. In reply. *JAMA* **287**:453.

32. **Swain F.** 23 July 2019. Is it right to use Nazi research if it can save lives? BBC Future. https://www.bbc.com/future/article/20190723-the-ethics-of-using-nazi-science.

33. **Higgins WC, Rogers WA, Ballantyne A, Lipworth W.** 2020. Against the use and publication of contemporary unethical research: the case of Chinese transplant research. *J Med Ethics* **46**:678–684. http://dx.doi.org/10.1136/medethics-2019-106044.

34. **Sand M, Bredenoord AL, Jongsma KR.** 2019. After the fact-the case of CRISPR babies. *Eur J Hum Genet* **27**:1621–1624. http://dx.doi.org/10.1038/s41431-019-0459-5.

35. **Angelski C, Fernandez CV, Weijer C, Gao J.** 2012. The publication of ethically uncertain research: attitudes and practices of journal editors. *BMC Med Ethics* **13**:4. http://dx.doi.org/10.1186/1472-6939-13-4.

36. **Halpin RW.** 2010. Can unethically produced data be used ethically? *Med Law* **29**:373–387.
37. **American Medical Association**. Code of Medical Ethics. Release of data from unethical experiments. Opinion 7.2.2. https://code-medical-ethics.ama-assn.org/ethics-opinions/release-data-unethical-experiments .
38. **Moe K.** 1984. Should the Nazi research data be cited? *Hastings Cent Rep* **14**:5–7. http://dx.doi.org/10.2307/3561733.
39. **Schafer A.** 1986. On using Nazi data: the case against. *Dialogue Can Philos Assoc* **25**:413–419. http://dx.doi.org/10.1017/S0012217300020862.

Chapter 30 – Plague Science

1. **Defoe D.** 1722. *A Journal of the Plague Year*. Nutt, Roberts, Dodd and Graves, London, United Kingdom.
2. **Beall RF, Kesselheim AS, Hollis A.** 2022. Premarket development times for innovative vaccines—to what extent are the coronavirus disease 2019 (COVID-19) vaccines outliers? *Clin Infect Dis* **74**:347–351. http://dx.doi.org/10.1093/cid/ciab389.
3. **Casadevall A.** 2021. The mRNA vaccine revolution is the dividend from decades of basic science research. *J Clin Invest* **131**:e153721. http://dx.doi.org/10.1172/JCI153721.
4. **Hvistendahl M, Mueller B.** 2023. Chinese censorship is quietly rewriting the Covid-19 story. *New York Times*. 23 April 2023. https://www.nytimes.com/2023/04/23/world/europe/chinese-censorship-covid.html.
5. **Welsh JM.** 2022. International cooperation failures in the face of the COVID-19 pandemic: Learning from past efforts to address threats. American Academy of Arts and Sciences. Cambridge, MA. https://www.amacad.org/publication/internation-cooperation-failures-covid-19-pandemic.
6. **Mueller B, Lutz E.** 1 February 2022. U.S. has far higher COVID death rate than other wealthy countries. *New York Times*. https://www.nytimes.com/interactive/2022/02/01/science/covid-deaths-united-states.html.
7. **Independent Panel for Pandemic Preparedness and Response.** 2021. *COVID-19: Make It the Last Pandemic*. World Health Organization, Geneva, Switzerland. https://theindependentpanel.org/wp-content/uploads/2021/05/COVID-19-Make-it-the-Last-Pandemic_final.pdf.
8. **Casadevall A, Fang FC.** 2016. Rigorous science: a how-to guide. *mBio* **7**:e1902-16. http://dx.doi.org/10.1128/mBio.01902-16.
9. **Gao J, Tian Z, Yang X.** 2020. Breakthrough: chloroquine phosphate has shown apparent efficacy in treatment of COVID-19 associated pneumonia in clinical studies. *Biosci Trends* **14**:72–73. http://dx.doi.org/10.5582/bst.2020.01047.
10. **Colson P, Rolain JM, Lagier JC, Brouqui P, Raoult D.** 2020. Chloroquine and hydroxychloroquine as available weapons to fight COVID-19. *Int J Antimicrob Agents* **55**:105932. http://dx.doi.org/10.1016/j.ijantimicag.2020.105932.
11. **Todaro J, Rigano GJ.** 2020. An effective treatment for coronavirus (COVID-19). https://www.semanticscholar.org/paper/An-Effective-Treatment-for-Coronavirus-(COVID-19)-Todaro-Rigano/435423cc2cf4a8b37b14e05feec6452837ddbb65.
12. **Gautret P, Lagier JC, Parola P, Hoang VT, Meddeb L, Mailhe M, Doudier B, Courjon J, Giordanengo V, Vieira VE, Tissot Dupont H, Honoré S, Colson P, Chabrière E, La Scola B, Rolain JM, Brouqui P, Raoult D.** 2020. Hydroxychloroquine and azithromycin as a treatment of COVID-19: results of an open-label non-randomized clinical trial. *Int J Antimicrob Agents* **56**:105949. http://dx.doi.org/10.1016/j.ijantimicag.2020.105949.
13. **Bik E.** 24 March 2020. Thoughts on the Gautret et al. paper about hydroxychloroquine and azithromycin treatment of COVID-19 infections. *Science Integrity Digest*. https://scienceintegritydigest.com/2020/03/24/thoughts-on-the-gautret-et-al-paper-about-hydroxychloroquine-and-azithromycin-treatment-of-covid-19-infections/.
14. **Million M, Lagier JC, Gautret P, Colson P, Fournier PE, Amrane S, Hocquart M, Mailhe M, Esteves-Vieira V, Doudier B, Aubry C, Correard F, Giraud-Gatineau A, Roussel Y, Berenger C,**

Cassir N, Seng P, Zandotti C, Dhiver C, Ravaux I, Tomei C, Eldin C, Tissot-Dupont H, Honoré S, Stein A, Jacquier A, Deharo JC, Chabrière E, Levasseur A, Fenollar F, Rolain JM, Obadia Y, Brouqui P, Drancourt M, La Scola B, Parola P, Raoult D. 2020. Early treatment of COVID-19 patients with hydroxychloroquine and azithromycin: a retrospective analysis of 1061 cases in Marseille, France. *Travel Med Infect Dis* **35**:101738. http://dx.doi.org/10.1016/j.tmaid.2020.101738.

15. **Sayare S.** 12 May 2020. He was a science star. Then he promoted a questionable cure for COVID-19. *New York Times.* https://www.nytimes.com/2020/05/12/magazine/didier-raoult-hydroxychloro-quine.html.

16. **Axfors C, Schmitt AM, Janiaud P, Van't Hooft J, Abd-Elsalam S, Abdo EF, Abella BS, Akram J, Amaravadi RK, Angus DC, Arabi YM, Azhar S, Baden LR, Baker AW, Belkhir L, Benfield T, Berrevoets MAH, Chen CP, Chen TC, Cheng SH, Cheng CY, Chung WS, Cohen YZ, Cowan LN, Dalgard O, de Almeida E Val FF, de Lacerda MVG, de Melo GC, Derde L, Dubee V, Elfakir A, Gordon AC, Hernandez-Cardenas CM, Hills T, Hoepelman AIM, Huang YW, Igau B, Jin R, Jurado-Camacho F, Khan KS, Kremsner PG, Kreuels B, Kuo CY, Le T, Lin YC, Lin WP, Lin TH, Lyngbakken MN, McArthur C, McVerry BJ, Meza-Meneses P, Monteiro WM, Morpeth SC, Mourad A, Mulligan MJ, Murthy S, Naggie S, Narayanasamy S, Nichol A, Novack LA, O'Brien SM, Okeke NL, Perez L, Perez-Padilla R, Perrin L, Remigio-Luna A, Rivera-Martinez NE, Rockhold FW, Rodriguez-Llamazares S, Rolfe R, Rosa R, Røsjø H, Sampaio VS, Seto TB, Shahzad M, Soliman S, Stout JE, Thirion-Romero I, Troxel AB, Tseng TY, Turner NA, Ulrich RJ, Walsh SR, Webb SA, Weehuizen JM, Velinova M, Wong HL, Wrenn R, Zampieri FG, Zhong W, Moher D, Goodman SN, Ioannidis JPA, Hemkens LG.** 2021. Mortality outcomes with hydroxy-chloroquine and chloroquine in COVID-19 from an international collaborative meta-analysis of randomized trials. *Nat Commun* **12**:2349. http://dx.doi.org/10.1038/s41467-021-22446-z.

17. **Marx K.** 1852. *The Eighteenth Brumaire of Louis Bonaparte.* Die Revolution, New York, NY.

18. **Caly L, Druce JD, Catton MG, Jans DA, Wagstaff KM.** 2020. The FDA-approved drug iver-mectin inhibits the replication of SARS-CoV-2 *in vitro. Antiviral Res* **178**:104787. http://dx.doi.org/10.1016/j.antiviral.2020.104787.

19. **Elgazzar A, Hany B, Youssef SA, Hany B, Hafez M, Moussa H.** 13 November 2020. Efficacy and safety of ivermectin for treatment and prophylaxis of COVID-19 pandemic. Research Square preprint. https://www.researchsquare.com/article/rs-100956/v1.

20. **Popp M, Stegemann M, Metzendorf MI, Gould S, Kranke P, Meybohm P, Skoetz N, Weibel S.** 2021. Ivermectin for preventing and treating COVID-19. *Cochrane Database Syst Rev* **7**:CD015017.

21. **Robins-Early N.** 13 September 2021. Ivermectin frenzy: the advocates, anti-vaxxers and telehealth companies driving demand. *The Guardian.* https://www.theguardian.com/world/2021/sep/13/ivermectin-treatment-covid-19-anti-vaxxers-advocates.

22. **Lawrence JM, Meyerowitz-Katz G, Heathers JAJ, Brown NJL, Sheldrick KA.** 2021. The lesson of ivermectin: meta-analyses based on summary data alone are inherently unreliable. *Nat Med* **27**:1853–1854. http://dx.doi.org/10.1038/s41591-021-01535-y.

23. **Schraer R, Goodman J.** 6 October 2021. Ivermectin: how false science created a Covid "miracle" drug. *BBC News.* https://www.bbc.com/news/health-58170809.

24. **Jung RG, Di Santo P, Clifford C, Prosperi-Porta G, Skanes S, Hung A, Parlow S, Visintini S, Ramirez FD, Simard T, Hibbert B.** 2021. Methodological quality of COVID-19 clinical research. *Nat Commun* **12**:943. http://dx.doi.org/10.1038/s41467-021-21220-5.

25. **Bak-Coleman JB, Alfano M, Barfuss W, Bergstrom CT, Centeno MA, Couzin ID, Donges JF, Galesic M, Gersick AS, Jacquet J, Kao AB, Moran RE, Romanczuk P, Rubenstein DI, Tombak KJ, Van Bavel JJ, Weber EU.** 2021. Stewardship of global collective behavior. *Proc Natl Acad Sci U S A* **118**:e2025764118. http://dx.doi.org/10.1073/pnas.2025764118.

26. **Bella T.** 2022. A vaccine scientist's discredited claims have bolstered a movement of misinforma-tion. 24 January 2022. Washington Post. https://www.washingtonpost.com/health/2022/01/24/robert-malone-vaccine-misinformatino-rogan-mandates/

27. **COVID-19 National Preparedness Collaborators.** 2022. Pandemic preparedness and COVID-19: an exploratory analysis of infection and fatality rates, and contextual factors associated with

preparedness in 177 countries, from Jan 1, 2020, to Sept 30, 2021. *Lancet* **399**:1489–1512. http://dx.doi.org/10.1016/S0140-6736(22)00172-6.

28. **Lenton TM, Boulton CA, Scheffer M**. 2023. Resilience of countries to COVID-19 correlated with trust. *Sci Rep* **12**:75. https://www.nature.com/articles/s41598-021-03358-w.

29. **Kaufman J**. 10 September 2021. Science alone can't heal a sick society. *New York Times*. https://www.nytimes.com/2021/09/10/opinion/covid-science-trust-us.html.

30. **Casadevall A, Fang FC**. 2010. Reproducible science. *Infect Immun* **78**:4972–4975. http://dx.doi.org/10.1128/IAI.00908-10.

31. **Open Science Collaboration**. 2015. PSYCHOLOGY. Estimating the reproducibility of psychological science. *Science* **349**:aac4716. http://dx.doi.org/10.1126/science.aac4716.

32. **Fang FC, Steen RG, Casadevall A**. 2012. Misconduct accounts for the majority of retracted scientific publications. *Proc Natl Acad Sci U S A* **109**:17028–17033. http://dx.doi.org/10.1073/pnas.1212247109.

33. **Mehra MR, Desai SS, Ruschitzka F, Patel AN**. 2020. RETRACTED: Hydroxychloroquine or chloroquine with or without a macrolide for treatment of COVID-19: a multinational registry analysis. *Lancet* doi:10.1016/S0140-6736(20)31180-6

34. **Mehra MR, Desai SS, Kuy S, Henry TD, Patel AN**. 2020. Cardiovascular disease, drug therapy, and mortality in Covid-19. *N Engl J Med* **382**:e102. http://dx.doi.org/10.1056/NEJMoa2007621.

35. **Offord C**. October 2020. The Surgisphere scandal: what went wrong? The Scientist. https://www.the-scientist.com/features/the-surgisphere-scandal-what-went-wrong--67955.

36. **Piller C**. 8 June 2020. Who's to blame? These three scientists are at the heart of the Surgisphere COVID-19 scandal. *Science*. https://www.science.org/content/article/whos-blame-these-three-scientists-are-heart-surgisphere-covid-19-scandal.

37. **Rabin RC**. 14 June 2020. The pandemic claims new victims: prestigious medical journals. *New York Times*. https://www.nytimes.com/2020/06/14/health/virus-journals.html.

38. **Walach H, Klement RJ, Aukema W**. 2021. Retracted: The safety of COVID-19 vaccinations—we should rethink the policy. *Vaccines (Basel)* **9**:693. http://dx.doi.org/10.3390/vaccines9070693.

39. **Walach H, Weikl R, Prentice J, Diemer A, Traindl H, Kappes A, Hockertz S**. 2021. Experimental assessment of carbon dioxide content in inhaled air with or without face masks in healthy children: a randomized clinical trial. *JAMA Pediatr* doi:10.1001/jamapediatrics.2021.2659.

40. **Evidence-Based Medicine Working Group**. 1992. Evidence-based medicine. *A new approach to teaching the practice of medicine. JAMA* **268**:2420–2425. http://dx.doi.org/10.1001/jama.1992.03490170092032.

41. **Fang FC, Benson CA, Del Rio C, Edwards KM, Fowler VG, Jr, Fredricks DN, Limaye AP, Murray BE, Naggie S, Pappas PG, Patel R, Paterson DL, Pegues DA, Petri WA, Jr, Schooley RT**. 2021. COVID-19—Lessons Learned and Questions Remaining. *Clin Infect Dis* **72**:2225–2240. http://dx.doi.org/10.1093/cid/ciaa1654.

42. **Fang FC, Naccache SN, Greninger AL**. 2020. The laboratory diagnosis of coronavirus disease 2019—frequently asked questions. *Clin Infect Dis* **71**:2996–3001. http://dx.doi.org/10.1093/cid/ciaa742.

43. **Guerin P, McLean ARD, Rashan S, Lawal AA, Watson JA, Strub-Wourgaft N, White NJ**. 2022. Definitions matter: heterogeneity of COVID-19 disease severity criteria and incomplete reporting compromise meta-analysis. *PLoS Glob Public Health* **2**:e0000561. http://dx.doi.org/10.1371/journal.pgph.0000561.

44. **Casadevall A, Dragotakes Q, Johnson PW, Senefeld JW, Klassen SA, Wright RS, Joyner MJ, Paneth N, Carter RE**. 2021. Convalescent plasma use in the USA was inversely correlated with COVID-19 mortality. *eLife* **10**:e69866. http://dx.doi.org/10.7554/eLife.69866.

45. **Libster R, Pérez Marc G, Wappner D, Coviello S, Bianchi A, Braem V, Esteban I, Caballero MT, Wood C, Berrueta M, Rondan A, Lescano G, Cruz P, Ritou Y, Fernández Viña V, Álvarez Paggi D, Esperante S, Ferreti A, Ofman G, Ciganda Á, Rodriguez R, Lantos J, Valentini R, Itcovici N, Hintze A, Oyarvide ML, Etchegaray C, Neira A, Name I, Alfonso J, López Castelo R, Caruso G, Rapelius S,**

Alvez F, Etchenique F, Dimase F, Alvarez D, Aranda SS, Sánchez Yanotti C, De Luca J, Jares Baglivo S, Laudanno S, Nowogrodzki F, Larrea R, Silveyra M, Leberzstein G, Debonis A, Molinos J, González M, Perez E, Kreplak N, Pastor Argüello S, Gibbons L, Althabe F, Bergel E, Polack FP; Fundación INFANT–COVID-19 Group. 2021. Early high-titer plasma therapy to prevent severe Covid-19 in older adults. *N Engl J Med* **384**:610–618. http://dx.doi.org/10.1056/NEJMoa2033700.

46. **RECOVERY Collaborative Group**. 2021. Convalescent plasma in patients admitted to hospital with COVID-19 (RECOVERY): a randomised controlled, open-label, platform trial. *Lancet* **397**:2049–2059. http://dx.doi.org/10.1016/S0140-6736(21)00897-7.

47. **Klassen SA, Senefeld JW, Johnson PW, Carter RE, Wiggins CC, Shoham S, Grossman BJ, Henderson JP, Musser J, Salazar E, Hartman WR, Bouvier NM, Liu STH, Pirofski LA, Baker SE, van Helmond N, Wright RS, Fairweather D, Bruno KA, Wang Z, Paneth NS, Casadevall A, Joyner MJ**. 2021. The effect of convalescent plasma therapy on mortality among patients with COVID-19: systematic review and meta-analysis. *Mayo Clin Proc* **96**:1262–1275. http://dx.doi.org/10.1016/j.mayocp.2021.02.008.

48. Janiaud P, Axfors C, Schmitt AM, Gloy V, Ebrahimi F, Hepprich M, Smith ER, Haber NA, Khanna N, Moher D, Goodman SN, Ioannidis JPA, Hemkens LG. 2021. Association of convalescent plasma treatment with clinical outcomes in patients with COVID-19: a systematic review and meta-analysis. *JAMA* **325**:1185–1195. http://dx.doi.org/10.1001/jama.2021.2747.

49. Sullivan DJ, Gebo KA, Shoham S, Bloch EM, Lau B, Shenoy AG, Mosnaim GS, Gniadek TJ, Fukuta Y, Patel B, Heath SL, Levine AC, Meisenberg BR, Spivak ES, Anjan S, Huaman MA, Blair JE, Currier JS, Paxton JH, Gerber JM, Petrini JR, Broderick PB, Rausch W, Cordisco ME, Hammel J, Greenblatt B, Cluzet VC, Cruser D, Oei K, Abinante M, Hammitt LL, Sutcliffe CG, Forthal DN, Zand MS, Cachay ER, Raval JS, Kassaye SG, Foster EC, Roth M, Marshall CE, Yarava A, Lane K, McBee NA, Gawad AL, Karlen N, Singh A, Ford DE, Jabs DA, Appel LJ, Shade DM, Ehrhardt S, Baksh SN, Laeyendecker O, Pekosz A, Klein SL, Casadevall A, Tobian AAR, Hanley DF. 2022. Early outpatient treatment for Covid-19 with convalescent plasma. *N Engl J Med* **386**:1700–1711. http://dx.doi.org/10.1056/NEJMoa2119657.

50. **Focosi D, Franchini M, Pirofski LA, Burnouf T, Paneth N, Joyner MJ, Casadevall A**. 2022. COVID-19 convalescent plasma and clinical trials: understanding conflicting outcomes. *Clin Microbiol Rev* **35**:e0020021. http://dx.doi.org/10.1128/cmr.00200-21.

51. **Tonelli MR, Shirts BH**. 2017. Knowledge for precision medicine: mechanistic reasoning and methodological pluralism. *JAMA* **318**:1649–1650. http://dx.doi.org/10.1001/jama.2017.11914.

52. **Jefferson T, Dooley L, Ferroni E, Al-Ansary LA, van Driel ML, Bawazeer GA, Jones MA, Hoffmann TC, Clark J, Beller EM, Glasziou PP, Conly JM**. 2023. Physical interventions to interrupt or reduce the spread of respiratory viruses. *Cochrane Database Syst Rev* **1**:CD006207. doi: 10.1002/14651858.CD006207.pub6.

53. **Tufekci Z**. 2023. Here's why the science is clear that masks work. *New York Times* 10 March 2023. https://www.nytimes.com/2023/03/10/opinion/masks-work-cochrane-study.html.

54. **Zhou H, Ji J, Chen X, Bi J, Li J, Wang Q, Hu T, Song H, Zhao R, Chen Y, Cui M, Zhang Y, Hughes AC, Holmes EC, Shi W**. 2021. Identification of novel bat coronaviruses sheds light on the evolutionary origins of SARS-CoV-2 and related viruses. *Cell* **184**:4380–4391.e14. http://dx.doi.org/10.1016/j.cell.2021.06.008.

55. **Maxmen A**. 2021. Divisive COVID "lab leak" debate prompts dire warnings from researchers. *Nature* **594**:15–16. http://dx.doi.org/10.1038/d41586-021-01383-3.

56. **Dance A**. 2021. The shifting sands of "gain-of-function" research. *Nature* **598**:554–557. http://dx.doi.org/10.1038/d41586-021-02903-x.

57. **Holmes EC, Goldstein SA, Rasmussen AL, Robertson DL, Crits-Christoph A, Wertheim JO, Anthony SJ, Barclay WS, Boni MF, Doherty PC, Farrar J, Geoghegan JL, Jiang X, Leibowitz JL, Neil SJD, Skern T, Weiss SR, Worobey M, Andersen KG, Garry RF, Rambaut A**. 2021. The origins of SARS-CoV-2: a critical review. *Cell* **184**:4848–4856. http://dx.doi.org/10.1016/j.cell.2021.08.017.

58. **Worobey M, Levy JI, Malpica Serrano L, Crits-Christoph A, Pekar JE, Goldstein SA, Rasmussen AL, Kraemer MUG, Newman C, Koopmans MPG, Suchard MA, Wertheim JO, Lemey P, Robertson DL, Garry RF, Holmes EC, Rambaut A, Andersen KG.** 2022. The Huanan Seafood Wholesale Market in Wuhan was the early epicenter of the COVID-19 pandemic. *Science* **377**:951–959. http://dx.doi.org/10.1126/science.abp8715.

59. **Pekar JE, Magee A, Parker E, Moshiri N, Izhikevich K, Havens JL, Gangavarapu K, Malpica Serrano LM, Crits-Christoph A, Matteson NL, Zeller M, Levy JI, Wang JC, Hughes S, Lee J, Park H, Park MS, Ching Zi Yan K, Lin RTP, Mat Isa MN, Noor YM, Vasylyeva TI, Garry RF, Holmes EC, Rambaut A, Suchard MA, Andersen KG, Worobey M, Wertheim JO.** 2022. The molecular epidemiology of multiple zoonotic origins of SARS-CoV-2. *Science* **377**:960–966. http://dx.doi.org/10.1126/science.abp8337.

60. **Liu WJ, Liu P, Lei W, Jia Z, He X, Shi W, Tah Y, Zou S, Wong G, Wang J, Wang F, Wang G, Win K, Gao B, Zhang J, Li M, Xiao W, Guo Y, Xu Z, Zhao Y, Song J, Zhang J, Zhen W, Zhou W, Ye B, Song J, Yang M, Zhou W, Dai Y, Lu G, Bi Y, Tan W, Han J, Gao GF, Wu G.** 2023. Surveillance of SARS-CoV-2 at the Huanan Seafood Market. *Nature* https://doi.org/10.1038/s41586-023-06043-2.

61. **Cohen J.** 2023. New clues to pandemic's origin surface, causing uproar. *Science* **379**:1175-1176. doi: 10.1126/science.adh9055.

62. **Maxmen A.** 2022. Wuhan market was epicentre of pandemic's start, studies suggest. *Nature* **603**: 15-16. doi: 10.1038/d41586-022-00584-8.

63. **Alwine JC, Casadevall A, Enquist LW, Goodrum FD, Imperiale MJ.** 2023. A critical analysis of the evidence for the SARS-CoV-2 hypotheses. mBio 2023 Mar 28:e0058323. doi: 10.1128/mbio.00583-23.

64. **World Health Organization.** 15 July 2021. WHO press conference on coronavirus disease (COVID-19). https://www.who.int/multi-media/details/who-press-conference-on-coronavirus-disease-(covid-19)—15-july-2021.

65. **Goodrum F, Lowen AC, Lakdawala S, Alwine J, Casadevall A, Imperiale MJ, et al.** 2023. Virology under the microscope: A call for rational discourse. *mBio* **14**:e0018823. doi: 10.1128/mbio.00188-23.

66. **Falkow S.** 2012. The lessons of Asilomar and the H5N1 "affair." *mBio* **3**:e00354-12. doi: 10/.1128//mBio.00354-12.

67. **Prather KA, Marr LC, Schooley RT, McDiarmid MA, Wilson ME, Milton DK.** 2020. Airborne transmission of SARS-CoV-2. *Science* **370**:303–304.

68. **Morawska L, Milton DK.** 2020. It is time to address airborne transmission of coronavirus disease 2019 (COVID-19). *Clin Infect Dis* **71**:2311–2313. http://dx.doi.org/10.1093/cid/ciaa939.

69. **Chagla Z, Hota S, Khan S, Mertz D; International Hospital and Community Epidemiology Group.** 2021. Re: It is time to address airborne transmission of COVID-19. *Clin Infect Dis* **73**:e3981–e3982. http://dx.doi.org/10.1093/cid/ciaa1118.

70. **Casadevall A, Fang FC.** 2014. Specialized science. *Infect Immun* **82**:1355–1360. http://dx.doi.org/10.1128/IAI.01530-13.

71. **Casadevall A, Fang FC.** 2015. Field science—the nature and utility of scientific fields. *mBio* **6**:e01259-15. http://dx.doi.org/10.1128/mBio.01259-15.

72. **Greenhalgh T, Ozbilgin M, Contandriopoulos D.** 2021. Orthodoxy, *illusio*, and playing the scientific game: a Bourdieusian analysis of infection control science in the COVID-19 pandemic. *Wellcome Open Res* **6**:126. http://dx.doi.org/10.12688/wellcomeopenres.16855.3.

73. **Gale RP.** 2020. Conquest of COVID-19. Publish it to death? *Br J Haematol* **190**:358–360. http://dx.doi.org/10.1111/bjh.16905.

74. **Clark J.** 2023. How covid-19 bolstered an already perverse publishing system. *BMJ* **380**:689. http://dx.doi.org/10.1136/bmj.p689.

75. **Cross R.** 25 January 2021. Will public trust in science survive the pandemic? Chemical Engineering News. https://cen.acs.org/policy/global-health/Will-public-trust-in-science-survive-the-pandemic/99/i3.

76. **Prasad V.** 16 October 2020. Op-Ed: Great Barrington vs John Snow is a false choice. *MedPage Today*. https://www.medpagetoday.com/opinion/vinay-prasad/89177.

77. **Ball P.** 2021. What the COVID-19 pandemic reveals about science, policy and society. *Interface Focus* **11**:20210022. http://dx.doi.org/10.1098/rsfs.2021.0022.

78. **Sandman PM**. 9 December 2021. Commentary: 8 things US pandemic commentators still get wrong. Center for Infectious Disease Research and Policy. https://www.cidrap.umn.edu/commentary-8-things-us-pandemic-communicators-still-get-wrong.

79. **Bollyky TJ, Castro E, Aravkin AY, Bhangdia K, Dalos J, Hulland EN, Kiernan S, Lastuka A, McHugh TA, Ostroff SM, Zheng P, Chaudhry HT, Ruggiero E, Turilli I, Adolph C, Amlag JO, Bang-Jensen B, Barber RM, Carter A, Chang C, Cogen RM, Collins JK, Dai X, Dangel WJ, Dapper C, Deen A, Eastus A, Erickson M, Fedosseeva T, Flaxman AD, Fullman N, Giles JR, Guo G, Hay SI, He J, Helak M, Huntley BM, Iannucci VC, Kinzel KE, LeGrand KE, Magistro B, Mokdad AH, Nassereldine H, Ozten Y, Pasovic M, Pigott DM, Reiner RC Jr, Reinke G, Schumacher AE, Serieux E, Spurlock EE, Troeger CE, Vo AT, Vos T, Walcott R, Yazdani S, Murray CJL, Dieleman JL.** 2023. Assessing COVID-19 pandemic policies and behaviours and their economic and educational trade-offs across US states from Jan 1, 2020, to July 31, 2022: an observational analysis. *Lancet* **401**:1341-1360. doi: 10.1016/S0140-6736(23)00461-0.

80. **Algan Y, Cohen D, Davoine E, Foucault M, Stantcheva S.** 2021. Trust in scientists in times of pandemic: Panel evidence from 12 countries. *Proc Natl Acad Sci USA* **118**:e2108576118. https://doi.org/10.1073/pnas.2108576118.

81. **Weber L, Achenbach J.** 2023. Covid backlash hobbles public health and future pandemic response. 8 March 2023. Washington Post. https://www.washingtonpost.com/health/2023/03/08/covid-public-health-backlash/

82. **Sachs JD, Karim SSA, Aknin L, Allen J, Brosbøl K, Colombo F, Barron GC, Espinosa MF, Gaspar V, Gaviria A, Haines A, Hotez PJ, Koundouri P, Bascuñán FL, Lee JK, Pate MA, Ramos G, Reddy KS, Serageldin I, Thwaites J, Vike-Freiberga V, Wang C, Were MK, Xue L, Bahadur C, Bottazzi ME, Bullen C, Laryea-Adjei G, Ben Amor Y, Karadag O, Lafortune G, Torres E, Barredo L, Bartels JGE, Joshi N, Hellard M, Huynh UK, Khandelwal S, Lazarus JV, Michie S.** 2022. The Lancet Commission on lessons for the future from the COVID-19 pandemic. *Lancet.* **400**:1224-1280. doi: 10.1016/S0140-6736(22)01585-9.

83. **Kennedy B, Tyson A, Funk C.** 2022. Americans' trust in scientists, other groups declines. Pew Research Center. 15 February 2022. https://www.pewresearch.org/science/2022/02/15/americans-trust-in-scientists-other-groups-declines/

84. **Camus A.** 1947. *La Peste.* Gallimard, Paris, France.

85. **Osborn B.** 2022. Northwell names Sandra Lindsay to public health leadership role. Northwell Health Newsroom. 3 October 2022. https://www.northwell.edu/news/the-latest/northwell-names-sandra-lindsay-to-public-health-leadership-role.

Chapter 31 – Reforming Science

1. **Levitt S, Dubner SJ**. 2005. *Freakonomics: A Rogue Economist Explores the Hidden Side of Everything.* William Morrow, New York, NY.

2. **Carlyle T**. 1858. *The Collected Works of Thomas Carlyle.* Chapman and Hall, London, United Kingdom.

3. **Machiavelli N**. 1532. *The Prince.* Antonio Blado d'Asola, Florence, Italy.

4. **Kaiser J**. 2009. Biomedical research. Rejecting "big science" tag, Collins sets five themes for NIH. *Science* **325**:927. http://dx.doi.org/10.1126/science.325_927a.

5. **Dorsey ER, Van Wuyckhuyse BC, Beck CA, Passalacqua WP, Guzick DS**. 2009. The economics of new faculty hires in basic science. *Acad Med* **84**:26–31. http://dx.doi.org/10.1097/ACM.0b013e3181904633.

6. **Boron WF**. 2009. Managing the business of science. *Physiology (Bethesda)* **24**:2–3. http://dx.doi.org/10.1152/physiol.00044.2008.

7. **Fang FC, Casadevall A**. 2010. Lost in translation—basic science in the era of translational research. *Infect Immun* **78**:563–566. http://dx.doi.org/10.1128/IAI.01318-09.

8. **National Institutes of Health.** 2022. *NIH Data Book.* http://report.nih.gov/nihdatabook/.

9. **Rhodes R.** 1986. *The Making of the Atomic Bomb.* Simon and Schuster, New York, NY.

10. **Casadevall A.** 2021. The mRNA vaccine revolution is the dividend from decades of basic science research. *J Clin Invest* **131**:e153721. http://dx.doi.org/10.1172/JCI153721.

11. **Goulden M, Mason MA, Frasch K.** 2011. Keeping women in the science pipeline. *Ann Am Acad Pol Soc Sci* **638**:141–162. http://dx.doi.org/10.1177/0002716211416925.

12. **Barr DA, Gonzalez ME, Wanat SF.** 2008. The leaky pipeline: factors associated with early decline in interest in premedical studies among underrepresented minority undergraduate students. *Acad Med* **83**:503–511. http://dx.doi.org/10.1097/ACM.0b013e31816bda16.

13. **Committee on Underrepresented Groups and the Expansion of the Science and Engineering Workforce Pipeline.** 2010. *Expanding Underrepresented Minority Participation: America's Science and Technology Talent at the Crossroads.* National Academies Press, Washington, DC.

14. **Hrabowski FA, III.** 2011. Boosting minorities in science. *Science* **331**:125. http://dx.doi.org/10.1126/science.1202388.

15. **Tabak LA, Collins FS.** 2011. Sociology. Weaving a richer tapestry in biomedical science. *Science* **333**:940–941. http://dx.doi.org/10.1126/science.1211704.

16. **Haywood JR, Greene M.** 2008. Avoiding an overzealous approach: a perspective on regulatory burden. *ILAR J* **49**:426–434. http://dx.doi.org/10.1093/ilar.49.4.426.

17. **Gutowsky HS.** 1981. Federal funding of basic research: the red tape mill. *Science* **212**:636–641. http://dx.doi.org/10.1126/science.212.4495.636.

18. **Infectious Diseases Society of America.** 2009. Grinding to a halt: the effects of the increasing regulatory burden on research and quality improvement efforts. *Clin Infect Dis* **49**:328–335. http://dx.doi.org/10.1086/605454.

19. **Whitney SN, Schneider CE.** 2009. Was the institutional review board system a mistake? *Clin Infect Dis* **49**:1957, author reply 1957. http://dx.doi.org/10.1086/648496.

20. **James A.** 2011. Too many tasks. *Nature* **475**:257. http://dx.doi.org/10.1038/nj7355-257a.

21. **Kaplan D, Lacetera N, Kaplan C.** 2008. Sample size and precision in NIH peer review. *PLoS One* **3**:e2761. http://dx.doi.org/10.1371/journal.pone.0002761.

22. **Fang FC, Bowen A, Casadevall A.** 2016. NIH peer review percentile scores are poorly predictive of grant productivity. *eLife* **5**:e13323. http://dx.doi.org/10.7554/eLife.13323.

23. **Martinson BC.** 2011. The academic birth rate. Production and reproduction of the research work force, and its effect on innovation and research misconduct. *EMBO Rep* **12**:758–762. http://dx.doi.org/10.1038/embor.2011.142.

24. **Martinson B.** 25 June 2015. The Elizabeth Goodwin case study—a family tree, or: why cohort matters. Prezi. https://prezi.com/p75rojb0mt0z/goodwin_case_study.

25. **Couzin-Frankel J.** 28 June 2010. Scientist turned in by grad students for misconduct pleads guilty. *Science.* https://www.science.org/content/article/scientist-turned-in-grad-students-misconduct-pleads-guilty.

26. **Ioannidis JP.** 2006. Concentration of the most-cited papers in the scientific literature: analysis of journal ecosystems. *PLoS One* **1**:e5. http://dx.doi.org/10.1371/journal.pone.0000005.

27. **Bauerlein M, Gad el-Hak M, Grody W, McKelvey B, Trimble SW.** 13 June 2010. We must stop the avalanche of low quality research. *Chronicle of Higher Education.* https://www.chronicle.com/article/we-must-stop-the-avalanche-of-low-quality-research/.

28. **Jinha AE.** 2010. Article 50 million: an estimate of the number of scholarly articles in existence. *Learn Publ* **23**:258–263. http://dx.doi.org/10.1087/20100308.

29. **Colquhoun D.** 5 September 2011. Publish or perish: peer review and the corruption of science. *The Guardian.* http://www.guardian.co.uk/science/2011/sep/05/publish-perish-peer-review-science.

30. **Fanelli D.** 2010. Do pressures to publish increase scientists' bias? An empirical support from US States Data. *PLoS One* **5**:e10271. http://dx.doi.org/10.1371/journal.pone.0010271.

31. **Ioannidis JP.** 2005. Why most published research findings are false. *PLoS Med* **2**:e124. http://dx.doi.org/10.1371/journal.pmed.0020124.

32. **Fanelli D.** 2009. How many scientists fabricate and falsify research? A systematic review and meta-analysis of survey data. *PLoS One* **4**:e5738. http://dx.doi.org/10.1371/journal.pone.0005738.

33. **Ioannidis JP.** 2011. An epidemic of false claims. Competition and conflicts of interest distort too many medical findings. *Sci Am* **304**:16. http://dx.doi.org/10.1038/scientificamerican0611-16.

34. **Dowd M.** 27 September 2011. Decoding the God complex. *New York Times*. http://www.nytimes.com/2011/09/28/opinion/dowd-decoding-the-god-complex.html.

35. **Lee C.** 28 May 2007. Slump in NIH funding is taking a toll on research, p A06. *Washington Post*.

36. **Strevens M.** 2003. The role of the priority rule in science. *J Philos* **100**:55–79. http://dx.doi.org/10.5840/jphil2003100224.

37. **Merton RK.** 1968. The Matthew effect in science. The reward and communication systems of science are considered. *Science* **159**:56–63. http://dx.doi.org/10.1126/science.159.3810.56.

38. **Rablen MD, Oswald AJ.** 2008. Mortality and immortality: the Nobel Prize as an experiment into the effect of status upon longevity. *J Health Econ* **27**:1462–1471. http://dx.doi.org/10.1016/j.jhealeco.2008.06.001.

39. **The Editors of The Lancet.** 2010. Retraction—Ileal-lymphoid-nodular hyperplasia, non-specific colitis, and pervasive developmental disorder in children. *Lancet* **375**:445. http://dx.doi.org/10.1016/S0140-6736(10)60175-4.

40. **Kohut A.** 9 July 2009. Public praises science; scientists fault public, media: scientific achievements less prominent than a decade ago. Pew Research Center. http://www.people-press.org/2009/07/09/public-praises-science-scientists-fault-public-media/.

41. **Leiserowitz AA, Maibach WR, Roser-Renouf C, Smith N, Dawson E.** 2010. Climategate, public opinion, and the loss of trust. SSRN. http://dx.doi.org/10.2139/ssrn.1633932

42. **Gauchat G.** 2012. Politicization of science in the public sphere: a study of public trust in the United States, 1974 to 2010. *Am Sociol Rev* **77**:167–187. http://dx.doi.org/10.1177/0003122412438225.

43. **Kennedy B, Tyson A, Funk C.** 15 February 2022. Americans' trust in scientists, other groups declines. Pew Research Center. https://www.pewresearch.org/science/2022/02/15/americans-trust-in-scientists-other-groups-declines/.

44. **Wikipedia.** 2022. List of scientific priority disputes. http://en.wikipedia.org/wiki/List_of_scientific_priority_disputes.

45. **Watson JD.** 1968. *The Double Helix: a Personal Account of the Discovery of the Structure of DNA*. Atheneum, New York, NY.

46. **Vahlne A.** 2009. A historical reflection on the discovery of human retroviruses. *Retrovirology* **6**:40. http://dx.doi.org/10.1186/1742-4690-6-40.

47. **Tollefson J, Gibney E.** 2022. Nuclear-fusion lab achieves "ignition": what does it mean? *Nature* **612**:597–598. http://dx.doi.org/10.1038/d41586-022-04440-7.

48. **Witze A.** 2022. This spacecraft just smashed into an asteroid in an attempt to change its path. *Nature* **2022**:23. http://dx.doi.org/10.1038/d41586-022-03030-x.

49. **Imperiale MJ, Casadevall A, Goodrum FD, Schultz-Cherry S.** 2022. Virology in peril and the greater risk to science. *mBio* **14**:e0333922. http://dx.doi.org/10.1128/mbio.03339-22.

50. **Bilmes L.** 26 October 2010. How the wars are sinking the economy. *The Daily Beast*. http://www.thedailybeast.com/articles/2010/10/27/the-economic-crisis-and-the-hidden-cost-of-the-wars.html.

51. **Clemins PJ.** 2010. *Overview of Federal Research and Development Funding*. American Association for the Advancement of Science, Washington, DC.

52. **Adams J, Pendlebury D.** 2010. *Global Research Report, United States*. Thomson Reuters, Leeds, United Kingdom.

53. **Alpert D, Hockett R, Roubini N.** 2011. *The Way Forward: Moving from the Post-bubble, Post-bust Economy to Renewed Growth and Competitiveness*. New America Foundation, Washington, DC.

54. **Nocera J.** 10 October 2011. This time, it really is different. *New York Times*. https://www.nytimes.com/2011/10/11/opinion/this-time-it-really-is-different.html.

55. **Kaiser J.** 2011. U.S. science and austerity. Darwinism vs. social engineering at NIH. *Science* **334**:753–754. http://dx.doi.org/10.1126/science.334.6057.753.

56. **Drew C.** 4 November 2011. Why science majors change their minds (it's just so darn hard). *New York Times.* https://www.nytimes.com/2011/11/06/education/edlife/why-science-majors-change-their-mind-its-just-so-darn-hard.html.

57. **Kowitt B.** 4 October 2011. DuPont CEO: We need more U.S. scientists. *CNN Money.* http://management.fortune.cnn.com/2011/10/04/ellen-kullman-dupont/.

58. **Will GF.** 2 January 2011. Rev the scientific engine. *Washington Post.* http://www.washingtonpost.com/wp-dyn/content/article/2010/12/31/AR2010123102007.html.

59. **Committee on Prospering in the Global Economy of the 21st Century.** 2005. *Rising above the Gathering Storm: Energizing and Employing America for a Brighter Economic Future.* National Academies Press, Washington, DC.

60. **Costello LC.** 2010. Perspective: is NIH funding the "best science by the best scientists"? A critique of the NIH R01 research grant review policies. *Acad Med* **85**:775–779. http://dx.doi.org/10.1097/ACM.0b013e3181d74256.

61. **Langin K.** 13 June 2022. As professors struggle to recruit postdocs, calls for structural change in academia intensify. *Science.* https://www.science.org/content/article/professors-struggle-recruit-postdocs-calls-structural-change-academia-intensify.

62. **Katz JI.** 1999. Don't become a scientist! https://muratk3n.github.io/thirdwave/en/2011/10/dont-become-scientist.html.

63. **National Institutes of Health.** 14 June 2012. *Biomedical Research Workforce Working Group Report.* National Institutes of Health, Bethesda, MD. https://acd.od.nih.gov/documents/reports/Biomedical_research_wgreport.pdf.

64. **Justice AC.** 2009. Leaky pipes, Faustian dilemmas, and a room of one's own: can we build a more flexible pipeline to academic success? *Ann Intern Med* **151**:818–819. http://dx.doi.org/10.7326/0003-4819-151-11-200912010-00013.

65. **Villablanca AC, Beckett L, Nettiksimmons J, Howell LP.** 2011. Career flexibility and family-friendly policies: an NIH-funded study to enhance women's careers in biomedical sciences. *J Womens Health (Larchmt)* **20**:1485–1496. http://dx.doi.org/10.1089/jwh.2011.2737.

66. **Organization for Economic Cooperation and Development.** 2008. *Main Science and Technology Indicators.* OECD Publishing, Paris, France.

67. **Bush V.** 1945. *Science, the Endless Frontier—a Report to the President on a Program for Postwar Scientific Research.* United States Government Printing Office, Washington, DC.

68. **Thomas L.** 1974. *Lives of a Cell: Notes of a Biology Watcher.* Viking Press, New York, NY.

69. **Atkinson RD, Stewart LA.** 2011. *University Research Funding: the United States is Behind and Falling.* Information Technology and Innovation Foundation, Washington, DC.

70. **Wadman M.** 2010. Study says middle sized labs do best. *Nature* **468**:356–357. http://dx.doi.org/10.1038/468356a.

71. **Fang FC, Casadevall A.** 2009. NIH peer review reform—change we need, or lipstick on a pig? *Infect Immun* **77**:929–932. http://dx.doi.org/10.1128/IAI.01567-08.

72. **Pagano M.** 2006. *American Idol* and NIH grant review. *Cell* **126**:637–638. http://dx.doi.org/10.1016/j.cell.2006.08.004.

73. **Ioannidis JP.** 2011. More time for research: fund people not projects. *Nature* **477**:529–531. http://dx.doi.org/10.1038/477529a.

74. **Cole MW.** 2011. Numbers are not everything. *Academe* **95**:30.

75. **Monod E.** 2004. Einstein, Heisenberg, Kant: methodological distinction and conditions of possibilities. *Inf Organ* **14**:105–121. http://dx.doi.org/10.1016/j.infoandorg.2003.12.001.

76. **Casadevall A, Pirofski LA.** 2006. A reappraisal of humoral immunity based on mechanisms of antibody-mediated protection against intracellular pathogens. *Adv Immunol* **91**:1–44. http://dx.doi.org/10.1016/S0065-2776(06)91001-3.

77. **Casadevall A, Fang FC.** 2008. Descriptive science. *Infect Immun* **76**:3835–3836. http://dx.doi.org/10.1128/IAI.00743-08.

78. **Casadevall A, Fang FC.** 2009. Mechanistic science. *Infect Immun* **77**:3517–3519. http://dx.doi.org/10.1128/IAI.00623-09.

79. **Devezer B, Navarro DJ, Vandekerckhove J, Ozge Buzbas E**. 2021. The case for formal methodology in scientific reform. *R Soc Open Sci* **8**:200805. http://dx.doi.org/10.1098/rsos.200805.

80. **Schooley R.** Personal communication.

81. **Postma E.** 2011. Comment on "Additive genetic breeding values correlate with the load of partially deleterious mutations." *Science* **333**:1221. http://dx.doi.org/10.1126/science.1200996.

82. **Haynes AB, Weiser TG, Berry WR, Lipsitz SR, Breizat AH, Dellinger EP, Herbosa T, Joseph S, Kibatala PL, Lapitan MC, Merry AF, Moorthy K, Reznick RK, Taylor B, Gawande AA; Safe Surgery Saves Lives Study Group**. 2009. A surgical safety checklist to reduce morbidity and mortality in a global population. *N Engl J Med* **360**:491–499. http://dx.doi.org/10.1056/NEJMsa0810119.

83. **Anonymous.** 2013. Announcement: reducing our irreproducibility. *Nature* **496**:398. http://dx.doi.org/10.1038/496398a.

84. **McNutt M.** 2014. Journals unite for reproducibility. *Science* **346**:679. http://dx.doi.org/10.1126/science.aaa1724.

85. **Anonymous.** 2018. Checklists work to improve science. *Nature* **556**:273–274. http://dx.doi.org/10.1038/d41586-018-04590-7.

86. **Han S, Olonisakin TF, Pribis JP, Zupetic J, Yoon JH, Holleran KM, Jeong K, Shaikh N, Rubio DM, Lee JS**. 2017. A checklist is associated with increased quality of reporting preclinical biomedical research: A systematic review. *PLoS One* **12**:e0183591. http://dx.doi.org/10.1371/journal.pone.0183591.

87. **Stern MD**. 2010. Patrimony and the evolution of risk-taking. *PLoS One* **5**:e11656. http://dx.doi.org/10.1371/journal.pone.0011656.

88. **Bennett LM, Gadlin H, Levine-Finney S**. 2010. *Collaboration and Field Science: a Team Guide (draft)*. National Institutes of Health, Bethesda, MD.

89. **Wilson DS**. 2007. *Evolution for Everyone*. Random House, New York, NY.

90. **Alon U.** 2010. How to build a motivated research group. *Mol Cell* **37**:151–152. http://dx.doi.org/10.1016/j.molcel.2010.01.011.

91. **Begley CG, Ellis LM**. 2012. Drug development: raise standards for preclinical cancer research. *Nature* **483**:531–533. http://dx.doi.org/10.1038/483531a.

92. **Fang FC, Steen RG, Casadevall A**. 2012. Misconduct accounts for the majority of retracted scientific publications. *Proc Natl Acad Sci U S A* **109**:17028–17033. http://dx.doi.org/10.1073/pnas.1212247109.

93. **Bik EM, Casadevall A, Fang FC**. 2016. The prevalence of inappropriate image duplication in biomedical research publications. *mBio* **7**:e00809-16. http://dx.doi.org/10.1128/mBio.00809-16.

94. **Casadevall A, Fang FC**. 2012. Reforming science: methodological and cultural reforms. *Infect Immun* **80**:891–896. http://dx.doi.org/10.1128/IAI.06183-11.

95. **Fang FC, Casadevall A**. 2012. Reforming science: structural reforms. *Infect Immun* **80**:897–901. http://dx.doi.org/10.1128/IAI.06184-11.

96. **Bosch G, Casadevall A**. 2017. Graduate biomedical science education needs a new philosophy. *mBio* **8**:e01539-17. http://dx.doi.org/10.1128/mBio.01539-17.

97. **Bik EM, Fang FC, Kullas AL, Davis RJ, Casadevall A**. 2018. Analysis and correction of inappropriate image duplication: the *Molecular and Cellular Biology* experience. *Mol Cell Biol* **38**:e00309-18. http://dx.doi.org/10.1128/MCB.00309-18.

98. **Proofig.** https://www.proofig.com.

99. **Jackson S, Williams CL, Collins KL, McNally EM**. 2022. Data we can trust. *J Clin Invest* **132**:e162884. http://dx.doi.org/10.1172/JCI162884.

100. **Walker R, Rocha da Silva P**. 2015. Emerging trends in peer review-a survey. *Front Neurosci* **9**:169. http://dx.doi.org/10.3389/fnins.2015.00169.

101. **Higgins JR, Lin FC, Evans JP**. 2016. Plagiarism in submitted manuscripts: incidence, characteristics and optimization of screening-case study in a major specialty medical journal. *Res Integr Peer Rev* **1**:13. http://dx.doi.org/10.1186/s41073-016-0021-8.

102. **Powell K.** 2006. Your own desktop crime lab. *Nat Med* **12**:493. http://dx.doi.org/10.1038/nm0506-493.

103. **Nuijten MB, Hartgerink CH, van Assen MA, Epskamp S, Wicherts JM**. 2016. The prevalence of statistical reporting errors in psychology (1985-2013). *Behav Res Methods* **48**:1205–1226. http://dx.doi.org/10.3758/s13428-015-0664-2.

104. **Didier E, Guaspare-Cartron C**. 2018. The new watchdogs' vision of science: a roundtable with Ivan Oransky (Retraction Watch) and Brandon Stell (PubPeer). *Soc Stud Sci* **48**:165–167. http://dx.doi.org/10.1177/0306312718756202.

105. **Resnik DB, Dinse GE**. 2013. Scientific retractions and corrections related to misconduct findings. *J Med Ethics* **39**:46–50. http://dx.doi.org/10.1136/medethics-2012-100766.

106. **Gunsalus CK, Marcus AR, Oransky I**. 2018. Institutional research misconduct reports need more credibility. *JAMA* **319**:1315–1316. http://dx.doi.org/10.1001/jama.2018.0358.

107. **Collaborative Working Group from the conference "Keeping the Pool Clean: Prevention and Management of Misconduct Related Retractions."** 2018. RePAIR consensus guidelines: responsibilities of Publishers, Agencies, Institutions, and Researchers in protecting the integrity of the research record. *Res Integr Peer Rev* **3**:15. http://dx.doi.org/10.1186/s41073-018-0055-1.

108. **Williams CL, Casadevall A, Jackson S**. 2019. Figure errors, sloppy science, and fraud: keeping eyes on your data. *J Clin Invest* **129**:1805–1807. http://dx.doi.org/10.1172/JCI128380.

109. **Casadevall A, Fang FC**. 2014. Causes for the persistence of impact factor mania. mBio 5:e00064-14. http://dx.doi.org/10.1128/mBio.00064-14.

110. **Casadevall A, Fang FC**. 2016. Rigorous science: a how-to guide. mBio 7:e1902-16. http://dx.doi.org/10.1128/mBio.01902-16.

111. van **Assen MA, van Aert RC, Nuijten MB, Wicherts JM**. 2014. Why publishing everything is more effective than selective publishing of statistically significant results. *PLoS One* **9**:e84896. http://dx.doi.org/10.1371/journal.pone.0084896.

112. **Epstein D**. 2019. *Range: Why Generalists Triumph in a Specialized World*. Riverhead Books, New York, NY.

113. **Ciubotariu II, Bosch G**. 2022. Improving research integrity: a framework for responsible science communication. *BMC Res Notes* **15**:177. http://dx.doi.org/10.1186/s13104-022-06065-5.

114. **Chubin DE, Hackett EJ**. 1990. *Peerless Science: Peer Review and U.S. Science Policy*. State University of New York Press, Albany, NY.

Chapter 32 – Diseased Science

1. **Casadevall A, Fang FC**. 2014. Causes for the persistence of impact factor mania. *mBio* **5**:e00064-14. http://dx.doi.org/10.1128/mBio.00064-14.

2. van **Diest PJ, Holzel H, Burnett D, Crocker J**. 2001. Impactitis: new cures for an old disease. *J Clin Pathol* **54**:817–819. http://dx.doi.org/10.1136/jcp.54.11.817.

3. **Kerr NL**. 1998. HARKing: hypothesizing after the results are known. *Pers Soc Psychol Rev* **2**:196–217. http://dx.doi.org/10.1207/s15327957pspr0203_4.

4. **Casadevall A, Fang FC**. 2008. Descriptive science. *Infect Immun* **76**:3835–3836. http://dx.doi.org/10.1128/IAI.00743-08.

5. **Casadevall A, Fang FC**. 2010. Reproducible science. *Infect Immun* **78**:4972–4975. http://dx.doi.org/10.1128/IAI.00908-10.

6. **Prinz F, Schlange T, Asadullah K**. 2011. Believe it or not: how much can we rely on published data on potential drug targets? *Nat Rev Drug Discov* **10**:712. http://dx.doi.org/10.1038/nrd3439-c1.

7. **Begley CG, Ellis LM**. 2012. Drug development: raise standards for preclinical cancer research. *Nature* **483**:531–533. http://dx.doi.org/10.1038/483531a.

8. **Johnson VE**. 2013. Revised standards for statistical evidence. *Proc Natl Acad Sci U S A* **110**:19313–19317. http://dx.doi.org/10.1073/pnas.1313476110.

9. **Olsen CH**. 2014. Statistics in *Infection and immunity* revisited. *Infect Immun* **82**:916–920. http://dx.doi.org/10.1128/IAI.00811-13.

10. **Ploegh H.** 2011. End the wasteful tyranny of reviewer experiments. *Nature* **472**:391. http://dx.doi.org/10.1038/472391a.

11. **Williams EH, Carpentier PA, Misteli T**. 2012. Minimizing the "re" in review. *J Cell Biol* **197**:345–346. http://dx.doi.org/10.1083/jcb.201203056.

12. **Walbot V.** 2009. Are we training pit bulls to review our manuscripts? *J Biol* **8**:24. http://dx.doi.org/10.1186/jbiol125.

13. **Fang FC**. 2008. On rejection. *Infect Immun* **76**:1802–1803. http://dx.doi.org/10.1128/IAI.00315-08.

14. **Casadevall A, Fang FC**. 2009. Mechanistic science. *Infect Immun* **77**:3517–3519. http://dx.doi.org/10.1128/IAI.00623-09.

15. **San Francisco Declaration on Research Assessment**. 2012. DORA. https://sfdora.org/.

16. **Casadevall A, Fang FC**. 2013. Is the Nobel Prize good for science? *FASEB J* **27**:4682–4690. http://dx.doi.org/10.1096/fj.13-238758.

17. **Casadevall A, Fang FC**. 2009. Is peer review censorship? *Infect Immun* **77**:1273–1274. http://dx.doi.org/10.1128/IAI.00018-09.

18. **Casadevall A, Fang FC**. 2012. Winner takes all. *Sci Am* **307**:13. http://dx.doi.org/10.1038/scientificamerican0812-13.

19. **Casadevall A, Fang FC**. 2009. Important science—it's all about the SPIN. *Infect Immun* **77**:4177–4180.

Index

Note: Page numbers followed by *f*, *t*, and *b* indicate figures, tables and boxes respectively.

Thinking about Science: Good Science, Bad Science, and How to Make It Better, First Edition.
Ferric C. Fang and Arturo Casadevall.
© 2024 American Society for Microbiology.